Serengeti III

SERENGETI III

Human Impacts on Ecosystem Dynamics

Edited by

A. R. E. Sinclair, Craig Packer, Simon A. R. Mduma,
& John M. Fryxell

The University of Chicago Press
Chicago and London

A. R. E. Sinclair is professor in the Biodiversity Research Centre at the University of British Columbia. Craig Packer is professor of ecology, evolution, and behavior at the University of Minnesota. Simon A. R. Mduma is director of the Tanzania Wildlife Research Institute. John M. Fryxell is professor of integrative biology at the University of Guelph.

The University of Chicago Press, Chicago 60637
The University of Chicago Press, Ltd., London
© 2008 by The University of Chicago
All rights reserved. Published 2008
Printed in the United States of America

17 16 15 14 13 12 11 10 09 08   1 2 3 4 5

ISBN-13: 978-0-226-76033-9   (cloth)
ISBN-13: 978-0-226-76034-6   (paper)
ISBN-10: 0-226-76033-2   (cloth)
ISBN-10: 0-226-76034-0   (paper)

Library of Congress Cataloging-in-Publication Data

Serengeti III: human impacts on ecosystem dynamics / edited by A. R. E. Sinclair . . . [et al.].
    p. cm.
  Includes bibliographical references and index.
  ISBN-13: 978-0-226-76033-9 (cloth: alk. paper)
  ISBN-10: 0-226-76033-2 (cloth: alk. paper)
  ISBN-13: 978-0-226-76034-6 (pbk.: alk. paper)
  ISBN-10: 0-226-76034-0 (pbk.: alk. paper) 1. Animal ecology—Tanzania—Serengeti
National Park Region. 2. Nature conservation—Tanzania—Serengeti National Park
Region. 3. Nature—Effect of human beings on—Tanzania—Serengeti National Park
Region. I. Sinclair, A. R. E. (Anthony Ronald Entrican) II. Title: Serengeti 3.
  QL337.T3S424 2008
  577.4′80967827—dc22

                                                                            2007049914

⊗ The paper used in this publication meets the minimum requirements of the American National Standard for Information Sciences—Permanence of Paper for Printed Library Materials, ANSI Z39.48-1992.

Richard LeRoy Hay
1926–2006

Beginning in 1962 with his work at Olduvai Gorge, Richard Hay's geological research there and at Laetoli spanned 40 years, contributing in numerous ways to our knowledge of the paleoecology of the Serengeti.

CONTENTS

Preface and Acknowledgments   *ix*

1 Introduction: Understanding the Greater Serengeti Ecosystem   *1*
Craig Packer and Stephen Polasky

2 Historical and Future Changes to the Serengeti Ecosystem   *7*
A .R. E. Sinclair, J. Grant C. Hopcraft, Han Olff, Simon A.R. Mduma,
Kathleen A. Galvin, and Gregory J. Sharam

3 Paleoecology of the Serengeti-Mara Ecosystem   *47*
Charles R. Peters, Robert J. Blumenschine, Richard L. Hay, Daniel A. Livingstone,
Curtis W. Marean, Terry Harrison, Miranda Armour-Chelu, Peter Andrews,
Raymond L. Bernor, Raymonde Bonnefille, and Lars Werdelin

4 The Resource Basis of Human-Wildlife Interaction   *95*
Han Olff and J. Grant C. Hopcraft

5 Generation and Maintenance of Heterogeneity in the Serengeti Ecosystem   *135*
T. Michael Anderson, Jan Dempewolf, Kristine L. Metzger, Denné N. Reed,
and Suzanne Serneels

6 Global Environmental Changes and Their Impact On the Serengeti   *183*
Mark E. Ritchie

7 The Multiple Roles of Infectious Diseases in the Serengeti Ecosystem   *209*
Sarah Cleaveland, Craig Packer, Katie Hampson, Magai Kaare, Richard Kock,
Meggan Craft, Tiziana Limbo, Titus Mlengeya, and Andy Dobson

8 Reticulate Food Webs in Space and Time: Messages from the Serengeti   *241*
Robert D. Holt, Peter A. Abrams, John M. Fryxell, and T. Kimbrell

9 Spatial Dynamics and Coexistence of the Serengeti Grazer Community   *277*
John M. Fryxell, Peter A. Abrams, Robert D. Holt, John F. Wilmshurst,
A. R. E. Sinclair, and Ray Hilborn

10 Dynamic Consequences of Human Behavior in the Serengeti Ecosystem   *301*
Christopher Costello, Nicholas Burger, Kathleen A. Galvin, Ray Hilborn, and Stephen Polasky

11 Human Responses to Change:
Modeling Household Decision Making in Western Serengeti   *325*
Kathleen A. Galvin, Steven Polasky, Christopher Costello, and Martin Loibooki

12 Larger-Scale Influences on the Serengeti Ecosystem:
National and International Policy, Economics, and Human Demography   *347*
Stephen Polasky, Jennifer Schmitt, Christopher Costello, and Liaila Tajibaeva

13 Land-Use Economics in the Mara Area of the Serengeti Ecosystem   *379*
Mike Norton-Griffiths, Mohammed Y. Said, Suzanne Serneels, Dixon S. Kaelo,
Mike Coughenour, Richard H. Lamprey, D. Michael Thompson, and Robin S. Reid

14 Propagation of Change through a Complex Ecosystem   *417*
Ray Hilborn, A. R. E. Sinclair, and John M. Fryxell

15 Who Pays for Conservation? Current and Future Financing Scenarios
for the Serengeti Ecosystem   *443*
Simon Thirgood, Charles Mlingwa, Emmanuel Gereta, Victor Runyoro,
Rob Malpas, Karen Laurenson, and Markus Borner

16 Integrating Conservation in Human and Natural Ecosystems   *471*
A. R. E. Sinclair

Appendix: The Main Herbivorous Mammals and Crocodiles
in the Greater Serengeti Ecosystem   *497*
Simon A. R. Mduma and J. Grant C. Hopcraft

Contributors   *507*

Index   *513*

Each of the Serengeti volumes has been designed to address different issues and problems. The intent of the first volume (1979) was modest; simply to put in one place what was then known about the functioning of the ecosystem. The second volume (1995), apart from updating the information, addressed issues of conservation in the protected area and surrounding human-dominated agricultural and pastoral regions. This third volume focuses on changes to the Greater Serengeti ecosystem. We start with changes in the recent past and palaeohistory. We examine possible future changes. We then model the effects of these changes on the protected area and the interaction of protected and human-dominated areas, considering biological, social, and economic components.

To do this has required people from a wide range of disciplines to work together—ecologists, palaeontologists, economists, social scientists, mathematicians, and disease specialists—to name a few. Members have come from many different countries—Tanzania, Kenya, Holland, Germany, Sweden, France, Belgium, Italy, Britain, Canada, and the United States. This embodies the intent and spirit of the National Center for Ecosystem Analysis and Synthesis (NCEAS), at Santa Barbara, California, that has funded three workshops (2001–2003). Subsequently, the National Science Foundation (USA) has provided funds through a biocomplexity grant to continue the work and support a fourth workshop at Seronera, Serengeti National Park, in 2004. We thank Elizabeth Lyons for her help with funding. This book is the outcome of these meetings.

Many people have been involved in the project. The staff at NCEAS went out of their way to help bring members from around the world, in particular Jim Reichman, the director, and administrators Marilyn Snowball, Gail Stichler, Leslie Allfree, Alicia Hernandez, Ginger Gillquist, and Marlene Sassaman. The Tanzania National Parks gave us permission to hold the fourth workshop in Serengeti. We are particularly grateful to Mr. Justin Hando, chief warden of Serengeti National Park, and his staff, for their help in providing facilities. We also thank the Tanzania Wildlife Research Institute, who was our host in Tanzania and our partner in the biocomplexity research.

Reviewers who are experts in their fields but are not associated with Serengeti commented on each of the chapters, providing an unaffiliated perspective. They include C. Barrett, R. Boone, J. Cruikshank, S. Eby, T. Fox, K. Homewood, P. Hudson, R. L. Jefferies, B. Kaltenborn, T. McCabe, R. McCann, K. Metzger, E. J. Milner-Gulland, D. Spratt, P. Thornton, R. Turkington, R. Varkis, L. Warman, and three anonymous referees. We thank them all for their time and consideration.

We thank the Natural Sciences and Engineering Research Council of Canada and the Frankfurt Zoological Society (FZS) for supporting the editors. Markus Borner of FZS has consistently supported our work over many years. ARES was also supported by a Canadian Senior Killam Research Fellowship for 2 years and by the Peter Wall Institute for Advanced Studies at the University of British Columbia to produce this book.

The British Ecological Society has supported the long-term work in Serengeti through a major grant from the George Jackson Memorial Fund, to provide housing for Tanzanian ecologists. We thank Paul Hutchinson, Mike Hutchings, and John Lawton for their help and encouragement.

Anne C. Sinclair collated, edited, and formatted the chapters. Her help was invaluable.

The Editors
January 2007

ONE

Introduction:
Understanding the Greater Serengeti Ecosystem

Craig Packer and Stephen Polasky

The very name of the Serengeti conjures up a vast vision of open spaces and phenomenal abundance, of vast herds free to wander immense plains that stretch on forever. To anyone who has ever had the good fortune to spend time in the Serengeti, the intertwining of plants and animals, herbivores and carnivores, weave their way into your imagination. It is the intricacy of the Serengeti that captures your imagination and won't let go. The sheer size and complexity of this intricacy reveals itself only very slowly—even for those of us who have worked in the Serengeti for decades, this interplay can barely be glimpsed by our limited experiences of the place.

In the twenty-first century, however, such complex interplays are being disrupted in most large ecosystems. Worldwide, it is estimated that roughly 50% of useable lands have been converted to human-dominated use (Tilman et al. 2001), a conversion that has largely been driven by the doubling of the human population between 1950 and 2000. Human population is expected to increase by another 50 to 67% during the next 50 years (UN 2003), and population growth is especially rapid in developing countries like Kenya and Tanzania, whose populations have both tripled since 1965 (see fig. 1.1). Although the total fertility rate in Kenya has declined from 7.7 children per family in the 1960s to 4.7 children today, and Tanzania's fertility has fallen from 6.6 in the 1960s to 5.6 today (Hinde and Mturi 2000), these birthrates, combined with population momentum, will carry these countries to ever-higher numbers for some time, and it is uncertain

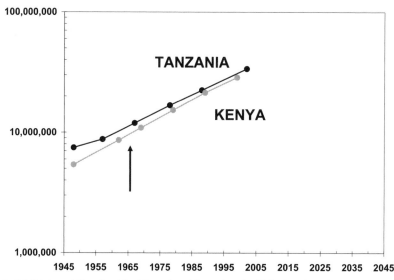

Fig. 1.1 Total population of mainland Tanzania and Kenya since the 1940s. Data from the official censuses of each country. Arrow indicates when the Serengeti Research Institute was established.

how they will respond to the impacts of the AIDS pandemic or to the economic/political unrest recently experienced in Kenya.

Thus, the Serengeti, despite its apparently endless bounty, is an ecological island in a rising sea of humanity. The core of the Serengeti ecosystem, within the Serengeti National Park in Tanzania and Maasai Mara National Park in Kenya, remains largely intact, but the northern reaches of the Serengeti/Mara system in Kenya have been converted to mechanized agriculture and the southwest boundary of the Maswa game reserve in Tanzania has been steadily eroded by cotton fields.

Worries about human impacts on the Serengeti are not new. After all, Bernard Grzimek's famous book, *Serengeti Shall Not Die* (1959), was prompted by the urgent need to define the limits of the wildebeest migration and to safely set aside the area as a national park. While the establishment of Serengeti National Park and its associated reserves has largely succeeded in maintaining the Serengeti's large wildlife populations, its increasing human influence is felt well inside these protected areas. Intensive poaching drove elephant and rhino to the verge of extinction in the late 1970s and early 1980s (see figs. 2.15, 2.19); Hilborn (1995) and Mduma, Hilborn and Sinclair (1998) showed how even the seemingly inexhaustible Serengeti wildebeest population could eventually collapse under rising pressure from illegal offtake when the migratory herds leave the confines of the protected areas. The human population around the Serengeti ecosystem now has enough domestic dogs to serve as a more-or-less permanent reservoir for rabies and canine distemper virus, which eliminated the Cape hunting dog between 1991 and 2001 and killed over a third of the Serengeti lion population in 1994.

It is abundantly clear that the fate of the Serengeti ecosystem rests largely in human hands. The future of this extraordinary place rests on the ways that people in the Greater Serengeti ecosystem use resources to earn a living, the extent to which foreign tourists find the Serengeti an attractive destination, and the policy choices by the governments of Kenya and Tanzania. For the Serengeti to survive, the people of Kenya and Tanzania must find conservation to be in their direct interests. Without economic development strategies that will allow some of the poorest people in the world to improve their lot in life in a way that is consistent with conservation, our grandchildren may only know Serengeti as a faintly exotic word, like Atlantis or Eden.

In this third volume of the Serengeti series, we develop a fully integrated research program that includes socioeconomic research on human activities and human decision making, in concert with ecological research. Our major goal is to begin to understand the complex feedback loops between natural and human components of the greater Serengeti ecosystem and how their disturbance propagates throughout the system. For example, a change in rainfall patterns, brought on by climate change, may have direct ecological impacts on vegetation and herbivore populations. But climate change may also lead to changes in human behavior that magnify these impacts by increasing the demand for bush meat at a time when herbivore populations are already stressed (Barrett and Arcese 1998). The presence of humans, therefore, will profoundly affect the dynamics of the entire system.

Because of the interacting complexity of the Serengeti system, we formed a working group to rally around a set of issues and develop a common set of research goals. Serengeti III is principally the outcome of a series of workshops held at the National Center for Ecological Analysis and Synthesis (NCEAS) in Santa Barbara, California, in 2001, 2002, and 2003. A secondary outcome of the workshops was the eventual award of an NSF grant on the Biocomplexity of the Serengeti, which in turn sponsored a workshop in the Serengeti National Park in 2004. The workshops and the grant explicitly linked together ecologists, economists, and anthropologists to work in an integrated fashion. Despite the usual centrifugal tendencies for scientists from different disciplines to cluster together, we enjoy a great degree of interdisciplinary camaraderie and a shared commitment to integrated research. The NCEAS workshops and the Biocomplexity grant produced a shared database and a related series of mathematical approaches to model the Greater Serengeti ecosystem, leading to this book.

We have attempted to integrate available data into a format that can be used to predict likely outcomes of perturbations to the current system. Understanding the interactions of such a complex ecosystem requires a comprehensive set of theoretical/simulation models that incorporate major processes of every important component of the system—from soils to human societies. Foci of these models include: (1) modeling of individual behavior, (2) expanding individual effects to the whole ecosystem, (3) connecting individuals with emergent properties of the ecosystem, and (4) predicting the behavior of a complex system.

Our intent is to address two general questions. First, how can new insights from complex ecological systems help improve the management of the greater Serengeti ecosystem? Second, can our knowledge of the Serengeti enhance the understanding and conservation of other ecosystems? More specific questions, which will appear in various chapters of this book, include:

- How does a disturbance propagate through a complex ecological system that includes both human and nonhuman components?
- What are the effects on ecosystem dynamics of spatial and temporal heterogeneity and the scales upon which they are expressed?
- Do emergent properties exist in the Serengeti? Are they useful indicators of ecosystem change? Do they have significant feedback on ecosystem components?
- How much detail/abstraction is required for models to analyze and predict the consequences of ecosystem change?
- What are the important abiotic, biotic, and socioeconomic constraints, and driving variables in the Serengeti ecosystem?
- *How do we evaluate the interplay of human and nonhuman components of ecological systems across different scales?*

Though we focus on the Serengeti, our research has broad implications. The alleviation of rural poverty is the greatest challenge to economic development in sub-Saharan Africa, but wildlife is the greatest long-term economic asset in several countries, such as Tanzania, Botswana, Namibia, and Mozambique (and, perhaps, on a local scale in Kenya). Yet international development and conservation agencies rarely consider the impact of specific economic projects on wildlife conservation, and vice versa. Local communities must be engaged as effective conservation partners—both to minimize their impact on wildlife resources and to cooperate with enforcement agencies to minimize illegal activities in the protected areas. The enormous capital that is being spent in protecting wildlife areas from local communities could far better be invested in the growth of the country's human capital, if villagers were to become full partners in conservation.

However, what does it mean to be "effective conservation partners"? The world is littered with unintended consequences—well-meaning efforts to improve agriculture through irrigation, for example, often come at a cost to the people living downstream and to the natural ecosystems that were sustained by scarce water in an arid environment. Only by understanding the intricacies of a system as complex as the Serengeti can decision makers possibly attempt to balance the demands of short-term economic growth with sustainable natural resource management.

REFERENCES

Barrett, C. B., and P. Arcese. 1998. Wildlife harvest in integrated conservation and development projects: Linking harvest to household demand, agricultural production, and environmental shocks in the Serengeti. *Land Economics* 74: 449–65.

Grzimek, B. 1959. *Serengeti darf nicht sterben.* Munich: Franke.

Hilborn, R. 1995. A model to evaluate alternative management policies for the Serengeti-Mara ecosystem. In *Serengeti II: Dynamics, management, and conservation of an ecosystem,* ed. A. R. E. Sinclair and P. Arcese, 617–37. Chicago: University of Chicago Press.

Hinde, A., and A. J. Mturi. 2000. Recent trends in Tanzanian fertility. *Population Studies* 54: 177–91.

Mduma, S., R. Hilborn, and A. R. E. Sinclair. 1998. Limits to exploitation of Serengeti wildebeest and implications for its management. In *Dynamics of Tropical Communities,* ed. D. M. Newbery, N. Brown, and H. H. T. Prins. *British Ecological Society Symposium* 37: 243–65. Oxford: Blackwell Science.

Tilman, D., J. Farigione, B. Wolff, C. D'Antonio, A. Dobson, R. Howarth, R. Schindler, W. H. Schlesinger, D. Simberloff, and D. Swackhamer. 2001. Forecasting agriculturally driven global environmental change. *Science* 292: 281–84.

United Nations. 2003. *World Urbanization Prospects: The 2003 Revision.* New York: United Nations, Department of Economic and Social Affairs, Population Division.

TWO

Historical and Future Changes to the Serengeti Ecosystem

A. R. E. Sinclair, J. Grant C. Hopcraft, Han Olff, Simon A. R. Mduma,

Kathleen A. Galvin, and Gregory J. Sharam

We document in this chapter what is known of recent historical changes in the Serengeti ecosystem and discuss possible future changes. The chapter sets the scene for the later analysis and modelling of the human-nature interactions. First, we describe the Serengeti in terms of geography, climate, soils, habitats, and animals, and place it in the human context of surrounding tribes and the larger region. Second, we give an overview of currently available information on the ecosystem. Third, we outline the various changes to the system that are either currently occurring or are expected in the future. We treat these as experimental perturbations, with which we alter the models in later chapters. We describe these changes, and why and how they may occur.

The overriding influence of the migratory wildebeest population on the dynamics of the system is the central theme of our story. This population is so large that it determines the structure and function of all components, both in the protected area and within the surrounding human areas. These effects have been reported in detail in the previous volumes, as well as many papers, and space allows only a brief synopsis and update. Further details are given in Schaller (1972), Sinclair (1977, 1979a), Sinclair and Norton-Griffiths (1979), Sinclair and Arcese (1995), Mduma, Sinclair, and Hilborn (1999), and Sinclair et al. (2007).

The explorers Speke and Grant passed through the Maswa country in 1861, describing the typical savanna and ungulates that are found there today (Speke 1863). Missionaries crossed the Serengeti plains in the 1860s and 1870s, following the ivory trade routes that passed Lake Lagarja (L. Ndutu), Simba kopjes, behind Mukoma hill, and Nata (Wakefield 1870, 1882; Farler 1882). In Kenya, Percival (1928) in July 1912 and White (1915) in August 1913 observed huge numbers of wildebeest during the dry season along the Mara River, although no one at the time knew where they came from. These tiny fragments of information are all that are reported until Johnson's three expeditions of 1926, 1928, and 1933, which provide the first references to the great wildebeest migrations and the first photographs of the area.

The Serengeti plains were known for their abundance of lions as far back as the 1910s, the area being a favorite of foreign hunting expeditions. Leslie Simpson pioneered the route for vehicles from Nairobi via Narok south through the northern Serengeti to what is now Seronera in 1920, and this has essentially remained unchanged to this day as the road from Kenya. Hunters Simpson, Klein, and Cunningham used the route in the 1920s, the photographer/hunters Martin and Osa Johnson used it in the 1920s and 30s, and the hunting firm of Syd Downey and Donald Kerr used it during the 1940s–1950s (Turner 1987).

An area of 2,286 km² was established in 1930 as a game reserve in what is now southern and eastern Serengeti. Warden Monte Moore was based at Banagi; he maintained a low level of stewardship from 1931 to 1939. Sport hunting was allowed until 1937, when all hunting was stopped. In 1940, Protected Area Status was conferred, and a national park was established in 1951, covering southern Serengeti and the Ngorongoro highlands. The park headquarters was based at Ngorongoro, with a western outpost situated at Banagi. After a commission of enquiry in 1956 (Pearsall 1957), the boundaries were realigned in 1959 to include the area between Banagi and the Kenya border (fig. 2.1). At the same time, the present Ngorongoro Conservation Area was excised from the Serengeti National Park and the new park headquarters was moved to Seronera, where it remained until 1998, when it moved to Fort Ikoma, outside the park. Both Chief Park Warden Gordon Harvey, and Deputy Warden Myles Turner moved to Seronera in 1959. In 1961 the Maasai Mara Reserve in Kenya was officially established, and in 1965 the Lamai Wedge between the Mara River and Kenya border was added, thus creating a continuous protected corridor for the wildebeest migration from the Serengeti plains in the south to the Loita plains in the north. A small area north of the Grumeti River in the corridor was added

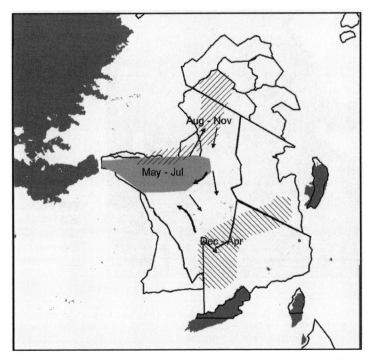

Fig. 2.1 Outline of the migration movements of the wildebeest, which define the outline of the Serengeti ecosystem. The different shaded zones indicate the main herds in different months of the year. The black lines are the boundaries of different protected areas (see fig 2.3; adapted from Maddock 1979).

in 1967. Subsequently, the Grumeti and Ikorongo game reserves were established in 1993, and the Wildlife Management Areas were decreed in 2003. The Serengeti National Park was one of the first areas to be proposed as a World Heritage site by UNESCO at the Stockholm conference of 1972, and it was formally established as such and as a Biosphere Reserve, along with the Ngorongoro Conservation Area, in 1981.

THE PRESENT SERENGETI-MARA ECOSYSTEM

The Area

The Serengeti-Mara ecosystem (fig. 2.1) is an area of some 25,000 km^2 on the border of Tanzania and Kenya, East Africa (34° to 36° E, 1° to 3° 30' S), defined by the movements of the migratory wildebeest. The eastern boundary is formed by the Crater Highlands and the Rift Valley (fig.2.2). An arm, called the *western corridor*, stretches west almost to Speke Gulf of Lake Vic-

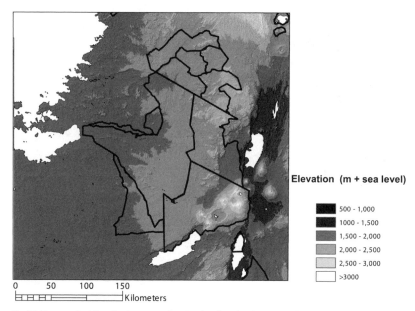

Fig. 2.2 Topography (elevation in meters above sea level) and main rivers in the Serengeti-Mara ecosystem. Elevation data from the Shuttle Radar Topography Mission (SRTM), USGS. Source for this dataset is the Global Land Cover Facility, http://www.landcover.org.

toria. The remaining western boundary is formed by dense cultivation. The northern boundary is formed by the Isuria escarpment and the Loita Plains in Kenya. The southern and southwestern boundary runs along an area of rocks, dense woodland, and cultivation.

Apart from the Serengeti National Park (14,763 km²), the ecosystem includes several other conservation administrations (fig. 2.3). The Ngorongoro Conservation Area (NCA; 8,288 km²) southeast of the park covers roughly half of the short-grass plains, and includes Olduvai Gorge and the Gol Mountains. North of the NCA is the Loliondo Game Control Area, which forms the eastern boundary of the Serengeti ecosystem and comprises plains in the south and savanna toward the north. It stretches east through the Gol Mountains, includes the semiarid Salai Plains, and ends at the Gregory Rift escarpment, which drops to Lake Natron. Wildebeest regularly use the Salai Plains in the wet season. The Maswa Game Reserve (2,200 km²) lies to the southwest of the park, and comprises dense *Acacia* woodland on flat alluvial soils. This is the refuge for wildebeest when there is a dry period during the rains, because it is the closest woodland area to the southern plains. At the southern end of Maswa the elevation drops and the terrain becomes a series of rocky hills and kopjes and the vegeta-

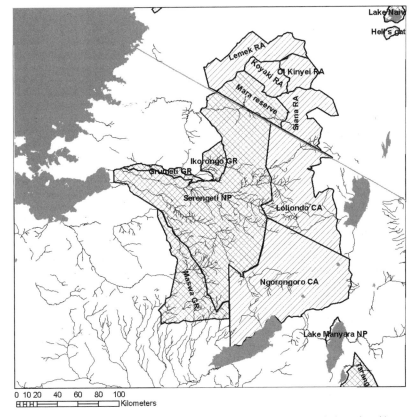

Fig. 2.3 Protection status of the area within which the Serengeti-Mara ecosystem is situated, and its surrounding districts in Kenya and Tanzania. Cross-hatched areas indicate national parks and game reserves managed solely in the interest of wildlife and ecotourism, within which agricultural use and livestock grazing are not permitted. Diagonal hatching indicates controlled, multiple-use pastoralism. Open areas are unprotected and do involve agriculture and rangeland.

tion includes baobabs (*Adansonia*). These habitat boundaries form the edge of the wildebeest range. North of the western corridor lie the Grumeti and Ikorongo Game Control Areas and the Grumeti Game Reserve. Adjacent to Serengeti, in Kenya, the Mara Reserve (1,672 km2) and adjoining group ranches in Kenya lie at the foot of the Isuria escarpment and eastward to include the Loita Hills. The Mara Reserve is largely grassland and relict *A. gerardii* savanna, but it includes the permanently flowing Mara River, which flows south into Serengeti and west to Lake Victoria at Musoma.

North of the Mara Reserve there occur the Loita plains and Loita Hills, which are the traditional lands of the Kenya Maasai, now devoted to enclosed ranching. In the northwest, above the Isuria escarpment, original highland forest with grassy glades (a century ago) has been transformed by pastoralists and, more recently, agriculturalists, into grassland. Wildebeest used to climb the escarpment in Kenya to graze these grasslands in the 1960s and early 1970s (pers. obs.) but they are now excluded. On the Tanzania side, dense settlement has precluded use by wildlife for at least the past 50 years, and probably much longer. These people, the Wakuria, are a tribe that stretches from inside Kenya, along the western boundary in Tanzania as far south as Mugumu. The Waikoma tribe takes over farther south—they stretch to the western-corridor boundary and west to Lake Victoria. In a line that runs through Ndabaka, at the tip of the corridor, the Waikoma give way to the Wasukuma, a very large tribe that stretches southward along the southwest boundary, through Maswa to Makao and then east to Lake Eyasi. These tribes are all agricultural, with smallholdings and small herds of livestock. The Maasai inhabit the eastern side of the ecosystem from the Narok district in Kenya south through the Loliondo area, Ngorongoro Conservation Area to the southern edge of the plains, where they meet the Wasukuma. Toward the southeast, the Maasai's pastoralist use extends farther into the Simanjiro plains across the Rift Valley, up to Dodma. The Maasai are more traditional pastoralists, without the use of fences, developing limited, small-scale agricultural settlements.

Topography and Geology

The Serengeti ecosystem is part of the high interior plateau of East Africa. It slopes from its highest parts on the crater highlands and the Gol Mountains toward Speke Gulf on Lake Victoria, at 920m (fig. 2.2). These highlands are the result of the recent volcanic activity related to the plate tectonics of the Rift Valley, which includes one still-active volcano (Ol Doynio Lengai). The Serengeti plains are situated at an elevation between 1,600 and 1,800 m above sea level. Several river catchments drain the area. The Mara River in the north flows from the Mau highland forests of Kenya south through the Mara Reserve, then west through northern Serengeti, out through the great Mara marshes and into Lake Victoria at Musoma. This is the only permanently flowing river in the Serengeti ecosystem. It supports dense riverine forest on its banks in the Mara Reserve and along its major tributaries in Serengeti. South of the Mara are the parallel catchments of the Grumeti and Mbalageti, which in the west form the western corridor extension of the park. Further south there are the much smaller Duma, Simiyu, and Semu rivers.

The area is gently undulating, heavily dissected by numerous small seasonal streams that drain into the main rivers. There are bands of hills that rise steeply from this relatively flat landscape (fig. 2.2). One band forms the eastern boundary of Serengeti National Park in the woodlands, running north from Grumechen to Kuka, then joining the Loita hills in Kenya. The Gol Mountains rise from the Serengeti plains east of the park. Another band stretches from Seronera west along the corridor to form the Central Ranges, and a third group of hills lies in the south, forming the Nyaraboro-Itonjo plateau. They have steep, stony sides, sometimes with flat tops.

West of a line going through Mgumu and Seronera (below about 1,500m elevation), the underlying rocks are very old (2,500 M years) Precambrian volcanic rocks, banded ironstones, and mineral-poor granites (see chapter 4, this volume). Late Precambrian sedimentary rocks unconformably overlie the shield and form the central and southern hills. A late Precambrian orogenic event produced outcrops of granitic gneisses and quartzite east of Seronera, forming the eastern hills and kopjes. The underground of the western corridor is of more recent geological history; it is a complex of colluvial and fluvial/alluvial formations, which form more nutrient-rich soils.

The Crater Highlands are volcanoes of Pliocene and Pleistocene age, and consist of basic igneous rocks and basalt. Aerial ash and debris from these volcanoes were blown westward to form the Serengeti plains, producing a lithology of basic, mineral-rich pyroclastic deposits (fig. 2.4). Eruptions of one volcano, Lengai, continue to the present, the last major one being in 1966. Peters et al. (chapter 3) give a detailed account of the geology and paleohistory of this area.

The granitic geology west of the Serengeti Park has produced low-nutrient soils (see chapter 4, this volume). The soils from the volcanic, fluvial, and collovial origin are higher in nutrients than those covering the rest of the system. Thus, the western boundary of the Serengeti parks marks a clear transition in soil fertility (see chapter 4, this volume). Since the western boundary was delineated partly by the migration movements of the wildebeest (fig. 2.1) it suggests wildebeest were responding to this low soil fertility in western Serengeti before agriculture took over in this region. Soils on the eastern plains are highly saline, alkaline, and shallow, due to the formation of a $CaCO_3$ hardpan, also known as *caliche*. This hardpan is formed under the influence of vegetation, but only on calcium-rich soils in dry climates. As water is removed by plant uptake, soluble substances precipitate. The caliche layer therefore shows a progressive development and cementation through time (Schlesinger 1991). The soils become progres-

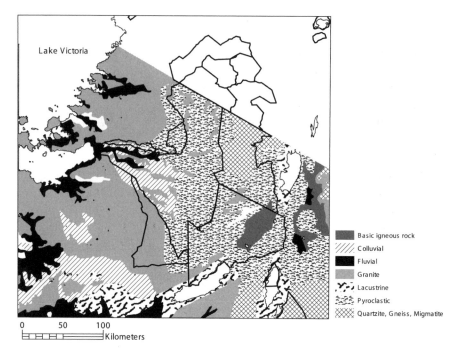

Lake Victoria

Basic igneous rock
Colluvial
Fluvial
Granite
Lacustrine
Pyroclastic
Quartzite, Gneiss, Migmatite

0 50 100
Kilometers

Fig. 2.4 The dominant lithology in the Serengeti-Mara ecosystem and its surroundings showing the seven dominant types. Map produced by aggregation of lithology categories, using data from the Soil and Terrain Database for Southern Africa (Dijkshoorn 2003). Data for Kenya were not available.

sively deeper (where the hardpan disappears) and less alkaline toward the northwest plains and into the woodlands, probably due to less calcium-rich soils and more precipitation. At levels of precipitation too high for hardpan formation, a characteristic soil catena is found. This is the gradient of soil types from ridgetop to drainage sump, characterized by shallow, sandy, well-drained soil at the top, changing to deep, silty, poorly drained soil at the bottom. These catenas form due to the long-term downwash of the finer soil particles downslope with surface runoff. Imposed upon this geological template, nutrient flow through the grazing food web determines the movement of migrants, the distribution of resident ungulates, and the biomass of herbivores. Herbivores are most likely especially important in the recycling and redistribution of nitrogen in the system, as this element is not present in any parent material from which soils are formed. Only nitrogen fixation by microorganisms (often associated with leguminous plants) and atmospheric deposition (which is naturally low, being so far from the sea) can bring N into the ecosystem.

Climate

The climate shows a relatively constant mean monthly maximum of 27° to 28°C at Seronera. The average maximum daily temperature changes mostly with elevation, from 15° C on the crater highlands to 30° C down by lake Victoria. Therefore, the Serengeti ecosystem is a relatively cool island in a much warmer region (fig. 2.5). The minimum temperature at Seronera varies from 16°C in the hot months of October–March to 13°C during May–August. Rain typically falls in a bimodal pattern, with the long rains during March–May and short rains in November–December (Griffiths 1958; Norton-Griffiths, Herlocker, and Pennycuick 1975; Brown and Britton 1980). However, the rains can fuse into one long period, particularly in the north, or the short rains can fail entirely, especially in the southeast. There is a rainfall gradient (fig. 2.6) from the dry southeast plains (500 mm per year) to the wet northwest on the Kenya border (up to 1,200 mm per year; Sinclair 1979c). The low rainfall at the Serengeti and Salai plains is caused by the rain shadow of the Crater highlands and the Meru-Kilimanjaro mountain range (fig. 2.2 and see chapter 4, this volume) as the prevailing

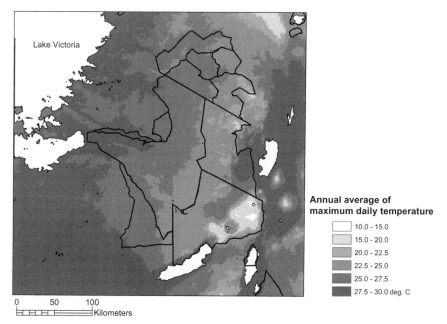

Fig. 2.5 Contours of annual mean daily maximum temperature for the Serengeti-Mara ecosystem and its surroundings, showing that temperature depends mainly on elevation (compare with fig. 2.2.). Data from the WORLDCLIM database (Hijmans et al. 2005).

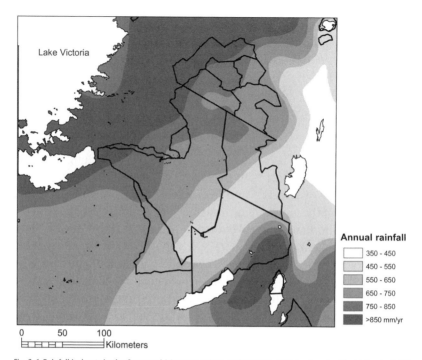

Fig. 2.6 Rainfall isohyets in the Serengeti-Mara ecosystem and its surroundings, showing the highest rainfall in northwest Serengeti and the lowest on the short-grass eastern plains, as caused by the rain-shadow effect of the Ngorongoro highlands and the influence of Lake Victoria. Map interpolated using all available data from local rainfall stations in the Serengeti Park (interpolation by M. Coughenour), combined with regional rainfall isohyets calculated by the FAO/UNEP Desertification and Mapping Project, as rasterized by UNEP/GRID.

winds carry moisture off the Indian Ocean from the south east. When the air rises towards these highlands, it cools off, holds less moisture and rain precipitates. However, shifting winds can carry moisture inland from Lake Victoria, counteracting this rainshadow effect, and inducing the rainfall gradient.

Vegetation

The southeast plains are treeless (fig. 2.7) except along Olduvai Gorge. The grasses are alkaline adapted, short, grazing tolerant, and rhiszomatous, and there are many forbs (herbaceous dicots). Dominant grasses include *Digitaria macroblephera* and *Sporobolus ioclades*. Due to the combination of water limitation and high soil fertility, the grasses of the short-grass plains produce high forage quality during the wet season, which attracts the migrant herbivores. Due to water limitation and intensive grazing, fuel for fires cannot build up, hence these grasslands never burn. Further to the west, deeper soils allow larger tussock grass species to establish, the dominants being *Themeda triandra, Pennisetum mezianum,* and the unpalatable *Cymbopogon excavatus;* and here fire becomes an important component of the ecosystem. These taller species also dominate in the *Acacia* woodlands. However, tall *Hyparrhenia* species dominate in the far northwest on nutrient-poor granitic soils.

The woodlands start at a sharp boundary running south of Seronera in one direction and east of Seronera in the other. The woodlands are dominated by various *Acacia* species in all areas except for a small region on sandy soils in the northwest of the system, where *Terminalia* and *Combretum* take over. Canopy cover of the trees varies between 2 and 30% (fig. 2.7). The woody cover shows strong small-scale heterogeneity, while showing little response to the larger-scale rainfall gradient above 600 mm/yr. This suggests that the local interplay of fire, herbivores, and soils is more important in regulating woodland cover. Along the main rivers, Mbalageti, Grumeti, and Mara and its tributaries, there is gallery forest with closed canopy.

Tree species, and in particular the Acacias, prefer particular edaphic conditions and locations on the soil catena (Mduma, Sinclair, and Turk-

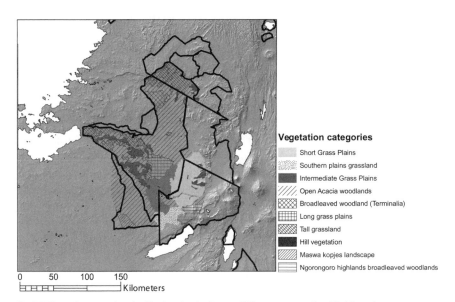

Vegetation categories

- Short Grass Plains
- Southern plains grassland
- Intermediate Grass Plains
- Open Acacia woodlands
- Broadleaved woodland (Terminalia)
- Long grass plains
- Tall grassland
- Hill vegetation
- Maswa kopjes landscape
- Ngorongoro highlands broadleaved woodlands

0   50   100   150
Kilometers

Fig. 2.7 The major vegetation classifications for the Serengeti-Mara ecosystem. Simplified from the vegetation map of Herlocker (1976a, b) and from data collected by the authors. Data for Loliondo are unavailable.

ington 2007). Shallow, stony, and sandy soils on the top of gentle ridges, medium-depth soils on the slopes and the foot of rocky hills, and deep, silty soils at the bottom near rivers and in areas of impeded drainage characterize the catena. There are many square kilometres of such poorly drained land in Serengeti and they vary in their extent of inundation during the rains, from merely damp on the edges to being completely inundated throughout the rainy season.

Two Acacias are found on well-drained ridges: the large umbrella tree, *A. tortilis,* which is found mostly around the edge of the Serengeti plains, and the smaller gum-arabic tree, *A. senegal,* which occurs on stony areas within the savanna proper. Midslope trees are the large stink-bark acacia *A. clavigera (= A. robusta),* probably the dominant tree of the savanna, and the smaller shrubby "wait-a-bit" thorn, *A. mellifera,* which lives on sandy areas of deeper soil. In the drier impeded drainage soils there occurs the wild date, *Balanites,* and the spindly gall-forming "whistling," *A. drepanolobium,* which forms large monospecific stands. In wetter areas this species gives way to the orange coloured *A. seyal,* also forming dense single-species thickets. Near the larger rivers in silty soil one finds the very large "yellow fever tree," *A. xanthophloea,* and finally on the banks and in the rivers, there occurs another large tree, *A kirkii.* In addition to the Acacias there occur two species of *Commiphora,* small trees that live on ridges often mixed with *A. senegal,* these being *C. schimpherii* and *C. africana.* A large broadleaved tree, *Terminalia mollis,* occurs on ridges only in the far northwest. Here the underlying geology changes to a granitic base and the savanna is mostly low-nutrient broadleaved trees and shrubs, but on midslope areas there occurs *A. gerrardii.* It appears that *A. gerrardii* and *A. clavigera* replace each other ecologically and that there is a narrow overlap in distribution. The "sausage tree," *Kigelia africana,* is usually solitary or in widely spaced groups of a few trees. They are found in drainage lines and near rivers. Fig trees, *Ficus sp.,* can be found on ridges or rock outcrops (kopjes) near a spring or seepage line, but otherwise are near rivers.

On the stony hills trees are sparse. *Combretum* is often found there along with *A. drepanalobium* and *A. nilotica.* Grasses are low-quality *Loudetia* and *Hypharrhenia.* On top of the higher Kuka hill there are remnant montane forest patches.

Fauna

Mduma and Hopcraft provide details of the larger mammal fauna in the appendix, tables A.1 and A.2. The Serengeti supports not only the largest herds of migrating ungulates but also one of the highest concentrations of large predators in the world. Estimates put wildebeest around 1.3 million, zebra at 200,000, and Thomson's gazelle at 440,000. Hyena are the most numerous of the large carnivores at about 7,500 with lion at about 2,800.

Long-distance migrants comprise wildebeest, zebra, Thomson's gazelle, and eland. Grant's gazelle also move some distance but little is known about where they go. The plains support the migrants in the wet season but only a few Grant's and Thomson's gazelle and ostrich live there in the dry season. Oryx occur on the Salai plains but their numbers are unknown. In the woodlands, topi occur throughout, but they form large herds on the wetter plains of the corridor and the Mara Reserve, and are absent in the east. In contrast, kongoni prefer the dry eastern woodlands and long-grass plains, and are absent in the corridor. Impala, steinbuck, dikdik, elephant, and buffalo are resident throughout the woodlands and avoid the plains. Giraffe are also found through the woodlands but some walk across the plains to the Gol Mountains. Waterbuck are confined to the larger rivers with riverine grassland. Bohor reedbuck are also found along rivers. They spread through the long-grass plains in the wet season but are only visible at night. Mountain (Chandler's) reedbuck are found only on top of the larger hills. Warthog are widespread but scarce in the woodlands, with a few on the plains. Their numbers have declined considerably in the past 20 years. Oribi are common in the northwest, characterised by tall *Hyparrhenia filipendula* grassland, with a few found northeast near Klein's camp spring. Grey duiker are also found in the northwest with a few on the hills elsewhere. Roan antelope occurred in two localities—the northwest, Lamai and Mara triangle in Kenya, and in the south through Maswa. Only the southern population remains, the northern one probably extinct by 1995. A few occurred in the central hills, the last being seen in 1981. The far south of Maswa also supports greater kudu. Bushpig are found in the riverine forests. The Mara River forests contain tree hyrax, whereas the Grumeti River supports black-and-white colobus. The endemic subspecies of red patas monkey has recently been found on the Mbalageti River. Both vervet monkeys and olive baboons are found through the woodlands close to water, and baboons are especially numerous along the corridor. Most of these species are found in the agricultural areas west of the park, particularly the smaller ones. However, poaching is removing many of them as by-catch in snarelines, or through night hunting with spotlights and dogs. Elephants and baboons cause problems with crop raiding.

Of the larger carnivores, lion, cheetah, and hyena are found in almost all habitats. Hyena form large groups on open areas such as the plains, but are solitary in much of the woodlands. Leopards are confined to the woodlands and along major rivers on the plains. Wild dog were also widespread

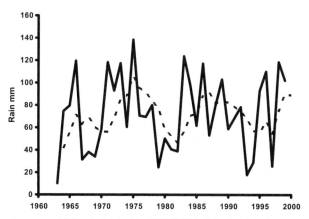

Fig. 2.9 Plains dry season rainfall (July–October) for 1963–1999 (solid line) and the 5-year running average (broken line), showing possible decadal oscillations.

a record from 1902–1910. Figure 2.8 shows a trend of increasing annual rainfall. Closer inspection shows that precipitation increased markedly to the peak rain of 1961–62, the highest of the century. Thereafter long-term mean rainfall has remained more constant or has declined slightly. However, there were floods in the wet season of 1997–98, one of the strongest El Nino events recorded, and there were severe droughts in 1960 and 1993. Within Serengeti the longest record is from Banagi, starting in 1938. This shows no long-term trend.

Dry-season rainfall (July–October) is important because it produces the critical new grass growth that ungulates feed on when resources are low. This rainfall exhibits a decadal oscillation (fig. 2.9).

Floods and droughts influence the biota. Thus, the floods of 1998 resulted in a high watertable, moist soils, and flowing springs well into the dry season, and unusually high biomass of grass. Ground-living birds such as helmeted guineafowl (*Numida meleagris*) experienced unusually high breeding success (probably from a combination of higher insect food abundance and reduced predation of chicks due to greater cover) compared to dry years, such as 2000. Droughts reduce survival of juveniles in all ungulate species except those of elephants. This was especially evident in the extreme drought of 1993, when drought affected both the survival of juvenile wildebeest and the conception rate of females (fig. 2.10).

Climatic change, apart from increasing overall rainfall, may also change the frequency of these floods and droughts. Richie (chapter 6, this volume) explores this topic in greater detail. Increasing wet-season rain results in higher fuel loads in the dry season and hence more extensive and hotter

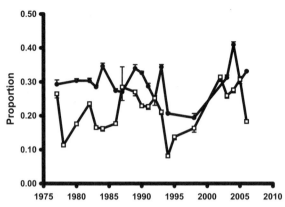

Fig. 2.10 Proportion of juveniles (< 1 year, closed circles) and yearlings (1–2 years, open squares) in the sum of juveniles and females (or yearlings and females) of wildebeest 1977–2006, showing the marked drop following the drought of 1993. Vertical bars = 1 SD.

fires. In contrast, increasing dry-season rain results in increasing fuel moisture and decreasing incidence of fires. At this stage it is not clear how climate change will affect these events. Other possible changes that might affect the system are increases in carbon dioxide, and shifts in the position of the Intertropical Convergence Zone (ITCZ). The ITCZ follows the sun south in November and north in April, delayed by some 6 weeks, and it brings the rain. Over the equator this results in the bimodal rainfall. Changes in how far this moves north or south, or changes in its width will affect the modality and duration of the rainfall.

Fire

Grass fires occur consistently each dry season, set by humans. No instance of lightning-caused fire has been recorded. It is generally understood that many fires are set during cattle-raiding excursions across the park and for purposes of poaching. In the early dry season grass regrows after a fire as a short, high-protein greenflush, which attracts grazers to the site and makes them available to snaring and ambush by poachers.

The extent of fires has been mapped intermittently through aerial surveys. The extent of burning was extremely high in the early 1960s; around 80% of the ecosystem was burnt each dry season. However, the extent of burning declined to low levels by the early 1980s and has remained below 25% (probably) into the 1990s (Norton-Griffiths 1979; Dublin 1986; Sin-

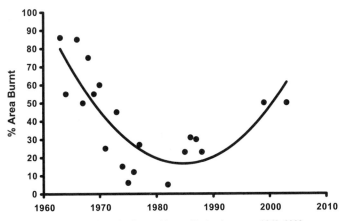

Fig. 2.11 Proportion of woodlands north of Banagi burned in the dry season 1963–2003.

clair et al. 2007). In recent years burning has increased as part of a policy of early burning by park managers. The decline of burning was caused by the increase of the wildebeest population (see the following), which reduced grass biomass and so limited the spread of fire (figs. 2.11, 2.12).

2. BIOTIC CHANGES

The Wildebeest Migration

The Johnsons first recorded the movements of large numbers of wildebeest, appearing from the east near Naabi and moving west through Moru, in July 1928. They guessed, incorrectly, that the population had migrated from the Athi plains near Nairobi past Lake Natron and were moving toward Lake Victoria. The Grzimeks documented the main migration routes in 1959–1960 (B. Grzimek and M. Grzimek 1960; M. Grzimek and B. Grzimek 1960), showing that the migrants used the eastern plains but not the Ngorongoro highlands or Crater. Talbot and Talbot in 1961 (1963) and Watson in 1963–65 (1967) showed that the northern extension from Banagi to the Kenya border was the essential route to the dry season refuge on the Mara River. Thus, by 1966 the migration routes were fully documented: the wet season (December–May) was spent on the Serengeti plains, animals moving to the far eastern Salai plains when conditions were very wet. When the rains stopped, in May–June, animals moved west through the Mbalageti and Grumeti valleys to the western corridor, and then northeast in July–August to the Mara River, the only permanently flowing river. The population remained in this refuge area of northwest Serengeti until the rains returned in November–December. At this time the wildebeest moved south to the plains again, waiting at the edge of the woodlands until the plains became green. If there were dry periods while the population was on the plains, then animals moved directly west into the Maswa woodlands, and this area was a refuge during the December–April period.

The migration routes of the 1960s have been modified as the population has increased (Maddock 1979). Thus, by the 1970s more of the population had moved into the Mara Reserve and beyond. Second, a segment of the population has moved directly north from the plains along the eastern side of the park, and even east of this in the Loliondo area, entering the Mara Reserve from the southeast. It is not known what proportion of the population uses this route. Third, it is known that historically wildebeest used the plains north of the western corridor and followed the Grumeti River upstream to the Mara drainage. Much of this area is now under human cultivation and, although the Grumeti route is still an essential conduit, it is being gradually cut off by settlement. The western corridor may be used less as it becomes a cul-de-sac. In addition, the shores of Lake Victoria at the far western end of the corridor were used as a dry season refuge in times of drought, as was seen in 1993. As settlement takes over the shoreline this refuge is becoming lost. Similarly, the progressive loss of land in the Maswa Reserve is compromising this area as a refuge.

The Great Rinderpest

Rinderpest, a viral disease of cattle that occurs naturally in Asia, was introduced into Ethiopia in 1887 by cattle brought from India by Italian invaders. The resulting epizootic killed over 95% of cattle during 1888–1889; in many cases complete herds died. Buffalo and many species of antelope, particularly hartebeest, were also decimated. By 1892 widespread famine occurred in Ethiopia, Somalia, southern Sudan, and other parts of eastern Africa. By 1896, rinderpest had spread throughout Africa to the Cape. Pankhurst (1966) provides detailed first-hand accounts of the origins of rinderpest, and Ford (1971) describes events in Tanzania. Spinage (2003) provides a review of the history of the event.

The repercussions from this panzootic have had a profound influence on the ecology of the Serengeti over the last century. Rinderpest circulated in the region until the early 1960s (Talbot and Talbot 1963), when it disappeared from the wildlife populations as a result of a cattle vaccination campaign (Sinclair 1977; Plowright 1982; Dobson 1995). By eliminating the disease from the domestic reservoir, the vaccination program effectively protected wildlife from infectious yearling cattle, and consequently the disease rapidly died out in wild populations. Only ruminants are affected by rinderpest, the greater morbidity being in species more closely related to cattle. Thus, buffalo were most affected, followed by wildebeest. Infections were also reported in giraffe and warthog, but other ruminants appear to have been less influenced by the disease (Scott 1981; Rossiter et al. 1987). Rinderpest reappeared in 1982 and caused a few deaths in buffalo but not wildebeest. It has not been an important mortality factor since that time (Rossiter et al. 1983).

The Eruption and Regulation of the Ungulates

A consequence of the disappearance of rinderpest in 1963 was the doubling of juvenile survival in wildebeest and buffalo. Both populations increased exponentially, with buffalo levelling out after 1973 at nearly 75,000 and wildebeest after 1977 at around 1.3 million (figs. 2.12 and 2.13). The removal of this disease provided the conditions for detecting the regulating mechanism operating in the two populations. Both were regulated through density-dependent adult mortality. This mortality took place in the dry season and was caused by a decline in per capita food supply (Sinclair 1977, 1979b; Mduma, Sinclair, and Hilborn 1999). The severe drought of 1993

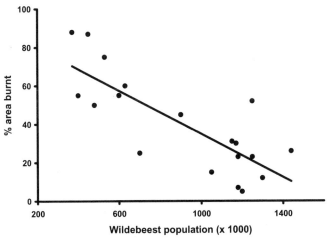

Fig. 2.12 Proportion of northern woodlands burnt relative to wildebeest population. Numbers interpolated from fig. 2.13.

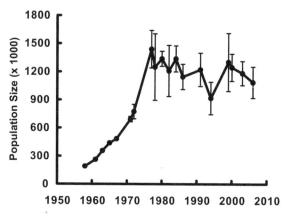

Fig. 2.13 Wildebeest population changes since 1961. Data from Mduma, Sinclair, and Hilborn 1999, and S. Mduma and G. Hopcraft, unpublished (see appendix).

reduced the food supply and caused a 30% drop in both populations. The wildebeest have subsequently rebounded to their previous level. We discuss what happened to the buffalo population in the following section.

Zebra, which as nonruminants are not affected by rinderpest, have remained at constant numbers, around 200,000, for the nearly 40 years 1966 to 2003. The regulation of this population is most likely due to predation (Grange et al. 2004). Also, smaller resident ungulates appear to be limited if not regulated by predation, whereas larger species are regulated by food supply (Sinclair, Mduma, and Brashares 2003).

Competition between Grazing Ungulates

The increase in wildebeest numbers provided circumstantial evidence for competition between grazing ungulates. In particular, the Thomson's gazelle migrant population have declined gradually since wildebeest reached their plateau in the late 1970s. In the Mara Reserve, buffalo numbers remained high until the great drought of 1993, when some 70% died of starvation. Whereas buffalo in other parts of the ecosystem rebounded from the drought (see the following) they did not do so in the Mara Reserve, staying at low numbers. This area is where they overlap with wildebeest during the dry season. Wildebeest consume much of the buffalo's riverine food supply (Sinclair 1977), and the inference is that wildebeest now keep buffalo numbers low. There has been a time delay (hysteresis effect) for compe-

Sinclair, Hopcraft, Olff, Mduma, Galvin, and Sharam

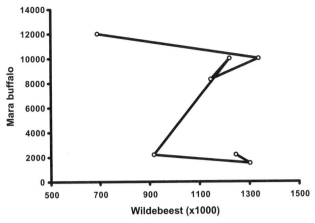

Fig. 2.14 Mara Reserve buffalo population relative to the wildebeest population, showing the decline of buffalo and wildebeest with the 1993 drought. Subsequently only wildebeest increased, suggesting a delayed competitive effect of wildebeest on buffalo.

tition to take effect, and it required the drought to precipitate the drop in numbers (fig. 2.14).

Changes in Savanna Trees

The impacts of wildebeest grazing on the savanna grasslands reduced the dry grass fuel for fires, and as mentioned earlier, the extent of burning declined in direct proportion to the wildebeest increase. Consequently, tree seedlings that had previously been unable to escape from fire were now able to grow. Their germination had probably started as a result of the several years (1971–1976) of unusually wet dry seasons, but until the fire was reduced few were able to survive.

Photographs were taken at established photopoints on woodland hills in 1980. These locations were subsequently rephotographed at intervals of 5 to 10 years. In addition, other photographs from known sites were obtained from the Johnson expeditions of the 1920s and 1930s, from Syd Downey in the 1940s (Dublin 1991), from Myles Turner in the 1950s, and from our own archives in the 1960s (A. R. E. Sinclair, personal observation). All of these sites were rephotographed later. Taking the density of trees in the most recent photos of the 2000s as the reference point, we see that trees have universally increased in density since the 1970s (see chapter 14, fig. 14.1, this volume) in the Tanzania part of Serengeti. This result applies to all

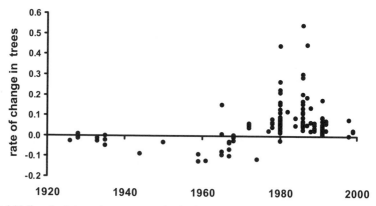

Fig. 2.15 The rate of change in savanna tree density, showing the decline during the 1920–1970 period, followed by a rapid increase in the 1980–1990 period.

species of savanna trees that we could identify, including *Acacia tortilis, A. robusta, A. senegal, A. mellifera, A. drepanalobium, A. seyal, A. polyacantha, A. gerrardiii,* and the broad-leaved trees of the northwest *Terminalia* and *Combretum.* By calculating the instantaneous rates of change in tree numbers between pairs of photos, we see that tree density declined from the 1920s to the 1970s before the rapid increase began (fig. 2.15).

The most widespread and abundant of the savanna trees is *A. robusta,* and ground measurements of the density of different ages (based on diameter and rings in known-age trees; see Sinclair 1995b) shows that there have been two relatively even-aged cohorts (fig. 2.16). One cohort started in the late 1970s in agreement with the photographic data, and the other began in the 1880s, coincident with the arrival of the Great Rinderpest (Sinclair 1995b). Evidence from Wakefield (1870), White (1915), and Johnson (1929) indicates that people were not living in the savanna part of the central Serengeti ecosystem as we now know it because of tsetse fly and the trypanosoma that it carries—this habitat appears to have been free of human habitation during the 1800s. However, humans were living on the boundary in the 1880s, as they are now, especially in Sukumaland and around Ikoma. When the Great Rinderpest arrived they abandoned the area, as the increase in tree thickets allowed the spread of tsetse from the Serengeti outward into previously cultivated areas (Ford 1971). The reduction of humans reduced the incidence of fire and so allowed the escape of tree seedlings in the 1890s (Pankhurst 1966). This cohort produced the extensive tree canopy seen in photos of the 1920s. It was reduced by the increase in burning when the human population returned after the 1930s—at that time there was active reduction of trees by the government as a measure to control the tsetse fly (Ford 1971).

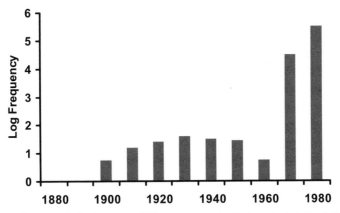

Fig. 2.16 The frequency distribution of trees of different ages, showing the two cohorts starting in the 1890s and 1980s. Before 1900 numbers were too low to be indicated.

By the 1960s *A. clavigera* was relatively uncommon, being present as widespread but low-density mature trees within a complex of small trees and bushes of other species such as *A. hockii* and *Commiphora spp.* (Herlocker 1976b). The second cohort of trees has changed the tree community to one where *A. clavigera* again dominates the habitat.

The Decline of the Riverine Forests

The extent of evergreen groundwater forests along the Mara and Grumeti rivers has been reduced since the 1950s, particularly on the Mara River (Sharam, Sinclair, and Turkington 2006). The width of the Mara River has almost doubled since 1967 (fig. 2.17), and consequently forest along the main riverbed of the Mara has been washed away. This bank erosion has been caused by more voluminous floods in the wet season, most likely due to the extensive deforestation of the Kenya highlands since 1975, because there has not been higher rainfall in that period. An important corollary of deforestation and flooding is the lower volume of water in the dry season. For example, in the Amazon, lower dry-season levels parallel higher wet-season levels (Gentry and Lopez-Parodi 1980, 1982; Nordin and Meade 1982). Since the Mara River is the most important source of water for migrants in the dry season, future reductions in water flow could have negative consequences on their distribution and survival.

On the edge of riverine forests fire, grass competition, and ungulate herbivory limit the regeneration of trees and shrubs (Sharam, Sinclair, and

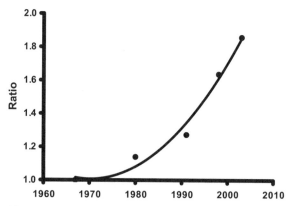

Fig. 2.17 Ratio of the width of the Mara River relative to that in 1967, showing the increase in the width due to floods and bank erosion. The river has almost doubled in width over 40 years.

Turkington 2006). In the Mara River, forests fire can reduce shrubs and trees to their rootstocks. Subsequently, grass grows in the openings caused by fire, so preventing new seedlings from germinating. In addition, browsers such as impala and bushbuck (*Tragelaphus scriptus*) can inhibit sapling growth (Sharam 2005). Thus, despite the overall reduction in burning over the past 40 years, the interactions of fire, grass competition, and browsing prevent the Mara forests from regenerating. There is a conservation concern from the future loss of forests because they support many animal species confined to this habitat. For example, they support some birds, such as Schalow's turaco (*Turaco schalowi*), endemic to the Serengeti region.

The Role of Elephants

As in other parts of savanna Africa, such as Kruger National Park, South Africa, and Chobe National Park, Botswana, the elephant populations of eastern Africa were decimated in the latter half of the 1800s as a consequence of the ivory trade (Spinage 1973; du Toit, Rogers, and Biggs 2003). Even Tsavo National Park, Kenya, now famous for its elephants, had none in the 1890s, when the railway was being built (Patterson 1907). We can assume, therefore, that the Serengeti elephants that supplied the ivory trade routes passing through the area were decimated at this time, because none were seen in the system until at least 1950 (fig 2.18). Elephants invaded the Serengeti as two populations in the early 1960s, one advancing from

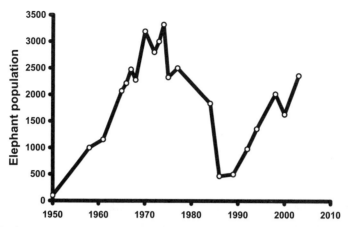

Fig. 2.18 Elephant population changes since 1950, showing the increase following the ivory trade of the 1800s, the decrease due to ivory poaching 1977–1988, and subsequent increase after the ivory trade ban.

Maswa in the south, the other from above the Isuria escarpment, through the Mara Reserve in the north (Lamprey et al. 1967; Watson and Bell 1969; Dublin and Douglas-Hamilton 1987). By 1965 the two populations had met at Seronera and the first elephants were recorded in the far west at Kirawira (A. Sinclair, pers. obs.).

Numbers increased until the mid-1970s, reaching 3,000 in the ecosystem. After the border closure in 1977 ivory poaching increased (see the following) and the population declined. The most severe poaching took place in the 1980s in Serengeti and some 80% of the population was lost. However, in the Mara Reserve little poaching occurred and the population remained constant or increased slightly. In 1988 ivory poaching ceased when elephants were given endangered species status and trade in ivory was banned by the Convention on International Trade in Endangered Species (CITES). Subsequently, the population has increased rapidly.

The role of elephants on the dynamics of savanna tree populations is described previously. In essence, during the dry season elephants depend on savanna *Acacia, Commiphora* trees, and riverine forest trees (Dublin 1986; Sinclair 1995a; Sharam 2005). With abundant savanna trees elephants cause only a small reduction in tree numbers. However, elephants can hold the vegetation in a grassland state by systematically removing seedlings and regenerating rootstocks, as they continue to do in the Mara Reserve since the 1970s, when they reached high densities. The new cohort of *Acacia* trees in Serengeti provides food for the increasing elephant population.

Over the past 40 years infectious diseases have been documented in ungulates (e.g., Sinclair 1977; Plowright 1982) and carnivores (Packer et al. 1999; Haas et al. 1996; Maas 1993), although very little is known about pathogens in other groups such as birds, small mammals, and reptiles.

The role of diseases on ecosystem dynamics in the Serengeti can be summarized in four categories. First, the impact of human populations at the boundaries of the park is reflected in the interaction of their livestock with wildlife. Several pathogens can be transmitted between wild and domestic ungulates, the most influential of which is rinderpest, described previously. Second, as human populations bordering the park have expanded (particularly during the last 40 years), there has been a concomitant increase in domestic dogs. Evidence suggests that this has led to increased incidence of canine distemper virus (CDV) and rabies in wild carnivores (Cleaveland et al. 2000). For example, wild dog numbers declined from around 100 in 1968 to extinction in 1993 (fig 2.19; in 2005 there was evidence of wild dogs returning to the ecosystem). Associated with this decline was the appearance of distemper outbreaks that reduced the population in the late 1960s, mid-1970s, mid-1980s, and early 1990s (Schaller 1972; Malcolm 1979; Creel 1992, MacDonald et al. 1992; Gascoyne et al. 1993; Alexander and Appel 1994). Periodic cases of distemper in jackals have also been recorded (Moehlman 1983). An extraordinary epidemic of CDV affected many carnivores in the mid-1990s (Harder et al. 1995; Roelke-Parker et al. 1996) and was responsible for a massive and rapid decline in lion numbers (Packer et al. 1999). Significant mortality was also recorded in hyenas (Haas 1996), with uncounted deaths in other less well-known carnivores. During the last 10 years, rabies has consistently been reported from dog populations bordering the national park and a range of carnivores inside the park, including jackals, hyenas, bat-eared foxes, and genets. During the late 1980s, rabies accounted for 90% of bat-eared fox mortality in one population (Maas 1993) and fluctuations in bat-eared foxes attributed to rabies outbreaks date back to the 1960s (Leakey 1969). It seems likely that rabies was involved in the final disappearance of wild dogs from the ecosystem. Rabies, therefore, represents a significant public health burden to local communities, and a conservation concern because of spillover into rare wild carnivores. Current research into these diseases of carnivores and vaccination of domestic dogs surrounding the park may curb these threats and increase our understanding of the role of infectious diseases in the Serengeti.

Third, infectious diseases of humans could have dramatic consequences on the ecosystem in the future. Hypothetically rising levels of HIV may

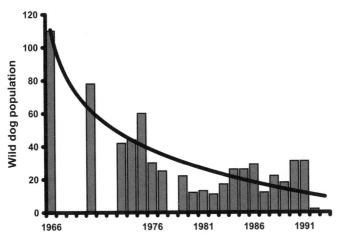

Fig. 2.19 Wild dog numbers on the Serengeti plains between 1966 and 1991.

even reverse population growth, lowering densities of human settlements and affecting their ability to farm and hunt. There is also the possibility of disease transmission from the human reservoir; tuberculosis was recently discovered in mongoose scavenging in refuse dumps at Chobe Park, Botswana (Alexander et al. 2002). Similar opportunities for disease transmission through inadequate waste disposal exist in settlements in and around Serengeti National Park.

Finally, outbreaks of disease appear to be triggered by extreme weather events such as droughts and floods. There are a number of mechanisms that could interact and facilitate disease outbreaks: under harsh conditions nutritional status is lowered, levels of stress are elevated, and immunity is weakened; such weather extremes may also facilitate contact between species as animals congregate at rare resources, and by forcing migratory species into marginal lands, interactions with domestic animals are more likely. During El Nino years, vector-borne (ticks, tsetses, mosquitos) pathogens also thrive. Thus, one of the worst droughts of the last century (1993) was closely followed by a severe outbreak of distemper in wild carnivores (chapter 6, this volume).

Wildebeest as a Keystone Species

We have already observed that the changes in wildebeest numbers have had direct effects by increasing some predator populations and decreasing grass food resources. There have also been indirect effects on ungulate competitors, and via grazing and fire, on savanna tree dynamics. We now see longer-chain indirect effects on the rest of the ecosystem (Sinclair 2003).

Wildebeest graze the eastern treeless plains during the wet season, and today they maintain a short-grass/dwarf-sedge/forb plant community (McNaughton and Sabuni 1988; Augustine and McNaughton 1998). However, when the short-grass plains west of Olduvai Gorge were protected from grazing for about 10 years by exclosure fences a very different long-grass plant community developed, similar to parts of the western plains today. Thus, when wildebeest numbers were low (before 1963), these eastern plains between Olduvai and Naabi supported long-grass communities (M. Leakey and A. Root personal communication; we have no knowledge of how the plains east of Olduvai and on the Salai looked). Thus, we can compare the long-grass and short-grass communities for the indirect effect of wildebeest on forb communities.

There are many more forb species on the grazed plains (some 70 species) relative to the long-grass plains (about 15 species). The short grassland herbs support a much higher density of butterflies. In contrast, grasshoppers are direct competitors with wildebeest, and both grasshopper species diversity and abundance have declined in the presence of wildebeest. This decline was observed during the period when wildebeest were increasing: before 1963, the number of grasshopper species was probably similar to that now found in the long-grass areas today (49 species). In 1972, thirteen species were found near Gol Kopjes in the eastern plains, and by 1986 only one species was found there (A. Harvey, personal communication). The community of small insectivorous and granivorous birds also changed with grazing intensity. Thus, of the 8 most abundant insectivorous bird species on the short-grass plains, 7 declined in abundance by 50 to 80% in the long-grass areas largely due to changes in the physical structure of the grass sward affecting nesting habitat or insect food supply. (A. R. E. Sinclair, personal observation).

HUMAN INFLUENCES

Livelihood Strategies of Peoples Who Live in the Greater Serengeti Ecosystem

The Greater Serengeti ecosystem is characterized by many different types of land uses that range from the Serengeti National Park (SNP), where people are prohibited from living, to highly intensive agriculture, conducted in some areas to the very edge of the park. In general, the southeast-northwest rainfall gradient affects land use such that in the drier areas livestock production prevails, while in the wetter zones to the west, agriculture is the predominant land use.

To the north and to the east of SNP are three multiple-use areas. The Mara Reserve in Kenya is controlled by many pastoral Maasai who live around the edges of the reserve. To the south and east of the Mara Reserve, Maasai live in the Loliondo Game Control Area, where herding and small-scale cultivation occur. Lack of roads and market towns in Loliondo result in a very local economy, except when livestock are sent to Kenyan markets. South of Loliondo is the Ngorongoro Conservation Area (NCA), a multiple-use area of Maasai pastoralists and resident wildlife that live side by side. Maasai in the NCA have been able to carry out limited agriculture since 1991, but they still rely on livestock as the basis of their economy. Although conservation policy has a direct effect on what people can and cannot do in this area, the authorization to allow limited agriculture has had the effect of more than doubling the human population in the NCA. In-migration of farmers from outside the NCA, such as from Arusha and Meru regions, has occurred in large numbers.

Densely settled villages of agriculturalists that also keep livestock characterize the west side of SNP. To the northwest and bordering Kenya are the Wakuria, once pastoralists, who now rely on intensive agriculture for their livelihoods. Maize is a cash crop but livestock are the main source of wealth. To the south and bordering the Serengeti Corridor are mixed groups of smallholder agriculturalists who use SNP resources to a great extent. To the south of the corridor are the Wasukuma, the largest ethnic group in Tanzania, who grow maize and rice as cash crops, but keep some livestock. Livestock owned by all populations that live on the west side of SNP are a source of "savings" for uncertain times.

One of the uncertainties that these populations must cope with is lack of available markets for their livestock and agricultural produce. People are forced to sell within villages for low prices, sell livestock for low prices, or move produce and livestock long distances to better markets in Kenya or to major towns in Tanzania such as Arusha or Musoma.

The increase of the human population is the most pervasive change in all the regions surrounding Serengeti National Park, especially in the regions with higher rainfall west of the park (fig. 2.20). Human population pressure on land use has caused diversification in livelihood strategies. For example, almost all Maasai pastoralists practice some agriculture where

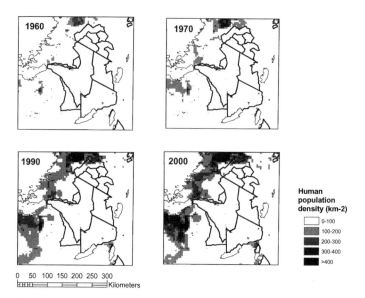

Fig. 2.20 Changes in human population density around the Serengeti-Mara ecosystem from 1960 to 2000. The highest population growth is observed in regions with the highest rainfall (compare with fig. 2.6). Data from the Africa Population Database, compiled by UNEP/GRID.

they have the opportunity. People on the west side of SNP have intensified agricultural production when possible, in some cases by manuring, mulching, and erosion control. This has led to an almost complete conversion of natural vegetation to agricultural land in the whole area between Serengeti and Lake Victoria (fig. 2.21). As agricultural pressures increase so do the pressures on the wildlife in and out of SNP. Illegal hunting supplements household incomes and provides food, especially by those who live on the west side of SNP. Conversion of land from rangeland to commercial agriculture has also impacted land-use strategies. This is most prevalent north of Maasai Mara, in Kenya, where pastoralists are forced to graze their livestock within a diminishing area.

The Border Closure

Political events have also had a major impact on the ecosystem in the past 30 years. In April 1977 the international border between Tanzania and Kenya was closed—and remained so until about 1986, when it was partially opened to tourism. However, the main tourist route between the Mara Re-

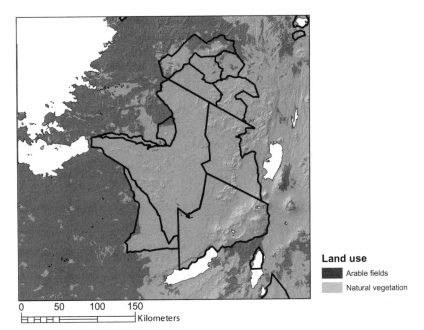

Land use
■ Arable fields
▒ Natural vegetation

0   50   100   150
Kilometers

Fig. 2.21 Map of the two dominant landuse types (2000) in the Serengeti-Mara ecosystem and its surroundings, showing the almost complete conversion of natural vegetation (green) to agricultural land (orange) in the region between Serengeti National Park and Lake Victoria. Primary data provided by the FAO AFRICOVER project, which were produced by supervised classification of Landsat images. Additional corrections for this region were made by checking all polygons within protected areas that were classified as agricultural in a 2000 Landsat image. Additional polygons of agricultural use within protected areas that were not picked up in the AFRICOVER classification were added to the map.

serve and Serengeti National Park (via the Sand River and Bologonja gates) has remained closed.

The immediate effect of the border closure was a precipitous drop in the number of foreign tourists, from around 70,000 in 1976 to 10,000 in 1977 (fig. 2.22). Visitor numbers remained at these very low levels throughout the 1980s; only in the late 1990s have they reached the pre-1977 levels. The Mara Reserve, in contrast, experienced an increase in tourist numbers during the same period of border closure.

One consequence of the border closure for the Serengeti National Park was a drop in income from visitors, with the consequent drop in operating budget. There was a net deficit in operating budget in the 1980s, so that antipoaching operations dropped to some 20% of that prior to border closure. These were the years when elephant poaching for ivory was at its peak (fig. 2.18).

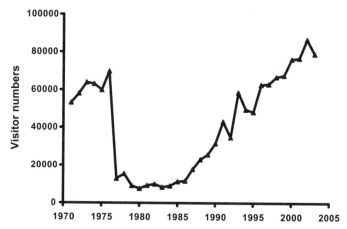

Fig. 2.22 Foreign tourist numbers entering Serengeti National Park, showing the collapse immediately following the border closure in 1977 and the gradual increase once the border re-opened in 1986 (data from Serengeti National Park archives).

Poaching

A marked increase in settlement along the western boundaries of both the Serengeti National Park and Mara Reserve has taken place since the 1960s (fig. 2.20; Dublin 1986; Campbell and Hofer 1995). On average, the annual increase in population has been 3%, but in some areas near the northwestern boundary it approaches 15% per year due to immigration, the supply of wildlife meat being an attraction.

The combined effects of the human population increase and the marked drop in antipoaching patrols resulted in an invasion of northern and western Serengeti by poachers (Hilborn et al. 2006). The first species to feel the effects of this was rhinoceros, which lost 52% of its population in one year alone, 1977, the first year of border closure (fig. 2.23). By 1980 the population was effectively extinct. However, in recent years a few animals have reinvaded the ecosystem and these are closely (and secretly) guarded.

Elephant and buffalo were the next two species to show the effects of poaching. Events concerning elephant are already described (fig. 2.18). Buffalo were almost absent throughout East Africa following the Great Rinderpest (fig. 2.24). Piles of bleached bones but no living animals were seen in 1892 (Pankhurst 1966), no buffalo were observed in 1900 (F.C. Selous in Lydekker 1908) and only a few herds were thought to exist in the whole of East Africa before 1910 (Roosevelt 1910). Numbers slowly increased through the 1920–1960 period (Percival 1924; Moore 1939; Sinclair 1977) in the presence of rinderpest, but then increased exponentially after rinderpest

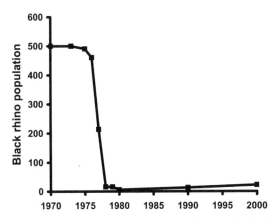

Fig. 2.23 Black rhino population changes showing the rapid decline following border closure.

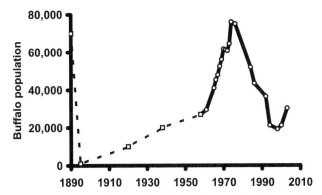

Fig. 2.24 Buffalo population trends, with the reported changes since the Great Rinderpest of 1890, assuming numbers started around the maximum observed in 1976 (broken line), followed by observed numbers since 1958 (solid line). This shows the rapid increase after the removal of rinderpest, the decline by poaching (1977–1992), the collapse in the drought of 1993, and the gradual recovery in areas distant from poachers.

disappeared, reaching a peak in 1976. Ten years later poaching had reduced numbers by some 40% over the park. However, poaching mortality was greatest in northwest and western Serengeti closest to human populations, and absent in the Mara Reserve. After 1986 numbers stabilized, except in the northwest where they continued to decline. In 1993 the great drought caused widespread density dependent decline of a further 50%. Numbers have finally begun to increase and by 2004 most areas were close to their 1986 levels. The exceptions are the northwest where numbers are so low that immigration is required, and in the Mara Reserve where numbers remain low due to possible competition from wildebeest (see above).

Other Possible Human Impacts

First, the increasing human population is likely to progressively take over more land around the boundaries of the protected areas. Wildebeest and other migrants currently move outside the protected areas, so human sequestering of land will reduce the resources for these populations. We have already mentioned this process in the NCA with changes in land tenure; it has been a continual process in Maswa and it will become a problem in the Loliondo area in future. Economic development will exacerbate this process, particularly if the road from Arusha via Endulen is constructed south of the plains to Maswa and Speke Gulf.

Second, pastoralism on the short-grass plains is determined by season, because the plains are only accessible when they are wet. Areas close to Olduvai are accessible year-round. However, placement of boreholes will allow year-round use of a much larger area. Experience in many areas has shown that such use will lead to overgrazing and eventual collapse of that habitat, to the detriment of both humans and wildlife.

Third, human expansion may cut off migration routes. One locality where this will occur is the corner, including Fort Ikoma and Robanda village, where migrants move from the corridor to the north, following the Grumeti River. The other involves the route along the east side of Kuka Hill into the Mara Reserve, where agriculture threatens to cut off access.

Fourth, there are repeated threats from Kenya to divert the headwaters of the Mara River to other river systems. If the Mara stops flowing in the dry season then wildebeest could run out of water, there would be a catastrophic collapse, and the whole ecosystem would change.

CONCLUSION

Long-term work in Serengeti has shown that:

1. Ecosystems involve long-term events related to the environment. Conservation needs to take into account infrequent and unpredictable events, such as floods, fire, and droughts. Long-term data are most necessary to measure these infrequent and unpredictable events, such as the Serengeti floods 36 years apart and droughts 33 years apart.

2. Infrequent natural disturbances provide insight into mechanisms of ecosystem regulation and stability, in this case to understanding wildebeest and buffalo population dynamics. These studies indicate that ecosystems can be self-regulating, and that both resource regulation and predator regu-

lation can occur in different parts of the food web. Thus, bottom-up processes take place through migrants and large herbivores. They supplement the predators that determine the top-down processes.

3. Long-term data are also required to measure slow trends due to environmental change, plant succession, and animal population fluctuation. These slow trends show an interaction between abiotic and biotic processes, each affecting the other, as illustrated by changes in grazing pressure that reduce the extent of burning.

4. Slow change can become an irreversible, rapid shift into a new state, as seen in the interaction of grasslands, woodlands, and elephants. Thus, the ecosystem can occur in multiple states. Some of these can be natural but others can be artifacts of human disturbance.

5. Ecosystems are not static and, therefore, management cannot aim to maintain the *status quo* or just one state—rather, it should allow natural change to take place. It is likely that these changes are oscillatory and will return to previous conditions after a time.

6. Keystone species, such as the wildebeest, can affect all components of the biotic system.

7. Human populations outside the park have increasing impacts on the natural system through land uses such as agriculture and hunting, as well as through the transmission of disease.

7. Protected ecosystems can then be compared with human ecosystems to detect slow human impacts. It is necessary to monitor baseline data to detect human impacts outside parks.

8. The timescales of both unpredictable, sudden events and slow change dictate the timescale for conservation planning. In this case, planning must be for 30- to 50-year periods, or longer.

9. Long-term baseline data are fundamental to conservation management because they provide the background to interpret causes of change and hence determine the course of conservation.

In general, the Serengeti ecosystem is constantly changing, and conservation and management must accommodate these changes.

ACKNOWLEDGMENTS

We thank Charles Peters, Katie Hampson, and several anonymous referees for their most helpful comments.

REFERENCES

Alexander, K. A., and M. J. Appel. 1994. African wild dogs (*Lycaon pictus*) endangered by a canine distemper epizootic among domestic dogs near the Masai Mara National Reserve, Kenya. *Journal of Wildlife Diseases* 30::481–85.

Alexander, K. A., E. Pleydell, M. C. Williams, E. P. Lane, J. F. C. Nyange, and A. L. Michel. 2002. Mycobacterium tuberculosis: An emerging disease of free-ranging wildlife. *Emerging Infectious Diseases* 8:598–601.

Augustine, D. J., and S. J. McNaughton. 1998. Ungulate effects on the functional species composition of plant communities: Herbivore selectivity and plant tolerance. *Journal of Wildlife Management* 62:1165–83.

Brown, L. H., and P. L. Britton. 1980. *The breeding seasons of East African birds.* Nairobi, Kenya: The East Africa Natural History Society.

Campbell, K., and H. Hofer. 1995. People and wildlife: Spatial dynamics and zones of interaction. In *Serengeti II: Dynamics, management and conservation of an ecosystem,* ed. A. R. E. Sinclair and P. Arcese, 534–70. Chicago: University of Chicago Press.

Channing, A., A. R. E. Sinclair, S. A. R. Mduma, D. Moyer, and D. A. Kreulen. 2004. Serengeti amphibians: Distribution and monitoring baseline. *African Journal of Herpetology* 53:163–81.

Cleaveland, S., M. G. J. Appel, W. S. K. Chalmers, C. Chillingworth, M. Kaare, and C. Dye. 2000. Serological and demographic evidence for domestic dogs as a source of canine distemper virus infection for Serengeti wildlife. *Veterinary Microbiology* 72: 217–27.

Creel, S. 1992. Cause of wild dog deaths. *Nature* 360:633.

Dijkshoorn, J. A. 2003. *SOTER database for Southern Africa (SOTERSAF).* Wageningen, The Netherlands: International soil reference and information centre (ISRIC).

Dobson, A. 1995. The ecology and epidemiology of rinderpest virus in Serengeti and Ngorongoro Conservation Area. In *Serengeti II: Dynamics, management and conservation of an ecosystem,* ed. A. R. E. Sinclair and P. Arcese, 485–505. Chicago: University of Chicago Press.

Dublin, H. T. 1986. Decline of the Mara woodlands: The role of fire and elephants. PhD diss., University of British Columbia.

———. 1991. Dynamics of the Serengeti-Mara woodlands: An historical perspective. *Forest and Conservation History* 35:169–78.

Dublin, H. T., and I. Douglas-Hamilton. 1987. Status and trends of elephants in the Serengeti-Mara ecosystem. *African Journal of Ecology* 25:19–33.

du Toit, J. T., K. H. Rogers, and H. C. Biggs, eds. 2003. *The Kruger experience.* Washington, DC: Island Press.

Farler, J. P. 1882. Native routes in East Africa from Pangani to the Masai country and the Victoria Nyanza. *Proceedings of the Royal Geographical Society,* new series 4:730–42.

Ford, J. 1971. *The role of trypanosomiases in African ecology.* Oxford: Clarendon.

Foster, R. 1993. *Dung beetle community ecology and dung removal in the Serengeti.* PhD diss., University of Oxford.

Gascoyne, S. C., M. K. Laurenson, S. Lelo, and M. Borner. 1993. Rabies in African wild dogs (*Lycaon pictus*) in the Serengeti region, Tanzania. *Journal of Wildlife Disease* 29: 396–402.

Gentry, A. H., and J. Lopez-Parodi. 1980. Deforestation and increased flooding of the upper Amazon. *Science* 210:1354–56.

———. 1982. Deforestation and increased flooding of the upper Amazon. *Science* 215:427.

Grange, S., P. Duncan, J.-P. Gaillard, A. R. E. Sinclair, P. J. P. Gogan, C. Packer, H. Hofer, and M. East. 2004. What limits the Serengeti zebra population? *Oecologia* 140: 523–32.

Griffiths, J. F. 1958. Climatic zones of East Africa. *East African Agriculture Journal,* 23: 179–85.

Grzimek, B., and M. Grzimek. 1960. *Serengeti shall not die.* London: Hamish Hamilton.

Grzimek, M., and B. Grzimek. 1960. A study of the game of the Serengeti plains. *Zeitschrift fur Saugetierkunde* 25:1–61.

Haas, L., H. Hofer, M. East, P. Wohlsein, B. Liess, and T. Barrett, 1996. Canine distemper virus infection in Serengeti spotted hyaenas. *Veterinary Microbiology* 49: 147-152.

Harder, T. C., M. Kenter, M. J. G. Appel, M. E., Roelke-Parker, T. Barrett, and A. D. M. E. Osterhaus. 1995. Phylogenetic evidence of canine distemper virus in Serengeti's lions. *Vaccine* 13:521–23.

Herlocker, D. J. 1976(a). Structure, composition and environment of some woodland vegetation types of the Serengeti National Park, Tanzania. PhD diss., Texas A&M University.

———. 1976(b). *Woody vegetation of the Serengeti National Park.* College Station: Texas A&M University.

Hijmans, R. J., S. E. Cameron, J. L. Parra, and A. Jarvis. 2005. Very high resolution interpolated climate surfaces for global land areas. *International Journal of Climatology* 25:1965–78.

Hilborn, R., P. Arcese, M. Borner, J. Hando, G. Hopcraft, M. Loiboki, S. Mduma, and A. R. E. Sinclair. 2006. Effective enforcement in a conservation area. *Science* 314:1266.

Johnson, M. 1929. *Lion.* New York: G. P. Putnam's Sons.

Lamprey, H. F., P. E. Glover, M. Turner, and R. H. V. Bell. 1967. Invasion of the Serengeti National Park by elephants. *East African Wildlife Journal* 5:151–66.

Leakey, L. S. B. 1969. *Animals of East Africa.* Washington, DC: National Geographic Society.

Lydekker, R. 1908. *The game animals of Africa.* London: Rowland Ward.

Maas, B. 1993. Bat-eared fox behavioural ecology and the incidence of rabies in Serengeti National Park. *Onderspoort Journal of Veterinary Research* 60:389–93.

MacDonald, D. W., M. Artois, M. Aubert, D. L. Bishop, J. R. Ginsberg, A. King, N. Kock, and B. D. Perry. 1992. Cause of wild dog deaths. *Nature* 360:633–34.

Maddock, L. 1979. The "Migration" and grazing succession. In *Serengeti: Dynamics of an ecosystem,* ed. A. R. E. Sinclair and M. Norton-Griffiths, 104–29. Chicago: University of Chicago Press.

Malcolm, J. R. 1979. Social organisation and communal rearing of pups in African wild dogs (*Lycaon pictus*). PhD diss., Harvard University.

McNaughton, S. J., and G. A. Sabuni. 1988. Large African mammals as regulators of vegetation structure. In *Plant form and vegetation structure,* ed. M. J. A. Werger, P. J. M. van der Aart, H. J. During, and J. T. A. Verhoeven, 339–54. The Hague, Netherlands: SPB Academic Publications.

Mduma, S. A. R., A. R. E. Sinclair, and R. Hilborn. 1999. Food regulates the Serengeti wildebeest population: A 40-year record. *Journal of Animal Ecology* 68:1101–22.

Mduma, S. A. R., A. R. E. Sinclair, and R. Turkington. Forthcoming. The role of rainfall and predators in determining synchrony in reproduction of savanna trees in Serengeti National Park, Tanzania. *Journal of Ecology.*

Moehlman, P. D. 1983. Socioecology of silverbacked and golden jackals (*Canis mesomeles, Canis aureus*). In *Recent advances in the study of mammalian behavior,* ed. J. F. Eisenberg and D. G. Kleiman, 423–52. Special publication no. 7. Lawrence, Kansas: American Society of Mammalogists.

Moore, A. 1939. *Serengeti.* London: Country Life.

Nordin, C. F., and R. H. Meade. 1982. Deforestation and increased flooding of the upper Amazon. *Science* 215:426–27.

Norton-Griffiths, M. 1979. The influence of grazing, browsing, and fire on the vegetation dynamics of the Serengeti. In *Serengeti: Dynamics of an ecosystem,* ed. A. R. E. Sinclair and M. Norton-Griffiths, 310–52. Chicago: University of Chicago Press.

Norton-Griffiths, M., D. Herlocker, and L. Pennycuick. 1975. The patterns of rainfall in the Serengeti ecosystem, Tanzania. *East African Wildlife Journal* 13:347–74.

Packer, C., S. Altizer, M. Appel, E. Brown, J. Martenson, S. J. O'Brien, M. Roelke-Parker, R. Hofmann-Lehmann, and H. Lutz. 1999. Viruses of the Serengeti: Patterns of infection and mortality in African lions. *Journal of Animal Ecology* 68:1161–78.

Pankhurst, R. 1966. The great Ethiopian famine of 1888–1892: A new assessment. Part 1. *Journal of the History of Medicine and Allied Sciences* 21:95–124.

Patterson, J. H. 1907 *The maneaters of Tsavo.* Repr., New York: Pocket Books, 1996.

Pearsall, W. H. 1957. Report on an ecological survey of the Serengeti National Park, Tanganyika. *Oryx* 4:71–136.

Percival, A. B. 1924. *A game ranger's notebook.* London: Nisbet & Co.

———. 1928. *A game ranger on safari.* London: Nisbet & Co.

Plowright, W. 1982. The effects of rinderpest and rinderpest control on wildlife in Africa. *Symposium of the Zoological Society, London* 50:1–28.

Roelke-Parker, M. E., L. Munson, C. Packer, R. Kock, S. Cleaveland, M. Carpenter, S. J. O'Brien, A. Pospischil, R. Hofmann-Lehmann, and H. Lutz. 1996. A canine distemper virus epidemic in Serengeti lions (*Panthera leo*). *Nature* 379:441–45.

Roosevelt, T. 1910. *African game trails.* London: John Murray.

Rossiter, P. B., D. M. Jessett, J. S. Wafula, L. Karstad, S. Chema, W. P. Taylor, L. Rowe, J. C. Nyange, M. Otaru, and Mumbala, M. G. R. 1983. Re-emergence of rinderpest as a threat in East Africa since 1979. *Veterinary Record* 113:459–61.

Rossiter, P. B., W. P. Taylor, B. Bwangamoi, A. R. H. Ngereza, P. D. S. Moorhouse, J. M. Haresnape, J. S. Wafula, J. F. C. Nyange, and I. D. Gumm. 1987. Continuing presence of rinderpest virus as a threat in East Africa 1983–1985. *Veterinary Record* 120:59–62.

Schaller, G. B. 1972. *The Serengeti lion.* Chicago: University of Chicago Press.

Schlesinger, W. H. 1991. *Biogeochemistry: An analysis of global change.* San Diego, CA: Academic Press.

Scott, G. R. 1981. Rinderpest. In *Infectious diseases of wild mammals,* ed. J. W. Davis, L. Karstad, and D. O. Trainer, 18–30. Ames: Iowa State University Press.

Sharam, G. J. 2005. The decline and restoration of riparian and hilltop forests in the

Serengeti National Park, Tanzania. PhD diss., University of British Columbia, Vancouver.

Sharam, G. J., A. R. E. Sinclair, R. and Turkington. 2006. Establishment of broad-leaved thickets in Serengeti, Tanzania: The influence of fire, browsers, grass competition and elephants. *Biotropica* 38:599–605.

Sinclair, A. R. E. 1977. *The African buffalo.* Chicago: University of Chicago Press.

———1979a. Dynamics of the Serengeti ecosystem: Process and pattern. In *Serengeti: Dynamics of an ecosystem,* ed. A. R. E. Sinclair and M. Norton-Griffiths, 1–30. Chicago: University of Chicago Press.

———1979b. The eruption of the ruminants. In *Serengeti: Dynamics of an ecosystem,* ed. A. R. E. Sinclair and M. Norton-Griffiths, 82–103. Chicago: University of Chicago Press.

———1979c. The Serengeti environment. In *Serengeti: Dynamics of an ecosystem,* ed. A. R. E. Sinclair and M. Norton-Griffiths, 31-45. Chicago: University of Chicago Press.

———1995a. Equilibria in plant-herbivore interactions. In *Serengeti II: Dynamics, management and conservation of an ecosystem,* ed. A. R. E. Sinclair and P. Arcese, 91–114. Chicago: University of Chicago Press.

———1995b. Serengeti past and present. In *Serengeti II: Dynamics, management and conservation of an ecosystem,* ed. A. R. E. Sinclair and P. Arcese, 3–30. Chicago: University of Chicago Press.

———2003. Mammal population regulation, keystone processes and ecosystem dynamics. *Philosophical Transactions of the Royal Society, London, B.* 358:1729–40.

Sinclair, A. R. E., and P. Arcese, eds. 1995. *Serengeti II: Dynamics, management and conservation of an ecosystem.* Chicago: University of Chicago Press.

Sinclair, A. R. E., and M. Norton-Griffiths, eds. 1979. *Serengeti: Dynamics of an ecosystem.* Chicago: University of Chicago Press.

Sinclair, A. R. E, S. A. R. Mduma, and J. S. Brashares. 2003. Patterns of predation in a diverse predator-prey system. *Nature* 425:288–90.

Sinclair, A.R.E, S. A. R. Mduma, J. G. C. Hopcraft, J. M. Fryxell, R. Hilborn, and S. Thirgood. 2007. Long term ecosystem dynamics in the Serengeti: Lessons for conservation. *Conservation Biology* 21:580–90.

Speke, J. H. 1863. *Journal of the discovery of the sources of the Nile.* London and Edinburgh: Harper & Brothers.

Spinage, C. A. 1973. A review of ivory exploitation and elephant population trends in Africa. *East African Wildlife Journal* 11:281–89.

——— 2003. *Cattle plague, a history.* Dordrecht, The Netherlands: Kluwer Academic/ Plenum.

Talbot, L. M., and M. H. Talbot. 1963. The wildebeest in western Masailand. Wildlife Monographs no. 12. Washington, DC: The Wildlife Society.

Turner, M. 1987. *My Serengeti years.* Ed. B. Jackman. London: Elm Tree Books/ Hamish Hamilton Ltd.

Wakefield, T. 1870. Routes of native caravans from the coast to the interior of East Africa. *Journal of the Royal Geographical Society* 11:303–38.

———. 1882. New routes through Masai country. *Proceedings of the Royal Geographical Society* 4:742–47.

Watson, R. M. 1967. The population ecology of the wildebeest (*Connochaetes taurinus albojubatus*) in the Serengeti. PhD diss., Cambridge University, Cambridge.

Watson, R. M., and R. H. V. Bell. 1969. The distribution, abundance and status of elephant in the Serengeti region of northern Tanzania. *Journal of Applied Ecology* 6: 115–32.

White, S. E. 1915. *The Rediscovered Country.* New York: Doubleday.

THREE

Paleoecology of the Serengeti-Mara Ecosystem

Charles R. Peters, Robert J. Blumenschine, Richard L. Hay, Daniel A. Livingstone,
Curtis W. Marean, Terry Harrison, Miranda Armour-Chelu, Peter Andrews,
Raymond L. Bernor, Raymonde Bonnefille, and Lars Werdelin

How and when did the current Serengeti-Mara ecosystem come into being? These are paleoecological questions that cannot be answered fully. We are limited in our understanding of the timing of geological changes that resulted in the landscape as we know it today, in fossil sites with well-preserved fauna and flora, and in our knowledge of how the system operated historically before the great rinderpest epidemic in the late nineteenth century, and before settlement, agriculture, and mining in the twentieth century cut off access to Lake Victoria and the lower Mara River wetlands to the northwest. Nonetheless, we can begin to outline some of the critical paleocomponents of the system, and summarize what is known about their evolution. The paleoperspective is an important one, for it reveals the ecological dynamics of the system on an evolutionary timescale. It raises long-term questions about the system that might otherwise be lost in the politics of short-term survival. For all of the deficiencies of its prehistoric record, the Serengeti-Mara is the modern, semiprotected African ecosystem for which the most comprehensive evolutionary synthesis can be attempted.

Our temporal scale is the past 4 to 5 million years. During this time, tectonic activity leading to the major faulting and volcanism that define the physical boundaries of the Serengeti-Mara ecosystem began. Our spatial scale, somewhat broader than that usually depicted for the current Serengeti-Mara ecosystem, spans northwest Tanzania and southwest Kenya. The geography of the modern ecosystem is depicted in fig. 3.1. To the east

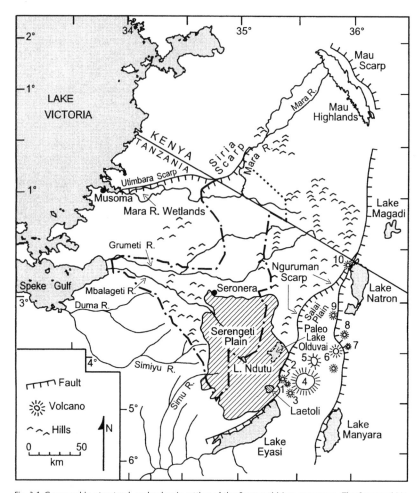

Fig. 3.1 Geomorphic, structural, and volcanic setting of the Serengeti-Mara ecosystem. The Serengeti National Park (Tanzania) is denoted by a dot-dash line, the Masai Mara National Reserve (Kenya) by a dotted line. These park and reserve boundaries, the Kenya/Tanzania border, and Lake Victoria are based on the USGS Digital Atlas of Africa (Hearn et al. 2001). Numbered volcanoes of the Crater Highlands and Rift Valley are: (1) Sadiman; (2) Lemagrut; (3) Oldeani; (4) Ngorongoro; (5) Olmoti; (6) Embagai; (7) Kerimasi; (8) Oldoinyo Lengai; (9) Mosonik; (10) Oldoinyo Sambu. The Rift Valley lakes of Eyasi, Manyara, Natron, and Magadi are shown for reference. Paleo-lake Olduvai, as depicted, was present from ca. 1.92–1.70 Ma (Hay and Kyser 2001). Rift Valley lakes, main river courses, the Mau Highlands (area above 9,000 ft [2,744 m]) and the locations of volcanoes are based on the 1982 Tactical Pilotage Chart M-5A (1:500,000 scale). The Mau, Siria, and Kenyan Nguruman escarpments are based on the 1987 Petroleum Exploration Promotion Project *Geological Map of Kenya with Structural Contours* (1:1,000,000 scale). The Utimbara Escarpment is based on Barth's (1990) Provisional Geological Map of the Lake Victoria Goldfields, Tanzania (1:500,000 scale). The Natron, Manyara, and Eyasi escarpments, plus the volcano representations, are based on Dawson (1992; fig. 1). Fault hachures are on the downthrown side of faults. Representation of hills is based on Talbot and Talbot (1963; fig. 2), in part. Mara River details are based on 1:250,000 scale topographic maps: Kisumu (Y503, SA-36-4, 1-GSGS) for the uppermost (NE) sections, Musoma (Y503, SA-37-7, I-TSO) for the lower (western) section and the Mara River wetlands.

and southeast are the Rift Valley and volcanic Crater Highlands. To the northeast is the Mau Escarpment, whose uplifted and back-tilted highland forms the upper catchment of the Mara River. To the west is Lake Victoria. In the northern Serengeti, the course of the Mara River, on its way to Lake Victoria, is confined first by the Siria Escarpment in Kenya, and then by the Utimbara Escarpment in Tanzania. The western corridor of the Serengeti National Park ends before the Grumeti and Mbalageti rivers reach Lake Victoria. These rivers are seasonal. The middle Mara River to the north, and the lower Mara River in Tanzania, along with some of their wetlands, are usually perennial. Southwest of the Serengeti National Park lie the woodlands of the Maswa Game Reserve.

Three aspects of the evolution of the Serengeti-Mara ecosystem are considered here, focusing on the possible origins of a broadscale migratory system. One, we examine the geoecology of the system, including the evolution of the eastern plain, the wet-season pasture of the wildebeest and zebra today, and the Mara River and Lake Victoria as potential drought refugia. Two, we sketch the evolution of large mammals, especially the larger herbivores and carnivores. Three, we assess the hominin factor, the use of the system by prehistoric hunter-gatherers and pastoralists, as well as the more dramatic historical human impacts.

GEOECOLOGICAL EVOLUTION OF THE SERENGETI PLAIN

Volcanic Ash Plains

The southeastern Serengeti today is a sediment plain, with soils developed largely from recent windblown volcanic ash and older consolidated ashes (tuffs). The east-central portion of the plain is most clearly influenced by recent volcanics (Anderson 1963; Anderson and Talbot 1965; de Wit 1978). This open, dry, almost treeless plain supports vast ungulate herds in the wet seasons of relatively wet years. It is a calving ground for wildebeest and zebra. In wet years the heavily grazed pasture is a mosaic of short grasses (esp. *Sporobolus*), dwarf sedges (esp. *Kyllinga*), and many species of forbs. Unlike the taller grasslands on the adjacent portions of the open plain to the west and northwest, which may be burned annually, this pasture is rarely, if ever, burned. Conspicuously lacking termite mounds, the soil is a shallow, alkaline, very fine sandy loam underlain by calcareous hardpan. Below the hardpan the subsoil is saline (de Wit 1978). In places there are stabilized dunes of dark, very fine volcanic material. In the Salai Plain, just to the northwest of the volcanic source, Oldoinyo Lengai, the dunes are mobile.

The southernmost Serengeti Plain, south of Lake Ndutu and west of Laetoli, is predominantly short-to-medium grassland on black cotton soils (silty clay loams) derived from calcareous tuff (Anderson and Talbot 1965). It is grazed heavily by livestock and, in wet years, by wildebeest and zebra from the north.

Laetoli (Eyasi) Sector

Laetoli is located in the Garusi Valley on the northeastern margin of the Eyasi Plateau in the southern Serengeti Plain, about 20 km north of Lake Eyasi (see fig. 3.1). The plateau is an uplifted fault block, with a general elevation of 1,700 to 1,800m, forming the divide between the Lake Eyasi and Olduvai Gorge drainage basins (Hay 1987). Outcrops of the Pliocene Laetolil and Ndolanya Beds are exposed over an area of ca.1,600 km², extending more than 20 km to the west and northwest of Laetoli. For a summary of the geological dates, see the appendix to this chapter. The Laetolil Beds, which are more than 123 m thick, are divisible into two major lithological units (Hay 1987). The lower unit, which rests unconformably on the Precambrian basement rock, consists mainly of eolian tuffs (i.e., wind-transported volcanic ashes) interbedded with primary airfall tuffs and some water-worked tuffs and conglomerates. Potassium-argon (K-Ar) dates indicate that the lower unit ranges in age from ca. 3.76 Ma (million years ago) to older than 4.3 Ma, and if we assume uniform sedimentation rates it might be as old as 4.5 Ma (Drake and Curtis 1987). The upper unit comprises a series of eolian and airfall tuffs, with localized water-worked tuffs deposited in small, ephemeral ponds. The top of the unit is delimited by the Yellow Marker Tuff, with other distinctive airfall tuffs throughout the sequence designated as marker tuffs (Tuffs 1–8, numbered from bottom to top). K-Ar dates constrain the age of the sediments to ca. 3.5–3.76 Ma (Drake and Curtis 1987).

The Laetolil Beds are derived from the now-extinct Sadiman Volcano (Hay 1978, 1981, 1986, 1987). Sedimentary evidence indicates that the volcanic ashes were deposited and reworked subaerially on a paleoland surface, and that there were no large permanent lakes or rivers on the Pliocene Eyasi Plateau. Vertebrate fossils occur in both the lower and upper Laetolil Beds, but are more common in the upper unit (Harris 1987). Fossil hominins, *Australopithecus afarensis,* are restricted to the upper unit (White 1977, 1980; Leakey 1987).

Overlying the Laetolil Beds disconformably are a series of massive eolian tuffs and calcretes comprising the Ndolanya Beds (15–20 m thick), subdivided into upper and lower units. Radiometric dating and faunal correlations are consistent with an age of ca. 2.6–2.7 Ma for the fossiliferous upper Ndolanya Beds (Beden 1976, 1987; Hooijer 1987; Harris 1987; Gentry 1987; Ndessokia 1990; Manega 1993). A diverse vertebrate fauna has been recovered (Harris 1987), including the fossil hominin, *Paranthropus aethiopicus* (Harrison 2002).

From ca. 4.5 to 3.5 Ma Sadiman (fig. 3.1) discharged large volumes of nephelinitic ash, most of which was transported by prevailing winds toward the southwest (Hay 1987). Sadiman ash covered much of the southern Serengeti Plain, judging from its 16 m thickness at Lake Ndutu 45 km to the northwest (Hay 1987). Sadiman ash was probably blown southwest more than 100 km to the Manonga Valley (based on a combination of evidence from Harrison and Mbago 1997; Verniers 1997; Mutakyahwa 1997; Harrison and Baker 1997). Natrocarbonatitic ash was ejected in some eruptions, and it contributed to cementing of the tuffs and preserving vertebrate remains and fossil footprints in the Laetoli area. (Natrocarbonatite is a carbonate rock of magmatic origin, composed chiefly of sodium, potassium and calcium carbonates. The calcium carbonate becomes a fine precipitate, useful in the preservation of footprints, etc. The sodium-carbonate component is readily dissolved in water, forming a highly alkaline solution that poisons the water for drinking [see historical notes on Oldoinyo Lengai, below].)

Interpretations of the footprinted natrocarbonatitic ashes bear on the possibility of a Serengeti-type migratory system in the middle Pliocene. Although composed mostly of reworked and heavily bioturbated eolian volcanic ash, some of the natrocarbonatitic tuffs in the upper unit of the Laetolil Beds are relatively undisturbed air-fall ashes similar to those erupted from Oldoinyo Lengai in 1966. A few of these finely laminated tuffs represent closely spaced series of ash falls that preserve footprints, and sometimes the impressions of raindrops, on bedding surfaces between the ash falls. These features are particularly abundant in the lower part of Tuff 7, and that part is termed the Footprint Tuff.

The Footprint Tuff is about 15 cm thick and has been subdivided into lower and upper units that differ in sedimentology and type of footprints. The lower and thicker unit comprises 14 thin ash-fall layers that were undisturbed after deposition, except for rainprints, found on nine tuff layers, and footprints, found on eight. Raindrop imprints were close together and well defined, having been produced by showers sufficient to dampen the ash but not to erode it. The upper unit comprises four ash-fall tuffs that were water-worked. Each layer was redeposited extensively by water, probably sheetwash from heavy showers. Rainprints were uncommon.

Footprints are particularly common on the basal ash of the upper unit of the Footprint Tuff, and at one horizon in the lower unit. The Footprint

Tuff was interpreted to represent a succession of ash falls covering a few weeks at the end of the dry season, extending into the early part of the following rainy season (Hay and Leakey 1982; Hay 1987). The ash falls were chiefly of natrocarbonatite and sand-size grains of silicate lava, but original proportions of these components are unknown.

Footprints of the lower unit of the Footprint Tuff are chiefly of lagomorphs, guineafowl, and rhinoceros, animals that today remain in grassland and savanna habitats throughout the dry season. Footprints of the upper unit of the Footprint Tuff include those of proboscideans, equids, baboons, and hominins, which are not found in the lower unit, and footprints of large bovids, which are much more common in the upper unit than in the lower unit. The apparent influx of large mammals was interpreted to record a rainy-season migration of the type seen in the Serengeti today (Hay and Leakey 1982; and Hay 1987). A combination of normal walking gait and the use of preexisting animal trails led to the conclusion that these movements of the large mammals at Laetoli were undisrupted by the soda ash fallout (Hay and Leakey 1982; Hay 1987). Judging from available historical notes on the impact on the local ecosystem of the eruptions of Oldoinyo Lengai in northern Tanzania (see the following), this interpretation is improbable.

Oldoinyo Lengai (fig. 3.1), the world's only active carbonatite volcano, is famous for its alkali-rich magma, natrocarbonatite (Krafft and Keller 1989). Explosive eruptions of this alkali-rich carbonatite in the form of soda ash dominated its volcanic activity during the twentieth century, although silicate igneous rocks make up the bulk of the volcano (Dawson et al. 1994). A number of published observations are relevant to interpreting the Laetoli ecosystem during the Pliocene.

In 1917, during an approximately 5-month eruption sequence (January until about June), ash from Oldoinyo Lengai fell over the nearby countryside to a distance of 40 to 50 km west and south (Hobley 1918). Large areas of grazing land were temporarily destroyed by the soda ash. Rainwater in pools was so strongly alkaline from the soda that it was undrinkable by cattle for 4 to 5 days. Many Masai cattle herds died drinking this water, where there was no other water to drink (springs were unaffected). In 1940–1941, during another 5-month eruption sequence (end of July until early January), fine ash was carried by prevailing winds to a distance of 100 km west-northwest, west, and southwest (Richard 1942). The air, hazy with fine ash, irritated the throat and had a salty taste. Nearer the volcano, especially on the western side, grazing land was completely ruined and waterholes were spoiled by the soda ash. The area was abandoned, and free of wildlife for several months. On a visit to OlBalbal and Olduvai Gorge, J. J. Richard was shown cattle that "had lost their hair through the action of the alkaline ashes" (Richard 1942, p. 96). During both the 1917 and 1940–1941 eruptions, the woody vegetation on the slopes of Oldoinyo Lengai apparently was also destroyed (1942, p. 96, and Richard's figure 11; also see Dawson 1962 for the 1917 eruption). In 1966, during a 3-month eruption sequence (early August to the end of October), fine ash was carried by the prevailing winds west to north-northwest, to a distance of 130 km (to Seronera in the NW corner of the Serengeti Plain) (Dawson, Bowden, and Clark 1968). Again, the Oldoinyo Lengai area was abandoned by both local Masai and wildlife.

Similarly, in the mid-Pliocene, during extended natrocarbonatitic eruptions of Sadiman, large mammals probably left the Sadiman-Laetoli area. Moreover, since these soda ash eruptions probably produced catastrophic effects for several kilometers downwind, these animals may have had to travel some distance to escape this temporarily inhospitable environment. Many of their movements may have been random searches for forage and drinkable water within the Laetoli area. Modern analogues indicate that it is more likely that the trails of footprints of large mammals were made by animals ultimately abandoning the area, rather than providing evidence of a modern-day Serengeti-type migration in the Pliocene.

Olduvai Sector

Several large volcanoes comprising the Crater Highlands were formed along the southeast margin of the Serengeti Plain between 2.5 and 1.7 Ma (Dawson 1992), during a time of minor faulting. The largest, Ngorongoro, was active from ca. 2.45 to 2.0 Ma (Hay 1976). The caldera rim ranges in elevation from ca. 1,400 to 2,100 m, and the summit of the volcano may well have had an elevation of 4,500 m prior to caldera collapse. Other volcanoes active during this time period include Olmoti, Embagai, and probably Lemagrut and Oldeani (fig. 3.1). These volcanoes created a rain shadow over most of the eastern Serengeti Plain.

Paleogeography of the Olduvai sector has varied over the past 2 million years, chiefly in response to faulting. The lower sedimentary deposits of Olduvai Bed I in the western part of the gorge (2.03–1.92 Ma, lowermost Bed I) accumulated on an alluvial plain that received fluvial detritus from both Ngorongoro and basement rocks to the west. A shallow closed-basin lake, termed Lake Olduvai (Hay and Kyser 2001), was formed ca. 1.92 Ma. It was saline-alkaline, and 15–20 km in average diameter at times of high lake level. At the end of Bed I, during deposition of Tuff IF (ca. 1.79 Ma), the lake was a dry playa. The lower Bed II lake quickly reached high levels again. The extent of the lake fluctuated considerably in a climate with average temperatures very likely higher than during the deposition of Bed I, judging from the common appearance of trona, an evaporative material found only in the Bed II lake deposits. Major deformation at ca. 1.70 Ma shifted the Bed II lake considerably to the east and greatly reduced its size. At ca. 1.2 Ma (in early Bed III times), the lake disappeared as a result of faulting. Beds III–IV (ca. 1.15–ca. 0.78 Ma [Hay 1976]) and the Masek Beds (ca. 0.78–ca. 0.50 Ma [Hay 1976]) were deposited on an alluvial plain. For a summary of the geological dates, see the appendix to this chapter. Major fault displacement at ca. 500 ± 100 ka produced the north-south rift valley at the western foot of the volcanoes Ngorongoro and Olmoti, lowering the base level in the Olduvai closed basin, and initiating erosion of the Olduvai Gorge, which continued through the deposition of the Ndutu Beds (ca. 0.5–ca. 0.1 Ma [modified from the dates in Hay 1976]).

Volcanic ash contributed to the formation of the eastern Serengeti Plain episodically through the Late Pliocene and Pleistocene. For example, eolian tuffs are common at Olduvai in the Lemuta Member of Bed II (Pickering 1958, 1960), which has an age of about 1.7 Ma. It is only in the later Pleistocene that natrocarbonatitic ash again plays a role. Eruptions of Kerimasi (ca. 0.78–0.50 Ma) and Oldoinyo Lengai (ca. 0.5 Ma to present) supplied large volumes of volcanic ash to the Olduvai Basin and eastern Serengeti Plain (fig. 3.1) in the middle and late Pleistocene. These volcanoes discharged principally nephelinite and phonolite tephra, but natrocarbonatite was erupted in the late stages of both volcanoes (Hay 1983; Dawson 1962). The modern C_4 vegetation of the semiarid short-grass plain apparently evolved during this time (see below).

Evolution of Vegetation on the Eastern Serengeti Plain

The eolian character of many of the Pliocene and Pleistocene sediments of the eastern Serengeti Plain at Laetoli and Olduvai Gorge suggests periods of bare ground after volcanic eruptions and succession of mobile to stabilized ash dunes vegetated with grasses, forbs, and scattered *Acacia,* as seen today in the Salai Plain and Serengeti short-grass plain near Olduvai Gorge. For example, in the Upper Laetolil Beds, sharp contacts between microlaminations in major marker tuffs, and extensive eolian reworking of tuffs, led Hay (1987) to infer that the volcanic ashes settled on a terrain with little or no vegetation. In contrast, the lowermost layer of the Footprint Tuff is associated with a rich and diverse assemblage of fossilized twigs, leaves, seeds, and tree boles in situ, that suggests woodland as deposition of Tuff 7 began. Deposits with fossilized termite mounds suggest brief periods of mound-free grassland succeeded by juvenile soil formation and mound construction over cycles of a few thousand years, during volcanically active periods (Sands 1987).

The fossil pollen record (Bonnefille and Riollet 1987) could be interpreted as indicating that open savanna, with scarce trees, existed over tens or hundreds of thousands of years, but the isotope signatures of the paleosol (ancient soil) carbonates and vertebrate fossils indicate predominantly woodland or grassy woodland vegetation (Cerling 1992; Kingston and Harrison 2001, 2002, 2007). Differing interpretations are possible of how modern in form the grassland was, especially when different lines of evidence are not synthesized to create an integrated reconstruction. Earlier interpretation tended to emphasize the possibility that the modern Serengeti system was present in the Pliocene and early Pleistocene 4.0 to 1.5 million years ago. Reevaluations provide evidence for less savanna and more woodland/bushland (Andrews 1989, 2006; Cerling 1992; Kingston and Harrison 2001, 2002, 2007; Kovarovic, Andrews, and Aiello 2002; Kovarovic 2004; Su 2005; Su and Harrison, 2007). This is not to say that grasses were an unimportant part of the ecosystem or to deny that grazing ungulates dominated the mammalian fauna, but rather to conclude that the Serengeti open grassland of today was not present.

Grasses have been an important part of the vegetation of Africa since at least the late Miocene, ca. 10 Ma (e.g., Morley 2000). However, stable carbon isotope analyses of paleosol carbonates from East African fossil sites spanning the past 9 million years indicate that the savanna grasslands we see in the modern landscapes of East Africa developed only in the last million years (Cerling 1992). Moreover, the virtually treeless short-grass plain of the Serengeti probably developed only in the mid to late Pleistocene, 500,000 to 100,000 years ago (Cerling 1992; Cerling and Hay 1986; see the following). Prior to the mid to late Pleistocene, a range of vegetation from grassy woodland to a wooded or shrub grassland mosaic may have dominated the plains, with local conditions of impeded drainage or volcanic ash dunes favoring grassland successions at finer spatial scales for hundreds to thousands of years. Deposits at Laetoli and Olduvai Gorge preserve some of this early record for the eastern Serengeti Plain.

Overview of 4 Million Years of Vegetation Change, Based on the Isotope Record for Paleosol Carbonates from the Eastern Serengeti Plain

Woody plants are typically C_3 in their mode of photosynthesis, whereas most of the low- to mid-elevation tropical grasses are C_4. The stable carbon isotopic composition of modern soil carbonate and paleosol carbonate is

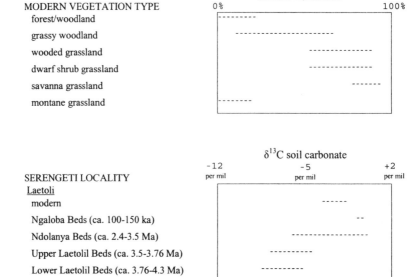

MODERN VEGETATION TYPE	fraction C$_4$ biomass 0%	100%
forest/woodland	----------	
grassy woodland	-----------------------	
wooded grassland	-----------------	
dwarf shrub grassland	----------------	
savanna grassland	--------	
montane grassland	--------	

SERENGETI LOCALITY	δ^{13}C soil carbonate −12 per mil	−5 per mil	+2 per mil
Laetoli			
modern		------	
Ngaloba Beds (ca. 100-150 ka)		--	
Ndolanya Beds (ca. 2.4-3.5 Ma)		-------------------	
Upper Laetolil Beds (ca. 3.5-3.76 Ma)	-----------		
Lower Laetolil Beds (ca. 3.76-4.3 Ma)	----------		
Olduvai Gorge			
modern		-------	
Ndutu (ca. 100-500 ka)		-------	
Masek (ca. 500-780 ka)		---------------	
Bed IV (ca. 780-950 ka)	------		
Bed III (ca. 950 ka - 1.15 Ma)	-----------		
Bed II (Lemuta, ca. 1.7-1.75 Ma)	---------------------		
Bed II (excl. Lemuta, ca. 1.15-1.79 Ma)	------		
Bed I (ca. 1.79-2.0 Ma)	--------		

Fig. 3.2 Estimates of C$_4$ biomass for modern east African vegetation types and the stable carbon isotope composition of modern soil carbonate and paleosol carbonate from Laetoli and Olduvai Gorge (from Cerling 1992, fig. 3, with some modifications). *Savanna grassland* here denotes grassland with virtually no tree cover. *Wooded* (or bushed) *grassland* denotes grassland with scattered or grouped trees, always conspicuous, but having a canopy cover of less than 20% (following Pratt, Greenway, and Gwynne 1966). *Dwarf shrub grassland* denotes arid land covered by grasses and shrubs not exceeding 1 m in height. We have added the geological dates to this figure: see the appendix to this chapter for the sources.

an indicator of the proportion of C$_4$ photosynthesis that occurred in the vegetation of the modern and ancient soils (Cerling 1992). Soil carbonate formed under pure C$_3$ vegetation has a delta 13C value of −10 per mil to −12 per mil, compared to ca. +2 per mil for the case of pure C$_4$ vegetation (see fig. 3.2). Modern soil carbonates from the Serengeti short-grass plain in the vicinity of Olduvai Gorge have an average delta 13C value of +0.5 per mil (± 0.5; Cerling 1992, table 1).

As can be seen in fig. 3.2, the carbon isotope values for the Laetolil Beds (ca. 4.3–3.5 Ma) fall within the range of grassy woodland. The land surfaces of the Ndolanya Beds (ca. 2.4–3.5 Ma) were overall probably drier than those of the Laetolil Beds, with less extensive tracts of woodland. Cerling (1992) suggests the possibility of a grassy bushland. Recent reanalysis of the fauna indicates that the vegetation of the upper Ndolanya Beds (ca. 2.6–2.7 Ma) was semiarid bushland (Kovarovic, Andrews, and Aiello 2002; Andrews, 2006). The Ngaloba Beds at Laetoli appear to have a strong grassland signature, but they may be as young as 100,000–150,000 years BP (Hay 1987).

For the Olduvai Gorge sequence, Cerling and Hay (1986) provide information on depositional environments important for the ecological interpretation of the pedogenic carbonates. For the period from ca. 1.9–2.0 Ma (lower Bed I), three samples of pedogenic carbonate are shown in their figure 4A, but not identified as such in their table 1. These are from eolian deposits (cemented by calcite) in the western paleobasin, and their isotope values indicate vegetation that was a mixture of C_3 and C_4 plants, suggesting a grassy woodland. From ca. 1.79 Ma (upper Bed I) there are two pedogenic carbonate samples originally thought (Hay 1976) to have been from lower Bed II. These are from alluvial fan deposits in the eastern basin prior to the eruption of Tuff IF. Their isotope values are similar to (though somewhat higher than) those of lower Bed I. Altogether they are interpreted as representing a range of C_4 biomass of 40–60%. From ca 1.70–1.75 Ma (the Lemuta Member, in lower Bed II), eolian deposits overlie alluvial fan deposits in the eastern part of the paleobasin. Three tongues of eolian tuff interfinger westward, with lake-margin claystones (Hay 1976, fig. 19). These eolian units were formed during relatively arid episodes in lower Bed II. Pedogenic carbonates of varied types in the upper unit (ca. 1.72–1.70 Ma) give isotope values indicating vegetation with 20–90% C_4 biomass (their table 2), a broader range than depicted in fig. 3.2. The lowest C_4 estimates very likely reflect diagenetic (or groundwater) carbonate formed at times of lake expansion. Pedogenic carbonates are rare for about 350,000 years following deposition of the Lemuta Member. Possible mixing of distinct sources of carbon in the lake-margin fluvial-lacustrine deposits make isotopic interpretation of analyses of groundwater nodules and nodular beds from this period problematic, but the climate was very likely moister and vegetation more like that preceding the Lemuta Member. From ca. 1.18–0.8 Ma (uppermost Bed II and Beds III and IV), pedogenic carbonates from fluvial deposits indicate a C_4 biomass range of 50–70% (their table 2), or 50–80% (p. 75), or 60–80% (p. 76), probably corresponding to the difference between Bed III and IV depicted in Cerling's (1992) figure.

The next period, from ca. 0.50–0.78 Ma (the Masek Beds), has somewhat higher isotope values in the lower part of the section, suggesting warmer and drier conditions. Overall, carbon isotope values from Masek eolian tuff and fluvial carbonates indicate a C_4 biomass of 50–80%. Following this, by at least 100,000 years ago (in the Ndutu Beds), modern carbon isotope values are reached, with C_4 biomass estimates of 80–90%. The terminal Cenozoic paleosol carbonates provide isotopic values indicating 80–90% C_4 biomass, similar to modern values (Cerling 1984; Cerling and Hay 1986). Cerling and Hay (1986) point out that this C_4 shift in the last 500,000 years was accompanied by an increase in the pH and salinity of the soils and the input of volcanic natrocarbonatitic ash, as well as an apparently drier climate. Thus, the modern eastern Serengeti short-grass plain around Olduvai is a special type of volcanic-ash edaphic grassland in a semiarid rain shadow created by the volcanic highlands immediately to the east.

The Laetoli Area Today

The vegetation of the Laetoli area is not as well studied as that of the Serengeti plain surrounding Olduvai. The overall impression provided by the paleontological monograph on Laetoli (Leakey and Harris 1987) is that the environment in the Pliocene was a dry savanna similar to that of today. However, today the area is not uniformly open savanna grassland. In fact, it is a mosaic of microhabitats, with vegetation ranging from grassland to bush, to open and closed canopy woodland, and what appears to be relict forest. The region is dominated by three major savanna-woodland associations: *Acacia tortilis–Combretum molle* woodland, varying from open to closed, on well-drained volcanic soils on hill slopes; *Acacia drepanolobium–Balanites aegyptiaca* woodland/bushland, varying from open to closed, on poorly drained vertisols in areas of low relief; and *Acacia kirkii–Vangueria infausta* woodland along water courses, with *Acacia xanthophloea–Euclea divinorum–Albizia gummifera* woodland where the water table is higher. Today, these associations range from open to closed depending on the degree of disturbance by fire, clearing, and overgrazing, all of which are prevalent in the area. There are also remnant elements of dry forest in wetter areas, evidenced by *Euclea divinorum, Diospyros mespiliformis, Croton macrostachys, Cassia usambarensis,* and *Markhamia lutea.* This last association is found on the eastern fringes of the Serengeti (i.e., Naisuri Gorge) and is similar to that found in the northern Serengeti woodlands (Sinclair 1979;

Jager 1982), which are thought to be remnants of forest mostly destroyed by fire and clearing (Sharam 2005; Sharam, Sinclair, and Turkington 2006).

The fossils from the Upper Laetolil Beds are dated to between ca. 3.5 to 3.76 Ma. It is a highly diverse fauna, comprising invertebrates (more than 20 species), tortoises (2 species), snakes (4 species), birds (7 species), and mammals (more than 70 species).

This Pliocene fauna has its greatest ecological similarities with present-day faunas from the more heavily wooded parts of the Serengeti in the northern and western sections of the ecosystem. Terrestrial gastropods, well represented in the Laetoli fauna, are particularly useful as paleoenvironmental indicators because they have relatively narrow ecological limits and a limited capacity for dispersal or postmortem transportation. The gastropod assemblage, therefore, can often reflect subtle, localized ecological conditions. In the Upper Laetolil Beds, the predominance of *Euonyma* and *Subulona* below Tuff 7 indicates relatively mesic conditions (with annual rainfall exceeding 1,000 mm) and the presence of forested habitats, since these two taxa are found in East Africa today in upland and riverine forests (Verdcourt 1987; Pickford 1995). Above Tuff 7, *Edouardia* is most common, which suggests that conditions at Laetoli became somewhat drier, and open woodland was the predominant vegetation. Moreover, the ubiquitous occurrence throughout the Laetoli sequence of a diverse guild of urocyclid slugs implies a dense litter of fallen leaves and tree trunks in a wooded environment that offered suitable habitats for estivation during the long dry season.

The fossil bird fauna, dominated by guineafowl and two species of francolins, is typical today of savanna and open woodland habitats: for example, guineafowl require trees to roost in at night (Watson 1987; Harrison 2005). Although small mammals are underrepresented in the Upper Laetolil Beds (Andrews 1989; Soligo and Andrews 2005), the community structure of the overall mammalian fauna confirms the presence of woodland at Laetoli during the Pliocene (Andrews 1989, 2006; Reed 1997; Musiba 1999; Kovarovic 2004; Su 2005). The numbers of mammal species with arboreal, frugivorous, and browsing adaptations are inconsistent with open grassland, and are representative of the full range of habitats seen today in the entire Serengeti (Sinclair and Norton-Griffiths 1979). This is emphasized by the diversity of the mammalian fauna in the Upper Laetolil Beds, which is similar in species richness to that found today in the woodlands of the Serengeti. It is distinct from the modern Serengeti grassland fauna, which has fewer than 40 mammal species. In addition, the fossil fauna includes several species of mammals whose modern close relatives prefer woodland and forest, such as cephalophine bovids (duikers), *Tragelaphus* (kudu), *Potamochoerus* (bushpig), *Rhynchocyon* (elephant shrew), *Galago* (bushbaby), and *Paraxerus* (bush squirrel), as well as guilds of giraffids and cercopithecids that collectively indicate wooded habitats. Aquatic and hydrophilic vertebrates, such as crocodiles, turtles, hippopotamids, and reduncine bovids, common at other East African paleontological sites, are entirely lacking from the Mid-Pliocene Laetoli fauna. This supports the geological evidence that no large permanent bodies of water were available near Laetoli during the Pliocene.

It is possible that the fossil fauna owes its high diversity in part to 'time-averaging,' since the Upper Laetolil Beds were deposited over a period of more than 250,000 years. However, a recent analysis (Su and Harrison 2003; 2007) has demonstrated that there are no discernable differences in species composition, and only very minor differences in the proportionality of taxa, when faunas from different localities and stratigraphic horizons are compared. This indicates that high species diversity characterized the entire Upper Laetolil sequence, and that the general ecology remained uniform throughout this time. Based on the faunal evidence, Mid-Pliocene Laetoli appears to have been predominantly a complex vegetation mosaic similar to that of today, but with a higher proportion of woodland and forest. It lacked, of course, the anthropogenic effects of the past few millennia. This vegetation was interrupted intermittently by catastrophic inundations of volcanic ash from Sadiman. During these periods, the vegetation of Laetoli was apparently dominated by grasses. However, since there are no indications of specialized grassland communities in any of the faunal assemblages from the Upper Laetolil horizons, these periods of disruption in the ecosystem were relatively short; grasslands were quickly replaced by wooded vegetation. The general ecosystem during deposition of the Upper Laetolil Beds was a mosaic of different, more-or-less wooded habitats, with a taxonomically diverse fauna that remained remarkably stable despite the brief and localized volcanic inundations.

Pollen Studies

For the identification of fossil pollen at Laetoli and Olduvai, studies on vegetation (appendix A, Leakey and Harris 1987) and modern pollen rain were carried out in grassland plains, woodlands, forests, marshes, and lake edges from the Serengeti to the mountain highlands of Sadiman, Ngorongoro, and Embakai. These studies showed that the pollen composition of the modern surface samples agrees in general with the floristic composition of the vegetation, and that the ratio of tree to herbaceous pollen corresponds to the physiognomy of the vegetation. Grasslands show up quite well through high frequencies of herbaceous plants in which grasses are dominant (Bonnefille and Riollet 1987, their fig. 2.14). Wooded grassland from the Laetoli area today is clearly represented by the greater proportion of tree pollen, although *Acacia,* one of the most common trees, is insect pollinated and so is underrepresented. The high percentage of grass pollen in the few fossil pollen samples recovered from the Upper Laetolil Beds suggested that the Pliocene vegetation was an open savanna (trees were scarce; Bonnefille and Riollet 1987). Also, the fossil pollen indicated a very diverse herbaceous component. But the proportion of pollen from forbs and sedges was much less in the fossil record than in samples from the Serengeti short-grass plain today. Pollen from montane forest trees was significantly better represented in the Pliocene samples than in modern samples. Although this fossil pollen might represent an allochthonous input from dry evergreen forests on the slopes of Sadiman, 20–30 km to the east, the greater abundance of such pollen may imply that forested habitats were located closer to Laetoli in the Pliocene. They could have been a minor component of the vegetation in the immediate vicinity, under wetter climatic conditions than today.

Olduvai Gorge Bed I: Evidence from the Small Mammals for the Late Pliocene Vegetation

Collections during the 1960s (Leakey 1971) provided evidence of a sequence of small mammal faunas in middle and upper Bed I. These are represented by nine levels, which encompass ca. 60,000 years of the late Pliocene between Tuffs IB and IF (Fernández-Jalvo et al. 1998). The nine small mammal faunas have been analyzed in several ways (Fernández-Jalvo et al. 1998). We show here the results of one of these analyses. Scores for weighted averages ordination yield the Taxonomic Habitat Index (THI) shown in fig. 3.3. This adds together the habitat ranges of all species included in a fauna based on observations of present-day species. Species in the fossil record that are the same as living species are scored the same as these, but extinct species in extant genera are scored for the summed species ranges of all species included in that genus. The scores for each habitat type for all species are added together and divided by the number of species to produce an average score; these scores are summarized here as the percentage of open country scores.

Figure 3.3 shows a trend from low open-country scores and low gerbil/ murid proportions for the middle Bed I faunas (localities FLKNN and

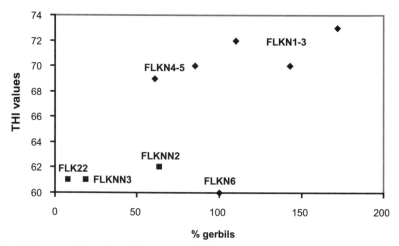

Fig. 3.3 Taxonomic Habitat Index (THI), plotted against the percentage of gerbils relative to murines for the rodent faunas of middle and upper Bed I, Olduvai Gorge. The THI is expressed as the proportion of mammalian habitat ranges in open country as opposed to closed woodland or forest (open country is defined here as those habitats encompassing semidesert, grassland, and bushland). The gerbil index is calculated by the ratio of gerbilline to murine rodents multiplied by 100. The three faunas from middle Bed I are indicated by squares, the six from upper Bed I by diamonds. Based on analyses in Fernández-Jalvo et al. (1998).

FLK22), to high open-country scores and high gerbil ratios in upper Bed I (locality FLKN), increasing through five of the six levels of upper Bed I. The low THI score for FLKN6, the stratigraphically lowest of the upper Bed I faunas, is anomalous and based on a small sample.

The middle Bed I faunas indicate a species-rich closed woodland habitat, and the diversity of arboreal and frugivorous species in the mammalian faunas indicate habitats richer than any part of the present-day savanna biome in Africa, possibly similar to the relict forests in northern Serengeti today. The upper Bed I faunas indicate more open habitats, with a trend through the five upper levels of upper Bed I toward increasingly open habitats. The most extreme values are from the topmost Bed I fauna, which is most similar to modern semiarid bushland like that of Tsavo in Kenya and the arid mixed environments of the Kalahari.

Olduvai Gorge Beds I and II: Evidence from the Larger Mammals for the Plio-Pleistocene Vegetation

Excavations by the Leakeys at Olduvai Gorge recovered over 10,000 large-mammal bones that provide a spatially and temporally broader indicator of vegetation than the small mammals. The excavations yielding large-mammal fossils span all of Beds I and II. Bovids account for approximately 70% of this fauna. The relative abundance of bovid taxa, based on teeth and postcranial remains (Gentry and Gentry 1978a,b; Kappelman 1984; Shipman and Harris 1988), combined with functional morphological indicators of bovid locomotor and dietary adaptations (Kappelman 1991; Kappelman et al. 1997; Plummer and Bishop 1994; Spencer 1997) permit reconstruction of vegetation physiognomy. Because bovids range widely, the spatial resolution of these reconstructions is lower than that from small mammals.

Bovid femur morphology reflects locomotion to avoid predators in environments with different degrees of woody cover. On this basis, the Bed I and Bed II bovids occupied habitats with a wide range of vegetation physiognomy, from unwooded open areas to "forest" (Kappelman 1991; Kappelman et al. 1997). The open areas were likely the lake margin flats inferred from the geology. These would support an edaphic grassland and sedgeland. Femora of bovids associated with open vegetation are the type most commonly found in the Bed I and Bed II sequence, indicating, along with their dental morphology, that they were grazers. These bovids include the reduncine *Kobus sigmoidalis,* the alcelaphine *Parmularis altidens,* and the antelope *Antidorcas recki* (Gentry and Gentry 1978b). In conjunction with stable carbon isotopes from soil carbonates (Cerling and Hay 1986; Sikes 1996), as well as the small-mammal results reported previously in this chapter, the bovids suggest an environmental mosaic of woodland with a grassy groundcover (at times a dense woodland with bushy understory), riparian forest, floodplain, and lake-flats pastureland. This mosaic would accommodate the grazing bovids and equids as well as the associated browsing herbivores (tragelaphines, *Aepyceros, Giraffa,* and elephant).

Parmularius and *Antidorcas* form a common species association throughout the sequence (Gentry and Gentry 1978b). The feeding preference of *Antidorcas recki* is the most equivocal of all the Olduvai bovids. Femoral measurements indicate affiliations with both "open" and "light cover" settings (Plummer and Bishop 1994; Kappelman et al. 1997). The feeding preferences of the extant springbok, *Antidorcas marsupialis,* suggest that dicots could have been an important component of the diet of *A. recki* (Gentry and Gentry 1978b). A modern comparative series of cranio-dental measurements suggests that *Antidorcas* was a mixed feeder eating a significant proportion of grasses (Spencer 1997). From isotopic analyses of early Pleistocene *Antidorcas recki* remains from Sterkfontein, South Africa, Luyt and Lee-Thorp (2003) concluded that this antelope was primarily or exclusively a browser. It appears *Antidorcas recki* possessed broad dietary tolerances, depending on setting. Several cranial and dental measurements support the hypothesis that *Parmularius* was a grazer (Spencer 1997). *Antidorcas* and

Parmularis may have preferred the drier settings of salt bush and grass at the upper margin of the lake flats.

Kobus sigmoidalis is considered ancestral to the waterbuck, *Kobus ellipsiprymnus,* with a transitional form identified from lower Bed II (Gentry and Gentry 1978a). Waterbuck are wetland grazers, seldom found far from water. The mandibular morphology of *Kobus sigmoidalis* is very similar to that of the modern waterbuck, suggesting that it occupied a similar niche.

The abundance of *Kobus* tends to vary inversely with the abundance of the *Parmularius/Antidorcas* association at individual sites through the Bed I and Bed II sequence. This pattern may reflect vegetation differences related to broad changes in climate, as is commonly assumed, or excavations having sampled different parts of a vegetation mosaic. As an example of the latter, *Kobus* represents over 50% of the bovids from all three levels of the Middle Bed I site of FLKNN. However, *Parmularius* and *Antidorcas* are dominant at level 22 of the site of FLK, which lies about 200 m to the southeast and was judged by Leakey (1971) to be the same stratum as level 3 at FLKNN, based on the similarity of sediments and faunal preservation. In this case, *Parmularius* and *Antidorcas* may have been drawn from the drier lake margin to watering sites in the wetland utilized by *Kobus.*

At a broad temporal scale, the bovids seem to reflect climatic changes inferred from other evidence. For example, *Kobus* is well represented in most assemblages within lower and middle Bed I. Less mesic conditions indicated by the small mammals for the assemblages in the lower portion of upper Bed I at the site FLKN are reflected in the dominance of *Parmularius* and *Antidorcas,* reaching 66% of all bovid individuals in levels 5 and 6 of that site.

The existence of some short grassland throughout the middle and upper Bed I sequence is indicated by the presence of *Connochaetes* sp., most specimens of which seem attributable to *C. gentryi.* The cranio-dental morphology of *C. gentryi* exhibits many of the short-grass feeding adaptations of the living wildebeest (Sinclair 1979, Harris 1991, Spencer 1997). Short grasslands, probably those supported by exposed lake flats and floodplains, are also indicated throughout the Bed I and II time period by the presence (and modern feeding habits) of hippopotamus and white rhinoceros.

Connochaetes gentryi and *Parmularius altidens* have been inferred to be the earliest bovid taxa in East Africa inhabiting secondary grassland, beginning ca. 2 Ma (Spencer 1997). This is based primarily on their having relatively short premolar tooth rows, similar to extant grazers characterized as preferring secondary succession grasslands (e.g., wildebeest, kongoni/hartebeest, Thomson's gazelle) in comparison to extant grazers preferring seasonal wetlands (e.g., reedbuck, kob, topi/tsessebe). An alternative interpretation is that this morphology is an unexplained preadaptation that appeared in the Pliocene species long before the advent of secondary succession grasslands in the Late Quaternary. This is reinforced by the conclusion, based on the microfauna, that mesic woodlands were present during most of the Bed I time period at Olduvai. In this setting, these fossil grazers were probably accommodated by grasslands on the dry upper-lake flats, including areas flooded seasonally by streams entering the lake (following Peters and Blumenschine 1995, 1996).

During the lower to middle and upper Bed II time period, relatively dry conditions are inferred from the rarity or absence of *Kobus,* the continued dominance of *Antidorcas* and *Parmularius* at many localities, and the first appearance of species adapted to drier, more open habitats, including *Camelus* sp. and *Damaliscus niro.* The numbers of bovines (*Pelorovis* and *Syncerus*) exceed that of tragelaphines for the first time in middle and upper Bed II (Gentry and Gentry 1978b). Moreover, the hyper-hypsodont grazer *Eurygnathohippus cornelianus* also becomes a prominent member of the fauna in upper Bed II (Bernor and Armour-Chelu 1999).

Olduvai Gorge Bed I: Evidence from Pollen for the Late Pliocene Vegetation

Fossil pollen assemblages were obtained from a few distinct stratigraphic layers in Bed I and Bed II (Bonnefille 1984). For Bed I they span the time between the deposition of Tuffs IB and IF. Lake-margin clays were selectively sampled and pollen was preserved in 10% of the processed samples.

The fossil pollen assemblages were dominated by herbaceous plants, particularly grasses, followed by sedges, Compositae (Asteraceae), Amaranthaceae, *Typha,* and so on. By comparison with the modern pollen rain of the Serengeti short-grass plain, sedge pollen was rare: less than 10%, compared with 30 to 40% in modern samples. *Kyllinga,* a dwarf C_4 sedge that sprouts quickly after brief rains, is very abundant today in the short-grass plain, but may have been rare during Pliocene times, before the short-grass plains existed. Alternatively, the fossil sedge pollen could be from wetland species such as *Cyperus laevigatus,* a sedge found today on the margins of small saline-alkaline lakes and streams in the eastern Serengeti.

Some of the fossil pollen types come from tree genera such as *Podocarpus, Olea, Syzygium,* and *Euclea. Podocarpus* grows in evergreen forest today, mostly at higher elevations (Eggeling 1952). *Olea* grows in vegetation today ranging from fresh water swamp and lowland rain forest to upland dry forest and grassland with scattered trees (Turrill 1952). *Syzygium* grows today in a variety of freshwater swamp, gallery and lakeside forest, woodland, and dambo (vlei) grassland. *Euclea* grows today in a range of vegetation types—

some moist, some dry—and extends to coastal elevations. Fossil pollen of Combretaceae was also found with the pollen of these four genera of trees, and may come from trees as well, but the family in East Africa includes shrubs, shrublets, and climbers. They grow in vegetation ranging from wooded grassland to bushland, thicket woodland, and evergreen forest.

The fossil pollen of these woody plants is relatively abundant, up to 25% of the total fossil pollen assemblage, in middle Bed I. In upper Bed I it is replaced by pollen of Capparidaceae and *Ximenia* (Olacaceae), and the total percentage of pollen from woody plants is lower (as low as 2%). The Capparidaceae include herbs, shrubs, lianas, and small trees (Elffers, Graham, and DeWolf 1964). They grow today in vegetation ranging from semi-desert scrub, disturbed land, grassland, bushland, thicket, and woodland to rain forest. *Ximenia* is represented in East Africa by two species of bushes to small trees growing in dry woodland, bushland, and wooded grassland. It is found today on the rocky footslopes of Naibor Soit near Olduvai Gorge, as well as along the rim of the gorge, and is abundant on the eastern side of Lake Eyasi, at a lower elevation. These fossil pollen changes are consistent with a shift from moist, tree-rich conditions to dryer vegetation with fewer trees, across the period of time represented by middle to uppermost Bed I.

In summary, the fauna and pollen record ecosystem-scale changes in the Olduvai area within a period of ca. 60,000 years during the Late Pliocene, beginning in middle Bed I with a tropical closed woodland and ending with an arid to semiarid bushland at the top of the Bed I sequence. This appears to have been reversed to some extent in lower Bed II, with the pollen record indicating a return to more wooded conditions—until deposition of the Lemuta Member, when drier conditions prevailed again. Following the Lemuta Member, another return to somewhat wetter conditions in Bed II occurred. These successional changes are embedded within an overall drift toward warmer and drier conditions across the past 2 million years (Cerling and Hay 1986).

EVOLUTION OF THE MARA RIVER AND LAKE VICTORIA AS POTENTIAL DROUGHT REFUGIA

Dry Season and Drought Refugia

In normal years the edges of papyrus marshes in the Amboseli National Park are late dry-season pasture for returning migrants, including wildebeest and zebra, and they provide an example of how marshlands could have been used in the Serengeti. During the severe Kenyan drought in 1973, elephants progressively pushed into the marshes, trampling and thinning out the tall, coarse sedges, enabling a secondary growth of smaller palatable sedges and grasses to proliferate. Buffalo, zebra, wildebeest, and gazelle were able to take advantage of the new fresh growth as the beaten-down margins of the marsh were expanded inward. Tens of thousands of animals were accommodated by relatively small areas of permanent marsh. They were emaciated, but most did not die of starvation (Western 1997).

The Serengeti is now cut off from two potential drought refugia within Tanzania. First, the Western Corridor stops short of Lake Victoria. Much of the intervening area near the lake edge is now settled and cultivated. Migrating animals are cut off from the lake and from papyrus marsh along Speke Gulf. In drought years, migrants move through the settled area to reach the water, as in the great drought of 1993 and also the drought of 2000. Probably, western-corridor migrants used the lake edge prehistorically, especially the Grumeti (Ruwana) River delta. The prehistory of Lake Victoria is probably an important part of the evolution of the Serengeti system.

A second potential drought refuge in Tanzania is the lower Mara River wetlands-complex below the Utimbara Escarpment. This appears to include a permanent marsh of ca. 100 km^2 (fig. 3.1). It is now difficult, because of intervening settlement (and before that, mining), to say how important this wetland might have been for the Serengeti. Now only the Masai Mara area in Kenya receives these migrants. Because of the importance of the Mara River in drought conditions, structural controls of this river are important to the evolutionary story. These include the Mau Highland (upper catchment), the Siria Escarpment (Masai Mara), and the Utimbara Escarpment (Tanzania). The Mau Highland and Lake Victoria appear to have originated in the later Pleistocene within the past 500,000 years. They are linked hydrologically (unpublished report cited by Wright 1967), contributing to the rainfall that sustains the flow of the Mara River.

Recently, many of the Serengeti migrants depend in the dry seasons and during droughts on forage and water in the Masai Mara National Reserve and the Trans-Mara Plateau above the Siria Escarpment in Kenya. Once every decade or so, serious drought results in the forage being depleted in Masai Mara N. R. and the grasslands north and east of the reserve that have not yet been converted to wheatfields (Thurow 1995). In these critical years the herds move up the escarpment onto Trans-Mara dambo grasslands. In 1995, this land was being converted to maize monoculture, up to the edge of the escarpment. Even the middle Mara River refuge in Kenya is threatened by anticipated further deforestation in the Mau Highlands, by water abstraction for irrigation along the river on the Loita Plains north of the Masai Mara N. R., and by Kenyan plans for water diversion for a hydroelectric scheme (Gereta et al. 2002).

The grassland-dominated plains of Masai Mara National Reserve as seen by Europeans a hundred years ago seem to have been created by pastoralist burning (see the section in this chapter on Prehistoric Humans as Part of the System). These grasslands probably did not exist prior to the introduction and subsequent development of pastoralism in East Africa. The relatively high dry-season rainfall received by the Masai Mara N. R. and the perennial Mara River offer late dry-season refuge to Serengeti wildebeest in this anthropogenic grassland.

Major Landform Events and Evolutionary Changes

The Serengeti lies on the Tanzanian Craton, an area tectonically stable from the Cambrian (ca. 550 Ma) until the African rift valleys formed in the mid-to-late Tertiary (< 15 Ma). This part of the craton consists largely of granitic and metamorphic rocks of Archean age (> 2.6 Ga; (Barth 1990). Metamorphic and granitic rocks in the southeastern part of the Serengeti Plain belong to the Mozambique Belt, metamorphosed at ca. 600 Ma (Mc-Connell 1972). These ancient rocks form the kopjes, hills, and ranges of low mountains seen to the west and northwest of the present-day Eastern Rift Valley (fig. 3.1).

The physical boundaries of the Serengeti ecosystem are defined in large part by Late Tertiary tectonic activity involving major faulting and volcanism. The oldest major faulting in northern Tanzania resulted in the uplift of several major blocks, including the Lake Victoria block (Dawson 1992). Major uplift has been dated at about 4.5 Ma on the basis of field relations in the Eyasi fault scarp bordering Lake Eyasi on the north (Pickering 1964; Dawson 1992). Early movement also occurred along the Nguruman Fault west of Lake Natron (Williams 1963). These movements of the Eyasi and Nguruman faults either raised or left elevated the eastern and southeastern parts of the Lake Victoria block. Volcanoes formed after the faulting, including Oldoinyo Sambu, Mosonik, and Sadiman (fig. 3.1). The Utimbara Fault, in northernmost Tanzania, is dominantly east-west, with the northern block uplifted about 500 m (Gray and Macdonald 1966). It joins the Siria Fault, mainly in Kenya, which is uplifted about 300 m on the northwest. The Siria Fault was probably active early in the Rift Valley formation (Williams 1963), and initial movement on both faults may have accompanied that of the Eyasi Fault ca. 4.5 Ma. The Siria and Utimbara escarpments determine the southward (middle reach) and westward flow (lower reach) of the Mara River (fig. 3.1). This structural control appears to be older than the Mau Highlands and Lake Victoria, the present beginning and endpoints of the Mara River.

Main rift faulting at 1.2–0.9 Ma (MacIntyre, Mitchell, and Dawson 1974) gave rise to the present-day Rift Valley. One result was the formation of the Lake Natron and Lake Manyara drainage basins along the foot of the east-facing escarpment (fig. 3.1). To the north in Kenya, the Mau Highlands were probably formed by elevation along faults in the late Middle Pleistocene ca. 400,000 years ago (Williams 1963), providing a new upper catchment for the Mara River. Lake Victoria also developed at this time.

Lake Victoria

Lake Victoria lies over a former drainage system that flowed from east to west. It was created following uplift of a fault block of the Western (Albert) Rift, which dammed the west-flowing drainage (Bishop and Trendell 1967). It is the largest of several lakes formed in the Pleistocene when the shoulder of the Western Rift rose faster than preexisting rivers could cut into it (Doornkamp and Temple 1966). Winam (Kavirondo) and Speke gulfs, though they are now part of Lake Victoria, occupy tectonic basins that may be older than the lake.

Seismic reflection data demonstrate a long and complex stratigraphic history for Lake Victoria (Johnson et al. 1996; Scholz et al. 1998). The stratified sequences are interpreted to be lacustrine sediments (Johnson et al. 1996). Using the known Holocene sedimentation rate and correcting for the lower water content of the more reflective older sediments indicates that the lake formed ca. 400,000 years ago (Johnson et al. 1996). The historical geography of the Victoria Basin (Kendall 1969) is consistent with this age estimate. The oldest exposed beds with a clear genetic relation to the modern lake are seen in the lower valleys of rivers entering along the western shore, especially the Kagera. Two sets of beds are recognized there, at elevations of 61–67 m and 33–35 m above the modern lake level (Bishop and Posnansky 1960). The upper and older set, the Nsongezi Beds, contains Acheulian and Sangoan lithic industries that suggest an age of a few hundred thousand years.

During this earlier phase of its existence, Lake Victoria lay further to the west, and in that location it held water for about 250,000 years. It seems to have been dry at least once during this phase. Since ca. 150,000 years ago the lake has existed episodically in something near its present position and shape. It has been largely or completely dry at least four times, most recently in the very late Pleistocene. The last episode of drying was much shorter than the wet episodes before and after it. High stands of the lake are recorded in undated raised beaches at 18–20 m, 12–15 m, and 3–4 m above the modern lake level (Temple 1969a, b).

Coring and seismic examination of terminal Pleistocene and Holocene sediments provide most of our information about changes in Lake Victoria. From at least 14,500 until 12,500 BP the lake was low and the water more concentrated than it is now. After 12,500 BP, the water was too dilute for precipitation of the sulfate and carbonate minerals formed earlier (Kendall 1969). It has been suggested that Victoria was a closed lake until 7,400 BP (Johnson et al. 1996) but much stronger evidence shows that the outlet was probably established by 12,500 BP, although the lake fell below it for about a millennium, centered on 10,000 BP (Kendall 1969). The characteristic strontium isotopic signature of Lake Victoria water appeared downstream in the White Nile by at least 11,500 BP (Talbot, Williams, and Adamson 2000).

It is not possible to demonstrate from shallow-water cores like those of Kendall that the lake dried up completely in the late Pleistocene, but later coring in deeper parts of the basin shows that the lake was dry in all places investigated so far. Seismic data demonstrate a hard reflector at a depth of 6–8 m in the sediment over the central part of the lake (Johnson et al. 1996; Scholz et al. 1998). This reflector was reached by coring in several places, and proved to be a buried soil coeval with the dry layers of lake mud described by earlier investigators (Kendall 1969; Stager 1982; Stager, Reinthal, and Livingstone 1986; Stager, Cumming, and Meeker 1997). It is possible, though increasingly improbable with each new investigation, that standing water persisted in some small part of the basin, not necessarily where the water is deepest today. Seismic profiling registers a clear signal of desiccation in all the places where it has been applied. The tectonic basins of Winam and Speke gulfs are too shallow to have held water during the dry late Pleistocene. Even if water remained in some small, as yet undiscovered, closed basin, it would probably have been too alkaline-saline for mammals to drink.

EVOLUTION OF THE LARGER-MAMMAL HERBIVORES AND CARNIVORES

Potential Migratory Species

Perhaps the migratory tendencies of living descendents and close living relatives of prehistoric large-mammal species may indicate that they were potentially migratory. If so, potential migratory species have been present for at least 1 to 2 million years. However, whether a prehistoric species was actually migratory would remain unknown, since no aspect of skeletal morphology can be uniquely related to the act of migration. Moreover, most migratory species today migrate only if they are in a system that facilitates or requires it. The Serengeti is such a system because of its combination of unusual geoecological circumstances. Even here, the migrations may have ceased intermittently for several hundred, or a few thousand, years when Lake Victoria dried up.

Pliocene and Pleistocene

All genera of larger-mammal carnivores and most genera of larger-mammal herbivores with body weights greater than ca.10 kg that are found in the Serengeti region today had appeared by the early part of the Olduvai sequence, in Bed I or lower Bed II (Gentry and Gentry 1978a,b; Harris and White 1979; Hooijer 1969; Churcher 1982; Leakey 1971). Modern genera absent from the early Olduvai record are represented by an equal or greater number of taxa from the same tribe or family, resulting in mammalian faunas that were substantially richer than those of the Serengeti today. Greater taxonomic richness is seen among the Plio-Pleistocene Olduvai giraffids, suids, and equids, and, among the bovids, in the alcelaphines. Among the larger-mammal herbivores, the only taxa that show greater species richness today are the bovid tribes Tragelaphini, Neotragini, and Cephalophini. Fossils of the latter two tribes, with small body sizes, may have suffered from relatively poor preservation. Alternatively, these bovids may not have utilized the unwooded lake-margin environments.

Among the larger Carnivora, the modern Serengeti has a greater number of feline and canine species than were present at Plio-Pleistocene Olduvai, again possibly due in part to the taphonomic underrepresentation of species with relatively small body size (e.g., jackal, small cats), as well as the tendency for mammalian carnivores to be rare elements of fossil vertebrate faunas. The latter factor prompts consideration of all East African Pliocene and Pleistocene fossil occurrences, from which it is clear that larger-carnivore taxonomic richness was greater between 4.0 and 1.0 Ma than it is today or at any time during the Holocene (fig. 3.4). The greater number of fossil species is accounted for mainly by the Felidae and Hyaenidae.

Part of the greater taxonomic richness of the fossil-mammal faunas might be an artifact of the amalgamation of a chronological series of less-speciose faunas. Such cumulative time effects for the fossil record might be expected, given the relatively long time periods it samples compared to that recorded historically in the Serengeti, and the great degree of climatic variability inferred across the timescale of the Plio-Pleistocene (e.g.,

Larger Plio-Pleistocene carnivora of eastern Africa

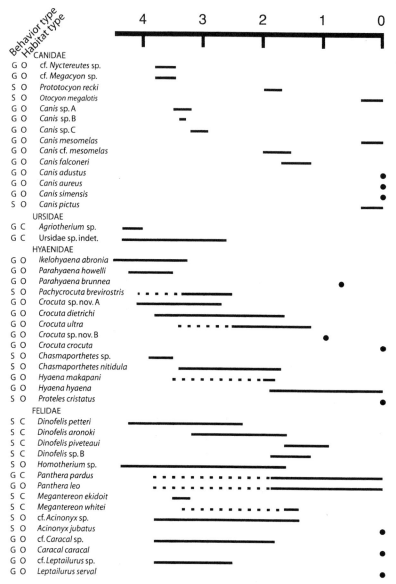

Fig. 3.4 Larger carnivora present during the Pliocene and Pleistocene in eastern Africa (based on Werdelin and Lewis 2005, plus work in progress). Horizontal scale: millions of years before present. Each taxon is classified as a generalist (G) or a specialist (S) with regard to inferred prey procurement, and as preferring open (O) or closed (C) habitats. Solid lines indicate a confirmed fossil record. Dashed black lines indicate an uncertain record. An internal dot indicates a taxon known only from a site with a very brief temporal extent. A terminal dot indicates a taxon with no known fossil record in eastern Africa.

deMenocal 1995). Nonetheless, the fossil faunas of the Serengeti region included a number of extinct forms in the Pliocene and early Pleistocene that imparted an ecological distinctiveness not seen today, as highlighted in the following.

Among the larger mammalian fossil faunas, three sets impart the greatest distinctiveness when compared with those of the modern Serengeti. One set is composed of a taxonomically diverse group of very large herbivores, larger than ca. 1,000 kg body weight. It includes the following series of extinct genera or species:

1. Two genera of proboscideans: *Elephas recki*, with a high-crowned lophodont dentition similar to *Loxodonta*, and *Deinotherium bosazi*, bearing a low-crowned bilophodont dentition suitable for shearing mainly C3 vegetation (Cerling, Harris, and Leakey 1999);

2. *Ancylotherium hennigi*, a large chalicothere (Perissodactyla) standing about 2 m at the shoulder, notable for its brachydont dentition (browser) and clawed digits;

3. *Hippopotamus gorgops*, substantially larger in body size than the modern hippopotamus (Coryndon 1970);

4. *Sivatherium marusium*, a very large giraffid, (probably a mixed feeder), stockier in build and shorter-necked than the modern giraffe; and

5. *Pelorovis oldowayensis*, a large buffalo with high-crowned teeth for grazing, and massive, curved horns spanning 1–2 m.

A second set is composed of extinct ungulates that were considerably larger than their closest modern relatives:

1. *Megalotragus kattwinkeli*, a large, grazing alcelaphine bovid, about 10% to 20% larger than the blue wildebeest, which is the largest of the extant alcelaphines;

2. *Kolpochoerus limnetes*, a bush-piglike suid larger than the extant forest hog (Harris 1983); and

3. *Metridiochoerus sp.*, an ancestor of the warthog, but larger than the extant forest hog.

The third set is a series of extinct larger carnivores (fig. 3.4), including:

1. Three genera of machairodont carnivores—the true and false saber-toothed cats *Homotherium*, *Megantereon*, and *Dinofelis*—which may have specialized in part on megaherbivore prey (see, e.g., Marean 1989 for a review);

2. Three extinct genera of hyaenids, including *Chasmaporthetes,* an actively hunting form with limited scavenging adaptations, *Pachycrocuta,* a powerful bone-cracking species that was present in East Africa until the later Pliocene, and *Parahyaena,* with the modern brown hyena now limited to southern Africa; and

3. *Canis falconeri,* a robust, wolfsized canid, probably related to the living African wild dog.

The former presence of these archaic carnivores underlies major changes in that guild over the last 4.5 Ma. Two parameters of guild structure, habitat preference and degree of prey-procurement specialization, do show some sensitivity to environmental restructuring. These parameters can be assessed by classifying each taxon into open versus closed habitat preference and into generalist versus specialist prey procurement on the basis of locomotor and gnathic skeletal anatomy (fig. 3.4). These classifications inevitably involve some subjectivity, but for the vast majority of larger Carnivora they are not controversial. The two classifications are not significantly associated (*G* test of independence; $G_{adj} = 3.08$; $p > 0.05$). An initial high percentage of specialist and closed-habitat taxa existed between 4.5 and 4.3 Ma, although this is based on a limited number of species. Around 3.9–3.6 Ma there is a drop in both—that is, a relative decrease in taxa with specialized prey procurement and a preference for closed habitats. The relative number of specialist and closed habitat taxa increases gradually subsequent to this until ca. 1.8–1.5 Ma. After this time, there is a steep drop: the relative numbers of generalist and open-habitat-adapted carnivores in the modern fauna are greater than at any time in the Plio-Pleistocene. Apparently, the climatic events of the later Pliocene (ca. 2.8–2.5 Ma) did not affect large-carnivore species richness negatively, as they did some other mammal groups (Behrensmeyer et al. 1997; Werdelin and Lewis 2005).

LATE QUATERNARY

Vegetation dynamics in the later Quaternary bear directly on the evolution of the large mammals. Unfortunately, the history of the vegetation of the Serengeti during the later Pleistocene and Holocene is not documented by paleobotanical evidence from the area itself. However, many late Quaternary pollen sequences (Jolly et al. 1997) have been extracted from East African lakes (e.g., Victoria, Tanganyika, Albert), volcanic and other mountainous settings (e.g., Mount Kenya, Embakai, Ruwenzori), and valley marshes and peat bogs (Burundi, Ruanda, and Uganda) . All of them indicate that significant changes affected the distribution of ecosystems on mountains and in lowlands. A significant reduction of areas occupied by forest and woodland occurred during the Last Glacial Maximum, from 18 to 20 ka BP (Jolly et al. 1998; Elenga et al. 2000). There was descent, reduction, and disappearance of the forest zone on mountains, replaced for the most part by grasslands and associated temperate plants (Bonnefille and Riollet 1988). Somewhat cooler, but much drier climatic conditions occurred in equatorial Africa (Bonnefille et al. 1992), as well as in the tropics worldwide (Farrera et al. 1999) at a time of global cooling and a lower concentration of atmospheric carbon dioxide. An opposite pattern is seen in the early Holocene from 10 to 6 ka BP, when forests reached their maximum development, occupying large belts on mountains and surrounding lakes that also had reached their maximum extension. This was the time of the wettest conditions in the last 10,000 years (Peyron et al. 2000). Suddenly, ca. 5 ka BP, a decline in the abundance of trees occurred simultaneously at many sites (Bonnefille and Chalie 2000). In forest sites there was a replacement of evergreen tree species by deciduous ones, attesting to greater seasonal contrast in annual precipitation. Abrupt drops in lake level indicate that the rainfall-to-evaporation ratio decreased strongly at that time. Although the period since 5 ka BP is marked by change toward much drier climate, several further declines of tree abundance at nonforest sites have been attributed to direct human impact, such as agriculture or pastoral burning.

East Africa has a small but growing sample of vertebrate faunal assemblages that provide only a glimmer of the probable changes in faunal communities that occurred during the Late Pleistocene and Holocene (Marean 1990; 1992b; 1997; Marean and Gifford-Gonzalez 1991). Three main clusters of sites occur: the central Rift Valley around the lakes Naivasha-Nakuru-Elmenteita in Kenya, the granitic inselberg of Lukenya Hill in the Athi-Kapiti Plains in Kenya, and some widely scattered sites in southwestern Kenya and in northern Tanzania. The last set includes the open-air site of Engaruani in Kenya, Nasera Rock and Loiyangalani in the Serengeti, Mumba Shelter near Lake Eyasi, and Kisese II Rockshelter at Kondoa in central Tanzania overlooking the Masai Steppe (Bower and Marean, n.d.; Deacon 1966; Inskeep 1962; Marean, Ehrhardt, and Mudida n.d.; Marshall 1986; Mehlman 1989).

The Late Quaternary mammalian fauna of sub-Saharan Africa is essentially modern. Unlike the North American record, no dramatic extinction marked the beginning of the Holocene (Klein 1984; Marean 1992b). Only a few specialized species became extinct just prior to or during the Holocene in Africa, and most of these are recorded only in South Africa. They consist of two sets: a suite of four very large-bodied species that seem adapted to arid conditions with low-quality forage, and a suite of three small-bodied species similarly adapted to arid conditions (Marean 1992b; Marean and Gifford-Gonzalez 1991). All are members of extant families, and some are members of an extant genus. They comprise a very large (draft-horse sized) zebra (*Equus capensis*); a very large (perhaps twice the body weight of *Syncerus*) long-horned buffalo (*Pelorovis antiquus*); a very large (eland-sized) hartebeest (*Megalotragus priscus*); a large suid (*Metridiochoerus compactus*); a small springbok (*Antidorcas bondi*); a small Grant's gazelle-sized alcelaphine, currently unnamed due to lack of a worthy holotype (Marean 1992b; Marean and Gifford-Gonzalez 1991); and a caprine (Brink 2000). Several of the large and small-bodied species likely formed a coassociation or grazing succession, much like modern species (Marean 1990, 1992b).

Most of the extinct species are known only from cave sites in South Africa. Only *Pelorovis* and the unnamed, small alcelaphine are known from Kenya and northern Tanzania. Both species display very high-crowned teeth, suggesting a coarse, grazing diet. These forms persist into the Last Glacial Maximum (ca. 21,000–12,000 years ago) of the terminal Pleistocene, but only the small extinct alcelaphine survives into the Holocene.

The modern grassland large herbivores and large carnivores of southern Kenya and northern Tanzania were present during the Late Pleistocene, but it is unclear if the modern community was dominant. Several sites show an abundance of species indicative of arid conditions more typical of northern Kenya today. A recurrent pattern is the presence of *Pelorovis* alongside Grevy's zebra and oryx. In addition, the extinct, small unnamed alcelaphine dominates the larger mammalian fossil assemblages of the Late Pleistocene arid-grassland community (Marean 1992b; Marean and Gifford-Gonzalez 1991).

The arid-adapted community of the Late Pleistocene gives way at the onset of the Holocene to mammal faunas more typical of today. More and better faunal samples are known from the south-central Kenyan and northern Tanzanian Holocene, most of which postdate the arrival of pastoralism, about 4,000 years ago. The largest is from Enkapune Ya Muto, at 2,400 m on the Mau Escarpment (Marean 1992a; Marean, Mudida, and Reed 1994). Prior to 4,800 years ago, the assemblage is dominated by forest species and bohor reedbuck, with the micromammals including water-dependent species that demonstrate the increased size of Lake Naivasha. After 4,800 years, the fauna registers a fall in the lake and the appearance of more grassy vegetation. Eventually, Enkapune Ya Muto is surrounded by more open vegetation and pastoralists occupy the site. At lower elevations in the Naivasha lake basin, a series of sites with very small faunal samples dating to the early and middle Holocene indicate a mix of grassland and woodland conditions (Gifford 1985; Gifford-Gonzalez 1998; Gifford-Gonzalez and Kimenengich 1984). By the late Holocene, most archaeological sites in the central Rift Valley display a significant pastoralist occupation and are dominated by cattle and caprines, while others preserve an abundant wild grassland fauna with substantial numbers of cattle and caprines (Gifford, Isaac, and Nelson 1980).

The grassland-dominated areas of southern Kenya have yet to yield early and middle Holocene faunal assemblages, but late Holocene assemblages are well known from the Lemek Valley in southwestern Kenya (Ngamuriak and Sambo Ngige). These differ from those of the central Rift Valley in that wild fauna is nearly absent and cattle and caprines dominate the assemblages (Marshall 1986, 1990a, b).

A small sample of early and middle Holocene assemblages is available for northern Tanzania. The Serengeti site of Gol Kopjes is not yet published, but preliminary reports indicate early to middle Holocene human occupation with an associated fauna that conforms to the modern Serengeti ecosystem, with the exception of rare occurrences of the extinct unnamed small alcelaphine (Marean and Gifford-Gonzalez 1991; Gifford-Gonzalez, personal communication). Nasera Rock in the Serengeti (north of Olduvai Gorge) preserves several Holocene occurrences, including a middle Holocene one for which the presence of domestic animals suggests mixing with more recent layers (Gifford-Gonzalez, in Mehlman 1989). The late Holocene layers at Nasera display less species diversity than the Pleistocene layers, and include domestic cattle and caprines, which dominate some layers, along with species typical of the modern Serengeti grasslands.

PREHISTORIC HUMANS AS PART OF THE SYSTEM

Hominins have been a part of the Serengeti large-mammal communities since at least the time of the Upper Laetolil Beds (ca. 3.76 Ma), where they are represented by *Australopithecus afarensis*. Despite earlier claims that even Pliocene hominins were hunters of large animals (e.g., Dart 1953), it is not likely that they became top carnivores in the Serengeti ecosystem until at least the advent of anatomically modern *Homo sapiens,* together with the appearance of the Middle Stone Age, around 250,000 years ago. Further, while hunting by Late Pleistocene humans and the Holocene advent of pastoralism (with possibilities of systematic burning) may have influenced the Serengeti system, it seems unlikely that human impact became sufficiently strong to disrupt the Serengeti migratory system until agricultural and mining practices became prevalent in the northern and western parts of the system during the last century. In the following section we highlight some behavioral innovations that are documented or alleged to have changed hominin niche characteristics and community interactions during the Quaternary, emphasizing the role of pastoralist burning in the creation of the Masai Mara subsystem.

Scavenging and Hunting in Early *Homo*

The Oldowan Industrial Complex, the world's first flaked-stone technology, appears ca. 2.5 Ma in Ethiopia (Semaw et al. 1997). Its advent was more-or-less coincident with the appearance of the genus *Homo.* The earliest stone tools in the Serengeti region date from the lower part of Bed I at Olduvai Gorge, during which time at least two genera (*Homo* and *Paranthropus* [*Australopithecus*]) and up to four species of hominins existed. The Oldowan's significance to the evolution of the Serengeti lies in its clear association with the beginning of hominin encroachment on the larger carnivore guild, seen in the stone-tool knife marks and hammerstone percussion marks on larger mammal bones associated with these early stone tool assemblages (e.g., Bunn 1981; Blumenschine 1995). Prior to this time, it is likely that hominin carnivory was limited to the type of small-prey (< 5 kg) predation seen among modern baboons and chimpanzees.

The nature of Oldowan hominin encroachment on the larger carnivore guild is debated. Hunting by hominins was long assumed to be the major mode of procurement of these animal foods (e.g., Washburn and Lancaster 1968; Schaller and Lowther 1969). More recent studies of the ecology and taphonomy of scavenging, based mainly on work conducted in the Serengeti, suggests that scavenging by hominins and, subsequently, hyaenids can explain the types and conditions of large-mammal remains preserved at the Oldowan occurrences (e.g., Blumenschine 1986). Whether hominin scavenging was restricted mainly to the exploitation of abandoned felid kills (e.g., Blumenschine 1987; Cavallo and Blumenschine 1989), or involved displacement of social felids from more fully fleshed carcasses (e.g., Bunn and Ezzo 1993; Dominguez-Rodrigo 1997) is a matter of debate. Nonetheless, the lack of obvious weaponry needed to take down large-prey animals during the Oldowan, and even the succeeding Acheulian times, suggests that hominins were subordinate carnivores throughout the Early and Middle Pleistocene, with minimal predatory or competitive impacts on the region's larger herbivores and carnivores.

Controlled Domestic Use of Fire

Claims for the advent of controlled fire for domestic use have been advanced for hominins that lived as far back as 1.5 Ma, based on evidence at archaeological sites in Ethiopia, Kenya, and South Africa (Bellomo 1994; Clark and Harris 1985; Brain and Sillen 1988). The evidence is scant and controversial, taking the form of discrete patches of oxidized (baked) clay thought to represent small campfires, or bone that had allegedly been burned by hominins prior to fossilization. No such traces of fire have been observed in the extensive excavations of similar-age deposits at Olduvai Gorge. Whether these early traces of fire were solely natural or were controlled by prehistoric hominins is the main uncertainty. Domestic fire use is speculated to have facilitated the Early Pleistocene colonization of highland regions of Ethiopia as well as temperate regions outside Africa (Clark and Harris 1985). Nonetheless, humanly controlled fire does not become common in the archaeological record until the Late Pleistocene, during the Middle Stone Age.

Middle and Later Stone Age Hunters

The Late Pleistocene saw the genesis and spread of anatomically modern humans throughout the world, and it is becoming increasingly clear that this event originated in Africa (Aiello 1993; Stringer 2003). Recent evidence suggests that these humans may also have been behaviorally modern (Henshilwood et al. 2001; Henshilwood et al. 2002). Certainly these people had a technology that was both sophisticated and adaptable enough to allow them to inhabit virtually every region of the world. In Africa, this period is referred to as the Middle Stone Age (MSA), beginning as early as 280,000 years ago (McBrearty and Brooks 2000).

Two features of MSA technology may bear on the impact of people on the ecosystems of Africa. First, MSA people apparently were equipped with stone-point-tipped weapons that were potentially effective projectiles (Shea, Davis, and Brown 2001). Specially made points, likely hafted, dominate many African MSA stone-tool assemblages. Analyses of faunal assemblages indicate that MSA people were effective hunters capable of taking the largest antelopes (Marean and Assefa 1999). Second, MSA people had controlled use of fire, as demonstrated by the many fireplaces present in African caves. Perhaps they used fire like hunter-gatherers—to modify vegetation so as to increase both hunting and plant-food collecting returns (Pyne 1991).

The MSA is replaced by the Later Stone Age (LSA) well before 46 ka (Ambrose 1998). Recent results on dating the lower part of the stone artifact-bearing Naisiusiu Beds in Olduvai Gorge (Skinner et al. 2003) suggest the LSA was present by at least 62 ka. Later Stone Age people were cognitively modern and equipped with projectile weapons, as well as controlled use of fire. They likely exhibited alpha predator capabilities and probably used fire as a technology to change the vegetation landscape. It is likely that hunter-gatherers faced larger carnivores as active competitors, and most likely they dominated them the way tropical, semiarid hunter-gatherers do today. National parks have provided a relatively safe haven for carnivores unique to our time. Because hunter-gatherers were pushed out of grassland-rich ecosystems by pastoralists prior to ethnographic documentation, we have no modern African analog for their lifeways in those ecosystems (Marean 1997). Current East African grasslands preserve the final overprint of pastoralists who modified the vegetation with their fire and grazing technologies, to one degree or another obscuring earlier conditions.

Holocene Pastoralists

Pastoralists and their domestic animals enter the grasslands-rich ecosystems of East Africa beginning 4,000 years ago, spreading rapidly south into Tanzania. This spread was associated with the expansion of grassland during the mid-Holocene dry phase, probably amplified by pastoral styles of burning(Marean 1990, 1992a, b). Most of the faunal assemblages dating to the earlier part of this event document large quantities of wild fauna alongside the domestic. In many cases, it is unclear if these represent pastoral or hunter-gatherer occupations. The specialized ('exclusive') pastoralism so common to East Africa in recent history appears to have evolved later. It is unlikely that the domestic-animal focus represented at these sites indicates the regional exclusion of wild fauna.

Historical analysis of vegetation changes in the Masai Mara National Reserve (Dublin 1991 and chapter 2, this volume) indicates that fires set by pastoralists played a major role in creating the current Mara component of the Serengeti ecosystem. At the end of the 1800s and into the early 1900s Masai Mara probably looked much as it does today: broad, open expanses of grassland and lightly wooded savanna predominated. Drought, followed by rinderpest in the late 1800s, resulted in catastrophic losses of cattle and consequently human life through famine and associated disease. Except for the ivory trade, the area was virtually devoid of human influence by the turn of that century and for many years thereafter. Buffalo and wildebeest herds also were reduced by the rinderpest virus. With the great reduction in ungulate, human, and elephant populations, woody vegetation began to colonize the plains. By the 1930s–1940s, *Acacia* woodlands occurred throughout the Masai Mara, and numerous belts of *Croton* thorn-bush thicket existed where before there was open grassland on the plains.

This succession of woody vegetation was reversed midcentury. By the 1950s both *Acacia* woodland and the *Croton* thickets were on the decline. Between 1950 and 1982 over 95% of the *Acacia* woodland disappeared. Across the same period, the *Croton* thickets were reduced significantly in size. The highest losses were in the early 1960s, when average rainfall in the district was ca. 50% higher than that of 1925–1950. Seasonal rainfall and the resulting grass growth were so high that the Masai were able to burn the plains two to three times per year. Large standing crops of dry grass fueled fierce hot fires that destroyed woody vegetation. Secondarily, elephant densities were on the increase in the reserve. Forced out of adjacent areas, they amplified woodland losses through ring-barking and heavy browsing, weakening trees and adding broken branches to the fuel loads. Thickets were thinned internally by structural damage from their movements and browsing, opening the thickets to invasion by grasses and fire. Elephants may not have contributed in a major way to the open grasslands of Masai Mara in the nineteenth century, but pastoralist burning practices over hundreds of years is a likely cause.

Drawing upon this historical analysis of the Masai Mara and fire-ecology research from the Serengeti (Norton-Griffiths 1979), a model of strong effects for long-term (prehistoric) pastoralist burning can be constructed around the following combination of factors/conditions: (1) seasonally dry subhumid climate (average annual rainfall ca. 800–1,000 mm); (2) exceptionally wet years (2 to 3 consecutive years), producing large standing crops of grass; (3) low grazing pressure (high fuel load); and (4) timing of fires to maximize the potential burn (human ignition). These effects could be amplified during wet years immediately succeeding periods of severe drought. Pastoralist short-term motivation for this style of ecological burning would presumably be that of removing low-quality forage across large areas within the higher-rainfall grassy woodlands, so that the scattered dry-season showers that occur there would have greater potential for producing accessible new-grass growth. Otherwise, the protein content of the dry-season standing crop of old grass is too low to provide a maintenance diet (Lamprey 1979). The result in the long term would be thinning of the woodlands and increased grassland. This fire regime also benefits migrating wildebeest (Talbot and Talbot 1963).

CONCLUSIONS

The Serengeti-Mara ecosystem is a relatively recent system, not an ancient one. Reexamination of earlier interpretations does not support the idea of a modern Serengeti-like system in the Pliocene or early Pleistocene. At that time, the vegetation of the area that is now the Serengeti Plain was probably dominated by multiple successions of mesic woodland and semiarid bushland, with localized edaphic grasslands, each succession spanning thousands of years. In the late Pliocene, for ca. 200,000 years, a shallow saline-alkaline lake, whose maximum diameter averaged 15–20 km, was located in what is now the short-grass plain. The fauna of the region included a variety of very large herbivores, hyaenids, and saber-toothed cats that became extinct by the mid-Pleistocene. Modern forms were also present, and their descendents are what remains of this once highly diverse fauna.

The enabling abiotic components of the modern ecosystem evolved in an additive fashion over a period of 3 to 4 million years, but only in the last 500,000 years have a number of the crucial structural components emerged. Lake Victoria (and its rainfall) plus the Mau catchment of the Mara River develop during this latter period. The Serengeti Plain also takes on its modern (sodic) volcanic-ash form within the past 500,000 years, and the C_4-dominated vegetation of the short-grass plain may have appeared as recently as 100,000 years ago.

In the aridity of the terminal Pleistocene, the ecosystem ceased to exist. Lake Victoria was a dry basin. The fauna of the region included species more typical of northern Kenya today.

Not only does the Serengeti sector of the ecosystem appear late in the history of human evolution, pastoralists probably created the Mara plains grasslands as a subsystem through their burning practices as recently as the past few hundred years. In the last century, most of the potential drought refugia in the northwestern portions of the ecosystem, including the shores of Lake Victoria and the lower Mara River wetlands, have been lost to development outside the protected areas of the Serengeti-Mara. This evolutionary and historical perspective highlights the extreme vulnerability of the current system, apparently now an isolated portion of what was once a much larger ecosystem.

ACKNOWLEDGMENTS

The authors, as follows, thank funding agencies and colleagues for their support for our work over the years: C. R. P. and R. J. B., the Boise Fund of Oxford University, the L. S. B. Leakey Foundation, the National Science Foundation, the Department of Anthropology and members of the Institute of Ecology at the University of Georgia, the Rutgers University Center for Human Evolutionary Studies; D. A. L., the National Science Foundation, the Arts and Sciences Research Council and Biology Department of Duke University, and Robert Wilbur for discussion; M. A.-C., for the help and advice of Alan Gentry, John Harris, and John de Vos; R. L. B., the National Science Foundation; T. H., the National Geographic Society, the L. S. B. Leakey Foundation, the National Science Foundation, the curators and staff at the National Museums of Tanzania in Dar Es Salaam, and Denise Su; C. W. M., the Boise Fund of Oxford University, the L. S. B. Leakey Foundation, the National Science Foundation, and Sigma Xi; L. W., the Swedish Research Council, and Margaret E. Lewis for long-term, fruitful collaboration. We also thank Jim Ebert for providing the base map for fig. 3.1 and Jim Abbott for drafting the composite product.

APPENDIX

Geological Dates and Estimated Ages of Stratigraphic Units in the Laetoli and Olduvai Gorge Areas

Stratigraphic unit	Dates and estimated ages	Technique(s) of dating and rationale for estimates[a]	References
Olduvai area			
Namorod Ash	ca. 1.25 ka	14C	Hay 1976
Naisiusiu Beds, upper part	ca. 14.6–22.5 ka, ca. 13 ka	14C	Hay 1976; Manega 1993
	39 ± 5 ka	esr	Skinner et al. 2003
Naisiusiu Beds, lower part	62 ± 5 ka	esr	Skinner et al. 2003
	> 42 ka	14C	Manega 1993
Ndutu Beds	ca. 100–500 ± 100 ka?	dating is questionable	Hay 1976; Manega 1993
Masek Beds	ca. 500–780 ka	pmag; in Brunhes Chron	C.C. Swisher, pers. comm.[b]
Bed IV	ca. 780–950 ka	pmag	Hay 1976; Tamrat et al. 1995
Bed III	ca. 0.95–1.15 Ma	pmag, tectonic history[c]	Hay 1976; Macintyre et al 1974
Bed II top	ca. 1.15 Ma	tectonic history[c]	Hay 1976; Macintyre et al 1974

Geological Dates and Estimated Ages of Stratigraphic Units in the Laetoli and Olduvai
Gorge Areas (*continued*)

Stratigraphic unit	Dates and estimated ages	Technique(s) of dating and rationale for estimates[a]	References
Bed II, Lemuta mbr.	ca. 1.70–1.75 Ma	thickness and alteration history	Hay 1976
Bed II, base	ca. 1.785 Ma	pmag	Blumenschine et al. 2003
Bed I, top[d]	ca. 1.785 Ma	pmag	Blumenschine et al. 2003
Tuff IC	1.839 ± 0.005 Ma	^{40}Ar/ ^{39}Ar	Blumenschine et al. 2003
Tuff IB[d]	1.845 ± 0.002 Ma	^{40}Ar/ ^{39}Ar	Blumenschine et al. 2003
Bed I, base[d]	ca. 2.0 Ma	^{40}Ar/ ^{39}Ar	Walter et al. 1992
Laetoli area			
Ngaloba Beds, upper	ca. 100–150 ka	U-Th from bone, correlation with upper Ndutu	Hay 1987
	≥ 205–290 ka	amino acid racemization of ostrich eggshells	Manega 1993
Olpiro Beds	ca. 2.14 Ma	^{40}Ar/ ^{39}Ar	Ndessokia 1990
	ca. 2.03 Ma	^{40}Ar/ ^{39}Ar	Manega 1993
Naibadad Beds	ca. 2.26 Ma	K-Ar	Drake and Curtis 1987
	ca. 2.15 Ma	^{40}Ar/ ^{39}Ar	Ndessokia 1990
	ca. 2.09–2.26 Ma	^{40}Ar/ ^{39}Ar	Manega 1993
Ogol Lavas	ca. 2.41 Ma	K-Ar	Drake and Curtis 1987
Ndolanya Beds	ca. 2.41–3.46 Ma	between lava and tuff dated by K-Ar	Drake and Curtis 1987
Ndolanya Beds, upper	ca. 2.58–2.66 Ma	^{40}Ar/ ^{39}Ar	Ndessokya 1990
Laetolil Beds, upper	ca. 3.46–3.76 Ma	K-Ar	Drake and Curtis 1987
Laetolil Beds, lower	ca 3.76–≥ 4.32 Ma	K-Ar	Drake and Curtis 1987

[a] Dating techniques are 14C, carbon 14; esr, electron spin resonance; pmag, paleomagnetic polarity; K-Ar, potassium-argon; ^{40}Ar/ ^{39}Ar, argon isotopes (single crystal laser-fusion); U-Th, uranium-thorium.
[b] C. C. Swisher III, 2006, personal communication of normal polarity for all samples of the Masek Beds.
[c] The disconformity between Beds II and III is correlated with the beginning of major rift-valley faulting to the east of Olduvai Gorge, dated by K-Ar at 1.15-1.2 Ma (MacIntyre Mitchell, and Dawson 1974). As further evidence, Bed III records displacements along several faults that had been inactive during the deposition of Beds I and II.
[d] Bed I is divided into three units: lower Bed I is from the base of the bed up to Tuff IB; middle Bed I is from Tuff IB up to Tuff ID (an undated tuff); upper Bed I is from Tuff ID to Tuff IF (the top of Bed I).

REFERENCES

Aiello, L. C. 1993. The fossil evidence for modern human origins in Africa: A revised view. *American Anthropologist* 95:73–96.

Ambrose, S. H. 1998. Chronology of the Later Stone Age and food production in East Africa. *Journal of Archaeological Science* 25:377–92.

Anderson, G. D. 1963. Some weakly developed soils of the eastern Serengeti plains, Tanganyika. *Sols Africains* 8:339–47.

Anderson, G. D., and L. M. Talbot. 1965. Soil factors affecting the distribution of grass-land types and their utilization by wild animals on the Serengeti plains, Tanganyika. *Journal of Ecology* 53:33–56.

Andrews, P. 1989. Palaeoecology of Laetoli. *Journal of Human Evolution* 18:173–81.

———. 2006. Taphonomic effects of faunal impoverishment and faunal mixing. *Palaeogeography, Palaeoclimatology, Palaeoecology* 241: 572–89.

Barth, H. 1990. Explanatory notes on the 1:500,000 Provisional Geological Map of the Lake Victoria Goldfields, Tanzania. *Geologisch Jahrbuch* Heft 72, Hanover.

Beden, M. 1976. Proboscideans from the Omo Group Formations. In *Earliest man and environments in the Lake Rudolf Basin,* ed. Y. Coppens, F. C. Howell, G. Ll. Isaac, and R. E. F. Leakey, 193–208. Chicago: University of Chicago Press.

———. 1987. Fossil Elephantidae from Laetoli. In *Laetoli: A Pliocene site in Northern Tanzania,* ed. M. D. Leakey and J. M. Harris, 259–94. Oxford: Clarendon.

Behrensmeyer, A. K., N. E. Todd, R. Potts, and G. McBrinn. 1997. Late Pliocene faunal turnover in the Turkana Basin of Kenya and Ethiopia. *Science* 278: 1589–94.

Bellomo, R. V. 1994. Methods for determining early hominid behavioral activities associated with the controlled use of fire at FxJj 20 Main, Koobi Fora, Kenya. *Journal of Human Evolution* 27:173–95.

Bernor, R. L., and M. Armour-Chelu. 1999. Toward an evolutionary history of African Hipparionine horses. In *African Biogeography, Climate Change and Early Hominid Evolution,* ed. T. Brommage and F. Schrenk, 189–215. Oxford: Oxford University Press.

Bishop, W. W., and M. Posnansky. 1960. Pleistocene environments and early man in Uganda. *Uganda Journal* 24:44–61.

Bishop, W. W., and A. F. Trendell. 1967. Erosion-surfaces, tectonics, and volcanic activity in Uganda. *Quarterly Journal of the Geological Society of London* 122:385–420.

Blumenschine, R. J. 1986. *Early hominid scavenging opportunities: Implications of carcass availability in the Serengeti and Ngorongoro ecosystems.* Oxford: British Archaeological Reports, International Series 283.

———. 1987. Characteristics of an early hominid scavenging niche. *Current Anthropology* 28:383–407.

———. 1995. Percussion marks, tooth marks, and experimental determinations of the timing of hominid and carnivore access to long bones at FLK *Zinjanthropus,* Olduvai Gorge, Tanzania. *Journal of Human Evolution* 29:21–51.

Blumenschine, R. J., C. R. Peters, F. T. Masao, R. J. Clarke, A. L. Deino, R. L. Hay, C. C. Swisher, I. G. Stanistreet, G. M. Ashley, and L. J. McHenry, et al. 2003. Late Pliocene *Homo* and hominid land use from western Olduvai Gorge, Tanzania. *Science* 299: 1217–21.

Bonnefille, R. 1984. Palynological research at Olduvai Gorge. *National Geographic Society Research Reports* 17:227–43.

Bonnefille, R., and F. Chalié. 2000. Long term time series of pollen inferred precipitation from equatorial mountains, Central Africa. *Global and Planetary Change* 26:25–50.

Bonnefille, R., F. Chalié, J. Guiot, and A.Vincens. 1992. Quantitative estimates of full glacial temperatures in equatorial Africa from palynological data. *Climate Dynamics* 6:251–57.

Bonnefille, R., and G. Riollet. 1987. Palynological spectra from the Upper Laetolil Beds. In *Laetoli: A Pliocene site in northern Tanzania,* ed. M. D. Leakey and J. M. Harris, 52–61. Oxford: Clarendon.

———. 1988. The Kashiru pollen sequence (Burundi). Palaeoclimatic implications for the last 40,000 yr B.P. in tropical Africa. *Quaternary Research* 30:19–35.

Bower, J., and C. W. Marean. n.d. The Loiyangalani Site (HcJd1): New observations on an MSA occurrence in the Serengeti National Park, Tanzania. Unpublished paper presented at 12th Biennial Conference of Africanist Archaeologists, Indiana University, April 28–May 1, 1994.

Brain, C. K., and A. Sillen. 1988. Evidence from the Swartkrans Cave for the earliest use of fire. *Nature* 336:464–66.

Brink, J. 2000. Preliminary report on a caprine from the Cape mountains, South Africa. *Archaeozoologia* 10:11–25.

Bunn, H. T. 1981. Archaeological evidence for meat-eating by Plio-Pleistocene hominids from Koobi Fora and Olduvai Gorge. *Nature* 291:547–77.

Bunn, H. T., and J. A. Ezzo. 1993. Hunting and scavenging by Plio-Pleistocene hominids: Nutritional constraints, archaeological patterns, and behavioral implications. *Journal of Archaeological Science* 20:365–98.

Cavallo, J. A., and R. J. Blumenschine. 1989. Tree-stored leopard kills: Expanding the hominid scavenging niche. *Journal of Human Evolution* 18:393–99.

Cerling, T. E. 1984. The stable isotopic composition of modern soil carbonate and its relationship to climate. *Earth and Planetary Science Letters* 71:229–40.

———. 1992. Development of grasslands and savannas in East Africa during the Neogene. *Palaeogeography, Palaeoclimatology, and Paleoecology* (Global and Planetary Change Section) 97:241–47.

Cerling, T. E., J. M. Harris, and M. G. Leakey. 1999. Browsing and grazing in elephants: The isotope record of modern and fossil proboscideans. *Oecologia* 120:364–74.

Cerling, T. E., and R. L. Hay. 1986. An isotopic study of paleosol carbonates from Olduvai Gorge. *Quaternary Research* 25:63–78.

Churcher, C. S. 1982. Oldest ass recovered from Olduvai Gorge, Tanzania, and the origin of asses. *Journal of Paleontology* 56:1124–32.

Clark, J. D., and J. W. K. Harris. 1985. Fire and its roles in early hominid lifeways. *African Archaeological Review* 3:3–27.

Coryndon, S. C. 1970. The extent of variation in fossil Hippopotamus from Africa. *Symposia of the Zoological Society of London* 26:135–47.

Dart, R. A. 1953. The predatory transition from ape to man. *International Anthropological Linguistics Review* 1:201–19.

Dawson, J. B. 1962. The geology of Oldoinyo Lengai. *Bulletin Volcanologique* 24:349–87.

———. 1992. Neogene tectonics and volcanicity in the North Tanzanian sector of the Gregory Rift Valley: Contrasts with the Kenyan sector. *Tectonophysics* 204: 81–92.

Dawson, J. B., P. Bowden, and G. C. Clark. 1968. Activity of the carbonatite volcano Oldoinyo Lengai, 1966. *Geologische Rundschau* 57:865–79.

Dawson, J. B., H. Pinkerton, D. M. Pyle, and C. Nyamweru. 1994. June 1993 eruption of Oldoinyo Lengai, Tanzania: Exceptionally viscous and large carbonatite lava flow and evidence for coexisting silicate and carbonate magmas. *Geology* 22:799–802.

Deacon, J. 1966. An annotated list of radiocarbon dates for sub-saharan Africa. *Annals of the Cape Provincial Museums* 5:5–84.

deMenocal, P. B. 1995. Plio-Pleistocene African climate. *Science* 270:53–59.

de Wit, H. A. 1978. *Soils and Grassland Types of the Serengeti Plain (Tanzania)*. PhD diss., University of Wageningen, The Netherlands.

Dominguez-Rodrigo, M. 1997. Meat-eating by early hominids at the FLK 22 Zinjanthropus site, Olduvai Gorge (Tanzania): An experimental approach using cut-mark data. *Journal of Human Evolution* 33:669–90.

Doornkamp, J. C., and P. H. Temple. 1966. Surface, drainage and tectonic instability in part of southern Uganda. *Geographical Journal* 132:238–52.

Drake, R., and G. H. Curtis. 1987. K-Ar geochronology of the Laetoli fossil localities. In *Laetoli: A Pliocene site in northern Tanzania*, ed. M. D. Leakey and J. M. Harris, 48–52. Oxford: Clarendon.

Dublin, H. T. 1991. Dynamics of the Serengeti-Mara woodlands: An historical perspective. *Forest & Conservation History* 35(4): 169–78.

Eggeling, W. J. 1952. *The indigenous trees of the Uganda Protectorate,* 2nd ed., revised and enlarged by I. R. Dale. Entebbe, Uganda: Government Printer.

Elenga, H., O. Peyron, R. Bonnefille, D. Jolly, R. Cheddadi, J. Guiot, V. Andrieu, S. Bottema, G. Buchet, J.-L. and de Beaulieu, et al. 2000. Pollen-based biome reconstruction for Southern Europe and Africa 18,000 years ago. *Journal of Biogeography* 27(3): 621–34.

Elffers, J., R. A. Graham, and G. P. DeWolf. 1964. *Capparidaceae: Flora of tropical east Africa.* London: Crown Agents for Overseas Governments and Administrators.

Farrera, I., S. P. Harrison, I. C. Prentice, G. Ramstein, J. Guiot, P.-J. Bartlein, R. Bonnefille, M. Bush, W. Cramer, and U. von Grafenstein, et al. 1999. Tropical climates at the Last Glacial Maximum: A new synthesis of terrestrial paleoclimate data. I. Vegetation, lake-levels and geochemistry. *Climate Dynamics* 15:823–56.

Fernández-Jalvo, Y., C. Denys, P. Andrews, T. Williams, Y. Dauphin, and L. Humphrey. 1998. Taphonomy and palaeoecology of Olduvai Bed-I (Pleistocene, Tanzania). *Journal of Human Evolution* 34:137–72.

Gentry, A. W. 1987. Pliocene Bovidae from Laetoli. In *Laetoli: A Pliocene site in northern Tanzania,* ed. M. D. Leakey and J. M. Harris, 378–408. Oxford: Clarendon.

Gentry, A. W., and A. Gentry. 1978a. Fossil Bovidae (Mammalia) of Olduvai Gorge, Tanzania. Part I. *Bulletin of the British Museum (Natural History),* Geology Series 29: 289–446.

———. 1978b. Fossil Bovidae (Mammalia) of Olduvai Gorge, Tanzania. Part II. *Bulletin of the British Museum (Natural History),* Geology Series 30:1–83.

Gereta, E., E. Wolanski, M. Borner, and S. Serneels. 2002. Use of an ecohydrology model to predict the impact on the Serengeti ecosystem of deforestation, irrigation and the proposed Amala Weir water diversion project in Kenya. *Ecohydrology and Hydrobiology* 2:135–42.

Gifford, D. P. 1985. The faunal assemblages from Masai Gorge Rock Shelter and Marula Rock Shelter. *Azania* 20:69–89.

Gifford, D. P., G. Isaac, and C. M. Nelson. 1980. Evidence for predation and pastoralism at Prolonged Drift: A pastoral Neolithic site in Kenya. *Azania* 15:57–108.

Gifford-Gonzalez, D. P. 1998. Early pastoralists in East Africa: Ecological and social dimensions. *Journal of Anthropological Archaeology* 17:166–200.

Gifford-Gonzalez, D. P., and J. Kimenengich. 1984. Faunal evidence for early stock-keeping in the Central Rift of Kenya: Preliminary findings. In *Origin and early development of food-producing cultures in north-eastern Africa*, ed. L. Krzyzaniak, 457–71. Poznan, Poland: Polish Academy of Science.

Gray, I. M., and A. S. Macdonald. 1966. North Mara (Tarima). *Geological Survey of Tanzania Quarter Degree Sheets 4 and 5, scale 1:125,000*. Dar es Salaam, Tanzania: Government Printer.

Harris, J. M. 1983. Family Suidae. In *Koobi Fora Research Project, volume 2. The fossil ungulates: Proboscidea, Perissodactyla, and Suidae*, ed. J. M. Harris, 215–302. Oxford: Clarendon.

———. 1987. Summary. In *Laetoli: A Pliocene site in northern Tanzania*, ed. M. D. Leakey and J. M. Harris, 524–31. Oxford: Clarendon.

———. 1991. Family Bovidae. In *Koobi Fora Research Project, vol. 3. The fossil ungulates: Geology, Fossil Artiodactyls, and Palaeoenvironments*, ed. J. M. Harris, 139–320. Oxford: Clarendon.

Harris, J. M., and T. D. White. 1979. Evolution of the Plio-Pleistocene African Suidae. *Transactions of the American Philosophical Society* 69:1–128.

Harrison, T. 2002. First recorded hominins from the Ndolanya Beds, Laetoli, Tanzania. *American Journal of Physical Anthropolology* Supplement 32:83 (abstract).

———. 2005. Fossil bird eggs from Laetoli, Tanzania: Their taxonomic and paleoecological implications. *Journal of African Earth Sciences* 41:289–302.

Harrison, T., and E. Baker. 1997. Paleontology and biochronology of fossil localities in the Manonga Valley, Tanzania. In *Neogene paleontology of the Manonga Valley, Tanzania: A window into east African evolutionary history*, ed. T. Harrison, 361–93. New York: Plenum.

Harrison, T., and M. L. Mbago. 1997. Introduction: Paleontological and geological research in the Manonga Valley, Tanzania. In *Neogene paleontology of the Manonga Valley, Tanzania: A window into east African evolutionary history*, ed. T. Harrison, 1–32. New York: Plenum.

Hay, R. L. 1976. *Geology of the Olduvai Gorge*. Berkeley, CA: University of California-Berkeley Press.

———. 1978. Melilitite-carbonatite tuffs in the Laetolil Beds of Tanzania. *Contributions to Mineralogy and Petrology* 17:255–74.

———. 1981. Palaeoenvironment of the Laetolil Beds, northern Tanzania. In *Hominid sites: Their geological settings*, ed. G. Rapp and C. F. Vondra, 7–24. Boulder, CO: Westview.

———. 1983. Natrocarbonatite tephra of Kerimasi volcano, Tanzania. *Geology* 11: 599–602.

———. 1986. Role of tephra in the preservation of fossils in Cenozoic deposits of East Africa. In *Sedimentation in the African rifts*, ed. L. E. Frostick, R. W. Renaut, I. Reid and J. J. Tiercelin, 339–44. Geological Society Special Publication no. 25. Oxford: Blackwell Scientific.

———. 1987. Geology of the Laetoli area. In *Laetoli: A Pliocene site in northern Tanzania,* ed. M. D. Leakey and J. M. Harris, 23–47. Oxford: Clarendon.

Hay, R.L., and T. K. Kyser. 2001. Chemical sedimentology and paleoenvironmental history of Lake Olduvai, a Pliocene lake in northern Tanzania. *Geological Society of America Bulletin* 113:1505–21.

Hay, R. L., and M. D. Leakey. 1982. The fossil footprints of Laetoli. *Scientific American* 246: 50–57.

Hearn, P., T. Hare, P. Schruben, D. Sherrill, C. LaMar, and P. Tsushima. 2001. *Global GIS database: Digital atlas of Africa.* Digital Data Series DDS-62-B. Reston, VA: U.S. Geological Survey.

Henshilwood, C. S., F. D'Errico, C. W. Marean, R. G. Milo, and R. J. Yates. 2001. An early bone tool industry from the Middle Stone Age, Blombos Cave, South Africa: Implications for the origins of modern human behaviour, symbolism and language. *Journal of Human Evolution* 41:631–78.

Henshilwood, C. S., F. D'Errico, R. Yates, Z. Jacobs, C. Tribolo, G. A. T. Duller, N. Mercier, J. C. Sealy, H. Valladas, and I. Watts, et al. 2002. Emergence of modern human behavior: Middle Stone Age engravings from South Africa. *Science* 295:1278–80.

Hobley, C. W. 1918. A volcanic eruption in East Africa. *Journal of the East African and Uganda Natural History Society* 13:339–43.

Hooijer, D. A. 1969. Pleistocene east African rhinoceroses. *Fossil Vertebrates of Africa* 1: 71–98.

———. 1987. Hipparion teeth from the Ndolanya Beds. In *Laetoli: A Pliocene site in northern Tanzania,* ed. M. D. Leakey and J. M. Harris, 312–15. Oxford: Clarendon.

Inskeep, R. 1962. The age of the Kondoa Rock paintings in the light of recent excavations at Kisese II Rock Shelter. *Actes du IV Congrès Panafricain de Préhistoire et de l'Étude du Quaternaire,* ed. G. Mortelmans and J. Nenquin, 249–56. Tervuren, Belgium: Musee Royale de l'Afrique Centrale.

Jager, T. 1982. *Soils of the Serengeti woodlands.* Wageningen, The Netherlands: Centre for Agricultural Publishing and Documentation.

Johnson, T. C., C. A. Scholz, M. R. Talbot, K. Kelts, R. D. Ricketts, O. Ngobi, K. Beuning, I. Ssemmanda, and J. W. McGill. 1996. Late Pleistocene desiccation of Lake Victoria and rapid evolution of cichlid fishes. *Science* 273:1091–93.

Jolly, D., S. P. Harrison, B. Damnati, and R. Bonnefille. 1998. Simulated climates and biomes of Africa during the late Quaternary: Comparison with pollen and lake status data. *Quaternary Science Reviews* 17:629–57.

Jolly, D., D. Taylor, R. Marchant, A. Hamilton, R. Bonnefille, G. Buchet, and G. Riollet. 1997. Vegetation dynamics in Central Africa during the late glacial and Holocene periods: Pollen records from the interlacustrine highlands of Burundi, Rwanda and western Uganda. *Journal of Biogeography* 24:495–512.

Kappelman, J. 1984. Plio-Pleistocene environments of Bed I and lower Bed II, Olduvai Gorge, Tanzania. *Palaeogeography, Palaeoclimatology, Palaeoecology* 48:171–96.

———. 1991. The paleoenvironment of *Kenyapithecus* at Fort Ternan. *Journal of Human Evolution* 20:95–129.

Kappelman, J., T. Plummer, L. Bishop, A. Duncan, and S. Appleton. 1997. Bovids as indicators of Plio-Pleistocene paleoenvironments in East Africa. *Journal of Human Evolution* 32:229–56.

Kendall, R. L. 1969. An ecological history of the Lake Victoria basin. *Ecological Monographs* 39:121–75.

Kingston, J., and T. Harrison. 2001. High-resolution middle Pliocene landscape reconstructions at Laetoli, Tanzania. *Journal of Human Evolution* 38:A11 (abstract).

———. 2002. Isotopically based reconstructions of early to middle Pliocene paleohabitats at Laetoli, Tanzania. *American Journal of Physical Anthropology* Supplement 34:95–96 (abstract).

———. 2007. Isotopic dietary reconstructions of Pliocene herbivores at Laetoli: Implications for early hominin paleoecology. *Palaeogeography, Palaeoclimatology, Palaeoecology* 243: 272–306.

Klein, R. G. 1984. The large mammals of southern Africa: Late Pliocene to recent. In *Southern African prehistory and paleoenvironments,* ed. R. G. Klein, 107–46. Rotterdam: A. A. Balkema.

Kovarovic, K. 2004. Bovids as palaeoenvironmental indicators. An ecomorphological analysis of bovid postcranial remains from Laetoli, Tanzania. PhD diss., University of London.

Kovarovic, K., P. J. Andrews, and L. Aiello. 2002. The paleoecology of the Upper Ndolanya Beds at Laetoli, Tanzania. *Journal of Human Evolution* 43:395–418.

Krafft, M., and J. Keller. 1989. Temperature measurements in carbonatite lava lakes and flows from Oldoinyo Lengai, Tanzania. *Science* 245: 168–70.

Lamprey, H. 1979. Structure and functioning of the semi-arid grazing land ecosystem of the Serengeti region (Tanzania). In *Tropical grazing land ecosytems: A state-of-knowledge report,* prepared by UNESCO/UNEP/FAO, 562–601. Paris: UNESCO.

Leakey, M.D. 1971. *Olduvai Gorge: Excavations in Beds I and II, 1960–1963.* Cambridge: Cambridge University Press.

———. 1987. The Laetoli hominid remains. In *Laetoli: A Pliocene site in northern Tanzania,* ed. M. D. Leakey and J. M. Harris, 108–17. Oxford: Clarendon.

Leakey, M. D., and J. M. Harris (eds.). 1987. *Laetoli: A Pliocene site in northern Tanzania.* Oxford: Clarendon.

Luyt, C. J., and J. A. Lee-Thorp. 2003. Carbon isotope ratios of Sterkfontein fossils indicate a marked shift to open environments c.1.7 Myr ago. *South African Journal of Science* 99:271–73.

MacIntyre, R. M., J. G. Mitchell, and J. B. Dawson. 1974. Age of the fault movements in the Tanzanian sector of the East African rift system. *Nature* 247:354–56.

Manega, P. 1993. *Geochronology, geochemistry and isotopic study of the Plio-Pleistocene hominid sites and the Ngorongoro volcanic highlands in northern Tanzania.* PhD diss., University of Colorado, Boulder.

Marean, C. W. 1989. Sabertooth cats and their relevance for early hominid diet and evolution. *Journal of Human Evolution* 18:559–82.

———. 1990. *Late Quaternary paleoenvironments and faunal exploitation in East Africa.* Unpublished PhD diss., University of California, Berkeley.

———. 1992a. Hunter to herder: Large mammal remains from the hunter-gatherer occupation at Enkapune Ya Muto rockshelter (Central Rift, Kenya). *African Archaeological Review* 10:65–127.

———. 1992b. Implications of late Quaternary mammalian fauna from Lukenya Hill (south-central Kenya) for paleoenvironmental change and faunal extinctions. *Quaternary Research* 37:239–55.

———. 1997. Hunter-gatherer foraging strategies in tropical grasslands: Model building and testing in east African middle and later Stone Age. *Journal of Anthropological Research* 16:184–225.

Marean, C. W., and Z. Assefa. 1999. Zooarchaeological evidence for the faunal exploitation behavior of Neandertals and early modern humans. *Evolutionary Anthropology* 8:22–37.

Marean, C. W., and D. Gifford-Gonzalez. 1991. Late Quaternary extinct ungulates of East Africa and paleoenvironmental implications. *Nature* 350:418–20.

Marean, C. W., N. Mudida, and K. E. Reed. 1994. Holocene paleoenvironmental change in the Kenyan Central Rift as indicated by micromammals from Enkapune Ya Muto rockshelter. *Quaternary Research* 41:376–89.

Marean, C. W., C. L. Ehrhardt, and N. Mudida. n.d. Late Quaternary mammalian fauna in eastern Africa: Its relevance for environmental change and faunal extinctions. Unpublished paper presented at Sixth International Conference, International Council for Archaeozoology, Washington, DC: July 1990.

Marshall, F. 1986. *Aspects of the advent of pastoral economies in East Africa.* Unpublished PhD diss., University of California-Berkeley.

———. 1990a. Cattle herds and caprine flocks. In *Early pastoralists of southwestern Kenya,* ed. P. Robertshaw, 205–60. Nairobi: British Institute of East Africa.

———. 1990b. Origins of specialized pastoral production in East Africa. *American Anthropologist* 92:873–94.

McBrearty, S., and A. S. Brooks. 2000. The revolution that wasn't: A new interpretation of the origin of modern human behavior. *Journal of Human Evolution* 39:453–63.

McConnell, R. B. 1972. Geological development of the Rift system in eastern Africa. *Geological Society of America Bulletin* 83:2549–72.

Mehlman, M. J. 1989. Later Quaternary archaeological sequences in northern Tanzania. PhD diss., University of Illinois at Urbana-Champaign.

Morley, R. J. 2000. *Origin and evolution of tropical rain forests.* New York: Wiley.

Musiba, C. M. 1999. Laetoli Pliocene paleoecology: A reanalysis via morphological and behavioral approaches. PhD diss., University of Chicago.

Mutakyahwa, M. 1997. Mineralogy of the Wembere-Manonga Formation, Manonga Valley, Tanzania, and the possible provenance of the sediments. In *Neogene paleontology of the Manonga Valley, Tanzania,* ed. T. Harrison, 67–78. New York: Plenum.

Ndessokia, P. N. S. 1990. *The mammalian fauna and archaeology of the Ndolanya and Olpiro Beds, Laetoli, Tanzania.* PhD diss., University of California, Berkeley.

Norton-Griffiths, M. 1979. The influence of grazing, browsing, and fire on the vegetation dynamics of the Serengeti. In *Serengeti: Dynamics of an ecosystem,* ed. A. R. E. Sinclair and M. Norton-Griffiths, 310–52. Chicago: University of Chicago Press.

Peters, C. R., and R. J. Blumenschine. 1995. Landscape perspectives on possible land use patterns for Early Pleistocene hominids in the Olduvai Basin, Tanzania. *Journal of Human Evolution* 29:321–62.

———. 1996. Landscape perspectives on possible land use patterns for Early Pleistocene hominids in the Olduvai Basin, Tanzania: Part II, expanding the landscape models. *Kaupia* 6:175–221.

Peyron, O., D. Jolly, R. Bonnefille A. Vincens, and J. Guiot. 2000. The climate of East Africa from pollen data, 6000 years ago. *Quaternary Research* 54(1): 90–101.

Pickering, R. 1958. *Oldoinyo Ogol (Serengeti Plain East).* Geological Quarter Degree Sheet 38 (12 SW), scale 1:125,000. Dodoma (Tanzania): Geological Survey Department.

———. 1960. *Moru (Serengeti Plain West).* Geological Quarter Degree Sheet 37, scale 1: 125,000. Dodoma (Tanzania): Geological Survey Division.

———. 1964. *Endulen.* Geological Quarter Degree Sheet 52, scale 1:125,000. Dodoma (Tanzania): Geological Survey Division, Ministry of Industries Mineral Resources and Power.

Pickford, M. 1995. Fossil land snails of East Africa and their palaeoecoogical significance. *Journal of African Earth Sciences* 20:167–227.

Plummer, T. W., and L. C. Bishop. 1994. Hominid paleoecology at Olduvai Gorge, Tanzania as indicated by antelope remains. *Journal of Human Evolution* 27:47–75.

Pratt, D. J., P. J. Greenway, and M. D. Gwynne. 1966. A classification of East African rangeland, with an appendix on terminology. *Journal of Applied Ecology* 3(2): 369–82.

Pyne, S. J. 1991. *The burning bush: A fire history of Australia.* New York: Holt.

Reed, K. E. 1997. Early hominid evolution and ecological change through the African Plio-Pleistocene. *Journal of Human Evolution* 32:289–322.

Richard, J. J. 1942. Volcanological observations in East Africa: I, Oldonyo Lengai; the 1940–1941 eruption. *Journal of the East Africa and Uganda Natural History Society* 16: 89–108.

Sands, W. A. 1987. Ichnocoenoses of probable termite origin from Laeloti. In *Laetoli: A Pliocene site in northern Tanzania,* ed. M. D. Leakey and J. M. Harris, 409–33. Oxford: Clarendon.

Schaller, G. B., and G. R. Lowther. 1969. The relevance of carnivore behavior to the study of early hominids. *Southwestern Journal of Anthropology* 25:307–41.

Scholz, C. A., T. C. C. Johnson, P. Cattaneo, H. Malinga, and S. Shana. 1998. Initial results of 1995 IDEAL seismic reflection survey of Lake Victoria, Uganda and Tanzania. In *Environmental change and response in east African lakes,* ed. J. T. Lehman, 47–57. Dordrecht: Kluwer.

Semaw, S., P. Renne, J. W. K. Harris, C. S. Feibel, R. L. Bernor, N. Fesseha, and K. Mowbray. 1997. 2.5-million-year-old stone tools from Gona, Ethiopia. *Nature* 385:333–36.

Sharam, G. J. 2005. The decline and restoration of riparian and hilltop forests in the Serengeti National Park, Tanzania. PhD diss., University of British Columbia, Vancouver.

Sharam, G., A. R. E. Sinclair, and R. Turkington. 2006. Establishment of broad-leaved thickets in Serengeti, Tanzania: The influence of fire, browsers, grass competition and elephants. *Biotropica* 38:599–605.

Shea, J., Z. Davis, and K. Brown. 2001. Experimental tests of Middle Palaeolithic spear points using a calibrated crossbow . *Journal of Archaeological Science* 28:807–16.

Shipman, P., and J. Harris. 1988. Habitat preference and paleoecology of *Australopithecus boisei* in eastern Africa. In *Evolutionary history of the 'robust' Australopithecines,* ed. F. E. Grine, 343–81. New York: Aldine de Gruyter.

Sikes, N. E. 1996. Hominid habitat preferences in Lower Bed II: Stable isotope evidence from paleosols. *Kaupia* 6:231–38.

Sinclair, A. R. E. 1979. The Serengeti environment. In *Serengeti: Dynamics of an ecosystem,* ed. A. R. E. Sinclair and M. Norton-Griffiths, 31–45. Chicago: University of Chicago Press.

Sinclair, A. R. E., and M. Norton-Griffiths. 1979. *Serengeti: Dynamics of an ecosystem.* Chicago: University of Chicago Press.

Skinner, A. R., R. L. Hay, F. Masao, and B. A. B. Blackwell. 2003. Dating the Naisiusiu Beds, Olduvai Gorge, by electron spin resonance: *Quaternary Science Reviews* 22: 1361–66.

Soligo, C., and P. Andrews. 2005. Taphonomic bias, taxonomic bias and historical non-equivalence of faunal structure in early hominin localities. *Journal of Human Evolution* 49:206–29.

Spencer, L. M. 1997. Dietary adaptations of Plio-Pleistocene Bovidae: Implications for hominid habitat use. *Journal of Human Evolution* 32:201–28.

Stager, J. C., 1982. The diatom record of Lake Victoria (East Africa): The last 17,000 years. In *Proceedings of the Seventh Diatom Symposium,* August 1982, Philadelphia, ed. D. G. Mann, 455–76. Koenigstein, Federal Republic of Germany: Otto Koeltz Science Publishers.

Stager, J. C., B. Cumming, and L. Meeker. 1997. A high-resolution 11,400-yr diatom record from Lake Victoria, East Africa. *Quaternary Research* 47:81–89.

Stager, J. C., P. R. Reinthal, and D. A. Livingstone. 1986. A 25,000 year history for Lake Victoria, East Africa, and some comments on its significance for the evolution of cichlid fishes. *Freshwater Biology* 16:15–19.

Stringer, C. 2003. Out of Ethiopia. *Nature* 423:692–95.

Su, D. 2005. The paleoecology of Laetoli, Tanzania: Evidence from the mammalian fauna of the Upper Laetolil Beds. PhD diss., New York University.

Su, D., and T. Harrison. 2003. Faunal differences in the sequence at Laetoli: Implications for taphonomy and paleoecology. *American Journal of Physical Anthropology* Supplement 36:203 (abstract).

———. 2007. The paleoecology of the Upper Laetolil Beds at Laetoli: A reconsideration of the large mammal evidence. In *Hominin environments in the East African Pliocene: An Assessment of the Faunal evidence,* ed. R. Bobe, Z. Alemseged, and A. K. Behrensmeyer, 279–313. Dordrecht: Springer.

Talbot, L. M., and M. H. Talbot. 1963. The wildebeest in western Masailand, East Africa. *Wildlife Monographs* no. 12. Washington, DC: Wildlife Society.

Talbot, M. R., M. A. J. Williams, and D. A. Adamson. 2000. Strontium isotope evidence for late Pleistocene reestablishment of an integrated Nile drainage network. *Geology* 28:343–46.

Tamrat, E., N. Touveny, M. Taieb, and N. D. Opdyke. 1995. Revised magnetostratigraphy of the Plio-Pleistocene sedimentary sequence of the Olduvai Formation (Tanzania): *Palaeogeography, Palaeoclimatology, and Palaeoecology* 114:273–83.

Temple, P. H. 1969a. Raised strandlines and shoreline evolution in the area of Lake Nabugabo, Masaka District, Uganda. *Quaternary Geology and Climate,* Publication 1701, 119–29. Washington, DC: National Academy of Sciences.

———. 1969b. Some biological implications of a revised geological history for Lake Victoria. *Biological Journal of the Linnaean Society* 1:363–71.

Thurow, T. 1995. Development in the TransMara. *Gnusletter* 14 (1): 17–18.

Turrill, W. B. 1952. *Oleaceae: Flora of tropical East Africa.* London: Crown Agents for the Colonies.

Verdcourt, B. 1987. Mollusca from the Laetolil and Upper Ndolanya Beds. In *Laetoli: A Pliocene site in northern Tanzania,* ed. M. D. Leakey and J. M. Harris, 438–50. Oxford: Clarendon.

Verniers, J. 1997. Detailed stratigraphy of the Neogene sediments at Tinde and other localities in the central Manonga basin. In *Neogene Paleontology of the Manonga Valley, Tanzania,* ed. T. Harrison, 33–65. New York: Plenum.

Walter, R. C., P. C. Manega, R. L. Hay, R. E. Drake, and G. H. Curtis. 1992. Tephrochronology of Bed I, Olduvai Gorge: An application of laser-fusion 40Ar/39Ar dating to calibrating biological and climatic change: *Quaternary International* 13/14:37–46.

Washburn, S. L., and C. S. Lancaster. 1968. The evolution of hunting. In *Man the hunter,* ed. R. B. Lee and I. DeVore, 293–303. Chicago: Aldine.

Watson, G. E. 1987. Pliocene bird fossils from Laetoli. In *Laetoli: A Pliocene site in northern Tanzania,* ed. M. D. Leakey and J. M. Harris, 82–84. Oxford: Clarendon.

Werdelin, L., and M. E. Lewis. 2005. Plio-Pleistocene carnivora of eastern Africa: Species richness and turnover patterns. *Zoological Journal of the Linnean Society* 144:121–44.

Western, D. 1997. *In the Dust of Kilimanjaro.* Washington, DC: Island Press.

White, T. D. 1977. New fossil hominids from Laetoli, Tanzania. *American Journal of Physical Anthropology* 46:197–230.

———. 1980. Additional fossil hominids from Laetoli, Tanzania: 1976–1979 specimens. *American Journal of Physical Anthropology* 53:487–504.

Williams, L. A. J. 1963. The geology of the Narok District. *Proceedings of the East African Academy* 1:37–49.

Wright, J. B. 1967. *Geology of the Narok area.* Report no. 80, Geological Survey of Kenya. Nairobi: Kenya Mines and Geological Department.

FOUR

The Resource Basis of Human-Wildlife Interaction

Han Olff and J. Grant C. Hopcraft

Throughout history, humans have always had to make choices about where to live, where to graze their livestock, and where to grow their crops. These spatial choices were originally based on the availability of food to hunter/ gatherers, then on opportunities for agriculture and livestock (particularly moist, fertile land), and now increasingly on the availability of jobs in commercial centers. Such resource transitions are a classic subject of study in ecological anthropology (White 1959; Hardesty 1975; Binford 2001; Borgerhoff, Mulder and Coppolillo 2005). In tropical savannas, soil fertility and rainfall have a major impact on the spatial distribution and land use of humans in the system. However, these same resources effect the spatial choices made by the abundant wild herbivores in savannas, with the potential for conflict in land use (Olff, Ritchie, and Prins 2002). In this chapter, we will discuss how current spatial patterns in land use by humans and the spatial distribution of native and domesticated herbivores are affected by resource availability (rainfall and soil fertility), and discuss the analogy between the current spatial patterns and the historical developments of agriculture and livestock husbandry in East Africa.

Within the savannas of East Africa, our relationship with the environment evolved through time, from scavenging meat and gathering seeds and tubers, to being pastoralists and agriculturalists, and eventually to being sedentary and commercially active with intensive agriculture and livestock husbandry (Bromage and Schrenk 1999; Reader 1999). The ecological consequences of these main resource transitions can still be studied in East

Africa, but now separated spatially rather than temporally. A comparison of the different ecological roles humans have played yields interesting lessons from our origins that are applicable to current-day conservation strategies, particularly in the light of increasing human-wildlife conflicts in savanna ecosystems (Prins, Grootenhuis, and Dolan, 2000; Prins 1992; Lamprey and Reid 2004; Ogutu, Bhola, and Reid 2005).

In this chapter, we analyze how current human population density and land use respond to environmental gradients, with an emphasis on rainfall and soil fertility, and compare this to the responses of large resident herbivores. We will identify historic shifts that led to intensified human land use in East Africa, identifying three distinct phases: hunter-gatherer, agripastoralist, and modern commercialized societies. Where chapter 3 deals with the long-term history (4 to 5 million years) we zoom in on the more recent history (10,000 years) since the end of the last ice age. These three historic phases of human land use are analogous to current main land use systems in the Serengeti-Mara ecosystem; (1) parks for wildlife and ecotourism, (2) protected multiple-use areas where people and wildlife coexist, and (3) the rural/village areas, with agricultural and livestock systems managed by a variety of more-or-less formal land tenure systems (Norton-Griffiths 1995). This sets the scene for a discussion of the resource basis of human-wildlife interactions in savannas, from which we can learn to manage these interactions better in the future.

HUMAN LAND USE CHANGES IN THE REGIONAL SETTING

Environmental Shifts and the Invasions of Agripastoralists into East Africa

During the last glaciation period, which extended from roughly 90,000 to 10,000 years ago, East Africa became significantly dryer and cooler (fig. 4.1a). The rain forest areas contracted while deserts and steppes expanded. For early humans foraging in this landscape, the drying process resulted in substantial range restrictions. As people were forced into smaller and smaller productive areas their resource demands on these confined areas grew. Sometime during this period, through several independent events around the world, humans started to foster the growth of their wild food species, probably at first by inadvertent reseeding. Not only did these cultures persist under the new environmental conditions, but as the climate warmed over the next several thousand years (fig. 4.1b, c) these new techniques advanced into organized cultivation and intensified food production (Diamond 1996; Ehret 2002; Gupta 2004). This was a major breakthrough for human evolution (Iliffe 1995).

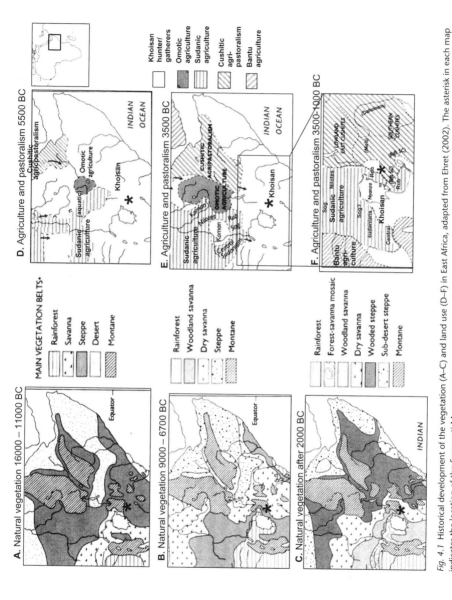

Fig. 4.1 Historical development of the vegetation (A–C) and land use (D–F) in East Africa, adapted from Ehret (2002). The asterisk in each map indicates the location of the Serengeti-Mara ecosystem.

Africa's largest and longest river, the Nile, enriches its floodplains seasonally, transporting abundant nutrients from the volcanic highlands of central and eastern Africa. The Nile floodplains were undoubtedly a cornucopia of food, with easy access to water, attracting early people to stay in these areas year round. Evidence suggests that Lake Victoria and the Nile were not connected before 12,500 years ago (chapter 3); the lake was much smaller and probably flowed westward before being blocked by tectonic faulting. The Nile probably became a much more stable and reliable source of water after this connection with Lake Victoria, and the more stable flow may well have encouraged the development of agriculture, particularly since the surrounding areas were arid during this time and humans were experiencing large range-restrictions. People cultivated grains such as millet in the southern Nile delta and supplemented their diet with fish. Similarly, during the same time period, humans retreating to the Niger delta in West Africa cultivated tubers and yams.

The independent development of agriculture in separate parts of Africa is important for understanding the Serengeti ecosystem because different cultures invaded the Lake Victoria and Rift Valley region in several repeating waves. This had significant consequences for land use and the ecology of the system. As the earth started warming 10,000 years ago, glaciers receded and Africa became wetter. Large forests and savannas grew in areas where there were previously deserts and steppe (fig. 4.1b), opening these areas once again to humans. However, by this time humans were no longer reliant only on gathering wild food. Archaeological evidence, from the use of tools and pottery as well as from linguistic lineages, suggests that agri-pastoralist cultures had begun to spread across the continent starting from several distinct sources. Starting 3,500 years ago, the collision of four very different cultures was taking place in the ecologically diverse crossroads of the Lake Victoria region (fig. 4.1d–f); the Khosian to the south were primarily hunter-gatherers, the Cushites from the Horn of Africa followed the Rift Valley with their livestock, the Southern Sudanic people (including the Rub, Sog, and later the Nilotes) followed the Nile, bringing a combination of grains (millet, sorghum), aquaculture (African wild rice), and livestock, and the Bantu people spreading from West Africa, bringing yams and other tubers. Although the main features of the Serengeti-Mara ecosystem have remained relatively unchanged for the last 6,000 years, despite being punctuated by catastrophic and pervasive volcanic events that transformed the landscape significantly for short periods of time, these diverse agripastoralist cultures—with very different demands on the landscape—influenced the ecology of the system significantly.

Analogous Systems to the Hunter-Gatherer Agripastoralist Transition

The Serengeti-Mara ecosystem is regionally situated at an intermediate position on the rainfall gradient that excludes the very driest and the very wettest areas in the region (fig. 4.2). The rainfall gradient is clearly pronounced from the southeast to the northwest due to the rain shadows of Mount Meru, Kilimanjaro, and the Ngorongoro highlands, and the influence of Lake Victoria (chapter 2, fig. 2.6). This particular rainfall gradient is however not unique: similarly strong transitions in rainfall (spatial variation in rainfall of more than 500 mm/yr over a distance of <300 km) also occur in eastern Kenya, southern Sudan, and southern Tanzania (fig. 4.2). Other explanations (such as geological features) are thus required to understand the extraordinary dominance of migrant herbivores in the Serengeti ecosystem.

Three types of past human land use (hunter/gatherer, pastoralist, agriculturalist) are analogous to the main current land use zones in the Greater

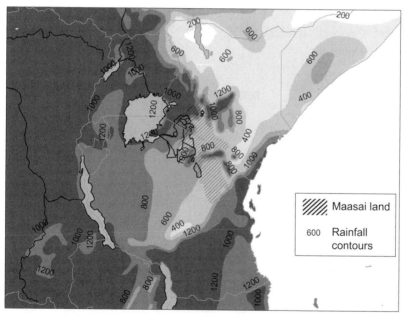

Fig. 4.2 The position of the protected areas in the Serengeti-Mara ecosystem (solid lines) within the regional rainfall gradient. Rainfall data from Food and Agriculture Organization /United Nations Environmental Program Desertification and Mapping Project, as rasterized by UNEP/GRID (1986). Labels refer to rainfall isohyets, not rainfall polygons. The current distribution of Maasai pastoralists (southern Maasailand) is indicated by the striped area (M. Said, ILRI, pers. comm.).

Serengeti-Mara ecosystem (fig. 2.3). The Serengeti Park and the Maasai Mara Reserve are core protected areas, managed only for wildlife and ecotourism, where humans have minimal impact on the landscape and which are probably similar to the environs of early hunter-gatherers. Maasai pastoralists, who move their livestock seasonally from xeric to mesic areas, generally occupy areas between 500 to 700mm of rainfall (fig. 4.2) on the eastern side of the ecosystem. The pastoralist connection with the land has remained remarkably stable over the last 4,000 thousand years, although historically the inhabitants may have been of Cushite origin rather than Nilotic (the Maasai migrated to this area only in the seventeenth century [Ehret 2002]). In many regions beyond the pastoral and core protected areas the land is cultivated by small-scale subsistence farmers, and subject to a variety of land use systems (figs. 2.3, 2.21). This is especially evident on the western side of the ecosystem.

THE ROLE OF HUMANS IN SAVANNAS: THE THREE PHASES EXPLAINED

Fossil remains of *Homo sapiens* in the Serengeti date back to 17,000 years ago, but with the largest ecological consequences for the Serengeti occur-

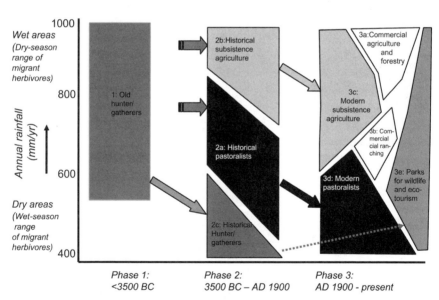

Fig. 4.3 Historical radiation of land use/ livelihood strategies in East African savannas as dependent on rainfall. The vertical extent of a polygon in each time period is proportional to the range of rainfall conditions used for the activity. The width of each polygon indicates the relative proportion of land used at different rainfall levels (Ehret 2002; Reader 1999; Iliffe 1995).

ring over the last 5,000 years. During this time, humans have altered the landscape, first through the use of fire, then through domesticating live-stock and developing agriculture. Despite this long history of human use, it is relatively short compared to the existence of the dominant large her-bivore groups in savannas. About 40 ma, all major orders of herbivores in savannas already existed. First the hyracoids, rhinoceroses, and elephants were evolutionarily successful and diverse, but only a few species of this group now remain. They were mostly replaced by other, more successful ungulates in recent radiations, which coincided with the radiation of the graminoids (grasses and sedges) about 8 ma.

The historical development of the role of humans in savannas is sum-marized in figure 4.3, and the main factors that regulated human densities in each phase are summarized in table 4.1. The figure shows how new forms of land use emerged through time, and how this modified the position of already-existing land use strategies along the rainfall gradient. This is a re-curring theme in our chapter that we develop in the following sections. We view the protected areas in which limited hunting, and no agriculture or livestock are allowed (i.e., the Mara Reserve, the Serengeti National Park, and the surrounding game reserves; fig. 2.3 and phase 3e in fig. 4.3) as an analogue for the land use where humans were both predators and prey (fig. 2.3 and phase 1 in fig. 4.3). We regard the Mara group ranches, the Loliondo game control area, and the Ngorongoro Conservation Area (fig. 2.3 and phase 2a, 3d in fig. 4.3) as an analogue for historical pastoralist land use. We take the remaining rural areas in the study area as an analogue to historic sedentary subsistence agriculture (phase 2b, 3c in fig. 4.3). Figure 2.3 indicates the current spatial distribution of these different land use types in the study area. In the next sections, we provide a further descrip-tion of the historical development of these land tenure systems and their ecological impacts.

Phase 1: From Prey to Predator

Early humans in Africa were foragers, actively searching for and collecting wild species of seeds, fruits, roots, and tubers (fig. 4.3). They probably sup-plemented their diet by hunting small animals under 5 kg and scavenging from larger carcasses (Blumenschine 1987; Blumenschine et al. 2003). The earliest stone tools emerged about 2 to 1.7 ma and provide the first evidence of hominids becoming large carnivores. Previous to this period, none of the tools were sufficient to hunt large prey, suggesting that hominids were still subordinate carnivores and, therefore, had minimal impact on populations

Table 4.1 Outline of the changing role of humans in savanna ecosystems.

Role of humans with respect to use of the land and natural resources	Period of emergence of new role in Serengeti-Mara ecosystem	Ecosystem impacts of humans	Current spatial analogue in the region	Relative importance of main factors regulating human population density in each phase (more plusses is more important)						Human population density in the non-urban areas	Proportion of population in nonurban areas	Spatial scale of regulation of human population	Local biodiversity (1–100 km²)	Regional biodiversity (100,000–10,000 km²)
				Predation	Rainfall, temperature	Soil fertility	Human and livestock diseases	Employment opportunities	Poverty					
Hunter/gatherers 1	<3500 BC	Hunting, scavenging, fires	National parks and game reserves, as Serengeti National Park, Mara Reserve	++	+	+				Low	All	Local	Medium	High
Pastoralists 2a	3500–1900 BC	Hunting, livestock grazing, fires	Controlled multiple use areas, as Ngorongoro Conservation area, Loliondo, Mara group ranches		++	++	+		+	Intermediate	All	Local to regional	Medium-high?	High

Sedentary subsistence agriculture with some livestock husbandry 2b	3500–1900 BC	Hunting, livestock grazing, conversion natural vegetation to cropland, modification of hydrology and nutrient cycling, erosion	Sedentary subsistence agricultural and rangeland areas as the Musoma, Bunda, Barradi, Narok, Nyaza districts outside protected areas	+++	+++	+++	+++	++	High	Intermediate	Local to regional	Low-medium	Medium
Commercial, intensive (irrigated) agriculture, fenced pastures, commercial ranches, feedlots 3a & 3b	>1900 AD	Livestock grazing, conversion natural vegetation to cropland, modification of hydrology and nutrient cycling, environmental pollution	Surroundings Arusha, Nairobi	+	++	++	++	+++	Intermediate	Low	Regional to global	Low	Low

of large herbivores and carnivores. Humans probably found themselves in the unfortunate position of being prey themselves, since large carnivores were plentiful at the time; a similar trophic position to modern baboons (Cowlishaw 1994). Weaponry such as spear heads and sharpened projectiles became more common in the Middle Stone Age (280,000 years ago), indicating a shift to hunting larger animals such as wildebeest and buffalo, and probably necessitating cooperative hunting techniques. This shift was combined with mastering fire (chapter 3), which became an essential tool as humans spread to cooler climates, while in warm climates it was used for protection and to attract prey. By as early as 62,000 years ago (Later Stone Age) hominids had become top predators, with fire and weaponry as integral tools in their repertoire. Undoubtedly they became a threat to larger herbivores such as elephant, buffalo, and hippo, which likely changed the perception of these species toward humans, increasing the risk of lethal confrontations (Brain 1981).

During this phase early human population densities were low. Human distributions were probably confined to areas within about 20km of water and with suitable foraging opportunities, because they had limited food and water storage strategies (Reader 1999; Bromage and Schrenk 1999). Rainfall and soil fertility most likely determined the areas that would have been occupied. Regulation of human population density was most likely by access to these resources as well as through predation and occasional disease episodes, particularly at high densities or during times of stress. Given the seasonal nature of the area, due to the Intertropical Convergence Zone (chapter 6), the hominid population was probably mobile, possibly to the point of being migratory. West African pastoralists in the Sahel zone still display such migrant behavior (Breman and De Wit 1983). Groups were probably not larger than 25 individuals. Therefore, local droughts and disease events most likely did not affect the entire population synchronously and could have occurred at more local scales. As these small groups of people moved across the landscape foraging they most likely depleted local resources, but regional biodiversity would have remained high.

The current national parks and game reserves in the region can be viewed as the modern analogue to this phase. Human populations in the Serengeti National Park, Mara National Reserve, Grumeti, Ikorongo, and Maswa Game Controlled Area are low. There are relatively few resources being extracted and human impacts are limited to some hunting and burning. In these areas, therefore, we can still observe the prehuman spatiotemporal choices made by wildlife in relation to resource heterogeneity.

The second phase started when domesticated livestock (sheep, goats, donkeys, and cattle) first arrived on the drier parts of the east African savannas about 4,000 years ago (Reader 1999; chapter 3; figs. 4.1, 4.3). Cushitic people, moving south from the Horn of Africa, first introduced cattle to East Africa sometime after 3,500 BC (fig. 4.1). The original cattle in East Africa probably originated from the Indian stock of drought-tolerant zebu strains. Sheep and goats were domesticated in the Mediterranean area and also brought to East Africa, most likely across the Red Sea into Ethiopia, where some of the earliest evidence of domestication in Africa exists. In an area stretching from the Narok district in Kenya to the Simanjiro and Kiteto districts in Tanzania, people have maintained a pastoral lifestyle based on cattle herding for at least the last 2,000 years (Lamprey and Reid 2004), although the Maasai, who are currently the largest pastoralist group in the area, have only been in the Serengeti ecosystem for the last 300 years (Ehret 2002). The pastoral way of life emerged simultaneously with sedentary agriculture, and was (and is) probably more profitable in the relatively drier regions, between 400 and 800 mm of annual rainfall, while both types of new land use 'pushed' the hunter/gatherers to the areas that were too dry for both (figs. 4.1, 4.3).

The ecological implications of early pastoralism differed from those of the hunter-gatherer lifestyle by a number of key attributes: (1) pastoralists, being less dependent on wild animal and plant populations, were buffered from ecological events beyond their control. Indeed, pastoralists may have viewed wild animals as grazing competitors and predators rather than as a food resource. (2) However, the degree of ecological decoupling due to pastoralism was still small because the livestock was still dependent on water and grazing. The time scale of lifestyle strategies would have shifted from thinking seasonally to thinking over the lifetime of a cow, maximizing yields and recruitment of livestock. Therefore, (3) the distributions of pastoralists were limited to areas where livestock could flourish, favoring grassland areas over heavily canopied woodlands. Compared to hunters and gatherers who could survive in a variety of different habitats, from moist, dense forests to dry, open semiarid landscapes, pastoralists became niche specialists, depending on grazing areas with strong seasonal movements. Evidence suggests that the expansion of grasslands about 4,000 years ago is synchronous with pastoralist invasion. Grassland may have developed partly from forests through an increase in the frequency of burning (Marean and Gifford-Gonzalez 1991; Marean 1992a; Marean 1992b;

Vincens et al. 2003). (4) Fire has always been an important management tool for pastoralists, who use it to improve the local forage conditions for their livestock. (5) Ecologically, fire stimulates fresh grass growth but also suppresses the establishment of denser woody vegetation, especially in areas with higher rainfall, and consequently reduces the abundance of tsetse flies (Reid et al. 2000), which transmit trypanosomes. The local occurrence of trypanosomiasis is thought to have set important boundaries on the historical distribution of the pastoralists (Prins 1996). The population density of pastoralists was therefore closely linked with the main livestock diseases, also including rinderpest (Sinclair and Norton-Griffiths 1979; Dobson 1995). Furthermore, diseases transferred from livestock to humans (zoonoses) may have played an important role in the evolution of culture as pastoralists moved into the traditional territories of hunter-gatherers, who until 5,000 years ago, had been little exposed to these diseases (Iliffe 1995; Diamond 1996). The densities of pastoralists would have been further regulated by largescale regional droughts causing catastrophic losses of livestock, but local droughts could have been avoided by a nomadic lifestyle. Large scale drought events would not have affected hunter-gatherers to the same extent because they could have switched food sources, concentrating their foraging in other habitats inaccessible to the pastoralists (forests, riverine, intertidal areas, etc.). (6) Livestock were not only a source of food and protein, but they were probably a sign of wealth and status in the community, similar to contemporary pastoralists. Therefore, there was motivation to increase the size of herds above that necessary for consumption. Large herds of livestock, particularly at high densities in confined areas, would have altered the grazing ecology and vegetation community of the landscape.

Historic shifts in land use to a more pastoralist lifestyle dependent on livestock are analogous to modern Maasai culture in and around the Serengeti ecosystem (fig. 4.2). The pastoral way of life has been generally viewed as compatible with nature conservation targets, leading to the establishment of multiple-use areas in which the interests of both wildlife and the pastoral Maasai are recognized (Norton-Griffiths 1995). These areas provide a good opportunity to explore how people and wildlife jointly respond to environmental gradients.

Phase 2b: Agriculturalists in Wetter Areas

The agriculturalists became even more decoupled from ecological events, but, relative to pastoralists, at the same time more specialized and constrained in the areas where they could live (figs. 4.1, 4.3). The latter could move with the seasons, maximizing their time on good grazing areas, whereas the agriculturalists developed areas that were most profitable (and therefore most evolutionarily advantageous) and became sedentary. Such areas had fertile soils with adequate rainfall throughout most of the year. Any areas beyond this could not support crops nor large densities of livestock year round and were left undeveloped. Agriculturists became less dependent on wild animals and plants, probably viewing them as a menace (pests and weeds). Their time frame of decision making became longer still; while pastoralists probably measured time by cohorts of livestock, agriculturalists probably thought seasonally on the short term but by family proprietorship of a tract of land in the long-term, which could have been for tens or even hundreds of years. Decisions included the management of soil fertility, pests, weeds, and seed sources, which require a longer-term (years to decades) perspective. Storage and provisioning of excess food meant that agriculturists could survive through lean periods, implying the population was regulated more by inherent soil fertility and long-term average rainfall rather than by seasonal or localized environmental events. Human, livestock, and crop diseases probably played a large role in regulating the density of people, as did social structures within the communities. Furthermore, a person's social status in the community provided incentives beyond the necessary acquisition of food for individuals to develop larger tracts of land or larger herds. Therefore the demands on the landscape by agriculturalists were in excess of what would be expected by a straightforward subsistence lifestyle, and even greater than pastoralists or hunter-gatherers. By relieving themselves from the bottlenecks of resource limitation, the agriculturalists developed higher population densities than the pastoralists and gained a competitive advantage over the pastoralists in the wetter areas. At the local scale, biodiversity in agricultural areas was most likely reduced, increasing at the regional level but still lower than that in nonagricultural areas (table 4.1).

Most current agriculture in the Serengeti-Mara region is still small-scale subsistence farming with an extensive use of hand implements. Single households maintain fields. Excess food is generally sold as a source of income and used to support the family's requirements. These include tools, medicine, education, or reinvestment into the property. Livestock are kept particularly as security for periods of adversity and as a status symbol. However, money and purchased luxury goods (technology, cars) are increasingly replacing livestock in this respect (chapter 13, this volume). Sedentary agriculture, currently employed by the Wakuria and Wasukuma people, dominates the western side of the Serengeti ecosystem (the high end of the rainfall gradient), while the eastern and northern sides are occupied by Maasai pastoralists (figs. 2.3, 4.2, 4.3). Most of the agricultural population in the region still lives in rural village areas, although towns such as Mugumu and Bunda are increasing in size as centers of trade and services.

Phase 3a and 3b: Commercialization of Agriculture, Livestock Husbandry and Urbanization in Wet Areas

The earliest evidence of expansive commercialized civilizations in Africa comes from Axum in northern Ethiopia and the Niger delta (Reader 1999; figs. 4.1, 4.3). These civilizations were large, complex societies, with elaborate social hierarchies where citizens specialized in a variety of skills. Tradesmen and artisans developed livelihoods that were not based solely on agriculture or selling or bartering artifacts that were not necessarily directly linked to daily survival. Commercial centers developed in rich agricultural areas, as this was the foundation of the economy, and humans became even less constrained by ecological events. Extensive provisioning and transport of goods meant these centers could persist through lean years, depending more on their local economy as a resource. In fact, modern commercial centers such as Arusha and Nairobi have grown to the point where they are dependent on the regional and global economy for the supply of goods, and they operate well beyond the confines of local ecological constraints. Food is produced by intensive large-scale mechanized agriculture and livestock are increasingly raised in high-density feedlots.

Although urban areas were originally dependent on soil fertility and rainfall in their direct surroundings, factors that regulate human density in these heavily populated centers are now mainly employment opportunities, poverty, and diseases (table 4.1). The connection between large commercial centers and their surrounding landscape has strongly weakened, abstracted to the point were it is even sold as a quaint holiday experience of "wilderness." Wild animals pose no threat to these commercial centers; ironically, wildlife are considered a commercial resource, which is a very different type of resource perceived by the original hunter-gatherers or the pests that plague agriculturists. The modern scale of time is now based on careers and measured in lifetimes and generations. Goods and credits are moved rapidly from one commercial center to the next. We seldom think about storage and provisioning of food in the same way that subsistence agriculturalists do, because we trust our local grocer and the miracles of supply and demand in the free market economy. Biodiversity in these urban centers and the surroundings areas has become strongly reduced due to intensive resource extraction and land use change.

In East Africa, the switch to intensive agriculture and urbanization was initiated by the colonial administrations over a century ago (fig. 4.3). Agriculture focused mainly on cash crops such as tea, coffee, sisal, wheat, and more recently, on fresh vegetables and flowers for export. This has lead to important changes in the spatial distribution of people: the employment per hectare in the rural areas has declined because of increased efficiency. Rural people are now inclined to sell their land (often to large agricultural businesses; chapter 13, this volume), and move to the cities in search of alternate employment in services or industries. If employment opportunities and economic growth are sufficient, then this phase can lead to an alleviation of poverty and reduced human population through internal mechanisms (more wealthy people generally decide to have fewer children; see chapter 12, this volume). Although this phase has yet to occur in the extended Serengeti-Mara ecosystem, urban centers such as Arusha and Nairobi influence local economies. Also, nearby towns such as Mwanza, Karatu, and Musoma are on the verge of urbanization. Intensive urbanization also serves as an interesting vantage point for comparison and analysis of trends in land use.

REGIONAL TRENDS IN RAINFALL, SOIL FERTILITY, AND HUMAN LAND USE

Following up on the previous considerations on human resource limitation, we will now quantify how rainfall and soil fertility interact in determining the distribution of both people and wildlife in the Serengeti-Mara ecosystem and its surroundings, to identify possible areas of conflict. We focus on the study area shown in figures 4.4 and 4.5, including parts of Kenya and Tanzania, with agricultural areas, pastoralist areas, and parks. Figure 4.4 also shows the rainfall gradient in the greater Serengeti-Mara ecosystem in more detail. The driest part lies in the east around Lake Natron, and the wettest part occurs in the northwest, around the Lemek group ranch (fig. 2.3). From a herbivore perspective, a rainfall gradient provides an interesting challenge. The amount of primary production generally increases with rainfall, which would drive herbivores to the wet end of the gradient, especially the large bulk feeders. However, the food quality for herbivores generally declines with rainfall (Breman and De Wit 1983; Olff, Ritchie, and Prins 2002), driving herbivores toward the dry end of the gradient, especially the smaller, more selective feeders. If areas with low and high annual rainfall are sufficiently close herbivores may migrate between dry and wet areas to profit from the high-quality food in the dry area during the wet season, while escaping its low biomass and lack of drinking water during the dry season by migrating to the area of high rainfall. This pattern is one possible explanation for the general migration patterns of wildebeest (fig. 4.4), zebra, Grant's gazelle, and Thomson's gazelle in the Serengeti-Mara ecosystem (Fryxell, Greever, and Sinclair 1988; Fryxell 1995; Fryxell et al. 2005). However, rainfall does not tell the whole story, because, if so, the migration should be more east-west than the observed north-south-oriented movements (fig. 4.4). In addition, we suggest that soil fertility also plays an important role in explaining the general location of the migration system (Anderson and Talbot 1965; McNaughton 1990; McNaughton 1985; East 1984; Bell 1982; McNaughton 1988).

The maps of lithology (fig. 2.4) and of total exchangeable bases (fig. 4.5) for the study area show that the highest fertility soils are most likely found in the western corridor, possibly explaining the detour that the migrants make on their way to the Maasai Mara (fig. 4.4).

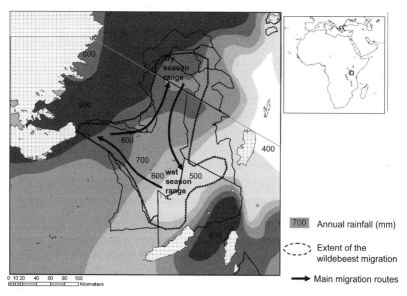

Fig. 4.4 Rainfall isohyets in the Serengeti-Mara ecosystem and its surroundings (shading), showing the highest rainfall in northwest Serengeti and lowest on the short-grass eastern plains, as caused by the rain-shadow effect of the Ngorongoro highlands. Solid lines indicate administrative boundaries; dashed lines indicate the extent of the wildebeest migration. Map interpolated from local rainfall stations in the Serengeti Park (interpolation by M. Coughenour, pers. comm.), combined with regional rainfall isohyets calculated by the Food and Agriculture Organization (FAO)/ United Nations Environmental Program (UNEP) Desertification and Mapping Project, as rasterized by UNEP/GRID.

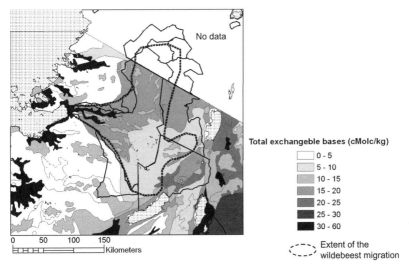

No data

Total exchangeble bases (cMolc/kg)
- ☐ 0 - 5
- ☐ 5 - 10
- ▨ 10 - 15
- ▨ 15 - 20
- ▩ 20 - 25
- ■ 25 - 30
- ■ 30 - 60

0   50   100   150
├─┼─┼─┼────────┤ Kilometers

⌐ ‾ ‾ ¬ Extent of the
⌊ _ _ _ ⌋ wildebeest migration

Fig. 4.5 Map of total exchangeable bases (cmol$_c$/kg of the sum of Ca^{2+}, Mg^{2+}, Na$^+$, and K$^+$, which is a proxy for soil fertility as set by the parent material) for the Serengeti-Mara ecosystem and its surroundings, calculated from the total cation exchange capacity (CEC) and the percentage base saturation (see box 4.1), based on the actual soil measurements and mapping as provided in the Soil and Terrain database for Southern Africa (SOTERAF), compiled by ISRIC (Eschweiler 1998; Batjes 2004; Dijkshoorn 2003). No data were available for Kenya.

In the 1950s, when the Serengeti National Park was being established, the government actively moved people out of some areas in the east, where wildebeest migrated (chapter 2, this volume). The Ngorongoro Conservation Area (a multiple-use zone) was established to the east of the park to preserve the Maasai pastoralist way of life and to protect wildlife, while agriculturists settled to the west. Analysis of the soils and rainfall patterns for these modern land use zones indicates that core protected areas have both higher annual rainfall and richer soils than the multiple-use pastoral and agricultural areas (fig. 4.6), suggesting the delineation of the wildebeest migration (fig. 4.4), and thus the establishment of the park, might be determined by soil nutrient patterns. In general, moister areas have higher soil fertility (fig. 4.6). However, the western edge of the migration system approximates a transition to lower soil fertility (fig. 4.5), originating from a geological transition. Beyond the western edge of the Serengeti National Park, the volcanic influence of the rift valley stops, and soils are derived from (less mineral-rich) granites (fig. 2.4). The soil fertility map shows that although the boundary of the park generally follows this soil fertility transition (fig. 2.4), it misses an area north of the western corridor (i.e., the western extension of the park), emphasizing the importance of

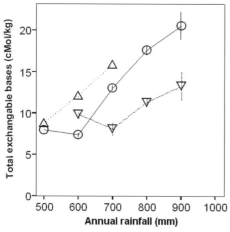

Land tenure type
⊖Parks for wildlife & ecotourism
△Controlled multiple use pastoralism
▽Unprotected, various land tenure

Error Bars show Mean +/- 1.0 SE

Fig. 4.6 Average total exchangeable bases (TEB) of the soil per 100 mm class of annual rainfall, for each of the land use types in the study area (fig. 2.3, this volume). Open circles (○) indicate controlled multiple use pastoralism areas, upright triangles (△) indicate parks for wildlife and ecotourism, and inverted triangles (▽) indicate unprotected areas with other land tenure. The x-values are the midpoints of 100-mm rainfall classes. The y-values are the mean TEB of all points falling within the rainfall class (with standard error bars).

the Grumeti and Ikorongo reserves in protecting the migration (Thirgood et al. 2004; figs. 4.4, 4.5). The eastern and the southeastern boundary of the migration seems to be set by low rainfall (fig. 4.4). Thus, in the regional setting, the migration system is located on generally rich soils with a strong rainfall gradient (figs. 4.4, 4.5).

To understand the seasonal wildebeest movements within this migration range (fig. 4.4), we need to explore how seasonal variation in rainfall interacts with soil properties in determining forage quality and risk. In general, forage quality for herbivores declines with rainfall, even when soil conditions remain constant, while quality increases with soil nutrient concentrations and salinity (Breman and De Wit 1983; Olff, Ritchie, and Prins 2002). The short-grass plains—the wet-season range for the migrant wildebeest and zebra—are situated at the lowest end of the rainfall gradient on alkaline soils (fig. 4.4). For the majority of the year these plains are an unsuitable habitat for larger grazers, due to lack of drinking water and completely desiccated vegetation. However, for a few months during the wet season (January to May), they produce a high-quality green flush that is much better than any other part of the ecosystem. This is probably due to four factors (McNaughton 1990; Fryxell et al. 2005; Wilmshurst et al. 1999; Fryxell 1991; Fryxell and Sinclair 1988): the accumulation of minerals in the soil during the dry season, the enhancement of the nutrient cycling (especially N and P) by the grazers (return in urine and dung), the inherently high base status of the pyroclastic (volcanic ash) parent material and the lower predation risk in this open habitat (Hopcraft, Sinclair, and Packer 2005). However, other habitats with inherently richer and deeper soils, and a less severe dry season, probably produce a better food quality and productivity in other seasons but are also more risky due to higher woody cover. For example, the fluvial soils of the western corridors are suggested to have a higher concentration of total exchangeable bases (fig. 4.5), but the higher rainfall in this area (fig. 4.4) may result in lower plant nutrient concentrations than on the plains. Despite these general insights, however, the relative importance of food quantity, food quality, and predation, and the detailed spatial interactions between geology, hydrology, soil formation, rainfall seasonality, and migrant herbivores still needs to be unraveled in more detail within this system.

Although agriculturalists on the western edge of the park first settled on the moister, more fertile soils, the majority of these areas lie within the protected areas and access is excluded. With further growth of the human population around the Serengeti National Park, people are becoming increasingly dependent on the poorer soils (fig. 4.5). The major part of southern Maasailand is located in areas where rainfall ranges from 500 to 700mm per year (fig. 4.2), an area too dry for dense tree cover and too dry for crops during the dry season, but still sufficiently productive for (seasonal) grazing opportunities.

Human population growth over the last 40 years has been fastest in areas with the highest rainfall (fig. 4.7), tripling in areas above 700mm of rainfall. There is an exponential relationship between human population density and rainfall where the slope depends on the land use system. Areas with the highest human population growth are also the areas with the most land converted to agriculture (fig, 4.8), indicating an intensifying resource extraction. Currently, up to 80% of the land in the study area with over 700 mm/yr of rain has become cultivated (fig. 4.8). As these prime agricultural areas become increasingly more occupied, human densities in marginal areas with lower rainfall (600mm) have doubled, especially since 1995 (fig. 4.7). Therefore, as human densities increase and agriculture becomes more intensive with larger economic gains, subsistence farming is pushed into marginal areas. Likewise, in the controlled pastoral areas (the Narok group ranches, Loliondo, Ngorongoro Conservation Area) human densities have also increased but are less pronounced, especially in areas above 800 mm/yr rainfall (fig. 4.7). It is only in these higher rainfall areas that pastoralists

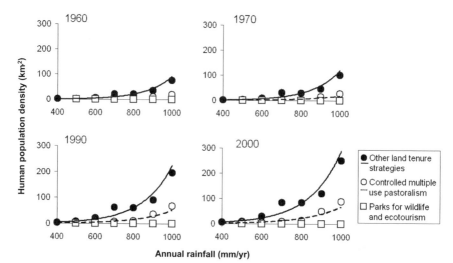

Fig. 4.7 Density of the human population across the rainfall gradient in the study area, separated by the three main land use strategies. The circles indicate the mean density within 100-mm rainfall classes, while the lines are fitted exponential regressions. The relationship was analyzed at a spatial resolution of 5 km. Data for the population density are from the Africa Population Database, compiled by UNEP/GRID. See fig. 4.4 for the source of the rainfall data.

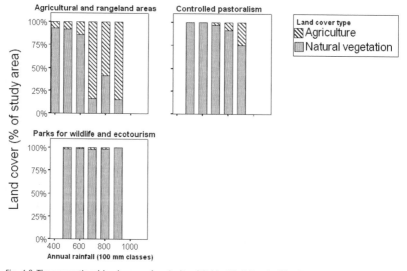

Fig. 4.8 The proportional land cover of agricultural fields (black hatched bars) versus natural vegetation (gray shaded bars) in the study area for 100-mm classes of annual rainfall, analyzed at a spatial resolution of 5 km, separated by the three main land use strategies. Land-use data were obtained from the Food and Agriculture Organization AfriCover project (www.africover.org)

are now starting to plant crops (fig. 4.8). It is probable that without additional legal restrictions, cultivation in pastoral areas will follow the same pattern as in agricultural areas—namely, that crops are grown in the wettest areas and then spread into dryer areas as human populations increase. This is a repeating trend over human history: new, more beneficial (and economically more viable) land use systems generally originate in the most productive areas, forcing old systems into lower rainfall areas (fig. 4.3). This context-dependency has important implications for protected areas: they only come under increasing pressure and isolation as land use patterns in the surrounding areas change. There is already evidence of some significant agricultural development in several Kenyan group ranches (chapter 13, this volume; Serneels and Lambin 2001; Lamprey and Reid 2004), and on the borders of protected areas (especially in the southern and western regions, such as Maswa's western boundary; fig. 2.21).

Resident Herbivore and Human Responses to Rainfall and Soil in the Serengeti-Mara Ecosystem

In addition to these human responses, we also studied the distribution of resident herbivores in the Serengeti through spatial autocorrelation, hot-spot analysis, and multiple regression (see text box 4.1 for methods), using aerial census data collected by the Conservation Information Monitoring Unit (under the Tanzania Wildlife Research Institute) and the Frankfurt Zoological Society. The distribution patterns of wildlife, livestock, and agriculture were related to rainfall (fig. 4.4) and soil gradients, using the soil concentration of total exchangeable bases (TEB) as a proxy for the soil quality, as determined by the parent material (see text box 4.1 for data sources). The distribution of different resident herbivores revealed strong differences in the degree of spatial aggregation, expressed as Moran's I (an index of spatial autocorrelation). The spatial autocorrelation of abundances predictably declined with body size (fig. 4.9a). This means that larger animals such as elephant and buffalo are less aggregated or spatially predictable, while smaller animals concentrate more in the same areas. This is also illustrated by spatial maps of the hot-spot statistic for the different-sized species (fig. 4.10).

The spatial distribution patterns of all the resident herbivores indicate clear segregation of most species during the wet season (fig. 4.10). The response analysis of the resident herbivores to rainfall and soil moisture indicates that smaller animals such as impala and warthog concentrate

BOX 4.1

DATA SOURCES AND ANALYSES

In order to study the relations between human population density, land use, wildlife distribution, and natural resources, we partitioned the study area in a 5×5 km grid. Using ArcGis 9.1, we created a GIS layer at this resolution containing all the attributes of climate, soil, herbivore, wildlife land use, and human population density data. The statistical distributions of the variables and the relations between them were analyzed by multiple regression using SPSS v. 12.0. The data used for this study came from a variety of sources.

Rainfall

A new rainfall map (fig. 4.4) was compiled by combining two data sources: an interpolated rainfall map created from historical weather station data in and around the Serengeti-Mara by M. Coughenour (unpub.) and a whole-Africa rainfall map created by UNEP/GRID. The latter data cover the regional patterns well, while the former are more suitable for local patterns within the Serengeti National Park. Both maps were combined by digitizing a contour map containing information from both sources. This contour map was then interpolated to a 5×5 km grid.

Soil Fertility

Soil nutrient availability (soil fertility) is a complex feature compounded by the interplay of parent material, weather, vegetation, and animals (Van Breemen 1993). It can be expressed in various dimensions and quantities, with differential importance of biotic and abiotic processes. In many temperate areas, nitrogen is an important limiting nutrient for primary productivity, especially in sandy soils. All nitrogen in soils is ultimately brought in through biotic processes, such as nitrogen fixation, and subsequently recycled by the interplay of plants, herbivores, and decomposers (Parton, Stewart, and Cole 1988; Aber and Melillo 1991; Schlesinger 1991; White 1993). Where N availability to plants is mostly set by the rate of litter input and subsequent microbial decomposition, the availability of other limiting plant nutrients is more complex. In many tropical ecosystems, nutrients other than N are often limiting plant growth (Schlesinger 1991; Proctor 1989). The flux and availability of P, Ca, Mg, Na, and K are determined by the interaction between the weathering of parent materials (their primary source), their absorption to the soil cation and anion exchange complexes, pH-dependent chemical equilibria, the seasonal water balance (salinity), and biotic processes (leaching from plants, decomposition of litter).

The total availability of base cations (Ca, Mg, Na, and K) is therefore strongly influenced by the basic geological template of an area, and reflects long-term, large-scale processes of geological transitions, climatic change, and soil formation, while N availability is set much more by short-term, smaller-scale biotic processes. Previous research has showed that the availability of exchangeable bases is a good predictor of worldwide hot spots of large herbivore diversity (Olff, Ritchie, and Prins 2002). This availability is set by the total potential of the soil to hold cations on the absorption complex, the so-called cation exchange capacity (CEC), *and* by the actual saturation of this complex by base cations (Ca^{2+}, Mg^{2+}, K^+, Na^+), the percentage base saturation (BS). Exchange sites that are not occupied by base cations will be occupied instead by H^+ or Al^{3+}, which never limit plant growth. Soils with a higher clay or loam percentage, with a higher organic matter content, and soils from parent material with higher mineral concentrations will all have higher CECs and a higher-percentage base saturation. For example, soils formed from volcanic ash, lava, or fluvial parent material will have higher total exchangeable bases than soils derived from granite. Because we are interested here in large-scale, stable patterns of soil fertility as set by the geological template and climate, we used total exchangeable bases as a proxy for soil fertility. We calculated the total exchangeable bases of the soil (TEB, $cmol_c/kg$) as TEB = BS/100 × CEC, where BS is the percentage base saturation (Ca^{2+}, Mg^{2+}, K^+, Na^+) of the soil adsorption complex, and CEC is the total cation exchange capacity (for the four base cations plus H^+ and Al^{3+}). Maps for CEC and BS were obtained from the Soil and Terrain database for Southern Africa (SOTERAF; Eschweiler 1998; Batjes 2004; Dijkshoorn 2003). SOTERAF is a unified soil map and typology for Southern Africa, with quantitative soil parameters for each polygon that are derived from a large soil reference collection. The main source of spatial information for the SOTER database for Tanzania has been the earlier work of de Pauw (1984). His map *Soils and Physiography (1983)* at a scale of 1:2 million and his report on "Soils, Physiography, and Agro-Ecological Zones of Tanzania" served as the basis for the delineation of the SOTER units and were combined with the large ISRIC soil reference collection for the determination of the physical and chemical properties of the soils.

Human Population Density

The changes in the spatial distribution of human population density were extracted for the study area from the UNEP/GRID human population database for the whole of Africa, and mapped for 1960, 1970, 1990 and 2000 for the study region.

Land Use

Maps of the distribution of land use types for the study area were derived from the Food and Agriculture Organization (FAO) AFRICOVER databases for Tanzania and Kenya (FAO 2002; FAO 2003). The land cover for these data was produced from visual interpretation of digitally enhanced LANDSAT Thematic Mapper (TM) images (Bands 4, 3, 2) acquired mainly in the year 1997. The land cover classes have been developed using the Food and Agriculture Organization/United Nations Environmental Program international standard land cover classification system (LCCS). Subsequently, we aggregated the land cover types to agricultural area (all croplands) versus natural vegetation. All polygons for reported agricultural areas within the protected areas in the study area were additionally checked by us against a LandSat Enhanced Thematic Mapper (ETM+) geocover mosaic S-36-00 (MDA Federal 2004). The images for this mosaic were collected between July 1999 and May 2002. Additional (small) agricultural use within protected areas was digitized from this LandSat image, and misclassifications within protected areas were removed.

Wildlife and Livestock Distribution and Response to Rainfall and Soil Fertility

The aerial wildlife census SE35 carried out by the Tanzanian Wildlife Research Institute (TAWIRI) was used to obtain insight into the spatial distribution of resident herbivores in the study area during the wet season. This census was conducted on April 3, 2001, using the Systematic Reconnaissance Flight (SRF) method (see Campbell [1995] for more details on the methods). The allometry of spatial aggregation of each resident herbivore species was assessed by calculating Moran's I per species (average overall distances) in ArcGis 9.1, and graphing this index against body size. An ecological response analysis (Huisman, Olff, and Fresco 1992; Manly et al. 2002) was then performed to test how rainfall and soil fertility interacted to determine the occurrence of the different species of large herbivores. To determine animal densities, the Getis-Ord GI* hot-spot statistic (Getis and Ord 1992) was calculated first from crude densities in ArcGis 9.1 to fill in the gaps between spots of observed high densities, at a resolution of 5 km. This computation accounts for the mobility of animals in the data, where local densities indicate that neighboring cells also have a high likelihood of having high densities at a slightly earlier or later time. It gives a good indication of which spots are preferred by animals (Boots 2002). Using the Getis-Ord GI* statistic as a response variable, we fitted a multiple regression model to explore its dependence on environmental conditions, with linear effects of rainfall and TEB, a quadratic effect of rainfall, and a linear rainfall × TEB interaction. A minimal model was selected, using forward selection of variables in the multiple regression procedure, using SPSS v. 12. In line with previously published theory (Olff, Ritchie, and Prins 2002), this model potentially accounts for an optimum of species against rainfall (significant quadratic effect of rainfall), while the position of this optimum, or the slope of the dependence on rainfall may depend on the soil fertility (significant rainfall × fertility interaction). For wildlife only, the data for the species distribution within the protected areas were used (Serengeti National Park, Grumeti, Maswa, and Ikorongo) in order to avoid interactions between resource conditions and land use. For example, the unprotected areas were generally lower in soil fertility, which would lead to confounding effects that cannot be separated using the current data. To explore how the dependence of different species on environmental conditions changed with their body size, we plotted the R^2 of the resulting regression models against their body mass, using the body mass data given by Prins and Olff (1998)

in areas with specific soil and moisture regimes (figs. 4.10, 4.11, table 4.2), suggesting they are constrained by food quality. Elephants occur all across the landscape, with a slight preference for moister areas over dryer areas but with no response to soil fertility, suggesting they are more constrained by food quantity and possibly access to drinking water. Kongoni and impala prefer the lower end of the rainfall gradient, additionally responding positively to soil fertility. For impala, a clear interaction between rainfall and soil fertility was found, where they prefer the most fertile, dry areas (fig. 4.11, table 4.2). Warthog and topi prefer the wetter end of the rainfall gradient, with an additional positive response to soil fertility (fig. 4.11, table 4.2). The R^2 of the regression models predicting the hot-spot index for each species from rainfall and soil fertility (i.e., the proportion of variation in abundance explained by environmental factors) declined with

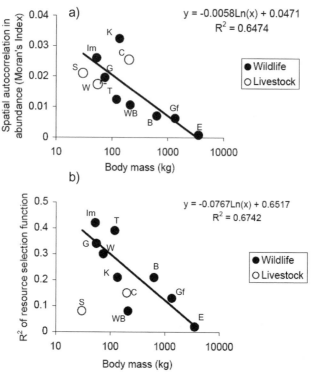

Fig. 4.9 (a) Change in Moran's I (degree of spatial autocorrelation) with body mass of resident herbivores and livestock. (b) The proportion of total variance in the Getis-Ord G statistic as explained by a multiple regression, with rainfall and soil fertility as predictors, as dependent on body size. The regression equation does not include livestock. E—elephant; Gf—giraffe; B—buffalo; WB—wildebeest; C—cattle; S—sheep and goats; K—kongoni; W—warthog; T—topi; G—Grant's gazelle; Im—impala.

body size (fig. 4.9b). This provides additional evidence that smaller animals are more responsive to rainfall and soil fertility, whereas these environmental factors are poor predictors of the spatial distribution of larger animals.

We found that the distribution of domestic animals such as sheep and goats (shoats) strongly deviates from the distribution of similar-sized wildlife (figs. 4.12, 4.13). For cattle, a buffalo-like response would be expected, while for sheep and goats an impala-like response would be expected. High livestock densities in the study area are found only at low and high rainfall, but not at intermediate levels (fig. 4.13). This distribution is probably caused by restricted access to protected areas, where most of the intermediate rainfall areas are. In addition, pastoralists move into low and high rainfall areas seasonally for only a few months at a time, to maximize their livestock's nutrient gain. The extent of agricultural area increased rapidly with annual rainfall (fig. 4.13). Also, areas with higher annual rainfall have a less severe

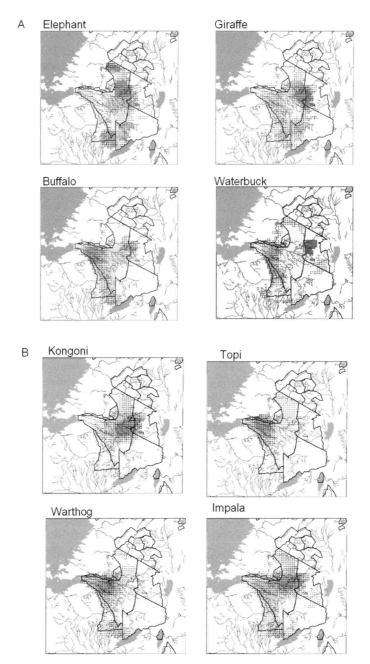

Fig. 4.10 The spatial distribution of different savanna herbivores resident in the study area during the wet season of 2001. The size of the dots is proportional to the value of the Getis-Ord hotspot statistic for that cell. (A) Displays the distribution of larger herbivores, and (B) displays that of smaller herbivores. The hotspot analysis was based on the Systematic Reconnaissance Flight (SRF) survey SE35, performed by the Tanzania Wildlife Research Institute and the Frankfurt Zoological Society (with permission).

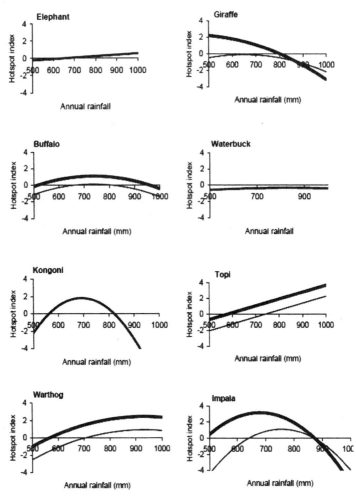

Fig. 4.11 Preference of different-sized resident herbivores to rainfall and the soil's total exchageable bases (TEB) as determined by forward multiple regression, predicting the hotspot statistic (see text box 4.1). Bold lines indicate the response to rainfall at high TEB (20 cMol$_c$/kg), and thin lines indicate the responses to rainfall at low TEB (5 cMol$_c$/kg). Figures with only one line indicate a significant response only to rainfall but not to TEB (see table 4.2). The parameter estimates and fit of the regressions are presented in table 4.2. Only data points within the land-use type "Parks for wildlife and ecotourism" were used in this analysis (see figure 2.3).

dry season, which is required for most crops. The extent of agriculture declined with soil fertility, which seems counterintuitive (we expect richest soils to be most selected). However, the richest soils occur on floodplains (such as the Mara delta), which may be unsuitable for crops during parts of the year. Additionally, most of the soils outside protected areas have lower total exchangeable bases than those inside the parks (fig. 4.5); therefore,

Table 4.2 Parameter estimates for multiple regression models predicting the hotspot indices for different-sized herbivores in the Serengeti ecosystem based on annual rainfall (mm) and soil fertility (Total Exchangeable Bases [TEB] in $cmol_c$/kg). For wildlife species the models were fitted using only observations within parks.

Species	Body mass (kg)	Regression model					
		Constant	Rainfall (mm)	Rainfall^2	TEB ($cmol_c$/kg)	Rainfall*TEB	R^2
Shoats	30.0	19.48655	−0.05435***	3.61E-05***			0.08***
Impala	52.5	−50.798	0.131799***	−8.44E-05***	0.751847***	−0.00086***	0.42***
Grant's gazelle	55.0	27.32158	−0.08311***	5.85E-05***	0.64161***	−0.00074***	0.34***
Warthog	73.5	−15.6683	0.034663***	−1.86E-05**	0.10115***		0.30***
Topi	119.0	−6.84087	0.008664***		0.093385***		0.39***
Kongoni	134.0	−50.2502	0.150532***	−0.00011***			0.21***
Cattle	200.0	19.72629	−0.07017***	5.70E-05***	0.48654***	−0.00063***	0.15***
Waterbuck	211.0	−1.89653	0.003843***	−2.30E-06**			0.08***
Buffalo	631.0	−12.1399	0.032444***	−2.20E-05***	0.064571***		0.21***
Giraffe	1350.0	−9.17307	0.024035***	−1.67E-05**	0.416087***	−0.00048***	0.13**
Elephant	3550.0	−1.05174	0.001652**				0.02***

*P < 0.05; **P < 0.01; ***P < 0.001

the concentration of agriculture on poorer soils is probably also a result of the protected area status and higher rainfall.

Migrant herbivores could not be quantitatively analyzed in the current study because wet-season counts for these species were restricted to the plains of the Serengeti, and recent dry-season counts are lacking. However, based on previous research (Maddock 1979; Fryxell 1995; Campbell and Borner 1995; Wilmshurst et al. 1999; Fryxell, Wilmshurst, and Sinclair 2004), we can formulate predictions on the distribution of migrants along the rainfall gradient, which strongly depends on season (fig. 4.14). Wildebeest and zebra concentrate in the low-rainfall areas during the wet season and shift to the wettest parts of the ecosystem during the dry season, with zebra being less constrained (i.e., showing less-pronounced curves). Drought-tolerant migrants such as gazelle and eland also concentrate on the low-rainfall areas during the wet season. However, they move to mid-

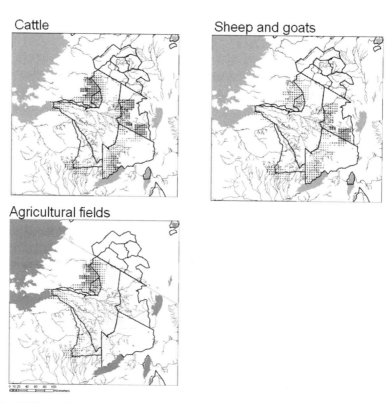

Fig. 4.12 The spatial distribution of livestock and agriculture in the study area during the wet season of 2001. The size of the dots is proportional to the value of the Getis-Ord hotspot statistic for that cell. The hotspot analysis was based on the Systematic Reconnaissance Flight (SRF) survey SE35, as for fig. 4.10.

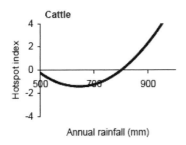

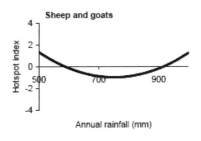

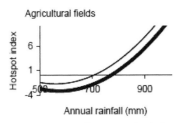

Fig. 4.13 Preference of livestock and agriculture to rainfall and soil total exchangeable bases (TEB) as determined by forward multiple regression predicting the hotspot statistic (preference). Bold lines indicate the response to rainfall at high TEB (20 cMol$_c$/kg) and thin lines indicate the responses to rainfall at low TEB (5 cMol$_c$/kg). Figures with only one line indicate a significant response only to rainfall but not to TEB (see table 4.2). The parameter estimates and fit of the regressions are presented in table 4.2. Data for all land use types were used (see fig. 2.3).

rainfall areas with the progression of the dry season, and do not necessarily go to the wettest parts of the ecosystem.

There are some parts of the protected areas where very few wild resident herbivores occur despite suitable rainfall and soil fertility. These areas are most notably along the northwestern boundary of the park, including the north end of Ikorongo Game Reserve, and the southwestern section of Maswa Game Control Area (figs. 2.3 and 4.10). These are also the areas with the highest illegal hunting activities (Arcese, Hando, and Campbell 1995; Campbell and Hofer 1995; Loibooki et al. 2002), which suggests a human cause for these low densities.

Consequences for Human-Wildlife Interactions

We have shown in this chapter that human land use has changed dramatically in the Serengeti-Mara ecosystem over the last 6,000 years, with widespread ramifications on natural resources and wildlife. The highest rate of land use change occurred in the last 50 years, due to rapidly in-

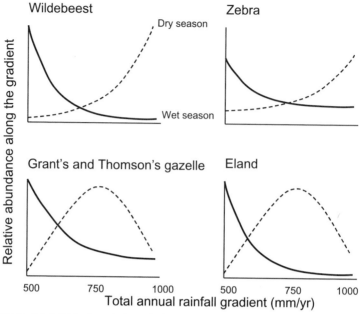

Fig. 4.14 Predicted distributions of migrant herbivores along the rainfall gradient in the Serengeti-Mara ecosystem, graphed separately for the wet (solid line) and the dry season (dashed line).

creasing human population densities. Whereas large-scale disturbances to the ecosystem and widespread human-wildlife conflicts were relatively uncommon with hunter-gatherers, pastoralists became increasingly demanding on the system. Similar to modern times, pastoralists most likely conflicted with wild animals with similar migratory grazing strategies as they concentrated their livestock in productive areas for short periods of time. Herbivores such as wildebeest, which compete for grazing but also pose threats to domestic stock in terms of transmissible diseases (malignant catarrhal fever, rinderpest, anthrax), present the largest potential for conflicts (fig. 4.14). Predation of livestock by large carnivores such as lions, leopard, hyena, and wild dog provide obvious additional grounds for animosity.

The transition to a more agricultural food base focused on areas with high rainfall (fig. 4.13). As with current subsistence agriculture, wildlife was regarded as a pest rather than a resource. The species most likely to conflict with agriculturalists were selective feeders in the smaller size classes, because they were restricted to areas with high rainfall and soil fertility. From the resident herbivores studied in the Serengeti-Mara ecosystem, topi and warthog prefer rainfall and soil fertility ranges that are now becoming permanently occupied by agriculture (figs. 4.10–4.13; chapter 2, fig. 2.20). Additionally, conflicts with migrant species are predicted to be most acute during the dry season, as agriculture now increasingly occupies the dry-season range of migrants (figs. 4.13, 4.14), especially for wildebeest and zebra. However, small animals are relatively easily killed or chased away and any conflict should be short-lived. Larger, less selective animals such as elephant and buffalo are more problematic. Although elephants survive in a wide range of niches and range over large areas, they maximize their time in productive pockets and are not easily killed (plus there is the added deterrent that it is a criminal offense to kill an elephant). Therefore, in areas with high contact rates between agriculturalists and wildlife, it is the small animals with a large degree of overlap (such as warthog and impala) that are the first to be exterminated or driven off, while large, difficult-to-kill animals such as elephant pose longer-term problems.

Species with large home ranges, such as elephants, may play an important role in the maintenance of buffer areas (Hoare and du Toit 1999). These are inherently risky areas for both wildlife and humans since they pose the most chronic sources of conflict. Long-term hard-to-resolve conflicts in these areas discourage both wildlife and humans from occupying them on an extended basis. In essence, they may form a "soft fence," based on two-way risk.

Most herbivores graze in areas with poorer soils only during times of higher rainfall and adjust to periods of lower rainfall by shifting to more fertile soils (dry-season refugia). These refugia in nutrient-rich areas are also selected for the cultivation of crops. The repercussions are that during the dry season, human-wildlife conflicts will be most acute. Small, selective wild herbivores, limited by local constraints, will be forced into these areas relatively quickly. Larger herbivores will most likely go to the closest profitable patch under all conditions, thus, they cause a more persistent conflict. During extended droughts, however, these dry-season refugia become depleted, forcing animals to move successively to mediocre, then poor areas. Therefore, we expect to see a pulse of human-wildlife conflict as the dry season progresses, increasing for some time, until the local resources are exhausted, then declining as animals are forced into suboptimal areas. However, as human populations expand into these less productive areas over time, dry season conflicts will only increase. Stochastic weather events, with unexpected periods of low and high rainfall, will further increase the contact rates and potential conflicts between humans and wildlife. Although these problems may be resolved in the future with new techniques, currently the most effective alleviation is to strictly enforce land use zoning and encourage the use of buffers.

CONCLUSION

Figure 4.3 illustrates an important lesson for human-wildlife conflicts and the role of parks and controlled areas. First, it shows the long-term effect that economic forces have on land use change, where new, more profitable strategies "push" older forms of land use down the rainfall gradient. New, more competitive land use strategies usually imply more intensive use of the natural resources and higher associated human population densities. This trend is generally at the expense of biodiversity, with the largest impacts occurring first in the wettest areas (fig. 4.3). The Serengeti-Mara system has experienced at least two new waves of major land use change in the last 4,000 years. The first incursion came with the introduction of subsistence agriculture in the wetter areas and pastoralism in the drier areas between 3,500 and 1,000 years ago, marginalizing existing hunter-gatherers. Agriculture and pastoralism coexisted for thousands of years, during which the human population grew very little, since they were limited by food availability and diseases. The second major invasion came in the twentieth century, with the introduction of new agricultural techniques and crops, improved health care, and globalized markets. This released local human populations from their historic limitations, leading to rapid population growth during the last 50 years, especially in the wet areas (fig. 4.7). Judging from historic patterns, traditional pastoralists such as the Maasai will be "pushed down" the rainfall gradient in forthcoming decades, similar to what happened to the hunter/gatherers 2,000 years earlier (fig. 4.3). Also, new commercial agricultural and livestock husbandry techniques, powered by their own economies, could force subsistence agriculture down the rainfall and soil fertility gradients, toward less favorable areas. This trend would lead to a serious deterioration in the livelihoods of subsistence farmers and has large implications for the majority of people living in rural Tanzania and Kenya. Without further measures, such changes in human land use will also accelerate biodiversity loss, especially in the high rainfall, high soil-fertility areas. Protected areas form an important bastion preventing biodiversity loss from occurring across the landscape (Sinclair, Mduma, and Arcese 2002, chapter 16, this volume), but increasing conflicts along park borders should be expected. Furthermore, new parks are now established less often in unprotected areas as land use intensifies. Therefore, current protected areas are of increasing importance to maintain our cultural and natural heritage.

As human pressure on land use increases, appropriate land use planning must develop creative conservation strategies that maintain ecosystem function while simultaneously reducing poverty through providing alternate job opportunities (Millennium Ecosystem Assessment 2005). This is the largest challenge we currently face—however, strong land use planning and zonation have proven to be effective at protecting both the cultural and natural heritage of the unique savannas environments of East Africa. Positive effects of parks on the biodiversity, ecosystem functioning, and local economies may extend far beyond their boundaries (du Toit, Walker, and Campbell 2004). If they are not maintained, the resulting loss of biodiversity and ecosystem function could have far-reaching consequences for people in the remaining agricultural and urban areas, due to the deterioration of soil fertility, primary productivity, climate, and lack of suitable drinking water (Wolanski and Gereta 2001; Loreau et al. 2001; Hooper et al. 2005; Chapin et al. 1998; Gereta 2004). But perhaps even more important, maintaining a legacy of protected areas for our grandchildren serves as a constant reminder of our evolutionary origins and our responsibilities for the future, especially now we have grown way too big to still fit in our East-African "cradle of mankind."

REFERENCES

Aber, J. D., and J. M. Melillo. 1991. *Terrestrial ecosystems.* Philadelphia: Saunders College Publishing.

Anderson, G. D., and L. M. Talbot. 1965. Soil factors affecting the distribution of the grassland types and their utilization by wild animals on the Serengeti plains, Tanganyika. *Journal of Ecology* 53:33–56.

Arcese, P., J. Hando, and K. Campbell. 1995. Historical and present-day anti-poaching efforts in Serengeti. In *Serengeti II: Dynamics, management and conservation of an ecosystem* ed. A. R. E. Sinclair and P. Arcese, 506–33. Chicago: University of Chicago Press.

Batjes, N. H. 2004. SOTER-based soil parameter estimates for Southern Africa (version 1.0). Wageningen, The Netherlands: International Soil Reference and Information Centre (ISRIC).

Bell, R. H. V. 1982. The effect of soil nutrient availability on community structure in African ecosystems. In *Ecology of Tropical Savannas,* ed. B. J. Huntly and B. H. Walker, 193–216. Berlin: Springer-Verlag.

Binford, L. R. 2001. *Constructing frames of reference: An analytical method for archaeological theory building using ethnographic and environmental data sets.* Berkeley: University of California Press.

Blumenschine, R. J. 1987. Characteristics of an early hominid scavenging niche. *Current Anthropology* 28:383–407.

Blumenschine, R. J., C. R. Peters, F. T. Masao, R. J. Clark, A. L. Deino, R. L. Hay, C. C. Swisher, I. G. Stanistreet, G. M. Ashley, and L. J. McHenry. 2003. Late Pliocene Homo and Hominid land use from western Olduvai Gorge, Tanzania. *Science* 299:1217–21.

Boots, B. 2002. Local measures of spatial association. *Ecoscience* 9:168–76.

Borgerhoff Mulder, M., and P. Coppolillo. 2005. *Conservation: Linking ecology, economics and culture.* Princeton, NJ: Princeton University Press.

Brain, C. K. 1981. *The hunters or the hunted: An introduction to African cave taphonomy.* Chicago: University of Chicago Press.

Breman, H., and C. T. De Wit. 1983. Rangeland productivity and exploitation in the Sahel. *Science* 221:1341–47.

Bromage, T. G., and F. Schrenk. 1999. *African biogeography, climate change and human evolution.* Oxford: Oxford University Press.

Campbell, K., and M. Borner. 1995. Population trends and distribution of Serengeti herbivores: Implications for management. In *Serengeti II: Dynamics, management and conservation of an ecosystem,* ed. A. R. E. Sinclair and P. Arcese, 117–45. Chicago: University of Chicago Press.

Campbell, K., and H. Hofer. 1995. People and wildlife: Spatial dynamics and zones of interaction. In *Serengeti II: Dynamics, management and conservation of an ecosystem,* ed. A. R. E. Sinclair and P. Arcese, 534–70. Chicago: University of Chicago Press.

Chapin, F. I., O. E. Sala, I. C. Burke, J. P. Grime, D. U. Hooper, W. K. Lauenroth, A. Lombard, H. A. Mooney, A. R. Mosier, and S. Naeem. 1998. Ecosystem consequences of changing biodiversity. *BioScience* 48:45–52.

Cowlishaw, G. 1994. Vulnerability to predation in baboon populations. *Behaviour* 131: 293–304.

de Pauw, E. 1984. Soils, physiography and agro-ecological zones of Tanzania. Dar es Salaam, Tanzania: Crop Monitoring and Early Warning Systems Project, FAO. GCPS/URT/047/NET, Ministry of Agriculture.

Diamond, J. 1996. *Guns, germs and steel: The fate of human societies.* London: W.W. Norton.

Dijkshoorn, J. A. 2003. SOTER database for Southern Africa (SOTERAF). Technical report. Wageningen, The Netherlands: International Soil Reference and Information Centre (ISRIC).

Dobson, A. P. 1995. The ecology and epidemiology of rinderpest virus in Serengeti and Ngorongoro Conservation Area. In *Serengeti II: Dynamics, management and conservation of an ecosystem,* ed. A. R. E. Sinclair and P. Arcese, 485–505. Chicago: University of Chicago Press.

du Toit, J. T., B. H. Walker, and B. M. Campbell. 2004. Conserving tropical nature: Current challenges for ecologists. *Trends in Ecology & Evolution* 19:12–17.

East, R. 1984. Rainfall, nutrient status and biomass of large African savannah mammals. *African Journal of Ecology* 22: 245–70.

Ehret, C. 2002. *The civilizations of Africa: A history to 1800.* Charlottesville: University of Virginia Press.

Eschweiler, J. A. 1998. SOTER database, Tanzania. Rome: FAO, AGLL Working paper no. 8.

Food and Agriculture Organization (FAO), Tanzania. 2002. Spatially Aggregated Multipurpose Landcover database (Africover)—version 2002-11-12. Rome: FAO.

———, Kenya. 2003. Spatially Aggregated Multipurpose Landcover database (Africover)—version 2003-03-28. Rome: FAO.

Fryxell, J. M. 1991. Forage quality and aggregation by large herbivores. *American Naturalist* 138:478—98.

————. 1995. Aggregation and migration by grazing ungulates in relation to resources and predators. In *Serengeti II: Dynamics, management and conservation of an ecosystem,* ed. A. R. E. Sinclair and P. Arcese, 257–73. Chicago: University of Chicago Press.

Fryxell, J. M., J. Greever, and A. R. E. Sinclair. 1988. Why are migratory ungulates so abundant? *American Naturalist* 131:781–98.

Fryxell, J. M., and A. R. E. Sinclair. 1988. Causes and consequences of migration by large herbivores. *Trends in Ecology and Evolution* 3:237–41.

Fryxell, J. M., J. F. Wilmshurst, and A. R. E. Sinclair. 2004. Predictive models of movement by Serengeti grazers. *Ecology* 85:2429–35.

Fryxell, J. M., J. F. Wilmshurst, A. R. E. Sinclair, D. T. Haydon, R. D. Holt, and P. A. Abrams. 2005. Landscape scale, heterogeneity, and the viability of Serengeti grazers. *Ecology Letters* 8:328–35.

Gereta, E. J. 2004. Transboundary water issues threaten the Serengeti ecosystem. *Ory* 38: 14–15.

Getis, A., and J. K. Ord. 1992. The analysis of spatial association by use of distance statistics. *Geographical Analysis* 4:189–206.

Gupta A. K. 2004. Origin of agriculture and domestication of plants and animals linked to early Holocene climate amelioration. *Current Science* 87:54–59.

Hardesty, D. L. 1975. *Ecological anthropology.* New York: Wiley.

Hoare, R. E., and J. T. du Toit. 1999. Coexistence between people and elephants in African savannas. *Conservation Biology* 13:633–39.

Hooper, D. U., F. S. Chapin, J. J. Ewel, A. Hector, P. Inchausti, S. Lavorel, J. H. Lawton, D. M. Lodge, M. Loreau, and S. Naeem. 2005. Effects of biodiversity on ecosystem functioning: A consensus of current knowledge. *Ecological Monographs* 75:3–35.

Hopcraft, J. G. C., A. R. E. Sinclair, and C. Packer. 2005. Planning for success: Serengeti lions seek prey accessibility rather than abundance. *Journal of Animal Ecology* 74: 559–66.

Huisman, J., H. Olff, and L. F. M. Fresco. 1992. A hierarchical set of models for species response analysis. *Journal of Vegetation Science* 7:37–46.

Iliffe, J. 1995. *Africans. The history of a continent.* Cambridge: Cambridge University Press.

Lamprey, R. H., and R. S. Reid. 2004. Expansion of human settlement in Kenya's Maasai Mara: What future for pastoralism and wildlife? *Journal of Biogeography* 31:997–1032.

Loibooki, M., H. Hofer, K. L. I. Campbell, and M. L. East. 2002. Bushmeat hunting by communities adjacent to the Serengeti National Park, Tanzania: The importance of livestock ownership and alternative sources of protein and income. *Environmental Conservation* 29:391–98.

Loreau, M., S. Naeem, P. Inchausti, J. Bengtsson, J. P. Grime, D. U. Hooper, M. A. Huston, D. Rafaelli, B. Schmid, and D. Tilman. 2001. Biodiversity and ecosystem functioning: Current knowledge and future challenges. *Science* 294:804–8.

Maddock, L. 1979. The migration and grazing succession. In *Serengeti: Dynamics of an ecosystem,* ed. A. R. E. Sinclair and M. Norton-Griffiths, 104–29. Chicago: University of Chicago Press.

Manly, B. F., L. McDonald, D. L. Thomas, T. L. McDonald, and W. P. Erickson. 2002. *Resource selection by animals: Statistical design and analysis for field studies.* Berlin: Springer.

Marean, C. W. 1992a. Hunter to herder: Large mammal remains from the hunter-gatherer occupation at Enkapune Ya Muto rockshelter (Central Rift, Kenya). *African Archeological Review* 10:65–127.

———. 1992b. Implications of late Quaternary mammalian fauna from Lukenya Hill (south-central Kenya) for paleoenvironmental change and faunal extinctions. *Quaternary Research* 37:239–55.

Marean, C. W., and D. Gifford-Gonzalez. 1991. Late Quaternary extinct ungulates of East Africa and paleoenvironmental implications. *Nature* 350:418–20.

McNaughton, S. J. 1985. Ecology of a grazing ecosystem: The Serengeti. *Ecological Monographs* 53:291–320.

———. 1988. Mineral nutrition and spatial concentrations of African ungulates. *Nature* 334:343–45.

———. 1990. Mineral nutrition and seasonal movements of African migratory ungulates. *Nature* 345:613–15.

MDA Federal Landsat GeoCover (ETM+) Edition Mosaics; Tile S-39-00. 2004. Sioux Falls, SD: USGS.

Millennium Ecosystem Assessment. 2005. *Ecosystems and human well-being: Biodiversity synthesis.* Washington, DC: World Resources Institute.

Norton-Griffiths, M. 1995. Economic incentives to develop the rangelands of the Serengeti: Implications for wildlife conservation. In *Serengeti II: Dynamics, management, and conservation of an ecosystem,* ed. A. R. E. Sinclair and P. Arcese. 588–604. Chicago: University of Chicago Press.

Ogutu, J. O., N. Bhola, and R. Reid. 2005. The effects of pastoralism and protection on the density and distribution of carnivores and their prey in the Mara ecosystem of Kenya. *Journal of Zoology* 265:281–93.

Olff, H., M. E. Ritchie, and H. H. T. Prins. 2002. Global environmental controls in diversity of large grazing mammals. *Nature* 415:901–4.

Parton, W. J., J. W. B. Stewart, and C. V. Cole. 1988. The dynamics of C, N, P, and S in grassland soils: A model. *Biogeochemistry* 5:109–31.

Prins, H. H. T. 1992. The pastoral road to extinction: Competition between wildlife and traditional pastoralism in East Africa. *Environmental Conservation* 19:117–23.

———. 1996. *Ecology and behaviour of the African Buffalo: Social inequality and decision making.* London: Chapman & Hall.

Prins, H. H. T., J. G. Grootenhuis, and T. T. Dolan. 2000. *Wildlife conservation by sustainable use.* Berlin: Springer.

Prins, H. H. T., and H. Olff. 1998. Species richness of African grazer assemblages: Towards a functional explanation. In *Dynamics of tropical communities,* ed. D. Newbery, H. H. T. Prins, and G. Brown, 449–89. Oxford: Blackwell Science.

Proctor, J., ed. 1989. Mineral nutrients in tropical forest and savanna ecosystems. *British Ecological Society Symposium, volume 9.* Oxford:Blackwell Science.

Reader, J. 1999. *Africa: Biogeography of a continent.* London: Vintage.

Reid, R. S., R. L. Kruska, U. Deichmann, P. K. Thornton, and S. G. A. Leak. 2000. Human population growth and the extinction of the tsetse fly. *Agriculture Ecosystems & Environment* 77:227–36.

Schlesinger, W. H. 1991. *Biogeochemistry: An analysis of global change.* San Diego, CA: Academic.

Serneels, S., and E. F. Lambin. 2001. Impact of land-use changes on the wildebeest migration in the northern part of the Serengeti-Mara ecosystem. *Journal of Biogeography* 28: 391–407.

Sinclair, A. R. E., S. A. R. Mduma, and P. Arcese. 2002. Protected areas as biodiversity benchmarks for human impact: Agriculture and the Serengeti avifauna. *Proceedings of the Royal Society of London, Series B* 269:2401–5.

Sinclair, A. R. E., and M. Norton-Griffiths, eds. 1979. *Serengeti: Dynamics of an ecosystem.* Chicago: University of Chicago Press.

Thirgood, S., A. Mosser, S. Tham, G. Hopcraft, E. Mwangomo, T. Mlengeya, M. Kilewo, J. Fryxell, A. R. E. Sinclair, and M. Borner. 2004. Can parks protect migratory ungulates? The case of the Serengeti wildebeest. *Animal Conservation* 7:113–20.

Van Breemen, N. 1993. Soils as biotic constructs favouring net primary productivity. *Geoderma* 57:183–212.

Vincens, A., D. Williamson, F. Thevenon, M. Taieb, G. Buchet, M. Decobert, and N. Thouveny. 2003. Pollen-based vegetation changes in southern Tanzania during the last 4200 years: Climate change and/or human impact. *Palaeogeography, Palaeoclimatology, Palaeoecology* 198:321–34.

White, L. A. 1959. *The evolution of culture.* New York: McGraw-Hill.

White, T. C. R. 1993. *The inadequate environment: Nitrogen and the abundance of animals.* Berlin: Springer-Verlag.

Wilmshurst, J. F., J. M. Fryxell, B. P. Farm, A. R. E. Sinclair, and C. P. Henschel. 1999. Spatial distribution of Serengeti wildebeest in relation to resources. *Canadian Journal of Zoology* 77:1223–32.

Wolanski, E., and E. Gereta. 2001. Water quantity and quality as the factors driving the Serengeti ecosystem, Tanzania. *Hydrobiologia* 458:169–80.

FIVE

Generation and Maintenance of Heterogeneity in the Serengeti Ecosystem

T. Michael Anderson, Jan Dempewolf, Kristine L. Metzger, Denné N. Reed, and Suzanne Serneels

An observer can gaze across the Serengeti grasslands and view a veritable sea of red oat grass (*Themeda triandra*), broken only by rolling hills and the occasional solitary *Acacia*. An unknowing spectator might assume the *Themeda*-dominated grassland represents a homogeneous community type, lacking complexity, and providing constant vegetation biomass to hungry, wandering ungulates. Likewise, in the northern woodlands, one may view seemingly uniform *Acacia*- or *Terminalia*-dominated woodland and assume a similar lack of complexity. While the grasslands and woodlands of Serengeti may give the illusion of modest variation, our observer would be incorrect to assume that the ecosystem is composed of different habitats (i.e., grassland and woodland) that are themselves homogeneous. Although heterogeneity has been acknowledged as a fundamental and conspicuous feature of other savanna ecosystems (Scholes 1990; du Toit, Rogers, and Biggs 2003), vegetation, soil, and landscape heterogeneity of the Serengeti has been largely overlooked because of its abundance of herbivore and carnivore species.

The purpose of this chapter is to describe the abiotic (e.g., soils, climate, landscape) and biotic (e.g., plant, animal) patterns of heterogeneity in Serengeti and identify processes that contribute to their generation and maintenance. In this chapter, patterns of heterogeneity in Serengeti are described in terms of the nature of their measurement—whether qualitative or quantitative—and the extent to which they included a spatial component. Furthermore, we identify how humans have impacted natural patterns of heterogeneity across the Serengeti ecosystem within the last century. Moreover, because abiotic and biotic sources of heterogeneity can interact in complex ways and can produce unpredictable results, we conclude the chapter by discussing three examples of heterogeneity generated and maintained by complex interactions between abiotic and biotic sources. Although we review processes germane to a wide range of organisms, much of the chapter deals with heterogeneity of soils and vegetation, because the majority of data on Serengeti heterogeneity are confined to those topics. While we acknowledge the importance of temporal heterogeneity, the chapter largely focuses on spatial heterogeneity, again because of data limitations. In this chapter, the Serengeti-Mara ecosystem will be collectively referred to as Serengeti. The Serengeti (see fig. 2.1) thus describes the national park in Tanzania, the Maasai-Mara National Reserve in Kenya, and the network of surrounding game reserves used by the wildebeest during their annual migration (Thirgood et al. 2004).

CONCEPTUAL ISSUES

Defining Heterogeneity

There is a vast technical and theoretical literature on concepts associated with heterogeneity, variation, and scale (Kolasa and Pickett 1991; Turner and Gardner 1991; Peterson and Parker 1998; Hutchings, Wijesinghe, and John 2000; Pickett, Cadenasso, and Benning 2003), and, except for a brief review of major concepts, the conclusions will not be reiterated here. Heterogeneity describes the degree to which elements or constituents of a system are different. This differs from variation, which describes different values of a variable of one kind (Kolasa and Rollo 1991). Under these definitions, differences in tree species composition represents heterogeneity, differences in total soil nitrogen represents variation. However, these terms, meant to make tractable a complex subject (Kolasa and Rollo 1991), may oversimplify the concepts.

In an attempt to quell the confusion associated with the terminology used to describe heterogeneity, Li and Reynolds (1995) and Weins (2000) recognized different types of heterogeneity that form a gradient from spatially implicit to spatially explicit. Weins (2000) defined four types of heterogeneity: spatial variance, patterned variance, compositional variance, and locational variance (box 5.1). These terms will be used in the following pages because they provide a framework for understanding how heterogeneity is measured and they allow an organized presentation of different types of heterogeneity as it applies to Serengeti. *Spatial variance* describes the simple statistical measure of dispersion associated with quantitative samples collected from different locations within a given area. This type of variance lacks explicit spatial information; the average deviation from the mean is indicated without reference to where samples occurred in space. *Patterned variance* is also a measure of dispersion among quantitative samples, but it incorporates a spatial reference. Patterned variance is still not spatially explicit, but it contains information about average differences among samples in relation to their proximity, that is, conveying that samples nearby one another tend to be similar, or patches of a particular size tend to be regularly spaced across a landscape. Spatial and patterned variance measure heterogeneity of quantitative data, but properties of a system can also vary qualitatively, such as when samples contain different species, vegetation, or soil types. *Compositional variance* describes qualitative differences among samples, but like spatial variance, does not contain spatial information regarding samples. Instead, compositional variance describes average qualitative properties of a system, such as association, dissociation, patchiness, or nestedness. Finally, *locational variance* describes qualitative differences among samples in which the spatial relations among all samples are explicit.

A final conceptual point requires clarification, and that is to differentiate between processes that generate heterogeneity and organisms that respond to heterogeneity. For example, variation in vegetation height, biomass, or composition may represent a significant source of heterogeneity for organisms in a community. However, in addition to being a source of heterogeneity, plants themselves respond to heterogeneity in climate, landscape, soil, or topographic variation. In a conceptual framework of heterogeneity (Pickett, Cadenasso, and Jones 2000; Pickett, Cadenasso, and Benning 2003), processes that generate heterogeneity are represented as chains that are linked through a series of interactions: agents → substrate → *heterogeneity* → recipient → *response.* Active components are underlined and those components that respond are in italics. Agents generate, modify, or sustain functional or structural properties of a system (Pickett, Cadenasso, and Benning 2003). We will use this term to refer to processes that generate heterogeneity; later in the chapter we will distinguish between biotic and abiotic agents. The substrate is the physical entity on which agents act, but it need not be inanimate; a grassland sward is the substrate for a herd of foraging Cape buffalo. Conditions not shown in the chain, called *controllers* (Pickett, Cadenasso, and Benning, 2003), can change the effect of an agent on a substrate. For example, the density of animals, water content of the soil, or time since previous defoliation will all change the effect that buffalo have on creating heterogeneity in the grassland sward. Finally, heterogene-

Box 5.1. Glossary of terms used in this chapter

TYPES OF HETEROGENEITY (FROM WEINS 2000)

Spatial variance—The simple statistical measure of dispersion associated with quantitative samples collected from different locations within a given area. Typical measures include variance, coefficient of variation, or variance/mean ratio.

Patterned variance—A measure of statistical dispersion among quantitative samples that contains information about average differences between samples in relation to their spatial proximity. Typical measurements include mean patch size, fractal dimension, lacuarity, correlograms, and semivariance.

Compositional variance—Describes qualitative differences between samples without including spatial information. Typical measurements are % similarity, evenness, patchiness index, and β-diversity.

Locational variance—Describes qualitative differences among samples in which the spatial relations among all samples are explicit. Typical measurements include mean nearest-neighbor distance, wavelet variance, and the proximity index.

Measures of Heterogeneity Used in This Chapter

Variance (s^2 or σ^2)—A statistical measure of dispersion among values in a population or a sample equal to the average squared deviation of values from the mean.

Coefficient of variance (CV)—A statistical measure of spatial variance used to compare data sets with different means. CV = (standard deviation/mean) × 100.

Fractal dimension (D)—Also known as the Hausdorff Dimension, it is the power ($0 < D < 3$) used to describe the space-filling properties of lines, surfaces, and volumes. In the case of Euclidean objects such as straight lines, squares, and cubes, D = an integer value (1, 2, and 3, respectively), while with fractal objects D is a noninteger value.

Semivariance (γ)—An autocorrelation statistic that estimates patterned variance by measuring the variance among samples as a function of the distance between them (h). Semivariance as a function of distance is: $\gamma(h)$ = $[1/2\,N(h)]\,\Sigma\,(z_i - z_{i+h})^2$; where $N(h)$ is the number of samples pairs used to calculate γ at distance class h, z_i is the measured value at point i, and z_{i+h} is the measured value at point $i + h$.

Beta-diversity (β)—A measure of among-sample species' compositional variance. In this chapter, $\beta = 1 - PS$, where PS is the % compositional similarity between plots. It should be noted that there are many ways of calculating β, and many theoretical issues associated with its estimation (see Velland 2001 for a discussion).

Heterogeneity as a Process
Agents—Abiotic or biotic processes that generate and maintain heterogeneity.
Substrates—The physical entities on which agents act.
Recipients—The organisms that perceive heterogeneity and in which there is elicited a response.

ity that is induced in the substrate is recognized by a recipient organism, in which there is elicited some response. So that there is a clear distinction between the sources that generate heterogeneity, the physical entities that display heterogeneity, and the organisms that respond to heterogeneity, we make the distinction among agents, substrates, and recipients throughout the chapter (box 5.1).

Heterogeneity across Scales

Different types of heterogeneity are relevant to different organisms (recipients) at different spatial and temporal scales. Natural processes themselves do not have a single scale at which they act, but rather the scale is defined by a recipient analyzing information at a given scale (McNaughton 1989; Allen and Hoekstra 1991). The scale at which organisms perceive resources, disturbances, threats, and so on, in their environment is proportional to their body size. Therefore, the importance of a particular agent in generating heterogeneity within a community is mediated by the body sizes of the recipients inhabiting it. In Serengeti, those processes range from deposition of urine and dung at small scales to gradients of climate and geomorphology at large scales.

On the temporal axis of fig 5.1, the position of an agent is determined by the relative amount of time its action persists on a substrate. For example, a termite colony and feeding ungulate exert influence on vegetation at similar spatial scales, a few square meters, but termites have effects on the physical and chemical properties of soil that can be long lasting (Jouquet et al.

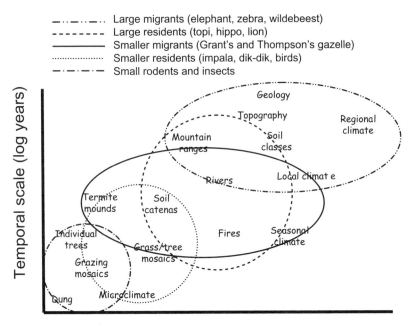

Large migrants (elephant, zebra, wildebeest)
Large residents (topi, hippo, lion)
Smaller migrants (Grant's and Thompson's gazelle)
Smaller residents (impala, dik-dik, birds)
Small rodents and insects

Spatial scale (log km)

Fig, 5.1 A schematic depicting the major agents of heterogeneity in the Serengeti ecosystem, plotted as a function of the temporal and spatial scale at which they have their primary influence. The temporal and spatial scales over which different organisms are primarily influenced are represented as circles and ellipses. Migratory species have elliptical regions of influence (wide on the spatial axis) because they encounter a much greater habitat area than resident species. The agents that overlap species regions of influence are hypothesized to be those most important in the generation of heterogeneity for those species.

2002), while the impact of defoliation can vanish within just a few months (Oesterheld and McNaughton 1988). The radius of influence of individual trees in savannas is between 4 and 12 m for a mature tree (Belsky et al. 1989), while fires can consume huge grassland swards (Stocks et al. 1996). Yet the effects of trees can be long lasting: *Acacia tortilis* can live more than 100 yr (Prins and van der Jeugd 1993), compared to the effects of fire, which can disappear after just 3 months (van de Vijver, Poot, and Prins 1999).

PATTERNS OF HETEROGENEITY IN SERENGETI

Across the Serengeti's gradient of soil fertility and rainfall, there are significant changes in woody vegetation (Herlocker 1976; Norton-Griffiths 1979) and grassland cover, composition, and structure (McNaughton 1983, 1985).

Dominant grasses change across the Serengeti plains, from short to intermediate, to tall from the southeast to the northwest, respectively (Anderson and Talbot 1965; Sinclair 1979). In the northern woodlands, tree height, density, and species richness increase with rainfall (Norton-Griffiths 1979; Metzger 2002). But what is the pattern of heterogeneity across the same gradient? To answer this we must first determine the best way to measure heterogeneity across Serengeti. As suggested in the previous section, the answer depends on the agents, substrates, and recipients of interest, the type of variance data collected (spatial, patterned, etc.), and scale. In the following summary, we group the available data for Serengeti based on whether they measure compositional, spatial, or patterned variance.

Compositional Variance

Two soil maps, one for the Serengeti plains (de Wit 1978) and another for the northern woodlands and western corridor (Jager 1982), revealed the complexity and heterogeneity of soil types across Serengeti (fig. 5.2, panel A). Likewise, the understanding of Serengeti landscape heterogeneity was improved tremendously by Gerreshiem's (1974) landscape classification (fig. 5.2, panel B); the map classified the Serengeti into areas of similar topography, geologic history, and climate, called Land Regions (hereafter regions). No spatially explicit statistics of association, variance, contiguity, and so forth, have ever been attempted with these maps, but current methods of landscape ecology provide a tremendous opportunity to link soil and landscape heterogeneity to the distribution and abundance of plants and animals.

In terms of herbaceous vegetation, Belsky (1988) measured among-quadrant β-diversity (box 5.1) at 16 sites, oriented north–south from the Serengeti plains to near Lobo. She concluded that plant compositional variation was greatest in the center of the national park and lower in the south and north; she had only one sample in the western corridor. Her data suggested that the presence of termite mounds and greater Na^+ availability were associated with greater β-diversity. McNaughton (1994) analyzed data collected from 103 sites (McNaughton 1983) and showed that β-diversity was greatest in the western corridor, but that it was also high in the eastern corridor and the central hills. McNaughton's analysis linked compositional variation to landscape position; β-diversity showed a unimodal response to topographic position. The results of β-diversity measured at 104 sites by Anderson (2004) largely supported the conclusions reached by Belsky (1988) and McNaughton (1983, 1994). However, a map of β-diversity across

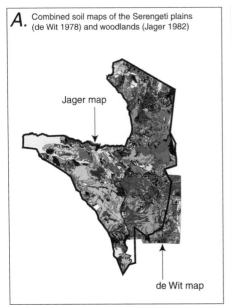

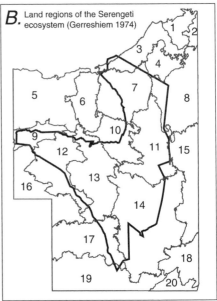

Fig. 5.2 Maps depicting patterns of compositional variance in soils and landscape features across the Serengeti ecosystem. (A) Soil maps from de Wit (1974) and Jager (1982); different shaded polygons from the Jager map (western and eastern corridor and northern extension) represent soil types grouped by soil depth, texture, and color; different shaded polygons from the de Wit map represent a hierarchical classification of soil type based on catena position, depth, texture, and geological parent type (note the finer-scale resolution of the de Wit map). (B) Serengeti Land Regions, as described by Gerreshiem (1974); Land Regions (grey lines) are main landscape types that share a common geologic history and climatic regime and have undergone comparable geomorphic influence (Gerreshiem 1974). Land Regions are themselves composed of successively smaller units called Sub-Regions, Land Systems, Land Elements, and Land Facets (not shown), which create a hierarchy of landscape classification. Smaller units in the hierarchy provide more specific information about landscape position, geologic parent material, vegetation, geologic features, and hydrology.

the Serengeti (fig. 5.3) suggests it may be an oversimplification to conclude that the greatest vegetation compositional variance exists in a band that transects the center of the Serengeti. First, high β-diversity clearly exists in the northern grasslands as well as in the central hills and the western corridor. Second, adjacent sites in the central hills and north often show great disparity in β-diversity, suggesting that within-region compositional variation may be as significant as among-region variation.

Compositional variance of woody vegetation in Serengeti was studied by Herlocker (1976) and more recently by Metzger (2002). Their studies revealed heterogeneity related to variation in topography, soil, and climate. Herlocker (1976) reported that 87.8% of the area supporting woody vegetation in Serengeti National Park was deciduous to semideciduous thorn tree woodland, but that this vegetation type was composed of a heterogeneous mix of 38 different dominant tree species combinations (species-types) of the genera *Acacia, Commiphora,* and very rarely, *Lonchocarpus* (Herlocker 1976). These species-types occur in a highly reticulated pattern that mirrors topographical variation across the landscape. The second most abundant woody vegetation type (4.7%) was semideciduous woodland composed of *Combretum molle* and *Terminalia mollis,* which occurs on sandy ridge tops and upper hill slopes in a large patch (~ 400 km²) in the northwest of the park. Semideciduous thorn tree wooded grassland occupied 2.7% of the woody vegetation and is dominated by *Balanites aegyptiaca;* it occurs in a few large patches to the west of Moru kopjes and in multiple small stands near

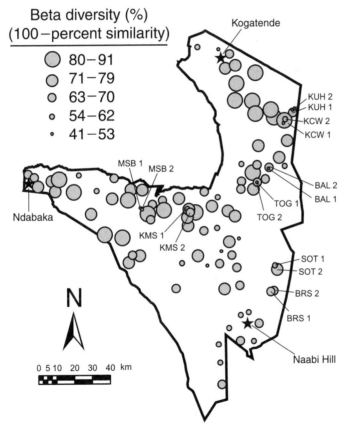

Fig. 5.3 Spatial distribution of plant species' β-diversity (measured as 100 − % similarity) in Serengeti National Park (dark line); large values of β indicate greater plant species dissimilarity among 1 m² subplots at a site (*n* = 104). Sites (*n* = 8) in which detailed sampling occurred (see text) are represented with three-letter codes; two plots were studied at each site, indicated by the number following the three-letter site code. Rain gauge locations, discussed later in the chapter, are indicated by stars.

the Grumeti River in the western corridor, and the Mara triangle. Semide-ciduous bushland and deciduous bushland (1.4%), composed primarily of *Acacia mellifera,* is found on bare, eroded, or disturbed soils throughout the western corridor or is associated with termite mounds in the north. At the time of Herlocker's survey, evergreen forests and evergreen to semidecidu-ous bushland occupied 2.9% of the total woody vegetation and occurred as narrow and often discontinuous vegetation bands that paralleled the Mara, Grumeti, Orangi, and Mbalageti rivers throughout the northern exten-sion and western corridor (Herlocker 1976). Inselberg vegetation, scattered throughout the Serengeti plains, Simiyu area and north of the Mara River, makes up a rather small (0.6%) but important part of the woody plant het-erogeneity in Serengeti (see the following).

Compositional variation was also measured by Folse (1982), who stud-ied the abundance of arthropods and birds at five sites from the Serengeti plains to the woodlands near Seronera. Arthropod abundance was greatest at the woodland site and generally decreased toward the short-grass plains. Many arthropod families showed site-dependent seasonal fluctuations in abundance, but total variation in arthropod abundance was negatively related to vegetation biomass and little variation was observed in the Serengeti plains. Bird species composition and abundance was highly vari-able in time and space throughout the study, but species showed significant habitat partitioning based on foraging behavior. Cursorial species required open, low-stature vegetation, while foliage-gleaners preferred dense, high-biomass vegetation. Bird species richness and abundance was greatest at sites with low vegetation height but a complex vertical structure because they could support both cursorial and foliage-gleaning feeding guilds. In general, bird densities did not track food availability (i.e., arthropods), but instead were correlated with vegetation greenness, which has a strong sea-sonal component and depends on the periodicity of local rain events.

Spatial Variance

We performed a reanalysis of the vegetation data collected by Metzger (2002), and found significant variation in herbaceous plant cover across the rainfall gradient, from completely bare soil to densely formed grass mats. The general trend for herbaceous vegetation was that the mean % cover decreased with rainfall (% cover = $-0.88 \times$ rainfall $- 26.3$; $F_{1,173} = 113.7$; $P < 0.001$; $r^2 = 0.40$). This was consistent with McNaughton (1985), who reported that grasslands transitioned from densely packed short-grasses in the southern plains to sparser, larger-stature vegetation in the northern Serengeti. To investigate spatial variance in plant cover among different regions of Serengeti, we compared the average CV of herbaceous plant cover in the Serengeti plains, the western corridor, and the northern corridor (Gerreshiem regions 14, 9, and 7, respectively; fig. 5.2, panel B). Average CV was calculated from 20 randomly located plots that were approximately equidistant within each region. Plots were composed of nine subsamples in an evenly spaced 100 m^2 grid. The mean CV of herbaceous plant cover was over twice as high in Serengeti plains than in either the western corridor or the north (region 7 = 6.6; region 9 = 4.2; region 14 = 16.7; $F_{2,57}$ = 15.2; $P < 0.001$; $r^2 = 0.34$). Moreover, when the CV of plant cover for each plot was regressed against mean annual rainfall, the relationship was linear and negative (CV plant cover = $-0.28 \times$ rainfall + 29.9; $F_{1,58}$ = 21.1; $P < 0.001$; $r^2 = 0.27$).

Unlike herbaceous vegetation, mean tree density increased linearly with rainfall (tree density = $0.009 \times$ rainfall – 0.39; $F_{1,58}$ = 81.1; $P < 0.001$; $r^2 = 0.31$). When the CV of tree density was compared among the Serengeti plains, western corridor, and the north, the Serengeti plains were almost an order of magnitude lower than the other regions (region 7 = 1.5; region 9 = 1.2; region 14 = 0.16; $F_{2,53}$ = 8.9; $P < 0.001$; $r^2 = 0.25$). In contrast to understory herbaceous vegetation, when the CV of tree density for each plot was regressed against rainfall, the relationship was positively linear (CV tree density = $0.02 \times$ rainfall – 0.61; $F_{1,54}$ = 7.0; $P = 0.01$; $r^2 = 0.12$).

To estimate the spatial variance of soil and plant characteristics in Serengeti, we conducted a reanalysis of data collected by Anderson (2004) that included eight variables from 16 plots across the rainfall gradient. Plant variables were above- and belowground biomass, leaf and root nitrogen, and soil variables were inorganic nitrogen, % sand, pH, and % water. Plots were paired at eight sites, and sites were nested within four of the ten Gerreshiem regions. As a measure of variation at different spatial scales, the among-sample variance was calculated within plots ($n = 9$), sites ($n = 18$), regions ($n = 36$), and the entire data set ($n = 144$), after which the variance was averaged at each scale. To control for the effects of different sample sizes, sample variance at each spatial scale was estimated by resampling the data in 10^4 plots using the freeware program PopTools (Hood 2004). Resampling is commonly used to reduce the bias created by different sample sizes and is a common procedure for hypothesis testing in ecology (Gotellie and Graves 1996).

Heterogeneity of soil and plant characteristics in Serengeti largely supports the ideas of fig. 5.1. The structure of variation across scales depends

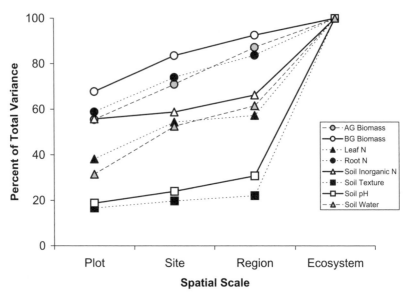

Fig. 5.4 Average variance of eight biotic and abiotic variables measured in 16 plots at eight sites across Serengeti (fig. 5.3). Variance was calculated within plots ($n = 9$), within sites ($n = 18$), within regions ($n = 36$), and among all plots ($n = 144$). To control for the different sample sizes among study scales, variance was calculated after the data were resampled 10^4 in PopTools (Hood 2004). Abbreviations in the legend are AG = aboveground, BG = belowground, N = nitrogen. Variables of the same symbol (circles, triangles, squares) have a similar pattern of variance across spatial scales. Only soil-inorganic N appears to not fall clearly into one of the groups.

on the variable considered, with the results generally conforming to three patterns (fig. 5.4). For the abiotic factors of soil texture and pH, most of the variation exists at the ecosystem level, with less of the total variation explained at the plot or regional scale. This is due to the geological origin of the parent material, as discussed later in the chapter. The remaining biotic factors vary over smaller spatial extents. For aboveground biomass, belowground biomass, and root %N, most of the total variation is explained at relatively small scales, with a small but consistent increase in the proportion of variance explained at larger scales (fig. 5.4). For leaf %N, soil inorganic N, and soil water, the situation is slightly more complex; these factors show some local and ecosystem variation, but little of the variance is explained by intervening scales. For these factors, there is a plateau in the proportion of variance explained between scales, such that increasing spatial scale has a small effect compared to the other scales; for soil inorganic N this occurs between plots and sites, for soil water and leaf N it occurs between sites and regions. Thus, organisms experience variation in leaf biomass and nutrients over smaller distances than factors such as pH or soil texture.

Anderson, Dempewolf, Metzger, Reed, and Serneels

The variation at these sites demonstrates the difficulty of describing heterogeneity based on a single variable. For example, at TOG 2, variation in aboveground biomass was the lowest of any of the 16 plots studied (s^2 TOG 2 = 1,446; mean s^2 without TOG 2 = 17,176), but variance in leaf nitrogen was the largest of any site (s^2 TOG 2 = 0.28; mean s^2 without TOG 2 = 0.06). So an herbivore foraging within this site is met with two very different types of plant variation, namely, relative constancy in plant quantity but variation in plant quality. To help visualize the structure of variation across the ecosystem, the within-site variance of the eight variables, plus a measure of within-site topographic variance, were ordinated with nonmetric multidimensional scaling (Minchin 1987) in version 4.01 of PCORD (McCune and Medford 1999). Ordination is informative in this case because sites that occur near each other in multivariate space have similar patterns of variation for the nine variables. Variation in % sand, soil water, and aboveground biomass best separated the sites along the first ordination axis, while variation in aboveground biomass, topographic relief, and belowground biomass were most correlated with separation on the second axis. However, a plot of the first versus the second ordination axis demonstrated a lack of consistent within-site or within-region variation (fig. 5.5): adjacent plots tended not to group together and in no case were plots within the same site

NMDS Ordination Results: Variance of Nine Variables

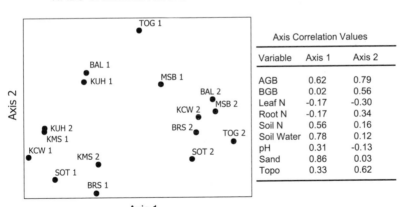

Variable	Axis 1	Axis 2
AGB	0.62	0.79
BGB	0.02	0.56
Leaf N	-0.17	-0.30
Root N	-0.17	0.34
Soil N	0.56	0.16
Soil Water	0.78	0.12
pH	0.31	-0.13
Sand	0.86	0.03
Topo	0.33	0.62

Fig. 5.5 Ordination results from nonmetric multidimensional scaling (NMDS) of variance measures within 16 plots at sites across Serengeti (see fig. 5.3). Variables used in the ordination are as in fig. 5.3, with the addition of topographic variation (Topo). Correlations between variables and axes one and two are shown to the right of the ordination plot. The results show a lack of within-site and within-region correspondence of heterogeneity; patterns of heterogeneity are often more similar between plots separated by hundreds of kilometers, compared to adjacent plots. AGB = aboveground biomass, BGB = belowground biomass, N = nitrogen.

nearest one another. Overall, patterns of variation between plots at distant sites were often more similar than plots separated by just a few kilometers.

Patterned Variance

Reporting a variance does not represent the spatial or temporal structure of heterogeneity experienced by organisms as they navigate through their environment (Weins 2000). An autocorrelation statistic, such as semivariance, is often used to represent variation in space or time because it incorporates spatial or temporal information into the calculation of variance (Anderson, McNaughton and Ritchie 2004). Neighboring points in space or time may be more similar than distant points, unless the spatial or temporal variation is random, in which case variation is constant as a function of distance or time. Data from Anderson, McNaughton and Ritchie (2004) allow a comparison between spatial and patterned variance of resin-extractable NO_3^- between paired plots at the eight study sites from fig. 5.5. From their data, it is clear that variance without a spatial reference may not adequately portray the data structure. For example, within-plot variation of resin-extractable NO_3^- in TOG 1 and TOG 2 were similar (s^2 TOG 1 = 226; s^2 TOG 2 = 187), yet the spatial structure differed significantly (fig. 5.6). The semivariogram for TOG 1 suggests a random NO_3^- distribution, while the semivariogram for TOG 2, shows a clear spatial pattern; resources show spatial autocorrelation between 0–18 m, with patches of an average size of ≈ 18 m arranged randomly on the landscape. Likewise, MSB 1 and MSB 2 had similar within-plot variance in resin-extractable NO_3^- (s^2 MSB 1 = 479; s^2 MSB 2 = 491), but again the spatial structure differed considerably (fig. 5.6). The semivariogram of MSB 1 suggests an average patch size of ≈ 14 m, with patches randomly arranged on the landscape. The semivariogram of MSB 2 suggests a constant increase in variation with distance, as would be observed in a unidirectional gradient of soil resources.

Some processes show spatial patterns that are self-similar across scales or scale invariant. This phenomenon is typical in natural landscapes, such as when the spatial distribution of vegetation cover or river networks is viewed at different scales (Milne 1992). Indeed, the pattern is evident for vegetation cover and river systems in satellite and aerial images taken of Serengeti at different spatial scales (fig. 5.7). When the distribution of a variable is self-similar across scales the variable is said to have a fractal or fractal-like distribution (Sugihara and May 1990). The fractal dimension (D) of a resource occurring in a two dimensional plane varies between 0 and 2; D = 0 is a single point, D ≈ 1 indicate highly clustered and self-similar

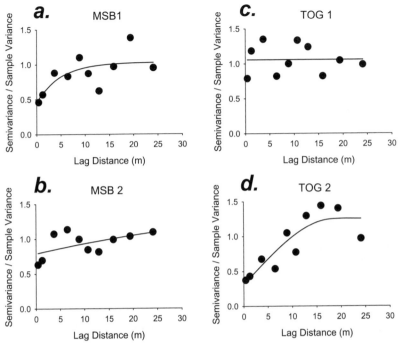

Fig. 5.6 Example semivariograms showing the average semivariance of soil NO_3^- availability as a function of separation distance. Soil NO_3^- availability was measured with 53 ion-exchange resin bags buried in each plot for one month after the beginning of the rainy season. Specific distance classes used to model semivariance are composed of multiple pairwise combinations of samples separated by a common distance; the number of sample pairs ranged between 30 at 0.4 m to 206 at 19.3 m. The examples shown compare adjacent plots at two sites, MSB and TOG (fig. 5.3). MSB 1 (A) shows spatial structure from 0–14 m, while MSB 2 (B) shows continuous increase in variance as the separation distance between points increases. At TOG 1 (C), the data showed no spatial structure, while at TOG 2 (D) there was spatial autocorrelation between 0–18 m. The distance at which the semivariance levels off, 14 m for MSB 1 and 18 m for TOG 2, is an estimate of the average NO_3^- patch size in the plot. TOG 1 shows no patch structure and MSB 2 suggests a continuous NO_3^- gradient. Data are from Anderson, McNaughton, and Ritchie 2004.

distributions, while D ≈ 2 more completely fill the plane and are likely indistinguishable from a random distribution (Milne 1997). The analysis of semivariance and fractals are linked because D can be obtained from a log-log plot of the semivariance versus distance, where a positive slope (*m*) indicates a fractal dimension with D = 2 − *m*/2. The analysis of soil-NO_3^- spatial distributions between 0.4–26 m at multiple sites in Serengeti demonstrated that fractal and random distributions were equally common and often occurred in adjacent plots (Anderson, McNaughton and Ritchie 2004). Thus, resources occur in a complex mosaic of random and fractal distributions embedded within a landscape that has self-similar properties (vegetation cover, rivers, topography) across scales (fig. 5.7).

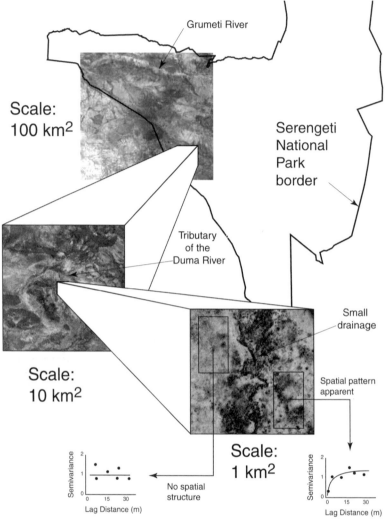

Fig. 5.7 Heterogeneity of vegetation cover and river systems across spatial scales. Three levels of spatial resolution are shown, 100, 10, and 1 km². 100 and 10 km² scales were extracted from Landsat 7 satellite images, 1 km² scale is from an aerial photograph. Although the images show disparate spatial scales, they share similarity in patterns of vegetation cover and river systems. The black rectangles embedded in the 1 km² image represent the different hypothetical patterns of heterogeneity, as were measured in adjacent plots in this study. The rectangle on the left lacks spatial heterogeneity, and therefore has a flat semivariogram. The rectangle on the right shows a self-similar pattern of heterogeneity across scales, and thus has a semivariogram that shows spatial structure. When taken together, one can see that heterogeneity exists in a complex continuum of random and structured features that exist on an equally complex template with self-similar characteristics across scales.

Summary of the Patterns

We now revisit the question put forth in the beginning of this section: what is the structure of heterogeneity across the Serengeti? In terms of compositional variation, heterogeneity in vegetation type (grasslands, woodlands, shrublands) increases with rainfall and topographic variation in the north and western corridor. Measured by plant β-diversity, the north, central hills, and western corridor are more heterogeneous than the regions dominated by plains to the south. Topography, geology, and grazing intensity also vary along the same gradient, and evidence from many sources suggests that compositional variation is linked to all these sources (see the following sections). In terms of spatial and patterned variance, the review emphasized grassland communities, so we limit our conclusions to those habitats. Our analysis revealed a common property of Serengeti grasslands: communities with substantial variation or complex spatial patterns were often adjacent to communities with modest variation and more-or-less random spatial patterns. Moreover, ordination results showed that sites on opposite ends of the ecosystem, separated by hundreds of kilometers, often displayed greater similarity in patterns of heterogeneity than sites within a few kilometers. Thus, heterogeneity is a universal property of the Serengeti, not restricted to regions that appear more complex because of differences in vegetation or topography.

AGENTS OF HETEROGENEITY IN SERENGETI

There are numerous processes that generate and maintain heterogeneity in African savanna ecosystems, but which are the most important in promoting the rich faunal and floral diversity observed in the Serengeti? In this section, we review the major agents of heterogeneity in Serengeti, focusing on those factors that have received the most attention and for which there are available data.

Abiotic Factors

Climate

At the largest scale in the Serengeti ecosystem, the influence of the inter-tropical convergence zone creates a bimodal pattern of rainfall: short rains, lasting from November to December, and long rains, lasting from March to May (chapter 2, this volume). However, other climate processes produce rainfall patterns and variability in Serengeti, including continental heat lows, orographic rainfall, convergence rainfall, and local convection rainfall (Bell 1979; Prins and Loth 1988; Swift 1996). Lake Victoria, the massive water body to the west of Serengeti, creates a small convergence zone that produces dry-season rainfall in the zone surrounding its eastern shores (Sinclair 1979; Swift, Coughenour, and Atsedu 1996). Localized rainfall is higher in regions of significant elevation change, such as the hills in the north and western corridor (Wolanski and Gereta 2001). The Serengeti plains west of the crater highlands typically receive < 500 mm yr^{-1} rainfall because they sit in a rain shadow created by the Ngorongoro highlands. Convection events create local thunderstorms that can be intermittent and highly isolated. Because of Lake Victoria and orographic effects, dry-season rainfall variability is lowest in the north, intermediate in the western corridor, and highest in the Serengeti plains (Sinclair 1979).

To characterize climate variability in different regions of Serengeti, we analyzed 77 monthly rainfall measurements between the years 1985 and 1993 (chosen because they were relatively complete) for three rain gauge stations: Kogatende in the north, Ndabaka gate in the west, and the Serengeti plains in the south. We calculated the CV by months, seasons, and years to examine if a change in temporal scale of measurement influenced the results (fig. 5.8). Regardless of the temporal scale, rainfall CV was always lowest at Kogatende in the north. When analyzed by month, the data supported the findings of Norton-Griffiths, Helocker, and Pennycuick (1975), that low-rainfall areas have greater variability than high-rainfall areas; Naabi Hill in the Serengeti plains had the highest CV and the lowest monthly mean precipitation. However, dry-season CV over this time period was the greatest for Ndabaka in the western corridor, which also received the most rainfall. Annual rainfall CV was also greater at Ndabaka but only slightly greater than at Naabi Hill. Thus, the transitional habitat in the western corridor not only experiences abundant rainfall, but also significant variability at times likely to influence migration patterns of zebra and wildebeest.

While seasonal variability (dry/wet season) is largely predictable, the Serengeti experiences enormous variation in annual rainfall that is not correlated with the Southern Oscillation Index, as might be expected; only during an extreme El Niño event does the Serengeti receive predictably high annual rainfall (Wolanski and Gereta 2001). Support is provided by a recent hypothesis that suggests that climate patterns in East Africa can be explained largely by the Indian Ocean climate system (Webster et al. 1999). Climate variability has important effects on primary production across the whole ecosystem, but the effects of climate variability on primary production are probably stronger in low-rainfall areas (fig. 5.9; Oesterheld

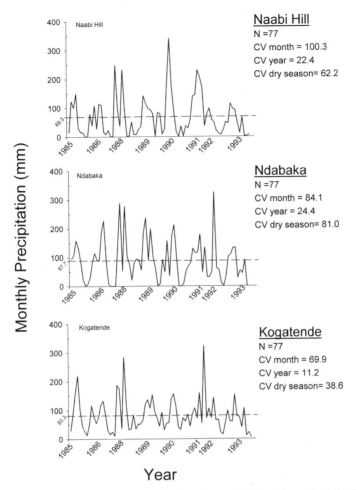

Naabi Hill
N =77
CV month = 100.3
CV year = 22.4
CV dry season= 62.2

Ndabaka
N =77
CV month = 84.1
CV year = 24.4
CV dry season= 81.0

Kogatende
N =77
CV month = 69.9
CV year = 11.2
CV dry season= 38.6

Fig. 5.8 Plots of monthly precipitation measures for three rain gauge locations in the south (Naabi Hill), the western corridor (Ndabaka), and the north (Kogatende) between 1985 and 1993 (see fig. 5.3 for rain gauge locations). Each plot has 77 months of data; when data were absent for one site, data were removed from all the sites for comparative purposes (note missing data between 1991 and 1992). Coefficients of variation (CV) were calculated by month, year, and dry season, and are listed for each site to the right of the plots. Mean monthly rainfalls are shown as dashed lines in each plot; the value is listed on the ordinate axis.

et al. 1999), such as the Serengeti plains. Behavioral models suggest that Serengeti grazers have adapted to stochastic temporal and spatial variation in rainfall and primary production by moving among grassland patches in ways that maximize their daily energy gains (Fryxell, Wilmhurst, and Sinclair 2004; Fryxell et al. 2005). Rainfall variability can affect plant species composition by influencing plant seedling emergence and recruitment (Veenendaal, Ernst, and Modise 1996a). Early rains can trigger multiple

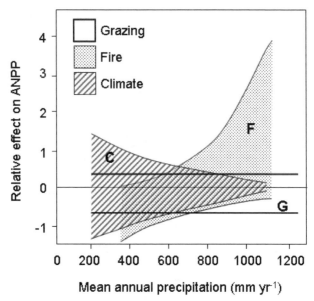

Fig. 5.9 The relative effects of fire (F), climate (C), and grazing (G) on annual net primary production (ANPP) across a range of mean annual precipitation in savanna-grassland ecosystems. Negative values represent a decrease in ANPP, while positive values represent stimulation of ANPP. Reprinted with permission from *Ecosystems of the World 16: Ecosystems of Disturbed Ground,* ed. M. Oesterheld, J. Loreti, M. Semmartin, and J. M. Paruelo. (1999). Grazing, fire, and climate effects on primary productivity of grasslands and savannas, pp. 287–306. The Netherlands: Elsevier.

emergence events, late rains can allow only a single emergence event, and early and late rains are associated with the emergence of different species (Veenendaal, Ernst, and Modise 1996b). Climate also has important effects on soil nutrient availability in Serengeti because the ability of soils to hold cations varies widely across the park. The soils of the western corridor and north are sandy, low in organic matter, and are easily leached of nutrients, while the soils of the Serengeti plains are silt-rich, have abundant organic matter, and are less easily leached (McNaughton, Ruess, and Seagle. 1988). Therefore, the rainfall patterns of Serengeti create a gradient of eutrophic to dystrophic soils typical of many African savanna systems (Huntley and Walker 1982).

Topography, Landscapes and Soil Composition

Soil heterogeneity in Serengeti is associated with parent material and landscape erosion processes. The broadest and most significant impact is historical and stems from Pleistocene and ongoing eruptions of natrocar-

Anderson, Dempewolf, Metzger, Reed, and Serneels

bonatitic volcanoes in the Ngorongoro highlands (chapter 3, this volume, and Dawson 1964, Dawson et al. 1994; Hay 1976). During these eruptions, the same prevailing winds that structure the precipitation gradients carried ejecta west and north from their source, blanketing what are now the plains with sodium-rich ash. Finer ash particles were carried farther, creating a gradient in soil texture (de Wit 1978). On a more local level, soil texture varies across shallow topographic gradients. Hydrodynamic activity in conjunction with gravity creates local variability in soil characteristics along the repeated pattern of hills and valleys formed by drainage lines. This edaphic pattern, termed a *catena* (Milne 1935; Pratt and Gwynn 1977), influences the vegetation structure and species composition along its profile (Bell 1970; Herlocker 1976; Vesey-Fitzgerald 1973). Well-drained eluvial soils tend to dominate hilltops, giving way to finer textured soils on lower slopes and valley bottoms (de Wit 1978; Jager 1982; Yair 1990; Gerrard 1990).

Landscape variation is associated with variability in soil fertility (Scholes 1990; Venter, Scholes, and Eckhardt. 2003) and soil alkalinity (Belsky 1988, 1992; Coughenour and Ellis 1993) which in turn influence plant water-use efficiency, morphology, chemistry, and rates of plant herbivory and growth (Scholes 1990). For example, in nearby Lake Manyara National Park, shrubs growing on volcanic soils had significantly higher diameter growth rates and height increases, 2.65 % and 18.9 %, respectively, compared to diameter increase of 1.78 % and height increases of 12.1 % for shrubs growing on nutrient-poor soils derived from basement complex (Prins and Van der Jeugd 1992). Thus, the results of landscape and soil complexity across Serengeti contribute to vegetation heterogeneity among vegetation types (Anderson and Talbot 1965; Belsky 1988).

Landscape and soil heterogeneity may also contribute to species coexistence within vegetation types. Across the 1,000 m^2 plots studied by Anderson, McNaughton, and Ritchie (2004), greater topographic variation was associated with greater among-sample variance in soil texture. In turn, the number of plant species in plots increased with variation in soil texture (fig. 5.10), explaining a relatively large proportion of the sample variance. Thus, if plant species are adapted to different soil types, topographic variation can influence plant species richness by promoting opportunities for coexistence through greater habitat heterogeneity (Shmida and Wilson 1985; Anderson, Metzger and McNaughton 2007). In another example, landscape heterogeneity contributes to coexistence between the Serengeti's two dominant grasses, *T. triandra* and *Digitaria macroblephara,* by influencing soil texture. When exposed to simulated grazing, soil texture has opposite effects on the two species; *T. triandra* acquired more N in low-sand soils

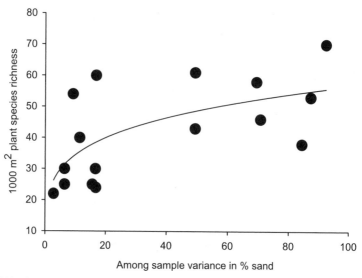

Fig. 5.10 Relationship of among-sample variance in % sand (from *n* = 9 subsamples from each plot) and corresponding plant species richness from 16 paired plots at 8 sites that span the Serengeti rainfall gradient (fig. 5.3). Samples were collected in 1000 m² plots at each site.

while *D. macroblephara* acquired more N in sandy soils (Anderson, Dong, and McNaughton 2006).

Across the Serengeti, variation among geographic land regions explained differences in soil microbial processes and nutrient cycling (Ruess and Seagle 1994). Soils with greater clay content were associated with greater electrical conductivity, pH, and water-holding capacity. As a result, soils with greater water-holding capacity and total carbon supported larger microbial populations with greater rates of respiration, N-mineralization, and carbon and N-turnover rates. As a consequence of regional landscape variation, soils in the Serengeti plains and western corridor had greater rates of biological activity and greater standing pools and nutrient fluxes than soils in the central hills and northern extension. Moreover, Ruess and Seagle (1994) linked the soil processes to higher trophic levels by demonstrating significant correlations between herbivore consumption and both soil microbial biomass and soil respiration rates, and between grazing intensity and rates of soil nitrogen mineralization.

Finally, granitic outcrops, such as kopjes, torrs, and inselbergs, provide unique habitat for organisms that cannot exist elsewhere, such as desiccation-tolerant vascular plants (Porembski and Barthlott 2000) and small-bodied ungulates, such as klipspringers. Moreover, kopjes provide protected foraging sites for elephants and other mammals (A. R. E. Sinclair,

pers. comm.). Bird diversity was significantly greater on kopjes than surrounding habitats and was composed of a unique community type that
included species that were rare elsewhere in Serengeti (Trager and Mistry
2003). For hyraxes, dispersal among the rock "islands" by members of small
populations decreases the colonies probability of local extinction (Gerlach
and Hoeck 2001).

Fire

Fire acts as an important disturbance regime and is major factor determining savanna structure. It directly affects vegetation and ecosystem processes
by creating a pattern of grass and woody biomass removal, destruction of
aerial portions or killing of woody plants, nutrient volatilization, and enhancement of above-ground net primary productivity. A classic theory suggests that the coexistence of woody plants and grasses is controlled by the
access of trees to moisture in deep-soil horizons and grasses to upper-soil
horizons (Walter 1971). While not mutually exclusive of this theory, Higgins, Bond, and Trollope (2000) suggested that trees/grass coexistence in
savannas is maintained instead because woody vegetation is more susceptible to high-intensity fires compared to quickly recovering herbaceous
vegetation.

Fire generates heterogeneity because it does not occur everywhere, and
where it does occur it does not burn uniformly. Heterogeneity that results
from fire is the result of fire type, frequency, and intensity. Savanna fires
are typically surface fires and either head fires spreading with the wind, or
back fires spreading against the wind. Under otherwise similar conditions,
head fires, which burn at high intensity and kill aerial portions of trees,
can have very different effects compared to back fires, which burn at lower
intensity but spread at a lower rate. Back fires threaten the grass sward more
than head fires because high temperatures close to the ground are maintained for prolonged periods of time, resulting in damage to grass apical
meristems (Trollope 1982).

Fire frequency depends on the availability of fuel, which increases
monotonically with rainfall (Oesterheld et al. 1999). Fires rarely occur
below 450 mm precipitation (Trollope 1974; Oesterheld et al. 1999); much
of the Serengeti plains (regions receiving < 500 mm rainfall) burn infre-
quently. The relationship between rainfall and fire frequency is particularly
strong in Serengeti ($r = 0.73$, $n = 150$; Norton-Griffiths 1979); moreover,
within-year fire frequency is highly correlated with wet-season rainfall (fig.
13.8 in Norton-Griffiths 1979). Fuel availability is also controlled by the
consumption of productivity by grazers; severe defoliation can reduce fuel

loads and lower fire frequency (Roques, O'Conner, and Watkinson 2001), an effect that is enhanced by high herbivore density (van Wilgen et al. 2003). Grazing increases with ANPP, but less so than biomass production, resulting in a positive correlation between unconsumed productivity or available fuel load and ANPP (Oesterheld et al. 1999).

Finally, intensity strongly influences the ecological impacts of fire (van Wilgen et al. 2003). Although fire intensity increases in plots with greater grass cover (Salvatori et al. 2001), it does not significantly influence the recovery of the grass sward (Trollope, Trollope, and Hartnett 2002). In contrast, fire intensity does have significant effects on woody vegetation and tree recruitment. Moreover, variance in fire intensity, created by variation in grass biomass, grazing, tree neighborhoods, and ambient conditions such as wind speed (Higgins, Bond, and Trollope 2000) produces variance in woody plant recruitment rates, which regulates tree/grass coexistence. Depending on intensity, fires can prevent tree recruitment by killing tree seedlings or seriously damaging the aboveground parts of shrubs and trees. Small trees can revert to a multistemmed, shrubby form that can be shaded by grasses and vulnerable to the next fire (Bond 1997). In Serengeti fires, Herlocker (sensu Norton-Griffiths 1979) reported that 92% of *Acacia* trees < 1 m were burned back to ground level, 68% of trees between 1–2 m, 28% of trees between 2–3 m, and 1% for trees > 3 m. Likewise, Norton-Griffiths (1979) found that in a *Combretum-Terminalia* woodland, fire burned back 94% trees < 1 m, 68% for trees between 1–2 m, and 45% for trees between 2–3 m. Even if young trees are not killed, they are unable to replace those lost to other factors, such as elephants, resulting in a woodland mosaic with patches of even-aged trees (Norton-Griffiths 1979).

For several months after the occurrence of fire, leaf nutrient concentrations are elevated in postfire vegetation. The increase results because of a greater ratio of leaf to stem, rejuvenation of plant material, and the distribution of similar nutrients in less aboveground biomass (van de Vijver, Poot, and Prins 1999). However, the effect of fire on nutrient heterogeneity has a complex interaction with other factors (van de Vijver 1999; Anderson et al., 2007); nitrogen and phosphorus concentrations in postfire vegetation depended on rainfall (wet vs. dry year), soil type (black cotton vs. lacustrine), and landscape position (ridge top vs. ridge slope). As a result of the increased live:dead ratio of leaves and increased plant nutrient concentrations, ungulates preferably forage in recently burned areas. But green flushes are not used equally by different species; there is a negative relationship between burn use and body size among ungulates (Wilsey 1996). This is because larger ungulates require large quantities of vegetation but can tolerate relatively poor-quality forage, such as occurs in unburnt vegetation.

Smaller ungulates, on the other hand, need less food volume but require relatively high quality forage (Illius and Gordon 1987), which is available in burnt grassland patches.

Biotic Factors

Grazing

Grazer effects on heterogeneity depend on the spatial pattern of grazing and the spatial pattern of the underlying vegetation (Adler, Raff, and Lauenroth 2001). On a regional scale, the dense wildebeest herds that graze intensively during the wet season in the Serengeti plains have created grazing lawns, a characteristic that results from both the reduction of the grass sward through defoliation and natural selection for prostrate, grazing-tolerant genotypes (McNaughton 1984). In contrast, less intensive and patchy grazing occurs in the mid- and tall grasslands that are utilized primarily during migration or localized rainfall (McNaughton and Sabuni 1988). On local scales, resident herbivores can maintain grazing lawns in heavy utilized areas called *hot spots* (see below) or around physical structures such as kopjes, trees, and termite mounds. Grazers can create spatial heterogeneity in grass sward height by selectively grazing patches of herbivore-tolerant species, such as *Cynodon dactylon,* and avoiding patches of unpalatable species, such as *Eleusine jaegeri,* as occurred in Arusha National Park (Vesey-Fitzgerald 1974).

In addition to effects on sward structure, Serengeti grazers can increase plant species richness (McNaughton 1983; Anderson, Ritchie, and McNaughton 2007), the mineralization rates of growth-limiting nutrients (McNaughton, Banyikwa, and McNaughton 1997), and primary production (McNaughton 1979, 1985). However, the generation of heterogeneity is increased because the modification of plant and soil characteristics by grazers often interacts with other factors that vary across the ecosystem. For example, overcompensation by plants in response to grazing that can lead to a stimulation of primary production requires a sufficient interval between defoliation events for regrowth to occur (Oesterheld and McNaughton 1991) and a threshold level of inorganic nitrogen (Hamilton et al. 1998). Evidence from other systems suggests that grazing alters plant demography in ways that apply to Serengeti, including effects on seed production (Anderson and Frank 2003), plant size (Butler and Briske 1988), age (Pfeiffer and Hartnett 1995), and density (O'Connor 1994). Herbivore density and body size appear to have important implications for the effects of grazers on heterogeneity in savanna ecosystems (Olff and Ritchie 1998;

Adler, Raff, and Lauenroth 2001; Bakker et al. 2004). Moreover, the interaction between plant tolerance and grazer forage selectivity can modify the influence of grazing on ecosystem processes and heterogeneity (Augustine and McNaughton 1998).

Various ungulate herbivores differentially promote heterogeneity. Hippos impact geomorphology by creating paths, aversions, levees, and swamps (McCarthy, Ellery, and Bloem 1998). In Serengeti, topi increase their efficiency and offtake of green leafy biomass by grazing selectivity in grassland swards in which reproductive stems have developed, whereas wildebeest graze to a lower height, thus increasing their efficiency and offtake in short vegetative grassland swards (Murray and Illius 2000). This effect can vary with season, phenology, and plant quality; selection of grass swards by roan antelope in South Africa shifted between high-quality forage in the late dry season and early wet season to high quantity in the late wet and early dry season (Heitkonig and Owen-Smith 1998).

Browsing

Landscapes represent a continuum of spatially heterogeneous resources across a hierarchy of scales for all organisms. For browsing ungulates, the hierarchy, from small to large, might be leaves, twigs, branches, trees, and woodland patches (Skarpe et al. 2000). The attributes of each level in the hierarchy can influence browsing selectivity, such as chemical defense of leaves, spines that occur on twigs, and species abundance in a woodland patch, attributes that influenced browsing selectivity in woodlands in Botswana (Skarpe et al. 2000). Small browsers, such as impala, Grant's gazelle, Thomson's gazelle, and dik-dik were implicated in a study of the effects of browsers on woodland regeneration in Serengeti (Belsky 1984). In the study, browsing significantly reduced tree heights at a mid-grass site and tall-grass site near Lobo and kept regenerating trees in the smallest size class for the entire three-year study.

Even though tree density varied widely in Serengeti woodland and riverine habitats, browsing and damage by elephant and giraffe was not related to the density of trees in a stand (Ruess and Halter 1990). Moreover, for the majority of tree species sampled, browsing damage occurred in direct proportion to their occurrence, suggesting that most species were selected more or less at random. However, *A. senegal* was significantly preferred in one stand, *Commiphora trothae* and *A. clavigera* (now *A. robusta*) were significantly avoided in several stands, while *Acacia tortilis, A. xanthophloea,* and *Albizia harveyi* were always damaged in proportion to their occurrence. As with the stimulation of herbaceous biomass by grazers, the removal of shoots by simulated giraffe browsing stimulated shoot production in *Acacia tortilis, A. xanthophloea,* and *A. hockii* (Pellew 1983). The impact of browsing on vegetation varies seasonally, with the most significant impact happening during the green flush that occurs approximately 1 month prior to the onset of the November rains (Pellew 1983).

Elephant have the greatest influence of any browsing mammal in Serengeti, and their behaviors have been at the heart of a controversy surrounding woodland decline for decades (Lamprey et al. 1967; Norton-Griffiths 1979). Their main direct effect on vegetation is to increase mortality by uprooting mature trees and stripping bark and to reduce recruitment of seedlings by consuming them. In the Maasai-Mara Reserve, elephants increase woodland fragmentation by removing branches and creating paths in woody thickets (Dublin 1995). However, their role in woodland degradation may be overstated; elephants largely consume grass (Croze 1974a, 1974b) and most moderately damaged trees survive (Sinclair 1995). Moreover, a study from a Kenyan savanna ecosystem demonstrated that *Acacia drepanolobium* seedling survival was lower in the absence of large mammalian herbivores (Goheen et al. 2004). Apparently, elephants and other large herbivores suppressed herbivory by small mammals and insects that decreased seedling mortality. In the same savanna, results of an exclosure study suggested that small browsers, such as dik-diks, had a major influence on suppressing shrub recruitment through selective browsing (Augustine and McNaughton 2004).

Termites

Unfortunately, termites themselves have gone largely unstudied in the Serengeti, perhaps because of the significant research emphasis on ungulates and carnivores. However, studies of vegetation (Glover, Trump, and Wateridge 1964; Belsky 1983) and soils (de Wit 1978) in the Serengeti region identified termites, termitaria, and abandoned mounds as significant determinants of vegetation pattern, species composition, and functional type (e.g., perennial, annual, short-grass, tall-grass). Total consumption of plant biomass by termites increases with rainfall, a pattern that has been observed among (Deshmukh 1989) and within (Buxton 1981) African savanna ecosystems. However, the proportion of total primary production that is consumed by termites is believed to decrease with increasing rainfall (Deshmukh 1989). Despite the presence of a strong rainfall gradient and abundant termite populations, whether these ecologically important relationships hold for Serengeti is not known.

The results of research from elsewhere in Africa have provided considerable insight into termite impacts in savanna ecosystems and have highlighted their role as ecosystem engineers. For example, in central Tanzania, the density and type of termites was one of the strongest indicators of soil depth, texture, clay mineralogy, drainage, and parent material at the regional level (Jones 1989). Different termite species have different life-history strategies, such as energetics, spatial distribution, nest-building, and habitat, which can have different effects on the surrounding environment (Eggleton and Tayasu 2001). Termite diets consist of dead plant material, leaf litter, woody debris, and dung (Dangerfield and Schuurman 2000) but can vary depending on species. The mounds of similar species, such as harvester ants (*Messor capensis*) in South Africa, can act as centers of plant germination and diversity and can improve seed production and growth rates of plants growing on compared to off mounds (Dean and Yeaton 1993). Moreover, harvester ant mounds disturbed by aardvarks contained a greater number of viable seeds for germination than soil between mounds, and seed germination on mounds varied significantly in time in a way that depended on rainfall (Dean and Yeaton 1992).

One conspicuous influence of termites is their effect on soil chemical and physical properties. First, termites mix soil layers within their nests by translocating small soil particles to the surface (Holt and Lepage 2000). Second, termites act literally as agents of weathering by increasing the expandable clay minerals in the soil used to build chamber walls in a more-or-less irreversible way (Jouquet et al. 2002). In terms of their effects on soil chemical properties, the presence of termite mounds results in the accumulation of bases in the surrounded soils (Malaisse 1978). In South Africa, soils of eroded termite mounds were more acidic and enriched in Mg, Ca, N, P, and total exchangeable cations compared to soil in between mounds. Moreover, the differences in soil nutrients translated into greater primary production and leaf % nitrogen in *T. triandra* that was grown on soils from eroded termite mounds compared to control soil (Smith and Yeaton 1998).

Individual Trees

Individual trees occurring in savannas change the light, nutrient, and water conditions in their immediate vicinity, as well as offer physical refuge for a wide variety of organisms such as ungulates, carnivores, birds, reptiles, amphibians, and insects. The influence of savanna trees on understory vegetation in Kenya was positive via effects of shading, reducing water stress and increasing nutrient availability (Weltzin and Coughenour 1990). Belsky et al. (1993) demonstrated that tree canopies decreased ambient light levels, soil temperatures, and soil C:N ratios, but increased organic matter and soil nutrient availability (total N, P, K, and Ca). Moreover, grassland production was higher under tree canopies at xeric sites compared to mesic sites, an effect believed to result from the higher benefit that reduced shade confers to grasses in xeric conditions compared to mesic sites. Belsky (1994) showed that herbaceous production in open grassland was nutrient limited compared to underneath trees, and that higher productivity resulted from increased nutrient inputs from trees. Furthermore, she suggested that water competition was reduced at arid sites because tree roots exploited soils farther from the tree and thus did not compete with grasses near the tree base, but competition for water limited herbaceous production at mesic sites. Amundson, Ali, and Belsky (1995) found that shade benefited crown species such as *Cynodon nlemfuensis* and *Panicum maximum* because they could close their stomata in response to shade and thus conserve water. On the other hand, grassland-zone species such as *D. macroblephara* and *Eustachys paspaloides* were unable to alter stomatal conductance in response to reduced light.

In Tarangire, trees effectively shifted nutrient limitation from N-limited in open grassland to P-limited under tree canopies (Ludwig et al. 2001, Ludwig, Dawson et al. 2004). Moreover, N:P ratios of grasses under small trees were intermediate to N:P ratios of grasses in open grassland and under large trees, suggesting that the shift from N- to P-limitation happens gradually (Ludwig, Dawson et al. 2004). In the mid-wet season nutrient concentrations of grasses were higher under tree canopies, suggesting that grass production was limited by light, when water and nutrients were abundant. However, canopy shade had positive effects on grass productivity in the dry season when water was scarce (Ludwig et al. 2001). Forb biomass and diversity were highest under tree canopies because of their tolerance for shade. Herbaceous vegetation was greatest under dead trees and was on average 60% more than under live trees, providing further evidence that trees and herbaceous vegetation compete for water. Finally, in support of Belsky's (1994) findings, Ludwig, de Kroon et al. (2004) showed that, even though hydraulic lift occurs under *Acacia tortilis* (Ludwig et al. 2003), competition between grass and trees overwhelms the positive effects of lift in African savannas.

Herbaceous Vegetation

While soils exert obvious effects on vegetation, plants also exert reciprocal effects on soil. Plants alter nutrient cycling (Wedin and Tilman 1990) soil microbial processes (Groffman et al. 1996; Hamilton and Frank 2001), and soil fertility (Ludwig et al. 2001). Small-scale variation in NO_3^- was correlated with local species diversity across grassland sites, suggesting that plant species directly influence nutrient concentrations in the small soil volumes surrounding plants roots (Anderson, McNaughton, and Ritchie 2004). Grasses in Serengeti are associated differentially with vesicular-arbuscular mycorrhizae fungi, which are more abundant at nutrient poor-sites and which may buffer plant nutrient quality against poor-quality soil (McNaughton and Oesterheld 1990). Plants with different life history strategies and growth forms (i.e., annuals, perennials, grasses, forbs, and shrubs) differ with respect to root and leaf tissue chemistry and elemental stoichiometry in ways that influence decomposition, soil mineralization, and microbial dynamics, all of which create plant-derived heterogeneity (Hobbie 1992). Belsky (1986) attributed small-scale vegetation patchiness in the Serengeti plains to the vegetative growth habits of dominant grasses, which form stable patches for long periods of time.

Recent studies reveal the profound impacts that temporal and spatial variation in primary production has on the movements and population dynamics of large-bodied Serengeti mammals. For example, the local movements of Thomson's gazelles among patches of herbaceous vegetation suggest they adaptively locate patches that maximize their energy intake (Fryxell, Wilmhurst, and Sinclair 2004; Fryxell et al. 2005). The population dynamics of Serengeti lions depend on variation in primary production because tall vegetation provides cover for hunting lions, thereby increasing the accessibility of prey (Hopcraft, Packer, and Sinclair 2005). Moreover, extreme climate events that cause substantial or sustained changes in herbaceous vegetation can trigger salutatory changes in the size of the lion population that remain stable on decadal time scales (Packer et al. 2005).

Human Impacts

People induce heterogeneity in vegetation and landscape patterns in Serengeti at different spatial and temporal scales and through a variety of processes. Fire is used to alter grassland/woodland mosaics or as a management tool. Grazing by domestic stock influences heterogeneity. At a more local scale, individual settlements, roads and paths, and cultivated land fragment the landscape. Written accounts from early explorers and settlers, and later on maps and aerial photographs allow for a reconstruction of the Serengeti landscape over the past 100 years. At the turn of the century, explorers, traders, and hunters described the Serengeti as open grassland with lightly wooded patches. The rinderpest pandemic of the 1890s devastated both domestic and wild ruminants and vegetation composition and landscape patterns changed drastically in the years that followed. By the time the colonial administrators arrived in the 1930s and early 1940s, the area had become densely wooded and infested with tsetse flies and trypanosomiasis (Lamprey and Waller 1990; Dublin 1995). From the 1950s onward, a combination of increased wildebeest numbers and high fire frequency (both natural and induced by the Maasai) led to the rapid disappearance of the woodlands, to make place for grazing lawns once more. Lamprey (1984) analyzed changes in woodland cover in three areas of the Mara from 1950 to 1983 and found that in all sites, woody cover had declined from 20–35% in 1950 to less than 10% by 1974. By the 1980s, woodland cover recovered at one of the sites, while it further declined at the second site and trees completely disappeared from the third site.

People also have a profound impact on the land through the development of permanent settlements and agriculture. Serneels, Said, and Lambin (2001) studied natural and anthropogenic changes in vegetation cover in and around the Serengeti between 1975 and 1995 using satellite imagery (fig. 5.11). The analysis demonstrated that the most important single type of land cover change was due to conversion to agriculture, ranging from small patches of subsistence cultivation (e.g., in NCA highlands) to large areas under mechanized farming (e.g., Loita plains; see Homewood et al. 2001). Although small, isolated patches of agriculture will at first increase the structural heterogeneity of the landscape, beta-diversity decreases when natural vegetation is replaced by one crop. As agriculture spreads, the structural heterogeneity is lost, too. This has implications for wildlife, as grazing/browsing resources are no longer available, and migration routes can become blocked. Mechanized farming in the Loita plains in Kenya spread from 4,875 ha in 1975 to 50,000 ha in 1995, resulting in the destruction of calving habitat for resident wildebeest population and a subsequent 70% population decline since the late 1970s (Ottichilo, de Leeuw, and Prins 2001; Serneels and Lambin 2001; Serneels, Said, and Lambin 2001). In recent years, the fields in the dryer parts have been abandoned, due to a combination of unfavorable weather conditions and land privatization. Activities are being shifted to irrigated agriculture on the banks of the Mara River. Since 2003, several new enterprises are drawing water from the Mara River, which may have consequences for wildlife in drought years (Gereta et al. 2002).

Other land changes in the northern Serengeti include the expansion of settlements of smallholders, mostly around the gates of the Maasai-Mara at Talek, Sekanani, and Aitong, including an increase in the number of Maasai settlements (Lamprey and Reid 2004) and their associated decrease in vegetation cover, and small-scale maize farming. The Maasai and similar nomadic peoples protect livestock in high animal density enclosures called

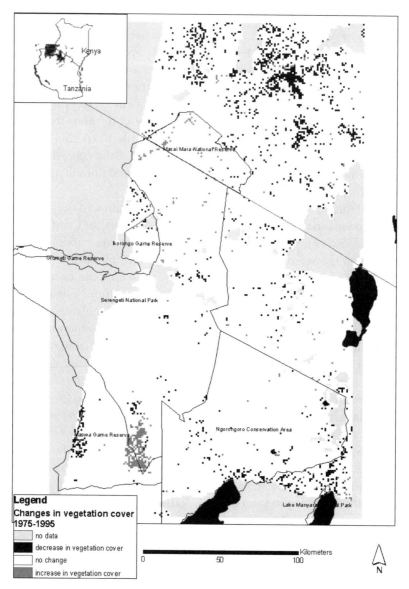

Fig. 5.11 Changes in vegetation cover in the Serengeti ecosystem as detected from a time series of Landsat images from 1975–1995. Pixel colors indicate land cover change: light grey = no available data; dark grey = increase in vegetation cover; black = decrease in vegetation cover.

bomas. While they are only temporary, significant dung and urine is deposited during occupation. People preferably settle in areas that present a number of advantages to them, such as proximity to permanent water (Reid et al. 2003), or with potential revenues from or employment in the tourism sector (Thompson 2002). These hot spots are also of key importance to wildlife, and a source of wildlife-people conflicts and habitat destruction. Several studies (Walpole 2002; Boydston et al. 2003) have shown changes in animal behavior in the Maasai-Mara, due to settlement expansion around the park boundaries. In 1970, the rangelands north of the Maasai-Mara were converted to group ranches under communal Maasai ownership and management. In 1999, Lemek Group Ranch was subdivided into private parcels, and currently the same process is ongoing in Koyaki Group Ranch, bordering the Maasai-Mara. An early manifestation of subdivision was the fragmentation of the traditional boma into one- or two-family units, and the proliferation of these new homesteads within their localities (Lamprey and Reid 2004). This increasing number of bomas effectively fragments the landscape, as wildlife behavior is altered by their presence.

CASE STUDIES: MULTIFACTOR INTERACTIONS AND HETEROGENEITY IN TIME AND SPACE

Elephants, Fire, Grazing, Humans, and Woodlands Dynamics

Over the past one hundred years, the Serengeti-Mara ecosystem has experienced major changes in woodland vegetation; grasslands transitioned into woodlands from 1890s to 1940s and back from woodlands to grasslands in the 1960s to 1980s, a trend that continued in the Maasai Mara but reversed in the northern Serengeti during the 1990s (Dublin 1995, and see chapter 2, this volume). The major factors attributed to woodland change have been fire, elephants, and grazing. However, assigning causation to any one factor has been problematic, because factors interact and are difficult to measure because of their dynamical nature. Between 1963 and 1972 the Serengeti lost 13% of its woody cover; the largest part of this occurred in northern Serengeti and Maasai-Mara (26%), compared to lower degradation in the central woodlands (7%). Moreover, the reduction was not uniform: woody vegetation cover showed a blanket decrease in the north, changes in woody cover showed a heterogeneous mosaic of increase and decrease in the central woodlands (Norton-Griffiths 1979).

Analysis suggested that fire was most strongly associated with reduction in woody cover in the north, while the dry-season elephant population size was most strongly associated with woody vegetation reduction in the central woodlands. Dublin, Sinclair, and McGlade (1990) showed by using a simple model derived from data on wildebeest, fire, elephants, and browsing antelopes, that the combined effect of elephants and fire had the greatest impact on Serengeti-Mara woodlands. They convincingly argued that high fire frequency was responsible for woodland decline during the 1960s, while high elephant densities maintained the grassland state throughout the 1980s. Apparently, human-induced fires were common during this period, and fires intensified because of the large fuel loads that followed the ungulate population declines during the rinderpest outbreak (Dublin 1995). However, after the eradication of rinderpest, the wildebeest population underwent a sixfold increase in just over 10 years (Sinclair 1995), resulting in the elimination of much of the fuel for fires, thus reducing the intensity and frequency of fires. By reducing fire damage to seedlings grazing may actually increase the potential for woodland regeneration.

Yet even after the drastic reduction in fire frequency, grassland states were maintained and existing woodlands continued to decline throughout the 1980s. Dublin, Sinclair, and McGlade (1990) showed that the best explanation lay in the burgeoning elephant population, especially in the northern Serengeti and Maasai-Mara. It was suggested that elephants act synergistically with fire, by opening up woodland thickets and canopies so that herbaceous fuel loads increase and fire damage is worsened, a scenario that was observed in Croton thickets in the Maasai-Mara reserve (Dublin 1995). Recently, trends in woodland cover in the Serengeti central hills region have reversed, while woodlands have continued to decline in the Maasai-Mara. Apparently, antipoaching efforts and high visitation rates have allowed high elephant densities in the Maasai-Mara, while the threat of poaching in the central hills keeps elephant densities relatively low.

Hot Spots: Formation and Maintenance

Resident herbivores in Serengeti are not distributed evenly across the ecosystem. Instead, they often occur in "hot spots," areas of high-density mixed herds, found in regions receiving greater than 700 mm yr^{-1} rainfall (McNaughton 1988). Hot spots are spatially and temporally stable for over 20 years, but they are heterogeneously distributed in the Serengeti ecosystem. Forage nutrient concentrations in hot spots are greater than in adjacent grasslands with low ungulate densities; magnesium, sodium, and phosphorus, important for ungulates during late pregnancy and lactation, occur in concentrations that meet dietary requirements in hot spots but not in adjacent unused grasslands (McNaughton 1988). Moreover, experiments showed that soil mineralization rates of sodium and nitrogen were higher in ungulate high-use areas, and that ungulate grazing actually promoted increased sodium mineralization rates by as much as an order of magnitude (McNaughton, Banyikwa, and McNaughton 1997). Therefore, grazing ungulates choose hot spots of greater leaf tissue nutrients and forage quality that occur heterogeneously across the landscape, and grazing acts to maintain high nutrient-rich forage and increases ungulate carrying capacity (McNaughton 1988, 1990; McNaughton, Banyikwa, and McNaughton 1997).

The factors responsible for causing hot spots are not known. Underlying soil differences cannot be implicated because total soil nutrient concentrations are not different between hot spots and adjacent control areas (McNaughton 1988). Several plausible mechanisms exist. One is that localized rainfall events create concentrated foraging areas by stimulating primary production, after which the high-density herds essentially fertilize large patches through urine and dung deposition, increasing nutrient mineralization and forage quality. A second hypothesis is that hot spots occur on old abandoned termite colonies. As discussed previously, termites change soil nutrient availability and increase forage quality in a way that could remain long after the colony disbands. A third hypothesis is that hot spots represent areas of historical and intense human use, especially behaviors of nomadic cattle herders of the Maasai tribe.

Only the third hypothesis has been adequately tested, with some support, although not in the Serengeti ecosystem. Abandoned cattle bomas create small areas of highly concentrated soil nutrients and forage with low C:N ratios that establish as long-term (> 40 years), grass-filled glades (Augustine 2003). These areas are preferentially used by ungulates and may potentially function as nutrient- and forage-rich hot spots in a new, stable state. The maintenance of nutrient-rich glades by the feeding behavior of ungulates was also identified in arid Kenyan grasslands; herbivores imported nutrients from surrounding woodlands to nutrient-rich glades composed of high-quality, grazing-tolerant grasses (Augustine, McNaughton, and Frank 2003). Calcium, nitrogen, and phosphorus were also at elevated levels in glades; phosphorus, in particular, was at levels high enough to support lactating livestock in the glade but below those levels in the surrounding woodlands (Augustine 2003). The connection between human use and hot spots has not been addressed in Serengeti, and the development of hot spots from abandoned bomas provides but one hypothetical mechanism for their occurrence.

A conceptual model by Osterheld et al. (1999) suggests that the relative importance of fire, climate, and grazing on primary productivity in savanna grassland ecosystems changes across a range of mean annual precipitation (fig. 5.9). At low rainfall, between 200 and 450 mm, interannual climate fluctuations have the greatest impact on production, while herbivore consumption rates and fire frequency are low. Between 450 and 700 mm precipitation, the importance of climate variation decreases, grazing still consumes a relatively small proportion of production, while fire frequency increases in importance but has mostly negative effects. Above 700 mm precipitation, interannual fluctuations in climate are relatively small compared to the effects of fire and grazing. Fire frequency is high and can increase productivity up to five times the mean. Grazing is substantial and can have beneficial effects because of compensatory regrowth. Interannual climate fluctuations can interact with fire and grazing, modifying their importance; this was observed in arid grasslands in Kenya, where overcompensation of primary production following grazing occurred during a wet year but not a dry year (Augustine 2002).

CONCLUSIONS

In summary, the heterogeneity of major vegetation types corresponded to the environmental gradient (e.g., rainfall) in a predictable manner, and the spatial scale at which variables expressed their heterogeneity closely matched the relationships depicted in fig. 5.1. However, a composite measure of heterogeneity demonstrated a significant lack of spatial consistency in heterogeneity across Serengeti grasslands. Regions were composed of adjacent sites that were often more dissimilar than distant sites, even those separated by over 100 km. Important abiotic sources of heterogeneity identified in the Serengeti are climate, fire, and geology, while ungulates (browsers and grazers), vegetation, termites, and humans are important sources of biotic heterogeneity. The agents of heterogeneity are dynamic in time and space and form a complex web of interactions (fig. 5.12). In the last section we attempted to represent some of that complexity by drawing on three examples in which multiple agents interact to affect heterogeneity.

To conclude, we highlight obvious gaps missing from this chapter and in general from studies of heterogeneity across the Serengeti ecosystem. In particular, we have identified four areas that, if elucidated through future research, would greatly enhance the understanding of heterogeneity

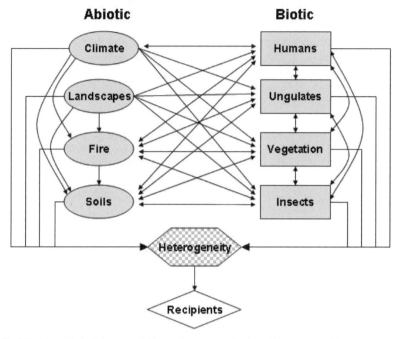

Agents of heterogeneity

Abiotic   **Biotic**

Fig. 5.12 Schematic depicting potential interactions among abiotic and biotic agents of heterogeneity in Serengeti National Park (substrate, controllers, and response not shown). Landscapes = landforms, kopjes, catenas, hills, mountains; ungulates = browsers and grazers; insects = especially termites, dung beetles, tsetse flies; vegetation = herbaceous and woody vegetation; soils = parent material and fertility; recipients include many of the biotic agents listed here so that organisms can modify the ecosystem heterogeneity in the way to which they themselves respond (i.e., positive/negative feedback loops). In these instances, controllers, substrates, and other agents modify the effects of the positive/negative feedback loops, so the system remains stable.

in Serengeti. First, maps of soil (de Wit 1978, Jager 1982), landscape (Gerreshiem 1974), and vegetation (Herlocker 1976) of Serengeti reveal considerable heterogeneity on their own. However, interactions among vegetation, soils, and landscape features across the ecosystem have not been considered. An attempt to explore spatial associations between soils, landscapes, and vegetation would provide a decent first approximation of the possibility that their interactions promote emergent forms of heterogeneity. Future work could include hydrological and land-systems models that incorporate feedbacks among biotic and abiotic components.

Second, the movements and behaviors of enormous herds of migratory ungulates are central and defining characteristics of Serengeti. However, an attempt to understand the causes and consequences of spatial and temporal variation in ungulate migration patterns has not been made. This is a daunting, if not overwhelming, task and will likely take a large collaboration of scientists to study the subject. A recent investigation by Thirgood et al. (2004) provided insight into the complexity of wildebeest migratory patterns. However, understanding the complexity of their movements and the potential to study their behavior as a self-organized, complex system represents one of the most compelling topics facing Serengeti researchers.

Third, as previously discussed, hot spots are key landscape features that provide sustained production of high-quality forage for resident ungulates. While their function is well known, little is known about how they are generated, where they occur, and how long they occupy the landscape. Understanding the spatial and temporal distribution of hot spots may help reveal the factors responsible for their generation and maintenance. The park would benefit greatly from a project to map and monitor hot spots, ungulate densities, and the properties of associated vegetation over time. Such a project would enhance the understanding of ecosystem functioning and enable conservationists and managers to better safeguard natural patterns of heterogeneity that sustain populations of resident herbivores.

Finally, insects, especially termites and dung beetles, have received glaringly little attention in Serengeti. Except for Folse (1982) and the mention of termite structures by those researchers studying plant diversity, information on the distribution, abundance, and effect of insects on ecological processes are all but absent from the Serengeti literature. In similar savanna ecosystems, insects account for a large proportion of animal biomass, and by comparison are likely responsible for the decomposition and redistribution of huge amounts of vegetation biomass and nutrients in Serengeti.

The consideration of these issues imposes a new layer of complexity to our understanding of the Serengeti ecosystems and savannas in general. However, expanding research efforts to incorporate these topics may reveal mechanisms that contribute to the dynamics and diversity that make the Serengeti one of the most singular and cherished ecosystems on earth.

ACKNOWLEDGMENTS

We thank Emilian P. Mayemba and Yusto B. Byarugaba for dedicated fieldwork and companionship while in Serengeti. We also thank A. R. E. Sinclair and two anonymous reviewers for their editorial comments and recommendations. Special thanks to Gericke Sommerville for providing the digitized map of the Gerreshiem regions, to Simon Mduma and Serengeti Ecological Monitoring Programme for providing rainfall data, and to Mike Coughenour for the use of his vehicle for fieldwork. Funding sources included an NSF doctoral fellowship (T. M. Anderson), NASA grant ESS NGT530459 (J. Dempewolf), NSF grant DEB-9711627 (K. Metzger), EU INCO-DC grant ERBIC18CT (S. Serneels), NSF grant DIG-9813706 (D. N. Reed), and an NSF postdoctoral fellowship (D. N. Reed). D. N. Reed acknowledges the Evolution of Terrestrial Ecosystems Program at the National Museum of Natural History (ETE pub. no. 96.).

REFERENCES

Adler, P. B., D. A. Raff, and W. K. Lauenroth. 2001. The effect of grazing on the spatial heterogeneity of vegetation. *Oecologia* 128:465–79.

Allen T. F. H., and T. W. Hoeekstra. 1991. Role of heterogeneity in scaling of ecological systems under analysis. In *Ecological heterogeneity,* ed. J. Kolasa and S. T. A. Pickett, 47–68. New York: Springer-Verlag.

Amundson, R. G., A. R. Ali, and A. J. Belsky. 1995. Stomatal responsiveness to changing light intensity increases rain-use efficiency of below-crown vegetation in tropical savannas. *Journal of Arid Environments* 29:139–53.

Anderson, G. D., and L. M. Talbot. 1965. Soil factors affecting the distribution of the grassland types and their utilization by wild animals on the Serengeti Plains, Tanganyika. *Ecology* 53:33–56.

Anderson, T. M. 2004. Determinants of plant species diversity across spatial scales in Serengeti National Park, Tanzania. PhD diss., Syracuse University, Syracuse, New York.

Anderson, T. M., Y. Dong, and S. J. McNaughton. 2006. Nutrient acquisition and physiological response of dominant Serengeti grasses to variation in soil texture and grazing. *Journal of Ecology* 94:1164–1175.

Anderson, T. M., and D. A. Frank. 2003. Defoliation effects on reproductive biomass: Importance of scale and timing. *Journal of Range Management* 56:501–16.

Anderson, T. M., S. J. McNaughton, and M. E. Ritchie. 2004. Scale-dependent relationships between the spatial distribution of a limiting resource and plant species diversity in an African grassland ecosystem. *Oecologia* 139:277–87.

Anderson, T. M., K. L. Metzger, and S. J. McNaughton. 2007. Multi-scale analysis of plant species richness in Serengeti grasslands. *Journal of Biogeography* 34:313–23.

Anderson, T. M., M. E. Ritchie, and S. J. McNaughton. 2007. Rainfall and soils modify plant community response to grazing in Serengeti National Park. *Ecology* 88:1191–1201.

Anderson, T. M., M. E. Ritchie, E. Mayemba, S. Eby, J. B. Grace, and S. J. McNaughton. 2007. Forage nutritive quality in the Serengeti ecosystem: the roles of fire and herbivory. *The American Naturalist* 170:343–57.

Augustine, D. J. 2002. Large herbivores and process dynamics in a managed savanna ecosystem. PhD diss., Syracuse University, Syracuse, New York.

———. 2003. Long-term, livestock-mediated redistribution of nitrogen and phosphorus in an East African savanna. *Journal of Applied Ecology* 40:137–49.

Augustine, D. J., and S. J. McNaughton. 1998. Ungulate effects on the functional species composition of plant communities: Herbivore selectivity and plant tolerance. *Journal of Wildlife Management* 62:1165–83.

———. 2004. Regulation of shrub dynamics by native browsing ungulates on East African rangeland. *Journal of Applied Ecology* 41:45–58.

Augustine, D. J., S. J. McNaughton, and D. A. Frank. 2003. Feedbacks between soil nutrients and large herbivores in a managed savanna ecosystem. *Ecological Applications* 13: 1325–37.

Bakker, E. S., H. Olff, M. Boekhoff, J. M. Gleichman, and F. Berendse. 2004. Impact of herbivores on nitrogen cycling: Contrasting effects of small and large species. *Oecologia* 138:91–101.

Bell, F. C. 1979. Precipitation. In *Arid land ecosystems: Structure, functioning, and management*, ed. D. W. Goodall and R. A. Perry, 373–92. Cambridge: Cambridge University Press.

Bell, R. H. V. 1970. The use of the herb layer by grazing ungulates in the Serengeti. In *Animal populations in relation to their food resources*, ed. A. Watson, 111–24. Oxford: Blackwell Scientific.

Belsky, A. J. 1983. Small-scale pattern in grassland communities in the Serengeti National Park, Tanzania. *Vegetatio* 55:141–51.

———. 1984. Role of small browsing mammals in preventing woodland regeneration in the Serengeti National Park, Tanzania. *African Journal of Ecology* 22: 271–279.

———. 1986. Population and community processes in a mosaic grassland in the Serengeti, Tanzania. *Journal of Ecology* 74:841–56.

———. 1988. Regional influences on small-scale vegetational heterogeneity within grasslands in the Serengeti National Park, Tanzania. *Vegetatio* 74:3–10.

———. 1992. Effects of grazing competition, disturbance and fire on species composition and diversity in grassland communities. *Journal of Vegetation Science* 3:187–200.

———. 1994. Influence of trees on savanna productivity: Test of shade, nutrients, and tree-grass competition. *Ecology* 75:922–32.

Belsky, A. J., R. G. Amundson, J. M. Duxbury, S. J. Riha, A. R. Ali, and S. M. Mwonga. 1989. The effects of trees on their physical, chemical, and biological environments in a semi-arid savanna in Kenya. *Journal of Applied Ecology* 26:1005–24.

Belsky, A. J., S. M. Mwonga, R. G. Amundson, J. M. Duxburg, and A. R. Ali. 1993. Comparative effects of isolated trees on their undercanopy environments in high- and low-rainfall savannas. *Journal of Applied Ecology* 30:143–55.

Bond, W. J. 1997. Fire. In *Vegetation of southern Africa*, ed. R. M. Cowling, D. M. Richardson, and S. M. Pierce, 421–46. Cambridge: Cambridge University Press.

Boydston, E. E., K. M. Kapheim, H. E. Watts, M. Szykman, and K. E. Holekamp. 2003. Altered behaviour in spotted hyenas associated with increased human activity. *Animal Conservation* 6:207–19.

Butler, J. L., and D. D. Briske. 1988. Population structure and tiller demography of the bunchgrass *Schizachyrium scoparium* in response to herbivory. *Oikos* 51:306–312.

Buxton, R. D. 1981. Change in the composition and and activities of termite communities in relation to rainfall. *Oecologia* 51:379–84.

Coughenour, M. B., and J. E. Ellis. 1993. Landscape and climatic control of woody vegetation in a dry tropical ecosystem: Turkana District, Kenya. *Journal of Biogeography* 20: 383–398.

Croze, H. 1974a. The Seronera bull problem. Part 1. The elephants. *East African Wildlife Journal* 12:1–28.

———. 1974b. The Seronera bull problem. Part 2. The trees. *East African Wildlife Journal* 12:29–48.

Dangerfield, J. M., and G. Schuurman. 2000. Foraging by fungus-growing termites (Isoptera: Termitidae, Macrotermitinae) in the Okavango Delta, Botswana. *Journal of Tropical Ecology* 16:717–31.

Dawson, J. B. 1964. Carbonatitic volcanic ashes in northern Tanganyika. *Bulletin of Volcanology* 27:81–91.

Dawson, J. B., H. Pinkerton, D. M. Pyle, and C. Nyamweru. 1994. June 1993 eruption of Oldoinyo Lengai, Tanzania: Exceptionally viscous and large carbonatite lava flows and evidence for coexisting silicate and carbonate magmas. *Geology* 22: 799–802.

Dean, W. R. J., and R. I. Yeaton. 1992. The importance of harvester ant *Messor capensis* nest-mounds as germination sites in the southern Karoo, South Africa. *African Journal of Ecology* 30:335–45.

———. 1993. The influence of harvester ant *Messor capensis* nest-mounds on the productivity and distribution of some plant species in the southern Karoo, South Africa. *Vegetatio* 106:21–35.

Deshmukh, I. 1989. How important are termites in the production ecology of African savannas? *Sociobiology* 15:155–68.

de Wit, H. A. 1978. Soils and grassland types of the Serengeti plains (Tanzania): Their distributions and interrelations. PhD diss., University of Wageningen, the Netherlands.

Dublin, H. T. 1995. Vegetation dynamics in the Serengeti-Mara ecosystem: The role of elephants, fire, and other factors. In *Serengeti II: Dynamics, management and conservation of an ecosystem*, ed. A. R. E. Sinclair and P. Arcese, 71–90. Chicago: University of Chicago Press.

Dublin, H. T., A. R. E. Sinclair, and J. McGlade. 1990. Elephants and fire as causes of multiple stable states in the Serengeti-Mara woodlands. *Journal of Animal Ecology* 59: 1147–64.

du Toit, J. T., K. H. Rogers, and H. C. Biggs. 2003. *The Kruger experience: Ecology and management of savanna heterogeneity*. Island Press, Washington.

Eggleton, P., and I. Tayasu. 2001. Feeding groups, lifetypes and the global ecology of termites. *Ecological Research* 16:941–60.

Folse, L. J. 1982. An analysis of avifauna-resource relationships on the Serengeti plains. *Ecological Monographs* 52:111–27.

Fryxell, J. M., J. F. Wilmshurst, and A. R. E. Sinclair. 2004. Predictive models of movement by Serengeti grazers. *Ecology* 85:2429–35.

Fryxell, J. M., J. F. Wilmshurst, A. R. E. Sinclair, D. T. Haydon, R. D. Holt, and P. A. Abrams. 2005. Landscape scale, heterogeneity, and the viability of Serengeti grazers. *Ecology Letters* 8:328–35.

Gereta, E., E. Wolanski, M. Borner, and S. Serneels. 2002. Use of an ecohydrology model to predict the impact on the Serengeti ecosystem of deforestation, irrigation and the proposed Amala Weir Water Diversion Project in Kenya. *Ecohydrology and Hydrobiology* 2:135–42.

Gerlach, G., and H. N. Hoeck. 2001. Islands on the plains: Metapopulation dynamics and female biased dispersal in hyraxes (*Hyracoidea*) in the Serengeti National Park. *Molecular Ecology* 10:2307–17.

Gerrard, A. J. 1990. Soil variations on hillslopes in humid temperature climates. *Geomorphology* 3:225–44.

Gerresheim, K. 1974. The Serengeti landscape classification. Serengeti Ecological Monitoring Programme. *Serengeti Research Institute Publication no.* 165.

Glover, P. E., E. C. Trump, and L. E. D. Wateridge. 1964. Termitaria and vegetative patterns on the Loita Plains of Kenya. *Journal of Ecology* 52:367–77.

Goheen, J. R., F. Keesing, B. F. Allan, D. Ogada, and R. S. Ostfeld. 2004. Net effects of large mammals on *Acacia* seedling survival in an African savanna. *Ecology* 85:1555–61.

Gotelli, N. J., and G. R. Graves. 1996. *Null models in ecology.* Washington, DC: Smithsonian Institution Press.

Groffman, P. M., P. Eagan, W. M. Sullivan, and J. L. Lemunyon. 1996. Grass species and soil type effects on microbial biomass and activity. *Plant and Soil* 183:61–67.

Hamilton, E. W., and D. A. Frank. 2001. Can plants stimulate soil microbes and their own nutrient supply? Evidence from a grazing tolerant grass. *Ecology* 82:2397–2402.

Hamilton, E. W., M. S. Giovannini, S. A. Moses, J. S. Coleman, and S. J. McNaughton. 1998. Biomass and mineral element responses of a Serengeti short-grass species to nitrogen supply and defoliation: compensation requires a critical [N]. *Oecologia* 116: 407–18.

Hay, R. L. 1976. *The geology of the Olduvai Gorge.* Berkeley: University of California Press.

Heitkonig, I. M. A., and N. Owen-Smith. 1998. Seasonal selection of soil types and grass swards by roan antelope in a South African savanna. *African Journal of Ecology* 36: 57–70.

Herlocker, D. 1976. *Woody vegetation of the Serengeti National Park.* Kleberg Studies in Natural Resources, the Texas Agricultural Experiment Station. College Station: Texas A&M University.

Higgins, S. I., W. J. Bond, and W. S. W. Trollope. 2000. Fire, resprouting and variability: A recipe for grass-tree coexistence in savanna. *Journal of Ecology* 13:295–99.

Hobbie, S. E. 1992. Effects of plant species on nutrient cycling. *Trends in Ecology and Evolution* 7:336–39.

Holt, J. A., and M. Lepage. 2000. Termites and soil properties. In *Termites: Evolution, sociality, symbiosis, ecology,* ed. T. Abe, D. E. Bignell, and M. Higashi, 389–407. Dordrecht: Kluwer Academic.

Homewood, K., E. F. Lambin, E. Coast, A. Kariuki, I. Kikula, J. Kiveli, M. Said, S. Serneels, and M. Thompson. 2001. Long-term changes in Serengeti-Mara wildebeest and land cover: Pastoralism, population, or policies? *Proceedings of the National Academy, USA* 98:12544–49.

Hood, G. M. 2004. PopTools version 2.6.2. Available at http://www.cse.csiro.au/poptools.

Hopcraft, J. G. C., C. Packer, and A. R. E. Sinclair. 2005. Planning for success: Serengeti lions seek prey accessibility rather than abundance. *Journal of Animal Ecology* 74: 559–66.

Huntley, B. J., and B. H. Walker. 1982. *Ecology of tropical savannas.* New York: Springer-Verlag.

Hutchings, M. J., D. K. Wijesinghe, and E. A. John. 2000. The effects of heterogeneous nutrient supply on plant performance: A survey of responses, with a special reference to clonal herbs. In *The ecological consequences of environmental heterogeneity,* ed. M. J. Hutchings, E. A. John, and A. J. A. Stewart, 91–110. Oxford: Blackwell.

Illius, A. W., and I. J. Gordon. 1987. The allometry of food intake in grazing ruminants. *Journal of Animal Ecology* 56:989–99.

Jager, Tj. 1982. *Soils of the Serengeti woodlands.* Centre for Agriculture Publishing and Documentation, Wageningen.

Jones, J. A. 1989. Environmental influences on soil chemistry in central semiarid Tanzania. *Soil Science Society of America Journal* 53:1748–58.

Jouquet, P., L. Mamou, M. Lepage, and B. Velde. 2002. Effect of termites on clay minerals in tropical soils: Fungus-growing termites as weathering agents. *European Journal of Soil Science.* 53:521–27.

Kolasa, J., and C. D. Rollo. 1991. Introduction: The heterogeneity of heterogeneity: A glossary. In *Ecological heterogeneity,* ed. J. Kolasa and S. T. Pickett, 1–23. New York: Springer-Verlag.

Lamprey, H. F., P. E. Glover, M. I. M. Turner, and R. V. H. Bell. 1967. Invasion of the Serengeti National Park by elephants. *East African Wildlife Journal* 5:151–61.

Lamprey, R. H. 1984. Masai impact on Kenya savanna vegetation: A remote sensing approach. PhD diss., University of Aston-in-Birmingham.

Lamprey, R. H., and R. S. Reid. 2004. Expansion of human settlement in Kenya's Masai Mara: What future for pastoralism and wildlife? *Journal of Biogeography* 31:997–1032.

Lamprey, R. H., and R. D. Waller. 1990. The Loita-Mara Region in historical times: Patterns of subsistence, settlement and ecological change. In *Early pastoralists of south-western Kenya,* ed. P. Robertshaw, 16–35. Nairobi: British Institute in Eastern Africa.

Li, H., and J. F. Reynolds. 1995. On definition and quantification of heterogeneity. *Oikos* 73:280–84.

Ludwig, F., T. E. Dawson, H. de Kroon, F. Berendse, and H. H. Prins. 2003. Hydraulic lift in *Acacia tortilis* trees on an East African savanna. *Oecologia* 134:293–300.

Ludwig, F., T. E. Dawson, H. H. Prins, F. Berendse, and H. de Kroon. 2004. Below-ground competition between trees and grasses may overwhelm the facilitative effects of hydraulic lift. *Ecology Letters* 7: 623-631.

Ludwig, F., H. de Kroon, F. Berendse, and H. H. Prins. 2004. The influence of savanna trees on nutrient, water and light availability and the understory vegetation. *Plant Ecology* 170:93–105.

Ludwig, F., H. de Kroon, H. H. Prins, and F. Berendse. 2001. Effects of nutrients and shade on tree-grass interactions in an East African savanna. *Journal of Vegetation Science* 12: 579–88.

Malaisse, F. 1978. High termitaria. In *Biogeography and ecology of southern Africa,* ed. M. J. A. Werger, 1279–1300. The Hague: W. Junk.

McCarthy, T. S., W. N. Ellery, and A. Bloem. 1998. Some observations on the geomorphological impact of hippopotamus (*Hippopotamus amphibius* L.) in the Okavango Delta, Botswana. *African Journal of Ecology* 36:44–56.

McCune, B., and M. J. Medford. 1999. *Multivariate analysis of ecological data,* v.4.01. Glenden Beach, OR: MjM Software.

McNaughton, S. J. 1979. Grazing as an optimization process: Grass-ungulate relationships in the Serengeti National Park, Tanzania. *American Naturalist* 113:691–703.

———. 1983. Serengeti grassland ecology: The role of composite environmental factors and contingency in community organization. *Ecological Monographs* 53:291–320.

———. 1984. Grazing lawns: Animals in herds, plant form, and coevolution. *American Naturalist* 124:863–86.

———. 1985. Ecology of a grazing ecosystem: The Serengeti. *Ecological Monographs* 55: 259–94.

———. 1988. Mineral nutrition and spatial concentrations of African ungulates. *Nature* 334:343–45.

———. 1989. Interactions of plants of the field layer with large herbivores. In *The biology of large African mammals in their environment,* ed. P. A. Jewell and G. M. O. Maloiy, 15–29. Oxford: Clarendon.

———. 1990. Mineral nutrition and seasonal movements of African migratory ungulates. *Nature* 345:613–15.

———. 1994. Conservation goals and the configuration of biodiversity. In *Systematics and Conservation Evaluation,* ed. P. L. Forey, C. J. Humphries, and R. I. Vane-Wright, 41–62. Oxford: Clarendon.

McNaughton, S. J., F. F. Banyikwa, and M. M. McNaughton. 1997. Promotion of the cycling of diet-enhancing nutrients by African grazers. *Science* 278:1798–1800.

———. 1998. Root biomass and productivity in a grazing ecosystem: The Serengeti. *Ecology* 79:587–92.

McNaughton, S. J., and M. Oesterheld. 1990. Extramatrical mycorrhizal abundance and grass nutrition in a tropical grazing ecosystem, the Serengeti National Park, Tanzania. *Oikos* 59:92–96.

McNaughton, S. J., R. W. Ruess, and S. W. Seagle. 1988. Large mammals and process dynamics in African ecosystems. *Bioscience* 38:794–800.

McNaughton, S. J., and G. A. Sabuni. 1988. Large African mammals as regulators of vegetation structure. In *Plant form and vegetation structure,* ed. M. J. A. Werger, P. J. M. van der Aart, H. J. During, and J. T. A. Verhoeven, 339–54. The Hague, Netherlands: SPB Academic.

Metzger, K. L. 2002. The Serengeti ecosystem: Species richness patterns, grazing and land-use. PhD diss., Colorado State University.

Milne, B. T. 1992. Spatial aggregation and neutral models in fractal landscapes. *American Naturalist* 139:32–57.

———. 1997. Applications of fractal geometry in wildlife biology. In *Wildlife and landscape ecology,* ed. J. A. Bissonette, 32–69. New York: Springer-Verlag.

Milne, G. 1935. Some suggested units of classification and mapping, particularly for East African soils. *Soil Research* 4:183–98.

Minchin, P. R. 1987. An evaluation of the relative robustness of techniques for ecological ordination. *Vegetatio* 69:89–107.

Murray, M. G., and A. W. Illius. 2000. Vegetation modification and resource competition in grazing ungulates. *Oikos* 89:501–8.

Norton-Griffiths, M. 1979. The influence of grazing, browsing and fire on the vegetation dynamics of the Serengeti. *In Serengeti: Dynamics of an ecosystem,* ed. A. R. E. Sinclair and M. Norton-Griffiths, 310–52. Chicago: University of Chicago Press.

Norton-Griffiths, M., D. Herlocker, and L. Pennycuick. 1975. The patterns of rainfall in the Serengeti ecosystem, Tanzania. *East African Wildlife Journal* 13:347–74.

O'Connor, T. G. 1994. Composition and population responses of an African savanna grassland to rainfall and grazing. *Journal of Applied Ecology* 31:155–71.

Oesterheld, M., and S. J. McNaughton. 1988. Intraspecific variation in the response of *Themeda triandra* to defoliation: The effect of time of recovery and growth rates on compensatory growth. *Oecologia* 77:181–86.

———. 1991. Effect of stress and time for recovery on the amount of compensatory growth after grazing. *Oecologia* 85:305–13.

Oesterheld, M., J. Loreti, M. Semmartin, and J. M. Paruelo. 1999. Grazing, fire, and climate effects on primary productivity of grasslands and savannas. In *Ecosystems of the world 16: Ecosystems of disturbed ground,* ed. L. R. Walker, 287–306. New York: Elsevier.

Olff, H., and M. E. Ritchie. 1998. Effects of herbivores on grassland plant diversity. *Trends in Ecology and Evolution* 13:261–65.

Ottichilo, W. K., J. de Leeuw, and H. H. T. Prins. 2001. Population trends of resident wildebeest (*Connochaetes taurinus* hecki [Neumann]) and factors influencing them in the Masai Mara ecosystem, Kenya. *Biological Conservation* 97:271–82.

Packer, C., R. Hilborn, A. Mosser, B. Kissui, M. Borner, G. Hopcraft, J. Wilmshurst, S. Mduma, and A. R. E. Sinclair. 2005. Ecological change, group territoriality, and population dynamics in Serengeti lions. *Science* 307:390–93.

Pellew, R. A. P. 1983. The impacts of elephant, giraffe and fire upon the *Acacia tortilis* woodlands of the Serengeti. *African Journal of Ecology* 21:41–74.

Peterson, D. L., and V. T. Parker. 1998. *Ecological scale.* New York: Columbia University Press.

Pfeiffer, K. E., and D. C. Hartnett. 1995. Bison selectivity and grazing response of little bluestem in tallgrass prairie. *Journal of Range Management* 48:26–31.

Pickett, S. T. A., M. L. Cadenasso, and T. L. Benning. 2003. Biotic and abiotic variability as key determinants of savanna heterogeneity at multiple spatiotemporal scales. In *The Kruger experience: Ecology and management of savanna heterogeneity,* ed. J. T. du Toit, K. H. Rogers, and H. C. Biggs, 22–40. Seattle, WA: Island Press.

Pickett, S. T. A., M. L. Cadenasso, and C. G. Jones. 2000. Generation of heterogeneity by organisms: Creation, maintenance and transformation. In *The ecological consequences of environmental heterogeneity,* ed M. J. Hutchings, E. A. John, and A. J. A. Stewart, 33–54. Oxford: Blackwell.

Porembski, S., and W. Barthlott. 2000. Granitic and gneissic outcrops (inselbergs) as centers of diversity for desiccation-tolerant vascular plants. *Plant Ecology* 151:19–28.

Pratt, D. J., and M. D. Gwynne. 1977. *Rangeland management and ecology in East Africa.* London: Hodder and Stoughton.

Prins, H. H. T., and P. E. Loth. 1988. Rainfall patterns as background to plant phenology in northern Tanzania. *Journal of Biogeography* 15:451–63.

Prins, H. H. T., and H. P. van der Jeugd. 1993. Herbivore population crashes and woodland structure in East Africa. *Journal of Ecology* 81:305–14.

Reid, R. S., M. Rainy, J. Ogutu, R. L. Kruska, M. McCartney, M. Nyabenge, K. Kimani, M. Kshatriya, J. Worden, and L. N'gan'ga. 2003. People, wildlife and livestock in the Mara ecosystem: The Mara Count 2002 Report. Nairobi: International Livestock Research Institute.

Roques, K. G., T. G. O'Conner, and A. R. Watkinson. 2001. Dynamics of shrub encroachment in an African savanna: Relative influences of fire, herbivory, rainfall, and density dependence. *Journal of Applied Ecology* 38:268–80.

Ruess, R. W., and F. L. Halter. 1990. The impact of large herbivores on the Seronera woodlands, Serengeti National Park, Tanzania. *African Journal of Ecology* 28:259–75.

Ruess, R. W., and S. W. Seagle. 1994. Landscape patterns in soil microbial processes in the Serengeti National Park, Tanzania. *Ecology* 75: 892–904.

Salvatori, V., F. Egunyu, A. K. Skidmore, J. de Leeuw, and H. A. M. van Gils. 2001. The effects of fire and grazing pressure on vegetation cover and small mammal populations in the Masai Mara national reserve. *African Journal of Ecology* 39:200–204.

Scholes, R. J. 1990. The influence of soil fertility on the ecology of southern African dry savannas. *Journal of Biogeography* 17:415–19.

Serneels, S., and E. F. Lambin. 2001. Impact of land-use changes on the wildebeest migration in the northern part of the Serengeti-Mara ecosystem. *Journal of Biogeography* 28: 391–407.

Serneels S., M. Y. Said, and E. F. Lambin. 2001. Land-cover changes around a major East African wildlife reserve: the Mara Ecosystem (Kenya). *International Journal of Remote Sensing* 22:3397–3420.

Shmida A., and M. V. Wilson. 1985. Biological determinants of species diversity. *Journal of Biogeography* 12:1–20.

Sinclair, A. R. E. 1979. The Serengeti environment. In *Serengeti: Dynamics of an ecosystem,* ed. A. R. E. Sinclair and M. Norton-Griffiths, 31–45. Chicago: University of Chicago Press.

———. 1995. Serengeti past and present. In *Serengeti II: Dynamics, management and conservation of an ecosystem*, ed. A. R. E. Sinclair and P. Arcese, 3–30. Chicago: University of Chicago Press.

Skarpe, C., R. Bergstroem, A. L. Braeten, and K. Danell. 2000. Browsing in a heterogeneous savanna. *Ecography* 23:632–40.

Smith, F. R., and R. I. Yeaton. 1998. Disturbance by the mound-building termite, *Trinervitermes trinervoides,* and vegetation patch dynamics in a semi-arid, southern African grassland. *Plant Ecology* 137:41–53.

Stocks, B. J., J. A. Mason, F. Weirich, A. L. F. Potgieter, B. W. Van Wilgen, W. S. W. Trollope, and D. J. McRae. 1996. Fuels and fire behavior dynamics on large-scale savanna fires in Kruger National Park, South Africa. *Journal of Geophysical Research* 101 D19: 23541–50.

Sugihara, G., and R. M. May. 1990. Applications of fractals in ecology. *Trends in Ecology & Evolution* 5:79–86.

Swift, D. M., M. B. Coughenour, and M. Atsedu. 1996. Arid and semiarid ecosystems. In *East African ecosystems and their conservation,* ed. T. R. McClanahan and T. P. Young, 243–72. New York: Oxford University Press.

Thirgood, S., A. Mosser, S. Tham, G. Hopcraft, E. Mwangomo, T. Mlengeya, M. Kilewo, J. Fryxell, A. R. E. Sinclair, and M. Borner. 2004. Can parks protect migratory ungulates? The case of the Serengeti wildebeest. *Animal Conservation* 7:113–20.

Thompson, M. 2002. Livestock, cultivation and tourism: Livelihood choices and conservation in Masai Mara buffer zones. PhD diss., University of London.

Trager, M., and S. Mistry. 2003. Avian community composition of kopjes in a heterogeneous landscape. *Oecologia* 135:458–68.

Trollope, L. A. 1974. Role of fire in preventing bush encroachment in the eastern Cape. *Proceedings of the Grassland Society of Southern Africa* 9:67–72.

———. 1982. Ecological effects of fire in South African savannas. In *Ecology of Tropical Savannas,* ed. B. J. Huntley and B. H. Walker, 292–306. New York: Springer Verlag.

Trollope, W. S. W., L. A. Trollope, and D. C. Hartnett. 2002. Fire behaviour a key factor in the fire ecology of African grasslands and savannas. In *Forest Fire Research and Wildland Fire Safety,* ed. D. X. Viegas, 1–16. Proceedings of IV International Conference on Forest Fire Research. Rotterdam: Millpress.

Turner, M. G., and R. H. Gardner. 1991. *Quantitative methods in landscape ecology.* New York: Springer Verlag.

van de Vijver, C. A. D. M. 1999. Fire and life in Tarangire: Effects of burning and herbivory on an east African savanna. PhD diss., Wageningen University, The Netherlands.

van de Vijver, C. A. D. M., P. Poot, and H. H. T. Prins. 1999. Causes of increased nutrient concentrations in post-fire regrowth in an East African savanna. *Plant and Soil* 214: 173–85.

van Wilgen, B. W., W. S. W. Trollope, H. C. Biggs, A. L. F. Potgieter, and B. H. Brockett. 2003. Fire as a driver of ecosystem variability. In *The Kruger experience: Ecology and management of savanna heterogeneity,* ed. J. T. du Toit, K. H. Rogers, and H. C. Biggs, 149–70. Seattle, WA: Island Press.

Veenendaal, E. M., W. H. O. Ernst, and G. S. Modise. 1996a. Effect of seasonal rainfall pattern on seedling emergence and establishment of grasses in a savanna in south-eastern Botswana. *Journal of Arid Environments* 32:305–17.

———. 1996b. Reproductive effort and phenology of seed production of savanna grasses with different growth from and life history. *Vegetatio* 123:91–100.

Velland, M. 2001. Do commonly used indicies of β-diversity measure species turnover? *Journal of Vegetation Science* 12:545–52.

Venter, F. J., R. J. Scholes, and H. C. Eckhardt. 2003. The abiotic template and its associated vegetation pattern. In *The Kruger experience: Ecology and management of savanna heterogeneity,* ed. J. T. du Toit, K. H. Rogers, and H. C. Biggs, 83–129. Seattle, WA: Island Press.

Vesey-Fitzgerald, D. 1973. *East African grasslands.* Nairobi: East Africa Publishing House.

———. 1974. Utilisation of the grazing resources by buffaloes in the Arusha National Park, Tanzania. *East African Wildlife Journal* 12:107–34.

Walpole, M. J. 2002. Factors affecting black rhino monitoring in Masai Mara National Reserve, Kenya. *African Journal of Ecology* 40:18–25.

Walter, H. 1971. *Ecology of tropical and subtropical vegetation.* New York: Van Nostrand Reinhold.

Webster, P. J., A. M. Moore, J. P. Loschnigg, and R. R. Leben. 1999. Coupled ocean-atmosphere dynamics in the Indian Ocean during 1997–98. *Nature* 401: 356–60.

Wedin, D. A., and D. Tilman. 1990. Species effects on nitrogen cycling: A test with perennial grasses. *Oecologia* 84:433–41.

Weins, J. A. 2000. Ecological heterogeneity: an ontogeny of concepts and approaches. In *The ecological consequences of environmental heterogeneity,* ed. M. J. Hutchings, E. A. John, and A. J. A. Stewart, 9–32. Oxford: Blackwell.

Weltzin, J. F., and M. B. Coughenour. 1990. Savanna tree influence on understory vegetation and soil nutrients in northwestern Kenya. *Journal of Vegetation Science* 1:325–34.

Wilsey, B. J. 1996. Variation in use of green flushes following burns among African ungulate species: The importance of body size. *African Journal of Ecology* 34:32–38.

Wolanski, E., and Gereta, E. 2001. Water quantity and quality as the factors driving the Serengeti ecosystem, Tanzania. *Hydrobiologia* 458:169–80.

Yair, A. 1990. The role of topography and surface cover upon soil formation along hillslopes in arid climates. *Geomorphology* 3:287–99.

SIX

Global Environmental Changes
and Their Impact on the Serengeti

Mark E. Ritchie

Human activity and natural environmental changes are expected to dramatically alter global climate and atmospheric chemical composition over the next 50 years. Such changes threaten major impacts on diverse, complex, and increasingly isolated natural ecosystems such as the Serengeti. The strongest global changes, such as elevated carbon dioxide (CO_2) and climate change, have the potential to alter, among other things, vegetation, hydrology, plant quality to herbivores, plant and animal species diversity, animal migration patterns, disease risk to wildlife and humans, and the success of different human agricultural and economic systems. A key element of climate change is uncertainty—both the uncertainty of the net effects of different global change factors, and the risks imposed if global changes lead to increased variability in climate. Risks apply to the productivity and diversity of the natural and human-dominated portions of the greater Serengeti ecosystem.

Predictions of the effects of global climate and atmospheric changes on ecosystems such as the Serengeti arise from two major sources. First, general-circulation models (GCMs; McCarthy et al. 2001) have been developed over the past 3 decades to assess differences in likely global changes in the atmosphere, especially temperature and precipitation, at relatively coarse scales ($1 \times 1°$ latitude and longitude). The advantage of GCMs is that they integrate large-scale data on atmospheric chemistry and climate in oceans and continents into their projections for local areas. Their disadvantage is their complexity: the large number of variables and multiple nonlinear relationships imbedded in GCMs creates large uncertainty in their projections. In addition, there is considerable uncertainty about whether the assumed functional relationships in GCMs apply locally for a region like the Serengeti (Coughenour and Chen 1997; McCarthy et al. 2001).

A second approach is to extrapolate current trends in global change factors into the future. The advantage of this method is that it is based on actual observations rather than simulations, and undeniably applies to the area where data are collected. However, extrapolating existing trends presumes that the past predicts the future, and without a really long time series, it becomes difficult to separate trends from natural fluctuations in climate. While neither method by itself is satisfactory, combining the two provide some confidence in projections if the trends revealed by both methods are similar.

Once the likely global environmental changes at a particular locality are determined, we can assess the consequences of these changes in several ways. Functional relationships between the global change factors and parameters of models of the dynamics of animal and plant populations, ecosystem processes, biodiversity, or human activities and welfare can be incorporated into these models. Examples of such models are in chapters 8 to 13 of this book. These models can explore different scenarios and can effectively include the effects of multiple factors that change simultaneously. This approach suffers from some of the same problems as modeling global changes themselves. However, these models can be tested with data on how the system has responded to past changes in global change factors, and to use these relationships to forecast future responses. For example, analysis of periodic animal censuses for how animal populations have responded to drought might indicate how animal populations would respond to long-term declines in rainfall and/or increased temperatures.

In this chapter, I use all of the methods mentioned to assess the likely global-level environmental changes that will affect the Serengeti and the consequences of these changes for the functioning of the Serengeti ecosystem and its member humans, habitats, flora, and fauna. In particular I focus on the major *global change factors* that are likely to impact the Serengeti: elevated atmospheric CO_2, climate changes in rainfall and temperature, and increased deposition of nitrogen (N). Other types of global changes, such as human population growth and land use change, are dealt with in other chapters in this volume. I consider both the projections of several different

GCMs and, where data are available, regional observed trends in climate over the past 50 years.

I also explicitly consider that climate factors, particularly rainfall, might change in their variability over time. Armed with these projections, I then speculate on the possible impacts to the Serengeti by using well-known functional relationships between global change factors and certain critical ecosystem processes in the Serengeti, such as photosynthesis, primary production, water-use efficiency, and plant chemical composition. I then speculate on the particular impacts global change factors might have on vegetation, soils, diseases, and animal densities and diversity, and the combined effects of these on human livelihood and activities. In particular, I evaluate the potential combined effects of different factors, such as carbon dioxide, temperature, and rainfall, since very rarely do these factors simply add in their effects.

Finally, I consider the risks imposed by changes in variability, particularly for human societies. These admittedly speculative exercises may raise some critical issues, identify potential management decisions, and better inform the management of the greater Serengeti ecosystem on how to adapt to global changes over the next decade.

EXPECTED ENVIRONMENTAL CHANGES IN THE SERENGETI

Carbon Dioxide

Atmospheric carbon dioxide (CO_2) is a critical global change factor because it is a greenhouse gas—that is, one that traps heat in the atmosphere and is a potentially limiting resource for plants. In the Serengeti, increases in CO_2 may increase productivity, plant species composition, and plant quality to the abundant and diverse array of herbivores. CO_2 has increased by 70% (from 220 ppm in 1910 to 375 ppm in 2000 at Mauna Loa, Hawaii, at 4,000 m elevation) in less than a century and a half, and by 18% from 1960 to 2000 (Keeling and Whorf 2002). Current rates of increase suggest that concentrations approaching 475–500 ppm may be achieved by 2050 at the 1,500 m-elevation characteristic of the Serengeti. These increases apply globally, even though CO_2 emissions are concentrated in industrialized countries, because CO_2 is readily mixed and transported in the atmosphere (Keeling et al. 1989). Given that global emissions of CO_2 are only now being regulated—and with relatively little success—the Serengeti therefore will almost certainly experience further dramatic increases in atmospheric CO_2 through the first half of the twenty-first century.

Rainfall

Rainfall is a critical climate variable, as it directly affects plant production and animal numbers and migrations in the Serengeti ecosystem (McNaughton 1985). Understanding possible rainfall change in East Africa and the Serengeti region requires a brief background on how climate functions in the region. Historically, the Serengeti experiences two annual wet seasons; from October to November and from March to April, as the intertropical convergence zone crosses the latitude of the Serengeti (2–3° south). This zone is the range of latitudes at which direct solar heating of the atmosphere drives a large convective pump, in which warm, moist air rises and condenses, pulling cooler air from higher latitudes, thus creating the trade winds and almost daily rainfall in the form of thunderstorms. In interior East Africa, these winds are amplified by additional convection between the hot land surface and the relatively cooler ocean, pulling moisture-laden air into the continent's interior. Orographic lift of moist air upward in elevation to the East African highlands also increases rainfall. The dry season occurs when the intertropical convergence zone shifts north of the equator during June–October and heating of the Indian subcontinent and southwest Asia creates the famous Indian monsoon, with winds blowing away from Africa. During this season, moisture off of Lake Victoria blows from west to east and rises and precipitates as rainfall over the highlands in the northwestern section of the Serengeti and the Mara region in Kenya, yielding fresh plant growth at a time when other areas are dry and plant production is minimal.

The majority of global climate models (GCMs) suggest that annual rainfall in East Africa should increase moderately, 25–50 mm (2–5%), by 2050 (Allali et al. 2001) with longer, wetter wet seasons. The primary driver of such predictions is the expected Indian Ocean warming of 2°C by 2050 (Webster et al. 1999). Warmer sea surface temperatures are expected to extend the wet season by intensifying the intertropical convergence and increasing the moisture of air reaching the interior of East Africa.

However, trends in rainfall since 1960, averaged across weather stations in the Serengeti National Park, do not agree with these predictions, despite demonstrated warming of the Indian Ocean (Clark, Cole, and Webster 2000). During the first half of the twentieth century, annual and wet-season rainfall in the Serengeti significantly ($P < 0.001$) increased (fig. 6.1a). However, since the 1960s, annual and wet-season (November–May) rainfall has declined significantly (fig. 6.1a,c), despite the fact that dry-season (June–October) rainfall increased slowly throughout the twentieth century (fig. 6.1b).

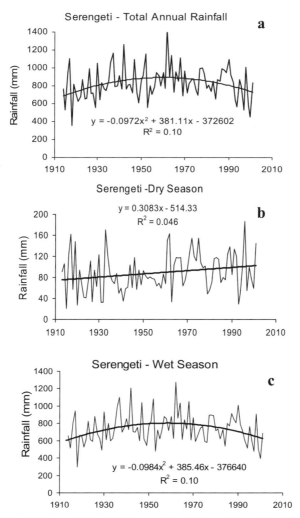

Fig. 6.1 Rainfall (mm) in Serengeti National Park, averaged over >20 reporting weather stations distributed throughout the park. All regressions are significant (*P* <0.01). A = Annual (Jan.–Dec.), B = Dry season (June–Oct.), C = Wet season (Nov.–May).

What could explain the discrepancy between the GCM predictions and recent long-term trends for the Serengeti? One possibility is that the GCMs overlook the effect of sea surface temperature on winds during the wet season. In the absence of cool water in the eastern Indian Ocean, monsoon winds are driven by the temperature gradient between land and sea (Saji and Yamagata 2003). If sea temperatures along the east African coast are warmer and thus closer to land temperatures, convection and winds will

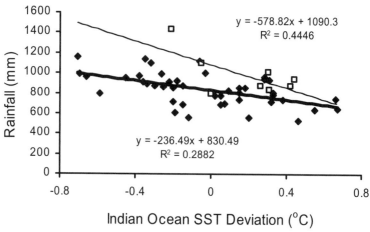

Fig. 6.2 Annual rainfall in Serengeti National Park (1946–1997) regressed negatively with average sea surface temperature (SST) in the Indian Ocean, expressed as the deviation from the average over 1946–1997 (Clark, Cole, and Webster 2000). Open squares are El Nino years, solid diamonds are non-El Nino years.

be weaker. Moister air, which results from warmer sea surface temperatures, will therefore penetrate less into the interior, resulting in increased rainfall along the coast but less rainfall inland. There is some evidence for this, as average annual rainfall in the Serengeti during the 50-year period 1946–1997 is negatively correlated with average Indian Ocean sea surface temperature (fig. 6.2). However, warmer ocean temperatures may intensify the intertropical convergence and extend its influence over a broader latitude range, which might explain why dry-season rainfall has steadily increased over the past 90 years. Warmer temperatures in Lake Victoria may also increase the June–October rainfall that arises from the orographic lift of moisture-laden southwesterly winds off the lake, over the Crater Highlands and Rift Valley in northern Tanzania and southwestern Kenya. This possibility is supported by an observed increase in annual rainfall at Musoma, on the shore of Lake Victoria, over the past 40 years (chapter 2, this volume).

Combining these analyses and model outcomes suggests that future wet-season rainfall will continue to decline as long as Indian Ocean sea surface temperatures continue to rise. This may be particularly likely for portions of the Serengeti ecosystem that lie in the rain shadow of the Rift Valley ranges, since these areas may be particularly sensitive to the strength of onshore winds driving moisture deep into the continent's interior.

Rainfall Variability

In addition to changes in mean annual rainfall, global climate change may result in changes in the variability of rainfall. Anomalous warming in the western and cooling in the eastern Indian Ocean, called the Indian Ocean Dipole, often but not always accompanies El Nino events (fig. 6.3), as it did with the 1997–1998 El Nino, and creates the most favorable conditions for rainfall in the Serengeti—very warm offshore waters in the Indian Ocean, combined with moisture from the intertropical convergence, plus a strong onshore wind generated by the gradient between warm water in the western Indian Ocean and cool water in the east. Evidence suggests that these anomalous climatic events have increased in frequency during the past 4 decades, and that their intensity varies with multiple periods (Saji and Yamagata 2003). Other major climate drivers, such as circulation patterns in the North Atlantic, which influences jet-stream position in southwest Asia and the Indian subcontinent, can also influence East African climate by altering the strength of monsoon winds. Each of these interdependent climate drivers varies on different time scales, from 3- to 5-year cycles (El-Nino Southern Oscillation, or ENSO) to decades (North Atlantic oscillation).

GCMs vary in their ability to predict variation in these drivers, and they seem unlikely to be able to accurately predict annual variability in rainfall. Perhaps the best option for evaluating future variability is to analyze long-

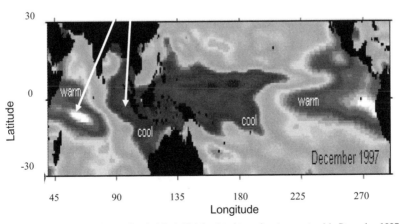

Indian Ocean Dipole, December 1997

Fig. 6.3 False-color map of sea surface height (which is related to surface temperature) in December 1997 reveals the sea surface temperature pattern of El Niño in the Pacific Ocean and a similar imbalance in the Indian Ocean, known as the Indian Ocean Dipole. Image credit: Tony Phillips. Source: NASA; http://science .nasa.gov/headlines/y2002/17apr_rvf.htm.

term trends in available data. We employed wavelet analysis (Torrence and Compo 1998; see box 5.1) to discern patterns of variability in annual, wet-season, and dry-season rainfall averaged across weather stations in the Serengeti from 1914 through 2001. We analyzed the Serengeti rainfall data for statistically significant patterns of variability in rainfall relative to that expected from a random walk. This analysis searches for correspondence between observed data and different modeled patterns of cycles, called wavelets, that have different periods and amplitudes, and evaluates this correspondence at different intervals along the time series of rainfall (fig. 6.4). We then compared the patterns of variability in rainfall in the Serengeti with those of Pacific sea surface temperatures (representing ENSO), and Indian rainfall (representing the intensity of the Indian Ocean monsoon). Areas of high correspondence, called *power,* between observed data and the wavelet (gray and light regions encircled by black line of fig. 6.4, $P < 0.05$) implies cycling with a particular cycle period in years (vertical axis) over a particular interval in the rainfall record (horizontal axis).

The results suggest that anomalous events, such as El Nino and the Indian Ocean Dipole, are major drivers of variability in Serengeti rainfall. Variation in wet-season rainfall was strongly associated with El Nino (3–5-year cycles surrounding known El Nino years) in the first half of the twentieth century, but not during the latter half of the century, except for 1997–1998. Elevated dry-season rainfall also was associated with El Nino events, but in decades 30 years apart (1930s, 1960s, and 1990s). Moreover, the period of the fluctuation, or length of time between peaks, in dry-season rainfall has increased during the past century. El Nino events were associated with both dry- or wet-season rainfall variability, but not both in the same decade. Third, the lack of El Nino signals in the wet-season rainfall variability in the last half of the twentieth century is consistent with the hypothesis stated earlier: that rainfall in the Serengeti is increasingly controlled by Indian Ocean warming and weakening of the monsoon wind during the months of intertropical convergence. Wavelet analysis of Indian Ocean sea surface temperatures (fig. 6.5) reveals that average sea surface temperatures of the Indian Ocean cycled with a period of 4 to 6 years prior to 1970, but varied in a manner not significantly different from a random walk since then, similar to patterns of the Indian monsoon (Torrence and Campo 1998). The increase in sea surface temperature seems even to be generally weakening the effect of El Nino or the Indian Ocean Dipole, despite the 1997–1998 El Nino event, as rainfall during years with these anomalous events declines significantly with Indian Ocean surface temperature (fig. 6.3) and through time (fig. 6.2).

These analyses suggest that variability in rainfall, particularly in the wet season, may decrease as Indian Ocean sea surface temperatures continue to

a

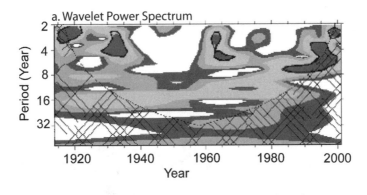

a. Wavelet Power Spectrum

b

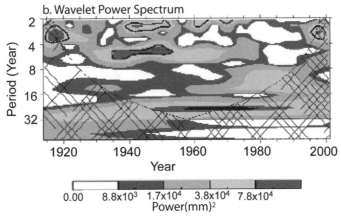

b. Wavelet Power Spectrum

0.00   8.8x10³   1.7x10⁴   3.8x10⁴   7.8x10⁴
Power(mm)²

Fig. 6.4 Graphical representation of wavelet analysis (Torrence and Compo 1998) of (A) dry- and (B) wet-season rainfall patterns in Serengeti National Park. Graphs present the power (a measure of variance explained) of fit of the time series of rainfall (see fig. 6.1) to a wavelet function (a combination of periodic functions, similar to Fourier functions, that can emulate cycles embedded inside larger cycles; see box 6.1). Darker interior areas, outlined in black, indicate regions of the time series whose dynamics are significantly explained (relative to a random walk time series) by a function with period (in years) equal to the corresponding value on the y-axis. Note the increasingly longer period in the variability in dry-season rainfall through the century and the lack of periodicity in variation in wet season rainfall over the last half of the century. Hatching indicates portions of the time series of rainfall that cannot be evaluated for pattern.

increase. There may be less difference between dry vs. wet season rainfall. The impacts of anomalous climatic events, such as El Nino and/or the Indian Ocean Dipole, on the Serengeti may lessen. As I discuss below, such changes in the patterns of variability have major implications for the fate of humans and biodiversity in the Serengeti ecosystem.

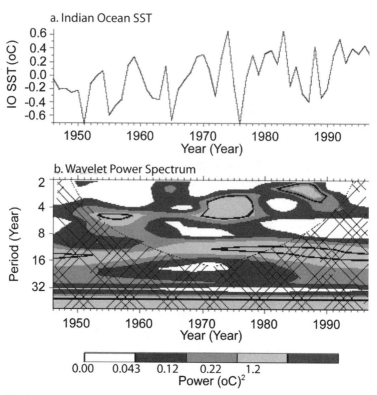

Fig. 6.5 (A) Trend in Sea Surface Temperature (SST) of the western Indian Ocean, 1945–2000, expressed as a deviation from the average SST over that period. (B) Wavelet analysis of the fluctuations in Indian Ocean SST (Torrence and Compo 1998). Hatching indicates portions of the time series of temperatures that cannot be evaluated for pattern.

Temperature

Average global temperature is expected to increase by 2–5°C by 2050 as a consequence of anthropogenic pollution of the atmosphere with greenhouse gases such as CO_2, methane, or nitrous oxide, or of natural changes in solar radiation striking the earth. However, these changes will not be distributed evenly around the globe. GCMs are very uncertain in their predictions of temperature change for interior East Africa (fig. 6.6a). The average outcomes of different models suggest that the East African highland region will experience little temperature change.

Long-term temperature trends (1976–2000) from nearby Amboseli National Park in Kenya suggest that temperatures are increasing (Altmann et al. 2002, fig. 6.6b). Daily maximum temperatures increased 3–5°C and

Box 6.1

Wavelet analysis is a method of finding patterns in long-term data, such as time series of annual rainfall. It is highly useful when patterns of variability change during the time series or when multiple factors cause variability at different timescales (Torrence and Compo 1998). For example, variability driven by events with a 2 to 5 year cycle, such as El Nino Southern Oscillation, may co-occur with events that vary on a decadal cycle, such as fluctuations in ocean currents. A computer algorithm attempts to fit a *wavelet,* or complex trigonometric function called a Fourier transform that describes fluctuations with different periods (time between peaks) in the data over a sequence, or n time intervals, to all possible sets of n intervals within the longer time series of k intervals ($n < k$). Different forms of wavelets, each with different sensitivities to detect multiple patterns of fluctuations within a time series, can be chosen to fit the data. The calculation of power, or correspondence between data and the wavelet, is made for different timescales, or sequences, in the data and for wavelets with different or multiple periods of oscillation. The results are typically displayed as a contour map of power across the length of the time series and for different periods up to $n = k/2$.

For the Serengeti rainfall data, I performed wavelet analysis interactively at the web site http://paos.colorado.edu/research/wavelets (Torrence and Compo 1998) using the Morlet wavelet, which is most sensitive in detecting multiple patterns of fluctuations. This wavelet is a normalized (each time step is expressed as a fraction of the total number of time steps in the dataset) function $\Psi_0(\eta)$:

$$\Psi_0(\eta) = \pi^{-1/4} e^{i\eta\omega_0} e^{-\eta^2/2}$$

where i is the square root of -1, an imaginary number, where e^{ai} (a is a function) has a known Fourier transform function (Boyce and DiPrima 1977). The variable η is the period of oscillations in the Fourier transform, and ω_0 is a constant fit to the data. Any patterns detected were compared with the patterns expected if rainfall followed a correlated random walk, or "red" noise, which I felt was the most appropriate null model for rainfall. The analysis produced a two-dimensional (cycle period versus year) pictorial representation of the power or correspondence of fluctuations in Serengeti rainfall to the Fourier transform of the wavelet function. This graph reveals sequences in the time series that exhibit significantly periodic fluctuations in rainfall and the period (in years) of these fluctuations.

a

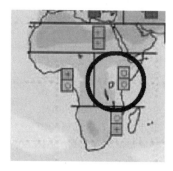

b

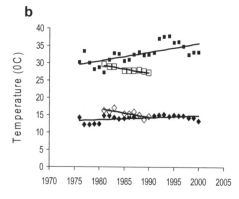

>> warming
≻Warming
≻< warming
Uncertain warming
Cooling

Fig. 6.6 (A) Temperature changes for Africa expected from several GCM's, where average global warming is 2–3 °C by 2050. Circled area is the region of East Africa, with uncertain changes in temperature predicted. (B) Trends in local mean annual maximum and minimum temperatures for Amboseli National Park, southern Kenya (Altmann et al. 2002)): maximum, filled squares; minimum, filled diamonds. Trends in Serengeti National Park, Tanzania (300 km from Amboseli) from 1982–1991: maximum, open squares; minimum, open diamonds. Trends in all four datasets are significant ($P<0.001$): Amboseli: $T°_{max} = 0.25$Year -473.5, $R^2 = 0.45$; $T°_{min} = 0.0615$Year -108.03, $R^2 = 0.25$; Serengeti: $T°_{max} = -0.23$Year $+494.4$, $R^2 = 0.83$; Serengeti Min: $T°_{min} = -0.24$Year $+495$, $R^2 = 0.63$.

daily minimum temperatures increased 1.5°C over this period. However, temperature records from the Serengeti, which are available only for the period 1981–1990, decreased (fig. 6.6b) during a period in which Amboseli temperatures steadily increased. The contrast in these two temperature trends is consistent with the GCM uncertainty.

A closer inspection, however, suggests that local factors may influence these trends. In contrast, both the Serengeti (fig. 6.1) and Amboseli, Kenya (Altmann et al. 2002) experienced increasing rainfall during 1981–1990, associated with a decade of declining Indian Ocean sea surface temperatures (fig. 6.5). Increased rainfall implies increased cloud cover during this period, which may have led to the cooler temperatures observed in the Serengeti. In Amboseli, however, increased albedo from the widespread decline in woody vegetation during the period (Koch et al. 1995) may explain why temperature, and daily maximum temperatures in particular, increased so dramatically, despite increasing rainfall and cooler ocean temperatures. Interestingly, woody plant cover increased during the decade in the Serengeti (Dublin 1995; Sinclair 1995).

The contrast in temperature trends for the Serengeti and Amboseli suggest the following. If rainfall, and thus cloud cover, continues to decrease as

Indian Ocean sea surface temperatures warm, then air temperatures in the Serengeti should *increase* in the long term. Nevertheless, local modification of habitat, particularly increases or decreases in woody plant cover, may modulate or even outweigh regional temperature changes. In the Serengeti, further increases in woody plant cover, if they occur, might reduce temperatures and balance the effects of expected Indian Ocean-driven regional climate changes.

CONSEQUENCES OF ENVIRONMENTAL CHANGE IN THE SERENGETI

The major global changes the Serengeti is likely to experience are decreased average wet-season rainfall, increased variability in rainfall with longer stretches of consecutive wet and dry years, and elevated CO_2. These changes have a number of likely consequences for primary production, the dynamics of key food web interactions (see chapters 7–9), biodiversity, and human welfare (chapters 10, 13, and 15) in the Serengeti. Here we explore these consequences for the dynamics of both the natural and human-dominated aspects of the Serengeti.

Plant Responses to Global Change Factors: Carbon Dioxide

A large literature on the effects of elevated CO_2 on plant physiology suggests that elevated CO_2 will increase rates of photosynthesis—but not dramatically—as most plants make internal physiological adjustments to down-regulate photosynthesis (Bazzaz, Miao, and Wayne 2003). However, these moderate rates of increase in photosynthesis can result in increases in primary productivity of up to 33% in grassland herbaceous plants, with a greater fraction of production occurring belowground (Coughenour and Chen 1997; Reich et al. 2001).

Elevated CO_2 will likely affect different plant species differently. For example, plants with the C_3 photosynthetic pathway (C_3 plants), such as woody plants and forbs, likely down-regulate photosynthesis under elevated CO_2 much less than those with the C_4 pathway (C_4 plants), such as grasses. C_3 plants thus may show much stronger increases in productivity and greater competitive ability under elevated CO_2. C_4 plants, in contrast, with their more limited gas exchange capacity and enzyme-limited carbon fixation pathway, are thought to respond less to elevated CO_2. If so, elevated CO_2 may favor the establishment of woody plants and may lead ultimately to dominance of woody plants and forbs in areas with sufficient rainfall. However, increased herbaceous production may provide more fuel for fires or support greater mega-herbivore (elephants, giraffes) densities, imposing greater mortality on woody plants. These nonlinear, opposing feedbacks of elevated CO_2 make it very difficult to predict the effects of CO_2 on the transition between woody plants and grasses in the Serengeti savannas (House et al. 2003)

The down-regulation of photosynthesis often results in decreased enzyme concentration in leaves and greater closure of stomata (openings in leaves that allow gas exchange with the atmosphere) in leaves to reduce evapotranspiration of water and increase efficiency in water use (Bazzaz, Miao, and Wayne 2003). The increased soil nutrient limitation that results from greater biomass production often reduces plant tissue nutrient concentration, with strong negative effects on herbivores and decomposers and cascading effects to other animals throughout the food web. Greater water-use efficiency would allow plants to be more tolerant of drought conditions. Consequently, elevated CO_2 alone may change the chemical composition of vegetation to make it more drought tolerant but lower in nutrient (N, Na, P, or Ca) concentration to herbivores and detritivores. One consequence of this is that consumers like large mammals and detritivores like termites may consume less of primary production as green leaves or litter. A greater standing crop of litter in the dry season may provide greater fuel for fires, thereby increasing fire frequency and intensity and balancing or outweighing any productivity advantage of woody plants.

Rainfall and Temperature

Multiple studies from the Serengeti and East Africa demonstrate increased productivity with increased rainfall (see McNaughton 1985; Fritz and Duncan 1994; Mills, Biggs, and Whyte 1995). Consequently, the projection of decreased rainfall, based on trends over 1960–2000 (fig. 6.1) suggests that productivity may decline. This may be particularly true since rainfall declines have occurred mainly in the wet season. A less-appreciated change that might result from decreased rainfall is an increase in plant nutrient concentration (Breman and DeWit 1983; Olff, Ritchie, and Prins 2002; Wright, Reich, and Westoby 2001) as more water-use efficient species replace less efficient ones. The mechanisms to explain why plants from drier sites have higher nutrient concentrations are not yet clear, but recent theoretical papers (Olff, Ritchie, and Prins 2002; Wright, Reich, and Westoby 2003) suggest that increasing nutrient concentrations correspond to increased enzyme concentrations that compensate for reduced CO_2 uptake when stomata are closed to conserve water. Whatever the mechanism, patterns of nutrient concentration of vegetation across rainfall gradients in Africa (Breman and DeWit 1983; Olff, Ritchie, and Prins 2002), Australia (Wright, Reich, and Westoby 2001), and North America (M.E. Ritchie, unpublished) all support this hypothesis. In the Serengeti, the short-grass plains, which are dominated by nutrient-rich, water-use efficient grass species growing on fertile volcanic soil, may expand westward and northward in the park as rainfall declines and wherever soils and topography allow. Likewise, hotter, drier conditions do not favor the establishment of woody plants, and such conditions might lead the distribution of woodlands in the park to shrink.

Variability in rainfall, if it continues to decline as indicated by wavelet analysis (fig. 6.5), would also tend to reduce establishment of woody plants. In many savannas around the world, woody plants often establish only when rainfall is extremely high, producing cohorts of saplings that grow fast enough to survive eventual fires and herbivory and grow to maturity. If such extreme events become increasingly infrequent, then fewer woody plant cohorts may establish over the next few decades, eventually leading to reduction of woody plants throughout the Serengeti. Extreme rainfall events may also be instrumental in maintaining plant diversity in the Serengeti, as species with high water requirements can recruit successfully during these extreme events and maintain their membership in the plant community despite one or more years of unacceptably dry conditions (Chesson and Huntly 1997). A decline in the frequency of extreme rainfall events may therefore cause plant diversity on the Serengeti to decline.

Combined Global Change Effects on Plants

In many respects, elevated CO_2 and temperature and decreased rainfall cause opposite effects on plant growth and ecosystem primary productivity. Elevated CO_2 is expected to enhance productivity and reduce plant nutrient content while lower rainfall should reduce productivity and increase plant nutrient content. An important question then is, what are the combined effects of changes in these factors? Do they simply add in their effects or do they interact in complex ways. Are the various ecosystem responses more sensitive to certain global changes than others, given their expected magnitude?

Coughenour and Chen (1997) explored these complex effects for tropical grassland in Kenya, similar to short-midgrass areas in the Serengeti. They found that on average, over 35-year simulations, changes expected from recent trends in temperature (increase 1–2°C), rainfall (decrease by

Table 6.1 Separate and combined effects of some expected global environmental changes in the Serengeti, a 1°C increase in mean temperature, a 20% decrease in rainfall, and a doubling of CO_2 from 350 to 700 ppm on % change in net primary productivity (NPP), water-use efficiency (WUE), and soil nitrogen mineralization (N_{min}), simulated for a Kenya grassland similar to areas in the Serengeti.

Variable	Change Factor			
	Temperature (+1°C)	Rainfall (−20%)	CO_2(+100%)	Combined
NPP	−5.83	−37.20	31.81	−18.38
WUE	−6.66	−10.47	48.57	29.52
N_{min}	−1.75	−46.20	11.69	−27.40

Note: Data taken from Coughenour and Chen, 1997.

Table 6.2 Sensitivity of net primary productivity (NPP), water-use efficiency (WUE), and soil net nitrogen mineralization (N_{min}) to expected changes (values in parentheses) in three global change factors, expressed as the absolute values of % change in response / % change in factor

Variable	Change Factor		
	Temperature (+1°C)	Rainfall (−20%)	CO_2(+100%)
NPP	1.63	1.86	0.32
WUE	1.86	0.52	0.49
N_{min}	0.49	2.31	0.12

Note: Data taken from Coughenour and Chen, 1996.

20%) should decrease productivity by up to 40%, while a doubling of atmospheric CO_2 to 700 ppm increased productivity by more than 30%. Collectively, the magnitude of expected change in CO_2, rainfall, and temperature for the Serengeti should result in nearly 20% less primary productivity.

The expected global changes also had equally dramatic but different simulated effects on plant water-use efficiency (WUE; table 6.1). WUE was most strongly positively affected by elevated CO_2 and only weakly negatively affected by increased temperature and decreased rainfall. In response to the combined expected environmental changes, WUE may increase by nearly 30%. This represents a large compensatory response to environmental changes and in part mitigates the expected decline in productivity.

Perhaps the most interesting outcome of Coughenour and Chen's

(1997) simulations was that plant responses were by far more sensitive to changes in rainfall and temperature than CO_2 (table 6.2). Plant responses to expected temperature were relatively small just because the expected change in temperature is small, only 1–2°C. In contrast, CO_2 levels may more than double over the next 50 years, but plant responses were much less than proportionate to the change in CO_2. Perhaps not too surprisingly, it is future changes in rainfall that are likely to have the greatest impact on plant production and species composition over the next half-century (insert fig. 6.7).

Another important question is how combined global environmental changes would affect the relative dominance of grasses versus trees. If plant production is most sensitive to rainfall, continued declines in rainfall would be expected to favor grasses over trees because trees might have difficulty establishing under dry conditions. However, this prediction is countered by an expected associated higher consumption of vegetation by

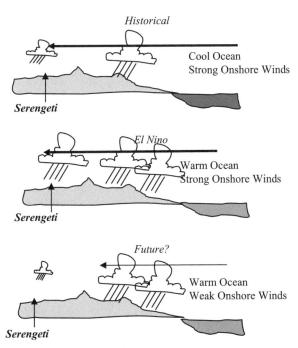

Fig. 6.7 Hypothetical mechanisms contributing to long-term future changes in rainfall in the Serengeti region. These can produce both an average decline in rainfall and an increase in annual variability. Compared to historical regional sea surface temperatures and winds, El Nino events bring warm seas and strong onshore winds, resulting in dramatically higher rainfall. However, warmer seas in the future, without the strong ocean temperature gradient, may produce weaker winds, which fail to transport moist air to interior regions like the Serengeti.

herbivores, reduction in competition and litter (see food web responses, in the following), and possible reduction in fire frequency and/or intensity. These changes would tend to reduce tree mortality and allow a higher frequency of saplings that did establish to reach maturity. Extremely high rainfall years, associated with Indian Ocean Dipole and/or El Nino events, will still occur, and woody plants may still establish in cohorts, especially if competition from herbaceous vegetation is reduced by heavier consumption by herbivores. Reviews by Scholes and Archer (1997) and House et al. (2003) suggest that the complex feedbacks between fire, water retention, herbivory, and other factors that govern the dynamics of grass and trees in tropical savannas make it difficult to confidently predict the impact of combined global environmental changes on grass versus woody dominance of vegetation in the Serengeti.

FOOD WEB RESPONSES TO GLOBAL CHANGE FACTORS

Elevated CO_2 and temperature are likely to have major effects on food webs in the Greater Serengeti, including those involving humans and their livestock and crops. A decrease in productivity accompanied by an increase in plant nutrient content is likely to benefit smaller grazers and cause declines in populations of larger herbivores, since smaller herbivores can tolerate lower abundance but more nutritious forage. Such changes may also cause shifts in diets from grazing to browsing or shifts in use of the landscape to track potential shifts in the distribution of woody plants. Elephants might become even more of a keystone species because they may become more dependent on woody vegetation for food if grass productivity is lower. Overall, herbivores may consume a greater fraction of available productivity than they do now. The net effects of change will ultimately depend on the magnitude of global changes and the sensitivity of plants (tables 6.1, 6.2), but Coughenour and Chen's simulations suggest that a decline in rainfall is likely to have the greatest effect on plant responses, and by inference, plant nutrient concentration.

Shifts in plant productivity, species composition, and nutrient concentration are likely to change the diversity of large mammalian herbivores in addition to their abundance and fraction of production consumed. Olff, Ritchie, and Prins (2002) show that globally, herbivore diversity is highest wherever rainfall is balanced by evapotranspiration (intermediate precipitation) and soil fertility is highest. The Serengeti currently produces one of the highest herbivore biomass and diversities in the world, given its rich soils, large areas with intermediate water availability, and the rich pool of herbivore species on the African continent available to occupy it. If production declines, as predicted by Coughenour and Chen's (1997) simulation (table 6.1), water availability may shift to a level that supports less herbivore diversity. Such a shift in species diversity would tend to favor smaller herbivores species and reduce numbers and diversity of larger herbivore species > 200 kg like buffalo, eland or elephants (Olff, Ritchie, and Prins 2002), since smaller herbivores can tolerate less abundant but more nutritious forage.

A review of herbivore abundance data from the Ngorongoro Crater just southeast of the Serengeti plains (Runyoro et al.1995) suggests that, in fact, rainfall may be a key driver of both herbivore biomass and diversity. For the period 1970–1992, after recovery from rinderpest, total herbivore biomass was surprisingly negatively associated with higher total annual rainfall (fig. 6.8a). This association was driven by declines in biomass of intermediate-sized wildebeest and zebra with higher rainfall, which, although resident in the Crater, are migratory in the Serengeti (fig. 6.8b). In addition, kongoni (hartebeest) and grant's gazelles also declined with increasing rainfall. Despite these relationships, these declines were not associated with decreased diversity at higher rainfall, as measured by the Shannon-Wiener index, H' ($R^2 = 0.03$, $P = 0.45$). One reason for this lack of diversity response is that numbers of large herbivores (> 500 kg), such as eland and buffalo, either did not respond to rainfall, as in the case of buffalo, or increased with rainfall, as in the case of eland. Elephant numbers were strongly associated with higher rainfall, lagged by a year (fig. 6.8c). These associations between rainfall and animal biomass, although not causal relationships, support the hypothesis that decreased rainfall leads to higher-quality plants and higher densities of smaller and intermediate-sized herbivores. Animal numbers in the Ngorongoro Crater could have changed for other reasons, such as migration, predation, or human hunting. As a preliminary step, these numbers suggest that different-sized herbivores may respond differently to current trends in rainfall, should they continue. In this way, the diverse ungulate community of the Serengeti ecosystem may be buffered against climate change.

These speculated shifts in herbivore abundance and species richness would also undoubtedly affect Serengeti predators. Reduced overall herbivore biomass, presumably caused by reduced rainfall, may have strong negative effects on carnivores. However, Sinclair, Mduma, and Brashares (2003) recently showed that herbivores < 150 kg are primarily limited by predators, whereas those > 150 kg in size are primarily limited by food. If plant production and N mineralization both decline while plant nutrient content increases, the biomass of smaller herbivores may increase relative

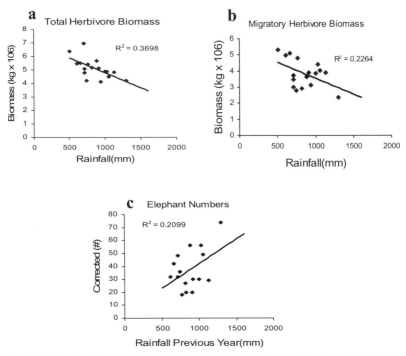

Fig. 6.8 Relationships between wet-season herbivore biomass, numbers, and diversity relative to rainfall in the Ngorongoro Crater 1970–1992. (A) Total herbivore live biomass; (B) Biomass of species that show large migrations in the Greater Serengeti Ecosystem: wildebeest, Grant's gazelle, and Thomson's gazelle; (C) Elephant numbers relative to rainfall in the previous year.

to that of larger herbivores (Prins and Olff 1998; Olf, Ritchie, and Prins 2002), providing more prey for carnivores, especially intermediate-sized predators like wild dogs and cheetahs. Recent trends in predator numbers in the Serengeti do not support this prediction, as wild dogs and cheetahs have declined while lions have increased in number (chapters 1, 2, and 14). However, these trends may be driven more by diseases, such as distemper in wild dogs, than by shifts in prey abundance (chapter 7).

A largely unappreciated effect of global change may be on decomposers and the belowground food web, dominated in the Serengeti by termites. If plant production is reduced and WUE increases, as suggested by Coughenour and Chen's (1997) simulations, then the ecosystem might shift toward one in which a greater fraction of production is consumed by herbivores or detritivores, leaving little standing biomass to fuel fires. If so, faster decomposition may decrease the likelihood of fire, which would lower mortality rates of woody plants. In contrast, if elevated CO_2 and increased temperatures outweigh changes in rainfall on plant quality, the opposite effect might result, leading to a detrital-based ecosystem with unconsumed low nutrient content production as standing biomass and a fire fuel (Hibbard et al. 2003). Declines in litter production, despite increased litter quality, may lead to lower nutrient mineralization, particularly of nitrogen, much of which is volatilized during fire (Ojima et al. 1991). This should result in reduced organic matter inputs into the soil and consequently less microbial activity and lower net nitrogen (N) mineralization rates (table 6.1). This effect would be balanced, however, by the likely greater N mineralization rates, resulting a greater consumption of production by herbivores and their indirect return of mineral N in dung and feces (McNaughton, Banyikwa, and McNaughton 1997; Augustine 2003).

LANDSCAPE RESPONSES TO GLOBAL CHANGE FACTORS

The hypothesized changes in plant production and species composition could result in potentially important shifts of vegetation across the landscape. The Coughenour and Chen (1997) simulations suggest that more water-use-efficient herbaceous plant species may replace less-efficient species in many areas. What is less clear is how woody plant cover would change across the landscape (see the previous section on combined global change effects on plants). This is a major unknown that warrants intensive study.

An overall decline in rainfall, despite any effects from elevated CO_2 and temperature, may alter the seasonal routes for the migratory herbivores, wildebeest, zebra, and Thomson's gazelle, which account for > 60% of herbivore biomass in the Serengeti. It's possible that their migrations might shift outside protected areas and make these species more vulnerable to illegal hunting. Shifts in herbivore prey distributions may also shift those of predators. Shifts in herbivore and carnivore populations closer to park boundaries would increase the threat to already vulnerable herbivores like sable antelope and topi and carnivores like cheetah and wild dogs. Such potential shifts would put a great premium on buffer areas surrounding the park, particularly those, like the Maasai Mara, in the lake effect belt that receives dry season rainfall from Lake Victoria. Such areas may be buffered from some of the changes expected for the regional climate.

A decline in mean and variability in rainfall may reduce the number of surface water points in the Serengeti, particularly in low-rainfall areas. Fewer water points are likely to cause higher local densities of animals near water and greater heterogeneity in animal distributions. Such congregations, particularly in the dry season, might be good for the tourist trade, as more animals in fewer spots makes for better public viewing and less total impact. However, high densities may make disease transmission more likely and would likely increase conflict between humans and wildlife over water points.

HUMAN WELFARE

Humans that depend on the Greater Serengeti ecosystem undertake an economy based on farming, livestock husbandry, and game hunting. The anticipated climate changes in the Serengeti, and their impacts on plants and animals, may have dramatic effects on livestock forage production, water distribution, and the success of crop agriculture. Although these could easily be seen as food web consequences of global change, we explore them separately in the following. Some of these outcomes are investigated in much more detail in other chapters; our goal is to highlight only some of the most critical impacts on humans.

A major question is whether livestock densities and production will change. The net effect of elevated CO_2, decreased rainfall, and increased temperature should result in little change in forage productivity (tables 6.1 and 6.2), but seems likely to result in higher average plant nutrient content. This means that livestock may have more nutritious forage available. Such changes would suggest that higher average livestock densities might be supported within the ecosystem. On the other hand, decreased rainfall may lead to fewer or less permanent surface water sources. Since cattle and goats must drink free water, they can only be herded within a certain radius of water holes. Thus, livestock distributions may become more restricted in space and herds may be unable to utilize the forage in major fractions of the landscape. In addition, wildlife and livestock may be forced to use the same water sources more often, leading to increased competition for forage and more human-wildlife conflicts. These conflicts may be exacerbated by bringing humans and livestock into greater risk of contracting diseases carried by wildlife, such as tuberculosis (chapter 7), rabies, malaria, or trypanosomosis (Leak 1999).

Decreased variability in rainfall and longer sequences of consecutive dry years suggest that there will be greater risks in "banking" with livestock. Current practices use livestock and their body tissues as a way of storing productivity during wet periods. In any given dry year when crop agriculture is likely to fail, human households can sell livestock for cash or harvest animals for meat. If dry periods begin to increasingly last longer than 3 years, households may lose livestock to starvation. If so, the pressure to harvest wildlife in the Greater Serengeti may become increasingly intense. If wildlife harvest intensifies mainly during dry periods, when wildlife populations have the least capacity to replace losses, human wildlife harvest may pose especially great risks for driving some wildlife populations extinct.

Crop agriculture in the Greater Serengeti is based on corn, which yields a high food energy return per area cultivated but also requires greater water inputs than lower-yielding but more water-use-efficient crops like millet (Jones and Thornton 2003). A critical step in responding to anticipated warmer, drier average conditions and longer sequences of consecutive dry years may be to switch to different crops under drier conditions, which would allow crop agriculture to occur in a greater number of years, placing less premium on livestock banks and less pressure on wildlife populations as potential food and cash sources.

Besides food security, the other major risk to human welfare is the myriad of lethal human and livestock diseases that occur in the Greater Serengeti. How will the risks and impacts of these diseases change in response to expected climate change? Drier conditions should make malaria less prevalent overall, as lower rainfall should produce less total area in permanent or seasonal standing water. However, humans may become increasingly exposed to other diseases like rabies, tuberculosis, and trypanosomosis, for which wildlife serve as alternate hosts, because of greater concurrent use of surface water sources by high densities of livestock and wildlife, as discussed previously. Factors that increase human hunting also increase contact between humans and wildlife, increasing the risk of contracting diseases such as rabies. Our understanding of the complex epidemiology of theses diseases is still poor (Grenfell and Dobson 1995, Leak 1999), so the precise magnitude of increased disease risks still needs further work.

CONCLUSION

Through a series of complicated interactions between ocean-atmosphere circulation patterns and land and sea surface temperatures, average and variability in rainfall in the Serengeti ecosystem should continue to decline over the next three decades. Temperature, which should track expected increases in Indian Ocean sea surface temperatures, is expected to increase by perhaps 1°C over this same time period. CO_2 concentrations are expected to increase by another 75–100 ppm. These global changes should result in somewhat lower aboveground productivity but increased nutrient concentrations, creating cascading changes through the diverse food web of the Serengeti. Changes in rainfall may also alter the distribution of grassland versus woodland and free surface water, which will further alter the distributions of wildlife and livestock.

These environmental changes will undoubtedly combine with increased human population growth land use and policy decisions to affect human welfare. Overall, humans in the Serengeti seem faced with somewhat worsened food security and health risk conditions, which will restrict their choices of livelihood (chapter 11) and the space in the landscape from which they can garner a living. Longer sequences of consecutive dry years may prevent the use of livestock banks and increase pressure to alter the choice of crops and to harvest wildlife, implying decreased food security. Increased contact between humans, livestock, and wildlife over spatially restricted free water sources and from hunting may expose humans to increased risks of certain diseases (rabies, tuberculosis, trypanosomosis), while drier conditions may reduce the risks of others (malaria). The expected combined global changes in climate and atmospheric CO_2 are likely to make human households more vulnerable to variation in climate and increase the pressure on conservation areas where human activities are restricted. Although these predictions are quite speculative, they illustrate some possible consequences and highlight the need for more research into the possible impacts of climate and atmospheric CO_2 changes, particularly in tropical coupled human and natural systems like the Serengeti.

REFERENCES

Allali, A., C. Basalirwa, M. Boko, G. Dieudonne, T. E. Downing, P. O. Dube, A. Githeko, et al. 2001. Africa. In *Climate change 2001: Impacts, adaptation and vulnerability,* ed. J. J. McCarthy, O. F. Canziani, N. A. Leary, D. J. Dokken, and K. S. White, Contributions of Working Group II of the Intergovernmental Panel on Climate Change. Retrieved from http://www.grida.no/climate/ipcc_tar/wg2/index.htm.

Altmann, J., S. C. Alberts, S. A. Altmann, and B. Roy. 2002. Dramatic changes in local climate in the Amboseli Basin, Kenya. *African Journal of Ecology* 40:248–51.

Augustine, D. 2003. Long-term, livestock-mediated redistribution of nitrogen and phosphorus in an East African savanna. *Journal of Applied Ecology* 40:137–49.

Bazzaz, F. A., S. L. Miao, and P. M. Wayne. 2003. CO_2-induced growth enhancements of co-occurring tree species decline at different rates. *Oecologia* 96:478–82.

Boyce, W. E., and R. C. DiPrima. 1977. *Elementary differential equations,* 3rd ed. New York: Wiley.

Breman, H., and C. T. deWit. 1983. Rangeland productivity and exploitation in the Sahel. *Science* 221:1341–47.

Chesson, P., and N. J. Huntly. 1997. The role of harsh and fluctuating conditions in the dynamics of ecological communities. *American Naturalist* 150:519–53.

Clark, C. O., J. E. Cole, and P. J. Webster. 2000. Indian Ocean SST and Indian summer rainfall: Predictive relationships and their decadal variability. *Journal of Climate* 13: 2503–19.

Coughenour, M. B., and D-X. Chen. 1997. Assessment of grassland ecosystem responses to atmospheric change using linked plant-soil process models. *Ecological Applications* 7:802–27.

Dublin, H.T. 1995. Vegetation dynamics in the Serengeti-Mara ecosystem: The role of elephants, fire and other factors. In *Serengeti II: Dynamics, management, and conservation of an ecosystem,* ed. A. R. E. Sinclair and P. Arcese, 71–90. Chicago: University of Chicago Press.

Fritz, H., and P. Duncan. 1994. On the carrying capacity for large ungulates of African savanna ecosystems. *Proceedings of the Royal Society of London* 256:77–82.

Grenfell, B.T., and A. P. Dobson, eds. 1995. *Ecology of infectious diseases in natural populations.* Cambridge: Cambridge University Press.

Hibbard, K. A., D. S. Schimel, S. Archer, D. S. Ojima, and W. Parton. Grassland to woodland transitions: Integrating changes in landscape structure and biogeochemistry. *Ecological Applications* 13:911–26.

House, J. I., S. Archer, D. D. Breshears, R. J. Scholes, and NCEAS participants. 2003. Conundrums in mixed woody-herbaceous plant systems. *Journal of Biogeography* 30: 1763–77.

Jones, P. G., and P. K. Thornton. 2003. The potential impacts of climate change on maize production in Africa and Latin America in 2055. *Global Environmental Change* 13: 51–59.

Keeling, C. D., and T. P. Whorf. 2002. Atmospheric carbon dioxide record 1958–2002. Retrieved from Mauna Loa Carbon Dioxide Research Group, Scripps Institution of Oceanography, at http://cdiac.esd.ornl.gov/trends/co2/sio-mlo.htm.

Keeling, C. D., R. B. Bacastow, A. F. Carter, S. C. Piper, T. P. Whorf, M. Heimann, W. G. Mook, and H. Roeloffzen. 1989. A three-dimensional model of atmospheric CO_2 transport based on observed winds: 1. Analysis of observational data. In *Aspects of climate variability in the Pacific and the Western Americas,* ed. D. H. Peterson, *Geophysical Monograph* 55:165–235.

Koch, P. L., J. Heisinger, C. Moss, R. W. Carlson, M. L. Fogel, and W. Behrensmeyer. 1995. Isotopic tracking of changes in diet and habitat use in elephants. *Science* 267: 1340–43.

Leak, S. G. A. 1999. *Tsetse biology and ecology: Their role in the epidemiology and control of trypanosomosis.* New York: CABI.

McCarthy, J. J., O. F. Canziani, N. A. Leary, D. J. Dokken, and K. S. White, eds. 2001. *Climate change 2001: Impacts, adaptation and vulnerability.* Contributions of Working Group II of the Intergovernmental Panel on Climate Change. Retrieved from http:// www.grida.no/climate/ipcc_tar/wg2/index.htm.

McNaughton, S. J. 1985. Ecology of a grazing system: The Serengeti. *Ecological Monographs* 55:259–94.

McNaughton, S. J., F. F. Banyikwa, and M. M. McNaughton. 1997. Promotion of the cycling of diet-enhancing nutrients by African grazers. *Science* 278:1798–1800.

Mills M. G. L., H. C. Biggs, and I. J. Whyte. 1995. The relationship between rainfall, lion predation and population trends in African herbivores. *Wildlife Research* 22:75–88.

Ojima, D. S., T. G. F. Kittel, T. Rosswall, and B. H. Walker. 1991. Critical issues for understanding global change effects on terrestrial ecosystems. *Ecological Applications* 1: 316–25.

Olff, H., M. E. Ritchie, and H. H. T. Prins. 2002. Global determinants of diversity in large herbivores. *Nature* 415:901–904.

Prins, H. H. T., and H. Olff. 1998. Species-richness of African grazer assemblages: Towards a functional explanation. In *Dynamics of Tropical Communities*, ed. D. M. Newberry, H. H. T. Prins, and N. Brown, 449–90. Oxford: Blackwell.

Reich, P. B., J. Knops, D. Tilman, J. Craine, D. Ellsworth, M. Tjoelker, T. Lee, et al. 2001. Plant diversity enhances ecosystem responses to elevated CO_2 and nitrogen deposition. *Nature* 410:809–812.

Runyoro, V., H. Hofer, E. B. Chausi, and P. Moehlman. 1995. Equilibria in plant-herbivore interactions. In *Serengeti II: Dynamics, management, and conservation of an ecosystem*, ed. A. R. E. Sinclair and P. Arcese, 146–68. Chicago: University of Chicago Press.

Saji, N. H., and T. Yamagata. 2003. Structure of SST and surface wind variability during Indian Ocean Dipole mode events: COADS observations. *Journal of Climate* 16: 2735–51.

Scholes, R. J., and S. R. Archer. 1997. Tree–grass interactions in savannas. *Annual Review of Ecology and Systematics* 28:517–44.

Sinclair, A. R. E. 1995. Equilibria in plant-herbivore interactions. In *Serengeti II: Dynamics, management, and conservation of an ecosystem*, ed. A. R. E. Sinclair and P. Arcese, 91–114. Chicago: University of Chicago Press.

Sinclair, A. R. E., S. Mduma, and J. S. Brashares. 2003. Patterns of predation in a diverse predator-prey system. *Science* 425:288–90.

Torrence, C., and G. P. Compo. 1998. A practical guide to wavelet analysis. *Bulletin of the American Meteorological Society* 79:61–78.

Webster, P. J., A. M. Moore, J. P. Loschnigg, and R. R. Leben. 1999. Coupled ocean-atmosphere dynamics in the Indian Ocean during 1997–1998. *Nature* 401:356–60.

Wright, I. J., P. B. Reich, and M. Westoby. 2001. Strategy shifts in leaf physiology, structure and nutrient content between species of high- and low-rainfall and high- and low-nutrient habitats. *Functional Ecology* 15:423–34.

———. 2003. Least-cost input mixtures of water and nitrogen for photosynthesis. *American Naturalist* 161:98–111.

SEVEN

The Multiple Roles of Infectious Diseases in the Serengeti Ecosystem

Sarah Cleaveland, Craig Packer, Katie Hampson, Magai Kaare, Richard Kock,

Meggan Craft, Tiziana Lembo, Titus Mlengeya, and Andy Dobson

Pathogens are important components of the Serengeti that have had a major role in shaping the ecosystem, with direct and indirect impacts on wildlife populations and community structure. As human and domestic animal populations continue to expand into areas surrounding the Serengeti National Park, the transmission of diseases between different populations has become an issue of growing concern, with implications not only for wildlife but also for livestock economies, human health, and land use strategies (fig. 7.1). Failure to resolve multiple-host disease concerns in these traditional systems is a major driver for pastoral people to seek alternative and/or agriculture-based livelihoods. As the vast majority of disease problems in the Serengeti arise from transmission between populations, generalist pathogens that can infect multiple hosts are a main focus of concern. Understanding the dynamics of these pathogens in complex ecosystems presents a considerable challenge, but it is essential to evaluating disease threats and designing effective and appropriate control strategies, where necessary.

A growing awareness of the need to develop integrated approaches toward human and animal health (Schwabe 1984) has recently led to the concept of *ecosystem health,* which extends the earlier focus on wildlife to include a broader socioecological definition that encompasses all the components of the ecosystem, including human and domestic animal populations (Kock 2005). Infectious disease research conducted in the Serengeti over the past 15 years has provided important contributions to this field.

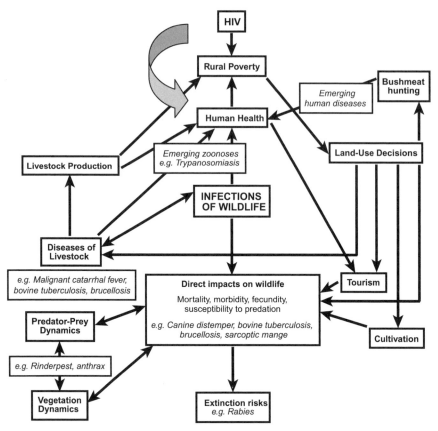

Fig. 7.1 Scheme showing the potential impacts and interacting factors relating to wildlife diseases in the Serengeti.

In this chapter, we describe research on specific pathogens that illustrates the multiple roles of infectious diseases in the Serengeti, including (a) impacts on individual hosts (canine distemper virus), (b) threats to endangered populations (rabies) and to the ecosystem (rinderpest), (c) impacts of disease on community structure and vegetation dynamics (rinderpest and anthrax), (d) wildlife as reservoirs of diseases that threaten livestock economies (malignant catarrhal fever) and human health (trypanosomiasis), (e) the consequences of animal diseases on rural poverty and land use decisions around the Serengeti, (f) the dynamics of multiple-host pathogens, and (g) the impact of major human diseases (Human Immunodeficiency Virus/Acquired Immunodeficiency Syndrome).

DIRECT IMPACTS ON INDIVIDUAL HOSTS

Mortality and Extinction Threats

Recent epidemics of canine distemper virus (CDV) and rabies in Serengeti carnivores have raised awareness about the role of pathogens as extinction risks for endangered wildlife. Rabies and CDV are RNA viruses that can infect a wide range of species and can be maintained in domestic animal populations. They both have short infection cycles and they both cause high mortality in infected individuals; CDV also produces lifelong immunity in individuals who survive infection. These traits prohibit the pathogens from persisting in small populations, because infection will repeatedly fade out due to a lack of new, susceptible hosts (Anderson and May 1991). Though the Serengeti is well known for its large populations of migratory ungulates, most Serengeti wildlife populations, especially carnivores, appear to be too small to host such virulent diseases by themselves. For diseases such as rabies and CDV, the major threat comes from large, continuous populations of domestic dogs that extend over the entire country. This threat will continue to grow as domestic animal and human populations expand in areas surrounding the Serengeti.

Canine Distemper Virus in Serengeti Lions

Canine distemper is usually caused by aerosol transmission and is a multisystemic, sometimes fatal viral disease that has been documented in a wide and expanding range of carnivore species (Harder and Osterhaus 1997), resulting in population declines in many wild carnivore populations (Funk et al. 2001). Mortality rates vary widely among species (Appel and Summers 1995), but can be as high as 90% in some populations (van de Bildt et al. 2002). In 1994, CDV caused a severe epidemic that affected lions (*Panthera leo*), spotted hyenas (*Crocuta crocuta*), bat-eared foxes (*Otocyon megalotis*), and domestic dogs (*Canis familiaris*) in both the Tanzanian and Kenyan regions of the Serengeti ecosystem (Roelke-Parker et al. 1996; Kock et al. 1998). Although infections with CDV had previously been confirmed in captive large cats (Blythe et al. 1983; Appel et al. 1994), the Serengeti outbreak was the first documented in free-living felids. Thirty percent of the lions died or disappeared in the Serengeti and Maasai Mara study areas (Roelke-Parker et al. 1996; Kock et al. 1998), leading to an estimate of 1,000 fatalities across the entire ecosystem.

Sequence analysis of the CDV isolates from Serengeti lions, spotted hyenas, bat-eared foxes, and domestic dogs, during the 1994 epidemic

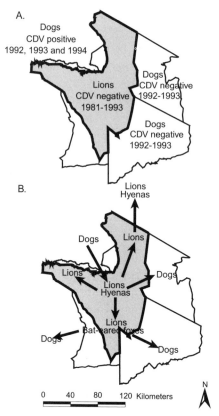

A.

Dogs
CDV positive
1992, 1993 and 1994

Lions
CDV negative
1981-1993

Dogs
CDV negative
1992-1993

Dogs
CDV negative
1992-1993

B.

Lions
Hyenas

Dogs   Lions

Lions

Lions   Dogs
Lions
Hyenas

Lions
Bat-eared foxes

Dogs

Dogs

0   40   80   120 Kilometers

N

Fig. 7.2 (panel A) Status of canine distemper virus (CDV) before the 1994 epidemic in Serengeti lions and (panel B) its spread during the epidemic as reconstructed using serological evidence and case-surveillance data (Roelke-Parker et al. 1996; Kock et al. 1998; Packer et al. 1999; Cleaveland et al. 2000; Harrison et al. 2004).

demonstrated that all four species were infected with very closely related viruses, suggesting that the Serengeti epidemic was caused by a single variant (Harder et al. 1996; Roelke-Parker et al. 1996; Carpenter et al. 1998). Age-seroprevalence patterns indicated that neither the lion population, nor low-density dogs of the Ngorongoro District, were exposed to infection in the years prior to the outbreak (Roelke-Parker et al. 1996; Packer et al. 1999). In contrast, the high-density dogs in the Serengeti district showed evidence of virus circulation in 1992 and 1993 (Cleaveland et al. 2000), suggesting that they were the source of the 1994 epidemic. The spatial and temporal pattern of cases indicated that CDV subsequently spread throughout the ecosystem in wild carnivores, re-emerging in domestic dogs in Ngorongoro in late 1994 and in the Shinyanga Region in 1995 (fig. 7.2).

Although the exact routes of transmission between dogs and wildlife are unknown, there are many opportunities for interspecific contact. Potential vectors of CDV, such as jackals (*Canis mesomelas, C. adustus, C. aureus*), spotted hyenas, small-spotted genets (*Genetta genetta*), African civets (*Civettictis civetta*), and bat-eared foxes and mongooses (*Ichneumia albicauda, Mungos mungo, Helogale parvula, Herpestes sanguineus*) have been recorded in close proximity to domestic dogs. From night-transect observations, white-tailed mongooses, small-spotted genets, and black-backed jackals are most commonly recorded in Serengeti District, and bat-eared foxes, white-tailed mongooses, and black-backed jackals in Ngorongoro District. Multiple opportunities for interaction between wild carnivores and domestic dogs can arise within villages—for example, when wild animals scavenge at slaughter slabs, rubbish dumps, or households, and during predatory attacks from mongoose or hyenas. Interactions may also occur in the buffer zone around the national park, when dogs accompany people during herding or hunting activities outside the villages. A link between domestic dogs and hyenas in the 1994 outbreak is also suggested by data from the Maasai Mara: low-ranking hyenas scavenge more frequently in villages than high-ranking hyenas, and subordinate individuals had a higher seroprevalence, consistent with increased exposure to CDV (Alexander et al. 1995). However, in a more recent episode in 2000–2001, although hyenas in the Maasai Mara were exposed to CDV, no effect of social rank could be detected (Harrison et al. 2004).

CDV is usually spread by aerosols, and as transmission requires only close proximity, frequent opportunities arise for both intra- and inter-specific transmission. For hyenas, jackals, and lions, intraspecific transmission may involve spread within social groups and between groups—for example, during territorial defence, kleptoparasitism, or long-distance movements. Long-distance movements have been documented in wide-ranging nomadic lions, dispersing jackals and commuting hyenas (Hofer and East 1995; M. Craft, personal communication). Interspecific transmission of CDV may also occur among lions, jackals, and hyenas during interactions at kills. For example, in 21% of kills, at least two carnivore species were simultaneously present around the carcass. Recent analyses of lion-kill data demonstrate the proximity between species at kills and the high potential for interspecific transmission (fig. 7.3).

Although CDV clearly has the potential to cause high mortality in lions and hyenas, the Serengeti lion population recovered extremely rapidly from the 1994 decline (fig 7.4; Packer et al. 2005). A single CDV epidemic in the Serengeti population is unlikely to result in extinction, but the disease could be an important factor if reduced density interacts with

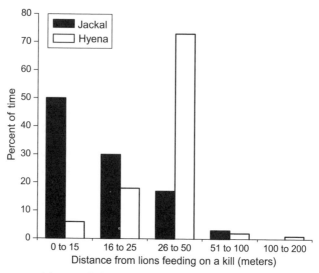

Fig. 7.3 Percentage of time spent by hyenas and jackals at various distances from carcasses at which lions were feeding. Observations were recorded from 111 carcasses between 1979 and 1981.

other density-independent factors. In smaller populations, CDV may pose a greater threat.

The impact of interacting cofactors on population dynamics and host pathogenicity has become an area of growing interest for research. For example, it is now clear that CDV is not invariably highly pathogenic in Serengeti lions, which have been exposed to CDV on several occasions (fig. 7.4), but have only suffered significant disease-associated mortality in 1994. Similarly, although CDV caused mortality in hyenas in the 1994 outbreak, exposure of juvenile hyenas to CDV in the Maasai Mara in 2000–2001 was not associated with mortality (Harrison et al. 2004). Many questions are thus raised about the determinants of CDV pathogenicity in Serengeti's wild carnivores and the role of genetic changes in virus variant, climatic factors, and other cofactors, including other concurrent infections that might affect host responses to CDV infection. For instance, in 1994, CDV appears to have caused particularly high mortality owing to high levels of coinfection with other pathogens, following the drought of 1993 (K. Terio and L. Munson, personal communication).

Rabies and African Wild Dogs

Since the 1980s, it has become clear that infectious diseases can represent a serious threat to endangered species (Dobson and May 1986; Funk et al.

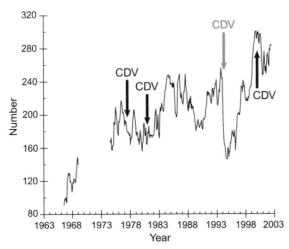

Fig. 7.4 Lion population size of the 2,000 km² long-term study area of the southeastern Serengeti National Park. Solid black arrows show approximate timing of episodes of canine distemper virus (CDV) exposure that were not associated with obvious morbidity or mortality. CDV exposure in lions was determined from age-seroprevalence patterns (Packer et al. 1999). The grey arrow shows the timing of CDV outbreak that caused 30–35% adult mortality.

2001; Woodroffe et al. 2004). This threat became apparent in the Serengeti with the deaths and disappearances of African wild dogs (*Lycaon pictus*) in the late 1980s and early 1990s. Between 1986 and 1991, rabies was recorded by laboratory diagnosis or clinical signs in 5 of 15 known packs of African wild dogs in the Serengeti-Mara ecosystem. A further two packs disappeared following undiagnosed signs, including lethargy and weakness, and the remaining eight packs all disappeared unobserved (Woodroffe 1997). Thus, by 1992, the wild dog population had become locally extinct (see fig. 2.16) and no breeding packs were documented in the ecosystem for the next 10 years. During the peak period of wild dog mortality (1987 and 1988), rabies epidemics also affected bat-eared foxes in the central Serengeti, killing 60% of all adult females and 20% of males and cubs (Maas 1993).

Rabies was apparently eliminated in the Serengeti region between 1954 and 1978, as a result of intensive dog vaccination and culling programs (Rweyemamu et al. 1973; Magembe 1985), indicating that domestic dogs at that time were the sole reservoir of infection. Rabies was re-introduced to the western Serengeti in the late 1970s during an epidemic that spread from an endemic focus in southern Tanzania, and the disease has persisted in domestic dogs in that area ever since (Cleaveland 1996). The close similarity of viruses (the canid Africa 1-b variant) isolated from wild dogs and domestic dogs, coupled with epidemiological evidence for disease persistence in domestic dogs, suggests that domestic dogs were the likely reservoir and source of infection for wild dogs (Cleaveland and Dye 1995; Kat et al. 1995). Since the wild dog outbreak, the same Africa 1-b virus variant has been recovered from domestic dogs and cats (*Felis catus*), livestock (cattle [*Bos taurus*], goats [*Capra hircus*], and donkeys [*Equus asinus*]) and a range of wildlife species, including a black-backed jackal, a small-spotted genet, an aardwolf (*Proteles cristatus*), a wild cat (*Felis libyca*), spotted hyenas, white-tailed mongooses, a leopard (*Panthera pardus*), and, most commonly, bat-eared foxes (East at al. 2001; Lembo et al. 2007). Preliminary analyses support a hypothesis of domestic dog-to-wildlife transmission of this variant.

A feature of rabies in African wild dogs, both in eastern and southern Africa, has been the associated loss of entire packs (Woodroffe, Ginsberg, and Macdonald 1997; Hofmeyr et al. 2000). The intensely social nature of African wild dogs allows rapid transmission and high rates of infection within the group, so a single transmission event can have disproportionately large effects. Although it is unlikely that pathogens infect or kill 100% of individuals, even in tight social groups, outbreaks of disease can result in the loss of whole groups, because survivors may fail to reestablish packs due to Allee effects: African wild dogs require helpers to successfully rear pups, so pup survival rates drop below replacement levels at a threshold pack size (Courchamp, Clutton-Brock, and Grenfell 2000).

Nonfatal Impacts of Disease on Wildlife Populations

Although epidemics are the most visible and dramatic manifestation of infectious diseases, pathogens can have more subtle, but potentially more significant, impacts through reducing host condition (which can lead to lowered host fecundity) and/or increased vulnerability to predation and reduced competitive ability. Bovine tuberculosis (bTB), caused by the bacterial pathogen *Mycobacterium bovis,* has been detected in a wide range of Serengeti ungulate and carnivore species and has been present in the Serengeti lions for at least 20 years (Cleaveland at al. 2005). In a preliminary study, *M. bovis* was isolated from 10% of migratory wildebeest (*Connochaetes taurinus*) culled as part of local meat-cropping programmes in the Ikorongo-Grumeti Game Control Areas, but no signs of disease were seen in affected animals and no visible lesions detected on post-mortem examination.

In the Kruger National Park and Hluhluwe-Umfolozi Park, South Africa, bTB has spread widely throughout the lion population, where consumption of infected buffalo (*Syncerus caffer*) is considered the predominant route of infection (Keet et al. 1996; Caron, Cross, and du Toit 2003; D. Cooper, personal communication). Serengeti lions would seem to be at similar risk of widespread infection from consuming wildebeest with *M. bovis,* but the lions' rate of infection in Serengeti (< 10%) appears to be much less than that in southern African populations. Although preliminary evidence indicates that infected lions in the Serengeti have shorter life expectancies than noninfected lions (Cleaveland et al. 2005), these effects appear to be mild compared to those in southern African lion populations, in which bTB causes severe debilitation and death. Comparative ecological and phylogenetic studies of bTB in eastern and southern Africa would clearly be valuable in determining whether this reflects TB strains of different virulence.

Disease outbreaks in predators can be harmful to those herbivore populations that are regulated primarily by parasitic infections rather than by predation (Packer et al. 2003). Carnivores preferentially attack sick individuals and thereby reduce the chances of disease transmission to other hosts. Herds that are released from predation pressure are expected to show a higher incidence of infection by any type of parasite, as well as reduced numbers of healthy individuals in the population. The impact on overall prey population size will depend on the type of infection. Where recovered individuals can become reinfected with the same parasite (e.g., sarcoptic mange mite), a significant drop in predation rates can cause the host population to decline. On the other hand, if recovered individuals are immune to subsequent reinfection (as in rinderpest), a drop in predation allows host population size to increase, since the large class of recovered animals would only have been susceptible to predation. With macroparasitic infection (e.g., intestinal parasites), predator removal reduces host population size, providing that the predators are sufficiently selective in capturing infected prey (Hudson et al. 2002). Since the migratory herbivores in the Serengeti appear to be limited by food availability rather than by predation, we predict that generalized disease outbreaks in the Serengeti carnivores (such as the 1994 CDV epizootic) are likely to be followed by increased incidence in herbivore diseases.

Pathogen impacts on host behavior and predation have been demonstrated in several other systems (Dobson 1988; Lafferty 1992), but not yet in the Serengeti. Sarcoptic mange, caused by the burrowing mite *Sarcoptes scabies,* has been reported to affect a wide range of species in the Serengeti, including ungulates (buffalos and Thomson's gazelles [*Gazella thomsonii*]), canids (bat-eared foxes and jackals), primates (baboons [*Papio cynocephalus anubis*]), and felids (cheetahs [*Acinonyx jubatus*] and lions). Increased mortality in wild animals can arise as a direct result of infection, ranging from a few isolated cases to dramatic epidemics and population declines (e.g., red foxes [*Vulpes vulpes*] and coyotes [*Canis latrans*]; reviewed by Pence and Ueckermann 2002). Selective predation of animals debilitated by mange may also contribute to increased mortality and enhance transmission of infection from prey species to predators. The apparently high prevalence of mange in Thomson's gazelles may provide an explanation for high infection rates in cheetahs. Immunosuppression is known to increase susceptibility to disease and stress-related mechanisms play an important role in the epidemiology of the disease in some species. Physiological stress, associated with failure to establish territories, has been linked to higher rates of disease in cheetahs in the Serengeti (Caro et al. 1987), and stresses associated with tourist pressure have been implicated in outbreaks of mange in cheetahs in the Maasai Mara (Mwanzia et al. 1995). Tourist pressure may be exacerbated by factors that affect sward length, such as increased pressure on pasture through mixed livestock-wild herbivore grazing (R. Kock, personal observation).

INDIRECT IMPACTS OF DISEASE

Ecosystem-Level Impacts of Disease: Rinderpest and Anthrax

Although it is evident that epidemics can have immediate impacts on affected host populations, the long-term consequences of epidemics on predator-prey and vegetation dynamics can also be profound. Rinderpest in the Serengeti provides a striking example of a single pathogen dominating the factors shaping the Serengeti landscape in the twentieth century (Sinclair 1979; Dobson 1995). For example, the huge impact of rinderpest on herbivore abundance in the great pandemic of the 1890s not only reduced the prey base for the carnivores, but also reduced levels of herbivory and created a "pulse" of tree recruitment (Prins and Weyerhaeuser 1987; Prins and van der Jeugd 1993; Dobson and Crawley 1994). Thus, a single pathogen has major impacts on three different trophic levels in the system. Similar effects have been observed where localized outbreaks of anthrax have caused high mortality in impalas (*Aepyceros melampus*), again leading to pulses of tree recruitment (Prins and van der Jeugd 1993). The continuing repercussions of the wildebeest population expansion (as a result of vaccination of cattle against rinderpest) are similarly profound, and are discussed in detail in several chapters in this volume (chapters 2, 5, 7, 9, and 14).

Malignant Catarrhal Fever—A Factor Driving Human-Wildlife Conflict in Maasailand

Disease transmission from wildlife to livestock invariably creates conflict with human populations and can be a major factor driving patterns of land use. The viral disease, malignant catarrhal fever (MCF), is particularly important to Maasai pastoralists of the Ngorongoro Conservation Area (NCA). The NCA is inhabited by over 40,000 pastoralist Maasai, but rapid human population growth, coupled with static livestock numbers over the past 40 years, has resulted in a four-fold decrease in per capita livestock, which has coincided with a decline in livestock productivity (Thompson 1997). More than 50% of children in the NCA are malnourished (McCabe, Schofield, and Nygaard Pederson 1997), 58% of the population face acute poverty (NCAA/NPW 1994) and a growing number of households are defined as "destitute" (i.e., owning less than two tropical livestock units [TLU] per *engaji* or household; Potanski 1999). Pathogens such as MCF, which spill over from wildlife to domestic livestock, further reduce the health and welfare of these households.

MCF is a gammaherpes virus that is transmitted by asymptomatic infected wildebeest and causes fatal disease in cattle (Barnard, van der Lugt, and Mushi 1994), with widespread consequences for pastoralists in the Serengeti ecosystem (fig. 7.5). The Maasai consistently record losses of around 5% of cattle per annum to MCF, with mortality occurring mainly in adults (Cleaveland et al. 2001). As the disease is primarily transmitted by wildebeest calves less than 3 months of age, Maasai are effectively forced to avoid the prime grazing lands of the short-grass plains during the wildebeest calving season. The increase in wildebeest numbers and the expansion of the migration over the past 30 years has led to Maasai cattle becoming increasingly confined to nonproductive highland and woodland pastures each wet season. The consequences are profound. First, cattle production is reduced through lowered access to salt and to high-quality forage (Homewood, Rodgers, and Arhem 1987; Homewood and Rodgers 1991) and, second, because cattle must be confined in highland areas at relatively high densities, they acquire an increased burden of vector-borne and directly transmitted diseases (Cleaveland et al. 2001). Third, the increased concentration of cattle in the fragile highland ecosystems contributes to over-grazing, deforestation, uncontrolled burning, and soil erosion (Kijazi 1997; Misana 1997; Aikman and Cobb 1997). Poor livestock production has resulted in families becoming trapped in a cycle of poverty, with insufficient animals to meet food and subsistence demands and a growing

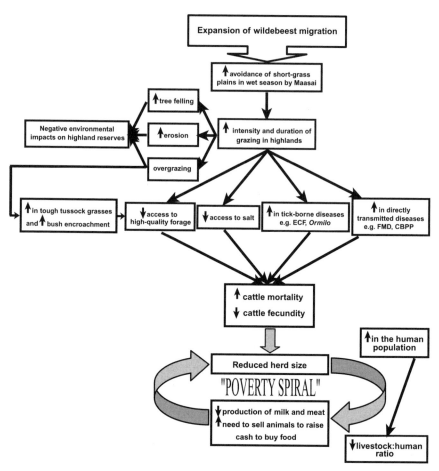

Fig. 7.5 Scheme illustrating the impact and consequences of malignant catarrhal fever (MCF) for pastoralists in the Serengeti ecosystem. ECF = East Coast Fever; FMD = Foot-and-mouth disease; CBPP = Contagious bovine pleuropneumonia.

dependence on grain. Inexorably, families are forced to sell their animals to raise cash, which further reduces their livestock resources and drives them further into destitution. Boone et al. (2002) developed a land-use model (Savanna) that suggested that the NCA could support 20,000 additional cattle if the Maasai were able to increase their wet-season access to the short-grass plains.

Although the decline in pastoralist livelihoods may not appear to be a factor of immediate concern for wildlife conservation, its implications may be serious. For example, concern about food security as a result of declining livestock production has been a major factor behind the recent expansion in crop cultivation in the NCA, a form of land use that is much less compatible with wildlife conservation than traditional pastoralism. While small-scale cultivation may arguably have relatively few impacts on Serengeti's wildlife, replacement of traditional pastoralism with large-scale cultivation is of great concern, with major declines in wild ungulate populations recorded in areas adjacent to the Maasai Mara as a result of conversion of rangelands to mechanised agriculture (Serneels and Lambin 2001; Homewood 2004).

ZOONOTIC DISEASES LINKED WITH WILDLIFE

Much recent interest in human infectious diseases has focused on emerging and reemerging diseases and the role of wildlife in the epidemiology of these diseases. In a recent survey, 75% of emerging human diseases were found to be zoonotic (caused by pathogens that can be transmitted from vertebrate hosts to humans; Taylor, Latham, and Woolhouse 2001) and 65% could infect wildlife hosts (Cleaveland, Laurenson, and Taylor 2001). Several other emerging diseases are also known to have originated in wildlife (e.g., Human Immunodeficiency Virus [HIV] 1 and 2; Gao et al. 1999; Hahn et al. 2000), even though the diseases are not classified as zoonoses (because infection now occurs as a result of human-to-human transmission). Thus, wildlife populations represent an important source of zoonotic pathogens. Increased contact between humans and wildlife (via human population growth, encroachment, habitat conversion, increasing human and wildlife movements) is likely to be the major factor in the emergence of established zoonoses (e.g., Ebola virus, Hantaviruses, Monkeypox) and diseases that have recently "jumped" into humans from wildlife (e.g., Severe Acute Respiratory Syndrome [SARS]).

The link between wildlife and human health poses several challenges. First, the lack of knowledge of transmission processes in wild animal populations limits the development of effective disease-control strategies. Detecting pathogens in free-living wildlife is notoriously difficult, hampered by the enormous practical problems of finding, collecting, and storing appropriate samples under field conditions, as well as the lack of species-specific diagnostic tests. Second, even where wildlife hosts and/or reservoirs have been identified, control options are limited. Public health officials generally advocate strategies that benefit people but are often harmful to wildlife, such as culling or erecting barriers. Vaccines and medications are often unavailable or untested for wildlife, and field delivery can be hindered by logistic, financial, ethical, and/or political considerations. For example, even though rabies vaccines had been successfully used to control rabies in wildlife throughout large parts of western Europe and North America, the rabies vaccination of African wild dogs in the Serengeti in 1990 was beset by debate and controversy that continues to affect wildlife disease control initiatives throughout Africa today (Macdonald et al. 1992; Burrows, Hofer, and East 1994, 1995; de Villiers et al. 1995; Creel, Creel, and Monfort 1997; Woodroffe 1997, 2001).

Trypanosomiasis—The Role of Wildlife and Domestic Animal Hosts in Human Sleeping Sickness

Trypanosomiasis is a vector-borne disease caused by protozoan parasites in the genus *Trypanosoma;* it exerts a profound impact on both livestock and public health throughout much of sub-Saharan Africa (Kristjanson et al. 1999; Schmunis 2004; Shaw 2004). *T. vivax* and *T. congolense* cause a debilitating disease of cattle, while *T. brucei* is less pathogenic to local breeds. A subspecies of *T. brucei, T. brucei rhodesiense,* is human-infective and the cause of sleeping sickness in East Africa, a potentially fatal disease of humans (Welburn, Fevre et al. 2001). The trypanosomes have profoundly affected the settlement and economic development of Africa, since they render vast areas of semiarid savanna unsuitable for cattle (Matzke 1979). In the Serengeti, the emergence of sleeping sickness in the wake of the great rinderpest pandemic (see chapter 2, this volume) was probably also a factor in preventing the resettlement of the Serengeti in the early twentieth century. While sleeping sickness had been present in the Serengeti in the 1960s, it reemerged as a concern for both human health and tourism in 2001. In this outbreak, people in local communities, staff working within the park, and 14 tourists were affected and four deaths were confirmed among local people (Jelinek et al. 2002; Mlengeya et al. 2002).

In a preliminary survey, 19% of warthogs (*Phacochoerus aethiopicus*) were infected with the human-infective *T. brucei rhodesiense* (Kaare et al. 2002). Although it is not known whether warthogs are the primary wildlife reservoir of human disease, they are thought to be preferred hosts for the tsetse vectors that predominate in this area (Moloo et al. 1971). Tsetses deposit their pupae in warthog burrows, with the result that the emergent pupae (which are most susceptible to infection) will feed on warthogs and sustain the cycle of infection. Clarifying the exact role of warthogs and other hosts, such as cattle (which act as reservoirs of infection in Uganda; Welburn, Fevre et al. 2001), is clearly important for designing tsetse and trypanosome control strategies in the Serengeti.

ANIMAL DISEASE, RURAL POVERTY, AND LAND USE

Rural poverty is a key factor underlying long-term threats to biodiversity. Communities adjacent to the Serengeti National Park with the highest rates of bush-meat hunting have the lowest access to livestock (Loibooki et al. 2002), suggesting that nutritional and economic requirements have been the driving forces behind local wildlife hunting. Livestock could provide alternative sources of protein to replace demand for wildlife meat in many villages, but livestock production is severely constrained by infectious diseases such as trypanosomiasis in cattle (Ministry of Agriculture 1995) and Newcastle Disease in poultry (Awan, Otte, and James 1994).

The establishment of effective veterinary services around the Serengeti has the potential not only to improve rural livelihoods but might also reduce illegal hunting. For example, it has been suggested that vaccination of poultry against Newcastle Disease, a highly virulent epidemic disease that causes major losses in Tanzania's poultry populations, may provide a cheap and effective means of increasing access to dietary protein and generating income for rural households. Research is currently underway to explore the links between Newcastle Disease vaccination, poultry productivity, and illegal meat hunting in villages on the western borders of the Serengeti.

DISEASE CONTROL

Integrated Disease Control—Rabies

The design of effective zoonotic control measures requires an understanding of how infections are maintained in reservoirs and transmitted from reservoirs to populations of concern (Haydon et al. 2002; Dobson 2004). Rabies provides a useful illustration of these issues; the virus has the potential to infect all mammals, including humans, but not all species are able to act as reservoirs. Identifying rabies reservoirs and critical transmission pathways in susceptible hosts is central to designing control measures that will reduce threats both to endangered wildlife and to human health. Domestic dogs are clearly a priority, since they are the most abundant carnivores in Tanzania, a likely reservoir of infection and the cause of the vast majority of human animal-bite injuries.

Rabies is a horrifying disease that not only kills thousands of people each year in Africa and Asia, but also causes major social and economic hardships (Knobel et al. 2005). The public health burden of rabies in Tanzania is routinely underestimated by several orders of magnitude (Cleaveland et al. 2002) and the economic and psychological impacts in local communities can be profound. For example, more than 400 people living on the periphery of Serengeti National Park were bitten by suspected rabid animals during an outbreak in the Serengeti ecosystem in 2003 and 2004. Bite victims frequently paid over half their yearly income for postexposure prophylaxis, and at least 10 people died. Livestock losses from rabies were substantial for some families, and the high costs of postexposure treatment were debilitating, not only at the family level but also for local governments.

Mass domestic dog vaccinations have been introduced around the Serengeti to identify the reservoirs of rabies and to reduce disease threats associated with the expanding domestic dog population. From 1997 to 2001, successive campaigns in the Serengeti district resulted in vaccination coverage of 61–75% and led to a significant decline, both in the incidence of dog rabies and in human dog-bite injuries (Cleaveland et al. 2003). As a result of mass dog vaccinations, demand for costly human rabies vaccine for postexposure rabies prophylaxis declined significantly (fig. 7.6), providing economic benefits that may help sustain any future control programmes. However, the disease rapidly re-emerged in 2003 following a drop in vaccination coverage in 2001 and 2002. The 2003-4 rabies outbreak also spread to low-density dog populations in Maasai areas east of the park. In the past, these dog populations appeared unable to sustain large-scale epidemics, but the rapid growth of the human population may have permitted the disease to reach its persistence threshold. As a result, dog vaccination campaigns

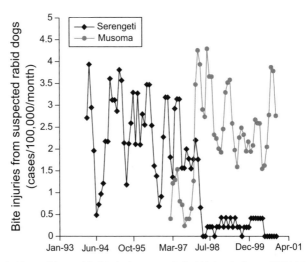

Fig. 7.6 The incidence of human bite injuries from suspected rabid dogs in Serengeti District (where dogs have been vaccinated since 1997) and Musoma District (where no large-scale dog vaccination has been undertaken).

Cleaveland, Packer, Hampson, Kaare, Kock, Craft, Limbo, Mlengeya, and Dobson

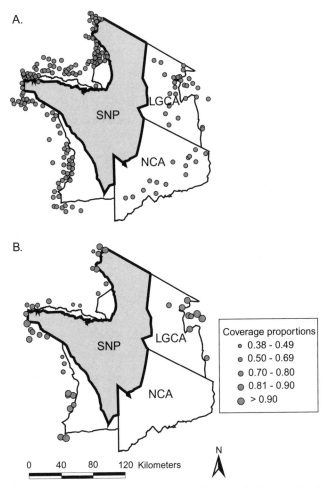

Fig. 7.7 (A) Location of villages included in the current domestic dog vaccination campaigns (each village is represented by solid circles). The vaccination coverage in 2003 is shown in (B) for randomly selected villages. SNP = Serengeti National Park; LGCA = Loliondo Game Control Area; NCA = Ngorongoro Conservation Area.

have now been extended to encompass all villages in Ngorongoro District, as well as all agropastoral villages within a 10 km zone bordering western boundaries of the Serengeti National Park (fig. 7.7), with marked declines in human rabies exposures.

These preliminary trials have demonstrated the feasibility of dog vaccination as a means of controlling rabies, but they also highlight the importance of maintaining vaccination coverage in a dog population with high birth and death rates (Cleaveland 1996). An important outcome of the program has been to demonstrate public health benefits arising from wildlife conservation and management initiatives, and efforts are now underway to further integrate public health, veterinary, and wildlife sectors into rabies control efforts around the Serengeti. Ultimately, community cooperation is essential for implementation of disease-control strategies targeting both people and parks, and the long-term conservation of the Serengeti ecosystem depends upon the support and compliance of local stakeholders. The issue of "all or nothing" is an important consideration for rabies control, and further work is needed to examine the impacts of partial control or low levels of vaccination coverage on the dynamics and virulence of rabies outbreaks.

Regional Control Programs and the Reemergence of Rinderpest

Although rinderpest was eradicated from most of Africa by the 1990s, a strain has persisted in the Somali ecosystem that causes only a mild disease in cattle. However, an epidemic in southern Kenya and northern Tanzania in 1993–1997 demonstrated the potential for the reemergence of the virus and the continuing threat to wildlife populations. Despite mild impacts in cattle, the virus caused over 50% mortality in buffalo and 80% mortality in kudu (*Tragelaphus strepsiceros*) in Tsavo (Kock et al. 1999). Wildlife in the Mkomazi Game Reserve were also infected, but the outbreak was controlled in northern Tanzanian by aggressive cattle vaccination before reaching the Serengeti. Although wildebeest did not show severe signs of disease in the 1997 outbreak, virulence (and possibly host specificity) can change, and another pandemic would have devastating consequences. This may explain why there was no evidence (clinical or serological) for transmission to wildebeest in Nairobi National Park in 1996, despite close contact between infected eland (*Tragelaphus oryx*) and buffalo with this species.

The mild impact in cattle makes the Somali strain difficult to detect, especially given the political instability in Somalia and northeastern Kenya. The Food and Agriculture Organization of the United Nations has organized a Global Rinderpest Eradication Program with the goal of complete eradication by 2010. Considerable effort is therefore being devoted to understanding the epidemiology of the virus and to developing effective eradication strategies. The East African savannahs are only a few hundred kilometers from the endemic zone, and hold large populations of susceptible wildlife and livestock. In order to identify the last traces of infection across the entire continent, vaccination has recently ceased in all regions of Africa, including the Somali outbreak zone. This is deemed necessary to identify

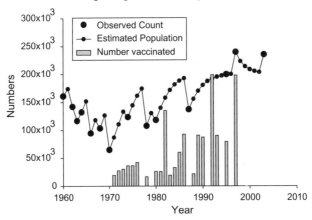

Ngorongoro Cattle Population

Fig. 7.8 The total size of the Ngorongoro and Loliondo cattle population from 1965 through to 2002 (solid circles). The vertical bars illustrate the numbers of cattle vaccinated for rinderpest each year and the solid line illustrates the estimated number of susceptible cattle in the region (from J. Miller and A. Dobson, in preparation).

areas that still retain rinderpest and can thus be targeted for intense vaccination campaigns that will hopefully lead to its final eradication. However, the growing size of the susceptible cattle population within the Serengeti ecosystem (and presumably in other areas) poses an enormous potential risk for wildlife should a new rinderpest outbreak spread, emphasizing the critical importance of sustained vigilance in wildlife populations (fig. 7.8).

DYNAMICS OF PATHOGENS WITH MULTIPLE HOSTS

A number of recent papers have examined the dynamics of pathogens with multiple hosts (Hudson and Greenman 1998; Woolhouse, Taylor, and Haydon 2001; Dobson 2004; Fenton and Pederson 2005). When the abundance of one host species significantly exceeds any of the others, it creates the potential to become a reservoir of infection that causes sporadic spillover infections into the less-abundant hosts. These events can be especially devastating when there are limited opportunities for the spillover host to build up "herd immunity" from endemic exposures. In these cases, old and young animals may be equally susceptible to infection, resulting in significant die-offs, particularly when virulence increases with host age. Many adult wildebeest and buffalo died in the initial rinderpest epidemics, and the deadly distemper outbreaks were characterized by abnormally high mortality in adult lions. This suggests that it will always be important to vaccinate rare species, such as wild dogs, when spillover outbreaks are likely to be frequent. Ultimately, the best policy may be to maintain adequate levels of vaccination in the domestic animals, as these are comparatively easy to locate; this will be particularly important for rinderpest and rabies, which also present major threats to human lives and livelihoods. Control efforts on domesticated reservoir species are also the best way to reduce the threat to wild species that are difficult to locate or capture (Fulford, Roberts, and Heesterbeek 2002)

Rinderpest As a Multi-Host-Species Pathogen

Rinderpest provides an important example of a pathogen that infects multiple host species. The dynamics of multi-host systems tend to be more stable (prevalence is relatively constant over time) than when a single host species is infected by a specific pathogen (Dobson 2004). Rinderpest was characterized by spectacular multispecies epidemics when it first appeared in east and southern Africa. However, in situations where a multi-host pathogen becomes endemic, interspecific differences in demographic rates should effectively desynchronize the underlying epidemic clockwork of regular, large-scale epidemics that characterize many single host-single pathogens, such as measles in humans, or rabies in European foxes. Instead, low rates of interspecific transmission would act to maintain the disease at a relatively constant prevalence in each host species (Dobson 2004). For carnivore diseases, such as CDV and rabies in the Serengeti, it is likely that intraspecific transmission within the large domestic dog population on the western borders of the Serengeti will dominate infection dynamics; the outbreaks observed in the much smaller wild carnivore populations will be caused by irregular spillover from the domestic reservoir, particularly at times when the epidemic peaks in the domestic dog population. In most cases, the spillover outbreaks in wild carnivores are likely to die out fairly quickly.

Multi-host pathogens have the potential to modify or even reverse competitive interactions between species while also tending to stabilize the potential for epidemic outbreaks (Dobson 2004; Hudson and Greenman 1998). The removal of rinderpest resulted in a dramatic increase in wildebeest and buffalo, while Thomson's gazelles and several other small herbivores subsequently declined—although this decline may have been due to either an overall increase in predator abundance or increased competition for forage (Dublin et al. 1990).

The Serengeti ecosystem is renowned for its extraordinary diversity of predators and prey, and therefore provides an exceptional opportunity for understanding the dynamics of multi-host pathogens in a complex community. A key question here is 'What is the role of species diversity in either amplifying, or buffering, infectious disease outbreaks?' Recent theoretical and empirical work on vector-transmitted pathogens such as Lyme disease in the United States and Louping ill in Scotland suggests that species diversity has the potential to create a "dilution effect" that reduces the magnitude of epidemic outbreaks (Ostfeld and Keesing 2000; LoGiudice et al. 2003). More theoretical work suggests that dilution effects may be strongest in vector-transmitted pathogens, while directly transmitted pathogens will only be buffered by increased host-species diversity when transmission is frequency dependent (Dobson 2004). For example, high levels of biodiversity may help buffer against epidemics of sleeping sickness. Because wildlife are preferred hosts for the tsetse vector of *T. brucei rhodesiense,* tsetse feeding rates and parasite transmission to livestock and humans are likely to be higher in areas with less wildlife. This may explain the observation that parasite prevalences in cattle are much higher in the wildlife-poor Soroti region of Uganda (18%; Welburn. Picozzi, et al. 2001) than in wildlife-rich areas surrounding the Serengeti (1%; Kaare et al. 2002). The decline of wildlife throughout Uganda in the 1980s may have contributed to the emergence of cattle as reservoirs for trypanosomiasis, which in turn led to new epidemics in the human population.

Human activities may also affect wildlife-infection dynamics. For example, cultivation and urbanization generally result in reduced biodiversity, and favor opportunistic species such as foxes, jackals, mongooses, and raccoons (*Procyon lotor*). These species may attain such high densities that they exceed the threshold for maintaining infectious diseases such as rabies, allowing persistence of the disease in areas previously free of endemic infection. High levels of biodiversity may therefore buffer against the establishment of rabies reservoirs and may explain the apparent absence of disease in protected areas, such as Hwange, despite the frequent occurrence of jackal rabies in areas surrounding the park (King 1992).

A second factor relates to the persecution of wildlife species, which may affect disease dynamics as a result of changing patterns of dispersal, contact, and aggressive encounters between individuals. Social disruption and perturbation as a result of culling have been suggested as factors contributing to the high incidence of jackal rabies in commercial lands of Zimbabwe (McKenzie 1993) and the failure of badger (*Meles meles*) culling to control bTB in the UK (Donnelly et al. 2003).

Social organization of the host species has an important effect on the impact of infectious disease. The key epidemiological effect of social organization is to fragment the host into a metapopulation, each subpopulation being a social group (Altizer et al. 2003; Dobson 2003). In species such as jackals, groups are small and relatively uniform in number: a breeding pair and their recent offspring, as well as one or two "helpers" from previous litters (Moehlman 1986). In lions and hyenas, groups are larger and more variable in size (Kruuk 1972; Hanby, Bygott, and Packer 1995) but are again based around a nucleus of breeding females, their mates, and offspring. The herbivores illustrate the largest and most variable groupings, ranging from the almost solitary dik-diks (*Madoqua kirkii*) to the huge aggregations of wildebeest.

In general, subdivision into social groups tends to increase the persistence time of a pathogen, compared to a homogenous population (Anderson and May 1986; Swinton 1998; Swinton et al. 1998). Relatively simple stochastic models for pathogens with similar levels of virulence and transmissibility as rabies, rinderpest, and CDV all suggest pathogen persistence times of several years in host populations of similar size and social organization as in the Serengeti (McCallum and Dobson 2006). However, the longer-term persistence of these pathogens would require either considerably larger host populations, or constant reintroduction from an alternate reservoir species, presumably domestic animals. To understand the ecology of infectious disease in the Serengeti we therefore need to look beyond the boundaries of the ecosystem. In the case of rabies, for example, large-scale epidemics apparently circulate over a much larger scale. In 2003, rabies was reported in domestic dogs living in 17 of Tanzania's 23 regions, whereas the norm is only one or two regions per year. Thus the frequency of infectious disease in the Serengeti may reflect the overall pattern across the entire country. Information generated at a local scale may therefore be critically important for future health planning in Africa.

HUMAN PATHOGENS IN THE SERENGETI

The current global epidemic of HIV/AIDS is likely to have a major impact on the human population surrounding the Serengeti. More than 90% of the world's AIDS cases are from sub-Saharan Africa, and a significant proportion of these are from East Africa. HIV prevalence is significantly higher among professional sex workers than among women visiting prenatal clinics, and prevalence is lower in rural rather than in urban areas of Tanzania. There is also some suggestion that prevalence levels in the rural community have stabilized at around 10%, which will have a significant impact on the local human demography and economy but will not be sufficient to reverse human population growth, as may occur in other parts of Africa. All of these findings underestimate the full impact of HIV, as uninfected relatives have to devote a considerable proportion of their income for the victim's treatment and funeral costs.

Finally, it is crucial to remember that malaria is still the worst killer in rural Tanzania. Malaria appears to interact synergistically with HIV, leading to relapses of disease and increased malaria transmission from adults who are HIV positive (Grimwade et al. 2004). The AIDS epidemic considerably frustrates attempts to assess the recent increase in malaria incidence. It is still impossible to determine whether changes in the distribution of malaria have arisen due to climate change or to increased vulnerability from HIV. Most likely, climate change, HIV, drug/insecticide resistance, and an expanding human population have all contributed to the observed increase in malaria cases. Reversing this trend and determining pragmatic ways of controlling malaria will increasingly require input from ecologists. Disease ecology has provided such important insights into the transmission of pathogens between wildlife and domestic animals that ecological research will eventually be viewed as fundamental to studying human disease.

CONCLUSIONS

Long-term studies in the Serengeti emphasize the role of infectious disease in determining the abundance, distribution, and variability of wildlife species. While pathogens have always played a subtle natural role in the system, the most dramatic epizootics have involved pathogens shared by wildlife and domestic animals. Rinderpest provides the classic example, but recent outbreaks of distemper and rabies illustrate that pathogen exchange between domestic animals and wildlife is a recurring consequence of human activities around the Serengeti. Pathogens shared by multiple-host species have serious implications for the management of the Serengeti. The importance of wildlife in emerging human diseases and as reservoirs for important livestock diseases is likely to require increased collaboration between wildlife managers, public health officials, and veterinary specialists. Decisions about whether and how best to intervene will require a coordinated and integrated approach that not only addresses the needs of human health and rural livelihoods but also minimizes negative impacts on Tanzania's wildlife resources.

ACKNOWLEDGMENTS

We are grateful to the Tanzanian Wildlife Research Institute and Tanzanian National Parks for permission to undertake research in the Serengeti and surrounding areas. Numerous colleagues have provided essential insights, particularly Tony Sinclair, Karen Laurenson, Dan Haydon, Robert Holt, Cassandra Nuñez, Christine Mentzel, Ernest Eblate, Eric Fèvre, and Julia Miller. SC was supported by a Wellcome Trust Fellowship in Tropical Medicine and by grants from the DFID Animal Health Programme (R5406, R7357, R7985). Material relating to carnivore viral transmission dynamics is based on work supported by the joint National Institute of Health/National Science Foundation Ecology of Infectious Diseases Program under grant no. 0225453. Any opinions, findings, conclusions, or recommendations expressed in this material are those of the authors, and do not necessarily reflect the views of the NIH or the NSF. We would like to thank two anonymous referees for extremely valuable comments on an earlier version of the manuscript.

REFERENCES

Aikman, D. I., and S. M. Cobb. 1997. Water development. In *Multiple land-use: The experience of the Ngorongoro Conservation Area, Tanzania,* ed. D. M. Thompson, 201–31. Gland, Switzerland: IUCN—The World Conservation Union.

Alexander, K. A., P. W. Kat, L. G. Frank, K. E. Holekamp, L. Smale, C. House, and M. J. G. Appel. 1995. Evidence of canine distemper virus infection among free-ranging spotted hyenas (*Crocuta crocuta*) in the Maasai Mara, Kenya. *Journal of Zoo and Wildlife Medicine* 26:201–6.

Altizer, S., C. L. Nunn, P. Thrall, J. M. Gittleman, J. Antonovics, A. A. Cunningham, A. P. Dobson, et al. 2003. Social organization and parasite risk in mammals: Integrating theory and empirical studies. *Annual Reviews of Ecology and Systematics* 34:517–47.

Anderson, R. M., and R. M. May. 1986. The invasion, persistence and spread of infectious diseases within animal and plant communities. *Philosophical Transactions of the Royal Society of London, Series B—Biological Sciences* 314:533–70.

———. 1991. *Infectious diseases of humans: Dynamics and control.* Oxford: Oxford University Press.

Appel, M. J. G., and B. A. Summers. 1995. Pathogenicity of morbilliviruses for terrestrial carnivores. *Veterinary Microbiology* 44:187–191.

Appel, M. J. G., R. A. Yates, G. L. Foley, J. J. Bernstein, S. Santinelli, L. H. Spelman, L. D.

Miller, et al. 1994. Canine distemper epizootic in lions, tigers and leopards in North America. *Journal of Veterinary Diagnostic Investigation* 6:277–88.

Awan, M. A., M. J. Otte, and A. D. James. 1994. The epidemiology of Newcastle disease in rural poultry: A review. *Avian Pathology* 23:405–23.

Barnard, B. H., J. van der Lugt, and E. Z. Mushi. 1994. Malignant catarrhal fever. In *Infectious Diseases of Livestock with special reference to Southern Africa*, ed. J. A. W. Coetzer, G. R. Thomson, and R. C. Tustin, 946–57. Cape Town, Oxford: Oxford University Press.

Blythe, L. L., J. A. Schmitz, M. Roelke, and S. Skinner. 1983. Chronic encephalomyelitis caused by canine distemper virus in a Bengal tiger. *Journal of the American Veterinary Medical Association* 183:1159–62.

Boone, R. B., M. B. Coughenour, K. A. Galvin, and J. E. Ellis. 2002. Addressing management questions for Ngorongoro Conservation Area, Tanzania, using the SAVANNA modeling system. *African Journal of Ecology* 40:138–50.

Burrows, R., H. Hofer, and M. L. East. 1994. Demography, extinction and intervention in a small population—the case of the Serengeti wild dogs. *Proceedings of the Royal Society of London, Series B—Biological Sciences* 256:281–92.

———. 1995. Population dynamics, intervention and survival in African wild dogs (*Lycaon pictus*). *Proceedings of the Royal Society of London, Series B—Biological Sciences* 262:235–45.

Caro, T. M., M. E. Holt, C. D. FitzGibbon, M. Bush, C. M. Hawkey, and R. A. Kock. 1987. Health of adult free-living cheetahs. *Journal of Zoology* 212:573–84.

Caron, A., P. C. Cross, and J. T. du Toit. 2003. Ecological implications of bovine tuberculosis in African buffalo herds. *Ecological Applications* 13:1338–45.

Carpenter, M. A., M. J. G. Appel, M. E. Roelke-Parker, L. Munson, H. Hofer, M. East, and S. J. O'Brien. 1998. Genetic characterization of canine distemper virus in Serengeti carnivores. *Veterinary Immunology and Immunopathology* 65:259–66.

Cleaveland, S. 1996. The epidemiology of rabies and canine distemper in the Serengeti. PhD Diss., University of London.

Cleaveland, S., M. J. E. Appel, W. S. K. Chalmers, C. Chillingworth, M. Kaare, and C. Dye. 2000. Serological and demographic evidence for domestic dogs as a source of canine distemper virus infection for Serengeti wildlife. *Veterinary Microbiology* 72:217–27.

Cleaveland, S., and C. Dye. 1995. Maintenance of a microparasite infecting several host species: Rabies in the Serengeti. *Parasitology* 111:S33–S47.

Cleaveland, S., E. M. Fèvre, M. Kaare, P. G. and Coleman. 2002. Estimating human rabies mortality in the United Republic of Tanzania from dog bite injuries. *Bulletin of the World Health Organization* 80:304–10.

Cleaveland, S., M. Kaare, P. Tiringa, T. Mlengeya, and J. Barrat. 2003. A dog rabies vaccination campaign in rural Africa: Impact on the incidence of dog rabies and human dog-bite injuries. *Vaccine* 21:1965–73.

Cleaveland, S., L. Kusiluka, J. ole Kuwai, C. Bell, and R. Kazwala. 2001. Assessing the impact of malignant catarrhal fever in Ngorongoro District, Tanzania. Department for International Development, Animal Health Programme. Edinburgh, UK: Centre for Tropical Veterinary Medicine, University of Edinburgh.

Cleaveland, S., M. K. Laurenson, and L. H. Taylor. 2001. Diseases of humans and their domestic mammals: Pathogen characteristics, host range and the risk of emergence.

Philosophical Transactions of the Royal Society of London, Series B—Biological Sciences 356:991–99.

Cleaveland, S., T. Mlengeya, R. R. Kazwala, A. Michel, S. L. Jones, E. Eblate, G. M. Shirima, and C. Packer. 2005. Tuberculosis in Tanzanian wildlife. *Journal of Wildlife Diseases* 41:446–53.

Courchamp, F., T. Clutton-Brock, and B. Grenfell. 2000. Multipack dynamics and the Allee effect in the African wild dog, *Lycaon pictus*. *Animal Conservation* 3:277–85.

Creel, S., N. M. Creel, and S. L. Monfort. 1997. Radiocollaring and stress hormones in African wild dogs. *Conservation Biology* 11:544–48.

de Villiers, M. S., D. G. A. Meltzer, J. van Heerden, M. G. L. Mills, P. R. K. Richardson, and A. S. van Jaarsveld. 1995. Handling induced stress and mortalities in African wild dogs (*Lycaon pictus*). *Proceedings of the Royal Society of London, Series B—Biological Sciences* 262:215–20.

Dobson, A. P. 1988. The population biology of parasite-induced changes in host behavior. *The Quarterly Review of Biology* 63:139–65.

———. 1995. The ecology and epidemiology of rinderpest virus in Serengeti and Ngorongoro crater conservation area. In *Serengeti II: Research, management and conservation of an ecosystem*, ed. A. R. E. Sinclair and P. Arcese, 485–505. Chicago: University of Chicago Press.

———. 2003. Metalife! *Science* 301:1488–90.

———. 2004. Population dynamics of pathogens with multiple hosts. *The American Naturalist* 164:S64–S78.

Dobson, A. P., and M. J. Crawley. 1994. Pathogens and the structure of plant communities. *Trends in Ecology and Evolution* 9:393–98.

Dobson, A. P., and R. M. May. 1986. Disease and conservation. In *Conservation biology: Science of diversity*, ed. M. E. Soulé, 485–505. Sunderland, MA: Sinauer.

Donnelly, C. A., R. Woodroffe, D. R. Cox, J. Bourne, G. Gettinby, A. M. Le Fèvre, J. P. McInerney, and W. I. Morrison. 2003. Impact of localized badger culling on tuberculosis incidence in British cattle. *Nature* 426:834–37.

Dublin, H. T., A. R. E. Sinclair, S. Boutin, E. Anderson, M. Jago, and P. Arcese. 1990. Does competition regulate ungulate populations? Further evidence from Serengeti, Tanzania. *Oecologia* 82:283–88.

East, M. L., H. Hofer, J. H. Cox, U. Wulle, H. Wiik, and C. Pitra. 2001. Regular exposure to rabies virus and lack of symptomatic disease in Serengeti spotted hyenas. *Proceedings of the National Academy of Sciences of the United States of America* 98:15026–31.

Fenton, A. and A. B. Pedersen. 2005. Community epidemiology framework for classifying disease threats. *Emerging Infectious Diseases* 11:1815–21.

Fulford, G. R., M. G. Roberts, and J. A. P. Heesterbeek. 2002. The metapopulation dynamics of an infectious disease: Tuberculosis and possums. *Theoretical Population Biology* 61:15–29.

Funk, S. M., C. V. Fiorella, S. Cleaveland, and M. E. Gompper. 2001. The importance of disease in carnivore conservation. In *Symposia of the Zoological Society of London*, 443–66. Cambridge: Cambridge University Press.

Gao, F., E. Bailes, D. L. Robertson, Y. L. Chen, C. M. Rodenburg, S. F. Michael, L. B. Cummins, et al. 1999. Origin of HIV-1 in the chimpanzee *Pan troglodytes troglodytes*. *Nature* 397:436–41.

Grimwade, K., N. French, D. D. Mbatha, D. D. Zungu, M. Dedicot, and C. F. Gilks. 2004. HIV infection as a cofactor for severe falciparum malaria in adults living in a region of unstable malaria transmission in South Africa. *AIDS* 18:547–54.

Hahn, B. H., G. M. Shaw, K. M. De Cock, and P. M. Sharp. 2000. AIDS—AIDS as a zoonosis: Scientific and public health implications. *Science* 287:607–14.

Hanby, J. P., J. D. Bygott, and C. Packer.1995. Ecology, demography and behavior of lions in two contrasting habitats: Ngorongoro crater and the Serengeti plains. In *Serengeti II: Dynamics, management and conservation of an ecosystem,* ed. A. R. E. Sinclair and P. Arcese, 315–31. Chicago: University of Chicago Press.

Harder, T. C., M. Kenter, H. Vos, K. Siebelink, W. Huisman, G. vanAmerongen, C. Orvell, T. Barrett, M. J. G. Appel, and A. D. M. E. Osterhaus. 1996. Canine distemper virus from diseased large felids: Biological properties and phylogenetic relationships. *Journal of General Virology* 77:397–405.

Harder, T. C., and A. D. M. E. Osterhaus. 1997. Canine distemper virus—a morbillivirus in search of new hosts? *Trends in Microbiology* 5:120–24.

Harrison, T. M., J. K. Mazet, K. E. Holekamp, E. Dubovi, A. L. Engh, K. Nelson, R. C. Van Horn, and L. Munson. 2004. Antibodies to canine and feline viruses in spotted hyenas (*Crocuta crocuta*) in the Maasai Mara National Reserve. *Journal of Wildlife Diseases* 40:1–10.

Haydon, D. T., S. Cleaveland, L. H. Taylor, and . K. Laurenson. 2002. Identifying reservoirs of infection: A conceptual and practical challenge. *Emerging Infectious Diseases* 8:1468–73.

Hofer, H., and M. East. 1995. Population dynamics, population size, and the commuting system of Serengeti Spotted Hyenas. In *Serengeti II: Dynamics, management and conservation of an ecosystem,* ed. A. R. E. Sinclair and P. Arcese, 332–63. Chicago: University of Chicago Press.

Hofmeyr, M., J. Bingham, E. P. Lane, A. Ide, and L. Nel. 2000. Rabies in African wild dogs (*Lycaon pictus*) in the Madikwe Game Reserve, South Africa. *Veterinary Record* 146: 50–52.

Homewood, K. M. 2004. Policy, environment and development in African rangelands. *Environmental Science and Policy* 7:125–43.

Homewood, K. M., and W. A. Rodgers. 1991. *Maasailand ecology.* Cambridge: Cambridge University Press.

Homewood, K., W. A. Rodgers, and K. Arhem. 1987. Ecology of pastoralism in the Ngorongoro Conservation Area, Tanzania. *Journal of Agricultural Science* 108:47–72.

Hudson, P. J., A. P. Dobson, I. M. Cattadori, D. Newborn, D. T. Haydon, D. J. Shaw, T. G. Benton, and B. T. Grenfell. 2002. Trophic interactions and population growth rates: Describing patterns and identifying mechanisms. *Philosophical Transactions of the Royal Society of London, Series B—Biological Sciences* 357:1259–71.

Hudson, P., and J. Greenman. 1998. Competition mediated by parasites: Biological and theoretical progress. *Trends in Ecology and Evolution* 13:387–90.

Jelinek, T., Z. Bisoffi, L. Bonazzi, P. van Thiel, U. Bronner, A. de Frey, S. G. Gundersen, P. McWhinney, and D. Ripamonti. 2002. Cluster of African trypanosomiasis in travelers to Tanzanian national parks. *Emerging Infectious Diseases* 8:634–35.

Kaare, M. T., K. Picozzi, S. C. Welburn, M. M. Mtambo, L. S. B. Mellau, S. Cleaveland, and T. M. Mlengeya. 2002. Sleeping sickness in the Serengeti ecological zone: The role of wild and domestic animals as reservoir hosts for the disease. *Proceedings of the Third Annual Scientific Conference of the Tanzania Wildlife Research Institute,* Arusha, Tanzania, December 3–5, 2002, 204–17.

Kat, P. W., K. A. Alexander, J. S. Smith, L. and Munson. 1995. Rabies and African wild dogs in Kenya. *Proceedings of the Royal Society of London, Series B—Biological Sciences* 262:229–33.

Keet, D. F., N. P. Kriek, M. L. Penrith, A. Michel, and H. Huchzermeyer. 1996. Tuberculosis in buffaloes (*Syncerus caffer*) in the Kruger National Park: Spread of disease to other species. *Onderstepoort Journal of Veterinary Research* 63:239–44.

Kijazi, A. 1997. Principal management issues in the Ngorongoro Conservation Area. In *Multiple land-use: The experience of the Ngorongoro Conservation Area, Tanzania,* ed. D. M. Thompson, 33–43. Gland, Switzerland: IUCN—The World Conservation Union.

King, A. 1992. African overview and antigenic variation. In *Proceedings of the International Conference on Epidemiology, Control and Prevention of Rabies in Eastern and Southern Africa,* ed. A. A. King, 57–68. Lusaka, Zambia, and Lyon, France: Editions Fondation Marcel Merieux.

Knobel, D. L., S. Cleaveland, P. G. Coleman, E. M. Fevre, M. I. Meltzer, M. E. G. Miranda, A. Shaw, J. Zinnstag, and F.-X. Meslin. 2005. Re-evaluating the burden of rabies in Africa and Asia. *Bulletin of the World Health Organization* 83:360–68.

Kock, R. 2005. "What is this infamous 'wildlife/livestock interface?'"—A review of current knowledge. In *Conservation and development interventions at the wildlife/livestock interface: Implications for wildlife, livestock and human health,* ed. S. A. Osofsky et al., 1–13. Gland, Switzerland: IUCN.

Kock, R. A., W. S. K. Chalmers, J. Mwanzia, C. Chillingworth, J. Wambura, P. Coleman, and W. Baxendale. 1998. Canine distemper antibodies in lions of the Maasai Mara. *Veterinary Record* 142:662–65.

Kock, R. A., J. M. Wambua, J. Mwanzia, H. Wamwayi, E. K. Ndungu, T. Barrett, N. D. Kock, and P. B. Rossiter. 1999. Rinderpest epidemic in wild ruminants in Kenya 1993–97. *Veterinary Record* 145:275–83.

Kristjanson, P. M., B. M. Swallow, G. J. Rowlands, R. L. Kruska, and P. N. de Leeuw. 1999. Measuring the costs of African animal trypanosomosis, the potential benefits of control and returns to research. *Agricultural Systems* 59:79–98.

Kruuk, H. 1972. *The spotted hyena: A study of predation and social behavior.* Chicago: University of Chicago Press.

Lafferty, K. 1992. Foraging on prey that are modified by parasites. *American Naturalist* 140: 854–67.

Lembo, T., D. T. Haydon, A. Velasco-Villa, C. E. Rupprecht, C. Packer, P. E. Brandoi, I. V. Kuzmin, et al. 2007. Molecular epidemiology identifies only a single rabies virus variant circulating in complex carnivore communities of the Serengeti. *Proceedings of the Royal Society, Series B,* 274 (1622): 2123–30.

Loibooki, M., H. Hofer, K. L. I. Campbell, and M. L. East. 2002. Bushmeat hunting by communities adjacent to the Serengeti National Park, Tanzania: The importance of livestock ownership and alternative sources of protein and income. *Environmental Conservation* 29:391–98.

LoGiudice, K., R. S. Ostfeld, K. A. Schmidt, and F. Keesing. 2003. The ecology of infectious disease: Effects of host diversity and community composition on Lyme disease risk. *Proceedings of the National Academy of Science of the United States of America.* 100: 567–71.

Maas, B. 1993. Bat-eared fox behavioral ecology and the incidence of rabies in the Serengeti National Park. *Onderstepoort Journal of Veterinary Research* 60:389–93.

Macdonald, D. W., M. Artois, M. Aubert , D. L. Bishop, J. R. Ginsberg, A. King, N. Kock, and B. D. Perry. 1992. Cause of wild dog deaths. *Nature* 360:633–34.

Magembe, S. R. 1985. Epidemiology of rabies in the United Republic of Tanzania. In *Rabies in the Tropics,* ed. E. Kuwert, C. Merieux, H. Koprowski, and K. Bogel, 392–98. Berlin: Springer-Verlag.

Matzke, G. 1979. Settlement and sleeping sickness control—a dual threshold model of colonial and traditional methods in East Africa. *Social Science and Medicine* 13D:209–14.

McCabe, J. T., E. C. Schofield, and G. Nygaard Pederson. 1997. Food security and nutritional status. In *Multiple land-use: The experience of the Ngorongoro Conservation Area, Tanzania,* ed. D. M. Thompson, 285–301. Gland, Switzerland: IUCN—The World Conservation Union.

McCallum, H., and A. Dobson. 2006. Disease and connectivity. In *Connectivity conservation,* ed. K. Crooks and M. Sanjayan, 479–501. Cambridge: Cambridge University Press.

McKenzie, A. A. 1993. Biology of the black-backed jackal *Canis mesomelas* with reference to rabies. *Onderstepoort Journal of Veterinary Research* 60:367–71.

Ministry of Agriculture. 1995. Mara Region Agricultural Development Project, United Republic of Tanzania. Dar es Salaam, Tanzania: Ministry of Agriculture.

Misana, S. B. 1997. Vegetation change. In *Multiple land-use: The experience of the Ngorongoro Conservation Area, Tanzania,* ed. D. M. Thompson, 97–109. Gland, Switzerland: IUCN—The World Conservation Union.

Mlengeya, T. D. K., C. Muangirwa, M. M. Mlengeya, E. Kimaro, S. Msangi, and M. Sikay. 2002. Control of sleeping sickness in northern parks of Tanzania. *Proceedings of the Third Annual Scientific Conference of the Tanzania Wildlife Research Institute.* Arusha, Tanzania, December 3–5, 2002, 274–81.

Moehlman, P. D. 1986. Ecology of cooperation in canids. In *Ecological aspects of social evolution: Birds and mammals,* ed. D. I. Rubenstein and R. W. Wrangham. Princeton, NJ: Princeton University Press.

Moloo, S. K., R. F. Steiger, R. Brun, and P. F. L. Boreham. 1971 Sleeping sickness survey in Musoma District, Tanzania. Part II. The role of *Glossina* in the transmission of sleeping sickness. *Acta Tropica* 28:189–205.

Mwanzia, J. M., R. Kock, J. M. Wambua, N. Kock, and O. Jarret. 1995. An outbreak of sarcoptic mange in the free-living cheetah (*Acinonyx jubatus*) in the Mara region of Kenya. *Proceedings of the American Association of Zoo Veterinarians and American Association of Wildlife Veterinarians Joint Conference.* Omaha, NB:105–12.

NCAA/NPW. 1994. Census results. Ngorongoro, Tanzania: Ngorongoro Conservation Area Authority.

Ostfeld, R. S., and F. Keesing. 2000. The function of biodiversity in the ecology of vector-borne zoonotic diseases. *Canadian Journal of Zoology* 78:2061–78.

Packer, C., S. Altizer, M. Ael, E. Brown, J. Martenson, S. J. O'Brien, M. Roelke-Parker, R. Hofmann-Lehmann, and H. Lutz. 1999. Viruses of the Serengeti: Patterns of infection and mortality in African lions. *Journal of Animal Ecology* 68:1161–78.

Packer, C., R. Hilborn, A. Mosser, B. Kissui, M. Borner, G. Hopcraft, J. Wilmshurst,

S. Mduma, and A. R. E. Sinclair. 2005. Ecological change, group territoriality, and population dynamics in Serengeti lions. *Science* 307:390–93.

Packer, C., R. D. Holt, P. J. Hudson, K. D. Lafferty, and A. P. Dobson. 2003. Keeping the herds healthy and alert: Implications of predator control for infectious disease. *Ecology Letters* 6:797–802.

Pence, D. B., and E. Ueckermann. 2002. Sarcoptic mange in wildlife. *Revue Scientifique et Technique del Office International des Epizooties* 21:385–98.

Potanski, T. 1999. Mutual assistance among the Ngorongoro Maasai. In *The poor are not us,* ed. D. M. Anderson and V. Broch-Due, 199–217. Oxford: James Currey.

Prins, H. H. T., and H. P. van der Jeugd. 1993. Herbivore population crashes and woodland structure in East Africa. *Journal of Ecology* 81:305–14.

Prins, H. H. T., and F. J. Weyerhaeuser. 1987. Epidemics in populations of wild ruminants: Anthrax and impala, rinderpest and buffalo in Lake Manyara National Park, Tanzania. *Oikos* 49:28–38.

Roelke-Parker, M. E., L. Munson, C. Packer, R. Kock, S. Cleaveland, M. Carpenter, S. J. Obrien, et al. 1996. A canine distemper virus epidemic in Serengeti lions (*Panthera leo*). *Nature* 379:441–45.

Rweyemamu, M. M., K. Loretu, H. Jakob, and E. Gorton. 1973. Observations on rabies in Tanzania. *Bulletin of Epizootic Diseases in Africa* 21:19–27.

Schmunis, G.A. 2004. Medical significance of American trypanosomiasis. In *The trypanosomiases.* ed. I. Maudlin, P. H. Holmes, and M. A. Mills, 283–302. Wallingford, UK: CABI International.

Schwabe, C.W. 1984. *Veterinary medicine and human health,* 3rd ed. Baltimore: Williams and Wilkins.

Serneels, S., and E. F. Lambin. 2001. Impact of land-use change on the Wildebeest migration in the Northern part of the Serengeti-Mara ecosystem. *Journal of Biogeography* 28:391–407.

Shaw, A. P. M. 2004. Economics of African trypanosomiasis. In *The trypanosomiases,* ed. I. Maudlin, P. H. Holmes, and M. A. Mills, 369–402. Wallingford, UK: CABI International.

Sinclair, A. R. E. 1979. The eruption of the ruminants. In *Serengeti: Dynamics of an ecosystem,* ed. A. R. E. Sinclair and M. Norton-Griffiths, 82–103. Chicago: University of Chicago Press.

Swinton, J. 1998. Extinction times and phase transitions for spatially structured closed epidemics. *Bulletin of Mathematical Biology* 60 (2): 215–30.

Swinton, J., J. Harwood, B. T. Grenfell, and C. A. Gilligan. 1998. Persistence thresholds for phocine distemper virus infection in harbour seal *Phoca vitulina* metapopulations. *Journal of Animal Ecology* 67:54–68.

Taylor, L. H., S. M. Latham, and M. E. J. Woolhouse. 2001. Risk factors for human disease emergence. *Philosophical Transactions of the Royal Society of London, Series B—Biological Sciences* 356:983-989.

Thompson, D. M. 1997. *Multiple land-use: The experience of the Ngorongoro Conservation Area, Tanzania.* Gland, Switzerland: IUCN—The World Conservation Union.

van de Bildt, M. W. G., T. Kuiken, A. M. Visee, S. Lema, T. R. Fitzjohn, and A. D. M. E. Osterhaus. 2002. Distemper outbreak and its effect on African wild dog conservation. *Emerging Infectious Diseases* 8:211–13.

Welburn, S. C., E. M. Fevre, P. G. Coleman, M. Odiit, and I. Maudlin. 2001. Sleeping sickness: A tale of two disease. *Trends in Parasitology* 17:19-24.

Welburn, S. C., K. Picozzi, E. M. Fevre, P. G. Coleman, M. Odiit, M. Carrington, and I. Maudlin. 2001. Identification of human-infective trypanosomes in animal reservoir of sleeping sickness in Uganda by means of serum- resistance-associated (SRA) gene. *Lancet* 358:2017-19.

Woodroffe, R. 1997. The conservation implications of immobilizing, radio-collaring and vaccinating free-ranging wild dogs. In *The African wild dog: Status survey and conservation action plan,* ed. R. Woodroffe, J. R. Ginsberg, and D. Macdonald, 124-38. Gland, Switzerland: IUSN/SSC Canid Specialist Group, IUCN.

————. 2001. Assessing the risks of intervention: Immobilization, radio-collaring and vaccination of African wild dogs. *Oryx* 35:234-44.

Woodroffe, R., S. Cleaveland, O. Courtenay, M. K. Laurenson, and M. Artois. 2004. Infectious disease in the management and conservation of wild canids. In *The biology and conservation of wild canids,* ed D. W. Macdonald and C. Sillero-Zubiri, 123-42. Oxford: Oxford University Press.

Woodroffe, R., J. R. Ginsberg, and D. Macdonald. 1997. *The African wild dog: Status survey and conservation action plan.* Gland, Switzerland: IUCN.

Woolhouse, M. E. J., L. H. Taylor, and D. T. Haydon. 2001. Population biology of multi-host pathogens. *Science* 292:1109-12.

EIGHT

Reticulate Food Webs in Space and Time:
Messages from the Serengeti

R. D. Holt, P. A. Abrams, J. M. Fryxell, and T. Kimbrell

Food webs—which characterize feeding relationships among species within defined temporal and spatial limits—are among the most complex entities studied by scientists. Yet understanding these complex networks is a prerequisite for forecasting the future impacts of environmental changes on the populations of any species within the web. Predicting the responses of individual species within a web to perturbations affecting a subset of those species has proven to be a particularly difficult problem. Yodzis' (1996, 1998, 2000) analysis of interactions in a complex marine ecosystem (the Benguela) is a well-known example of the difficulty of determining responses of population densities to environmental changes within a trophic web. In spite of a large body of studies on the interspecific interactions in the Benguela ecosystem, uncertainties about the strengths of these interactions made it impossible to predict even the direction of change in a commercially valuable fish stock following harvesting of fur seals. The problem in Yodzis' model arises because of long chains of interactions that result in complicated indirect effects between species, and complicated feedbacks on a single species. Uncertainties from such long chains of effects arise in most theoretical models of highly complex systems, as noted by Yodzis (1988), Schoener (1993), and Abrams et al. (1996), and can even arise in systems with just a few interacting species (Holt and Kotler 1987; Abrams 2002).

The magnitude of uncertainty will often depend on the temporal scale over which one is attempting prediction. For some management purposes, a short time horizon may suffice. Following a perturbation, the impact of indirect effects passing through several intermediary species can usually be ignored when the time frame for prediction is relatively short (Yodzis 1988). Over short time scales, the behavior of a food web may be dominated by transients, reflecting interactions among strongly interacting species. But short-term predictions are not sufficient for achieving most goals in conservation biology, where one is most concerned with the long-term viability of an ecosystem or the persistence of a set of species. At longer timescales, cascading indeterminacies due to indirect chains through the whole web can loom large.

There is a second class of uncertainties that arises from our lack of knowledge of the appropriate functional form to use when representing feeding relationships, and about the amount and role of between-individual variation in these relationships that must be included in models. These uncertainties have not been investigated explicitly, but they may be at least as much of a barrier for accurate prediction as are the intrinsic unpredictabilities that can emerge due to complex webs of interactions. The emergent effects of such dynamic uncertainties are often clear in management efforts. For instance, in Kruger National Park, there have been multiple unforeseen consequences of human attempts to manipulate the system via culling, fire management, and manipulation of water supplies (du Toit, Rogers, and Biggs 2003).

At the same time, there are also reasons for believing that some systems may have special features that make predictions more reliable. In theory, predictions should have greater reliability if a relatively small number of abundant species drive the behavior of the entire web, provided that the interactions between these few species is well known. In this case, analyses of community modules (sensu Holt 1997) could capture some of the essential dynamics of the full system. Alternatively, if the web consists of small sets of strongly interacting species that are weakly coupled to the rest of the web, this could also provide the basis for accurate predictions (although in practice, communities may not be organized in neat compartments; Melian and Bascompte 2002). There are a variety of other factors (e.g., life history, ecosystem constraints) that have been proposed as general explanations that could allow predictions in spite of complexity (Schoener 1993). Nevertheless, when accurate predictions have been required for practical applications, as in the case of most commercially important fisheries, they have either not been forthcoming, or have not been sufficiently certain to convince skeptical policymakers.

The Serengeti is a system that is both unusually well studied (Sinclair and Norton-Griffiths 1979; Sinclair and Arcese 1995), and unusually important. It is the last preserved ecosystem whose dynamics are driven by migratory herbivores, and it ranks among the world's largest national parks. Along with a few other systems (e.g., the Kruger and Yellowstone ecosystems), the Serengeti presents a unique opportunity for assessing the extent to which ecological research has, or can, lead to credible predictions of the future dynamics of major components of a terrestrial food web. A focus on food webs automatically means one is concerned with predator-prey (or plant-herbivore) interactions, and there is a large body of existing work that informs us about the processes that determine direct and indirect effects in models representing these interactions. Given what is already known about the Serengeti, it is clear that additional interspecific interactions such as exploitative competition, facilitation, and interference could also be critical determinants of food web dynamics. The question is whether both trophic and nontrophic interactions are known in sufficient detail to identify the direction of change in major species or species groups in response to a number of possible environmental changes. McNaughton (1992) suggested that strong trophic interactions in savannas could lead to the propagation of localized disturbances across the entire system, and buttressed this suggestion with retrospective analyses of historical disturbances. It is important to revisit this issue in the context of our contemporary understanding of the complexity of food web interactions.

Ecologists are increasingly being asked to provide predictions about the ecosystem consequences of various human activities, and some have argued that reliable predictions should soon be possible for a wide range of systems (Clark et al. 2001). It might seem that identifying the directions of changes in food web components is a rather limited goal, since many decisions about policy hinge on quantitative rather than qualitative responses. However, identifying the signs of the changes in the population size of particular species or species groups is still a difficult proposition, even for well-studied systems like the Serengeti. There is an increasing awareness that populations can respond in counterintuitive ways to change because of indirect pathways of effect within the food web (Yodzis 1988; Schoener 1993; Holt 1997; Abrams 1993, 1995, 2002; Abrams and Vos 2003). Moreover, it is likely that the sign of the change in a focal population in response to a perturbation imposed on another population will often depend on the magnitude of the original perturbation (Abrams 2001; Abrams 2004). Furthermore, effects that are propagated through the web can cause major changes in community composition. If a keystone predator is pushed to low numbers, a cascade of extinctions may ensue, because a mechanism of coexistence between prey species has been disrupted (e.g., Estes et al. 1998). Species may increase under the impact of increased mortality; predators may increase with increased mortality imposed on their prey, and prey may decrease because of mortality imposed on their predators (Abrams and Vos 2003). A reduction in spatial coupling via immigration and emigration (e.g., due to land use change, or mortality during transit) can lead to surprising effects on average abundance and population stability (Holt 2002). It is certainly not true that all indirect effects in food webs are indeterminate in sign, or that they produce either extinctions or counterintuitive responses. However, these possibilities simply illustrate our general inability to predict indirect effects in the absence of previous experimental manipulations or a mathematical model that incorporates accurate descriptions of the interspecific interactions in the web. Although poaching and disease have provided some "natural experiments" in the Serengeti ecosystem (e.g., Sinclair 1995; Sinclair, Mduma, and Brashares 2003), there is still considerable uncertainty about the mechanisms underlying the changes that those perturbations caused. This chapter will use some simple models to ascertain what kinds of predictions can be made about ecosystem responses to some plausible perturbations, and what additional information is needed to either enable or improve these predictions.

We begin by proposing a set of three potential perturbations that can be applied within a modeling context to determine our ability to predict the responses of the rest of the system: disruptions of populations of dominant vertebrate predators, or herbivores, and changes in plant productivity. We then evaluate our knowledge of the Serengeti food web and apply it to the task of determining the consequences of these perturbations. In addition to assessing how well we can predict the fate of major players within the ecosystem, this exercise can shed light on the potential ability to make similar types of predictions in other ecosystems. If we fall far short of sufficient knowledge for predictions in a system as well-studied as the Serengeti, it is likely that predictions will be even more difficult in other ecosystems.

POTENTIAL PERTURBATIONS TO THE SYSTEM

We will use three perturbations to assess our ability to predict food web responses. The Serengeti faces a range of potential threats, including a number of additional environmental changes whose effects are more difficult to classify as positive or negative. Here, we focus on the impacts of: (1) increased mortality of lions (arguably the best-studied predator), which could arise because of upsurges in disease such as canine distemper, or direct attacks from humans (see chapter 7, this volume); (2) increased mortality of wildebeest (which is commonly the largest single source of food supporting both lion and hyena population growth), which, for instance, could reflect an outbreak of rinderpest; and (3) an increase in the growth rate of the major grass species (a potential, although far from certain, outcome of global climate change). A variety of other perturbations could have been analyzed, but this set seems sufficient to assess our power to predict responses in the Serengeti.

Historically documented changes in the system have included major perturbations to the mortality rates of both carnivores and herbivores (Dobson 1995; Arcese, Hando, and Campbell 1995; Sinclair, Mduma and Brashares 2003, see chapters 1, 2, and 7, this volume). Although local changes in productivity are difficult to predict, it seems virtually certain that global climate change will have some significant impacts on plant productivity. Furthermore, there have been pronounced differences in rainfall at both annual and decadal timescales in the past (see chapters 2–6, this volume). The consequences of our three types of perturbations have been studied empirically in a wide range of other ecosystems (Power, Parker, and Wootton 1996). There have also been many analyses of simple models that have addressed the consequences of environmental changes affecting one species or level, both for the abundances of all trophic levels, and, to a lesser extent, for the diversity of species on each level (e.g., Abrams 1993; Holt, Grover, and Tilman 1994; Abrams and Roth 1994; Grover and Holt 1998). Thus, there is both an empirical and a theoretical basis for assessing what changes in abundance of selected species (or food web compartments) we might expect from each of these three hypothetical changes to the system. Predictions require some sort of model of the system.

SIMPLE DESCRIPTIONS OF THE SERENGETI FOOD WEB

At the most basic level, a food web consists of a number of nodes, representing species, connected by arrows, representing significant predatory interactions. As many authors have pointed out (e.g., Paine 1988; Schoener 1989), such qualitative descriptions allow very little in the way of prediction, although they do suggest possible pathways of indirect effects between species. A slightly more detailed description of a food web includes information on the amount to which each prey species contributes to the food intake of each predator, together with a rough idea of the extent to which changes in food density are reflected in changes in predator population densities. On the prey side, this second level of description includes the extent to which one or all predator species contribute to mortality in the prey. Such second-level models can be translated into Lotka-Volterra models, which can be regarded as linear approximations to a true underlying, nonlinear model. The Lotka-Volterra approximation may be adequate when a perturbation is sufficiently small in magnitude. A third level of description includes quantitative descriptions of the functional relationships that constitute the consumer-resource and other important interactions in the system. This includes descriptions of both functional and numerical responses of consumers, and should also include a reasonably quantitative description of any other major inter- or intraspecific interactions. Such a description is the minimum that might provide accurate predictions of the consequences of perturbations that are large in magnitude. The key set of questions addressed here are (1) the extent to which we approach this third level for the major players in the Serengeti; and (2) the reliability of predictions that can be made without such a complete description.

Unfortunately, a complete description of the entire food web, even in terms of the first-level description of nodes and arrows, is not possible for the Serengeti. A fully articulated food web for the Serengeti would require observations of thousands of interacting species, some of which are only present periodically. The raw data needed to construct such a food web (to generate a graph comparable to those of Winemiller 1990, Carpenter and Kitchell 1993, or Yodzis 1996, for example) does not exist, and it is uncertain whether even decades of well-funded research would change this situation. However, the Serengeti ecosystem does appear to be dominated to an unusual extent by the action of large mammals (although Sinclair [1975] documents high levels of herbivory by grasshoppers, suggesting the role of invertebrate herbivory may simply be poorly understood). A logical first step toward understanding biological interactions would therefore begin with the numerically dominant mammal species together with a model of vegetation that aggregates many distinct categories. Modeling this minimal food web can serve to suggest possible behaviors of a more complex web, as well as revealing quantities or processes that would need to be understood to make predictions about the fate of such a web.

The two earlier volumes in this series (Sinclair and Norton-Griffiths 1979; Sinclair and Arcese 1995) included simulation models describing the dynamics of several of the dominant ungulates, including interactions with predators and vegetation (Hilborn and Sinclair 1979; Hilborn 1995), and other chapters in those books contain a wealth of information relevant to constructing a minimal food web (e.g., Norton-Griffiths 1979; McNaughton and Banyikwa 1995; Scheel and Packer 1995). Most of the models we explored were based on this information. The "major players" web we attempt to model here is shown in figure 8.1. We emphasize again that this includes only a tiny fraction of the biodiversity present in the system. We focus on the part of the ecosystem (the open plains) where one can

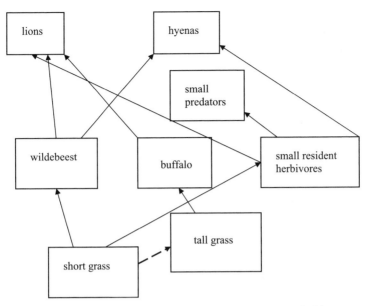

Fig. 8.1 A simplified web for a subset of abundant large mammals in the Serengeti. All solid arrows connect prey to predator. The bold dashed arrow denotes a growth relationship.

neglect trees and browsers (e.g., elephants and giraffes). Even a complete listing of a narrow taxonomic group on one trophic level—the herbivorous ungulates—would include 28 species (Sinclair 1995), and there are 10 mammalian predators on that set of ungulates (Sinclair, Mduma, and Brashares 2003). Plants have been aggregated into two functional groups, even though McNaughton and Banyikwa (1995) identify 17 different community types, each distinguished by a unique set of between one and three dominant species. Each of the predators in the system illustrated in fig. 8.1 includes in its diet to some degree prey species not depicted. Invertebrates and microbes are absent completely, although it is known that at least the latter group has played a key role in past population fluctuations of most of the species represented (Sinclair 1979; Dobson 1995; Cleaveland et al. chapter 7). Figure 8.1, and our models of that web, lack a substructure by size, age, spatial location, or any other variable.

Nevertheless, a quantitative description of the interactions between the major players represented in fig. 8.1 must be an important component of any more complete food web. Given the large body of published research on each of the species or groups represented in fig. 8.1, there is some reason to hope to achieve a reasonable model of this subset of the full web. An even simpler version of fig. 8.1 can be obtained by including only lions and hyenas on the top level, and wildebeest and buffalo on the herbivore level. This reduces the system to one in which there are two groups on each level, producing a six-variable system. This retains the essential feature of allowing shifts in species composition at each level. Abrams (1993) presents a complete catalogue of the qualitatively distinct possible responses of trophic-level abundances to fertilization in such a system, under a number of simplifying assumptions, which probably do not apply to fig. 8.1. Even this simplified system is capable of a wide variety of qualitatively different responses of the total abundances of each trophic level to fertilization. The ultimate in simplification would be to regard the system in fig. 8.1 as having three homogeneous trophic levels. Most theory about the responses of ecosystems to external perturbations has in fact been developed for this simplest of all food web models (starting with Oksanen et al. 1981). Even in this system, mortality at one level is capable of changing the abundances at other levels in either direction, provided there is some adaptive anti-predator behavior or evolution on the middle level (Abrams and Vos 2003). Density dependence due to causes other than food (Gatto 1991), and the presence of four or more trophic levels (McCarthy, Ginzburg, and Akçakaya 1994) can also alter the simple predictions of three-level models (Oksanen et al. 1981)

Once we have restricted ourselves to describing the small but important part of the ecosystem shown in fig. 8.1, we can achieve a second level of resolution, having some quantification of the relative strengths of links in the web. The first and second Serengeti volumes, as well as a body of previous literature (Fryxell, Greever, and Sinclair 1988; Scheel 1993; Sinclair, Mduma, and Brashares 2003), have included descriptions of the proportions of the diet of several major predators that are composed of different prey species, and, for several prey species, the fractions of total mortality due to predation. On the predator side, contributions to lion diet seem to be heavily influenced by prey availability (Rudnai 1974; Scheel 1993). Smaller predators such as cheetah specialize more on small gazelles (Caro 1994), and hyenas are more dependent on wildebeest than are lions (Hofer and East 1995). Herbivore mortality from predation is largely determined by herbivore size. Basically, large herbivores ($>1,000$ kg) experience almost no mortality from predators as adults, while smaller species (< 150 kg) have mortality rates that are almost entirely due to predation (Sinclair, Mduma, and Brashares 2003). Species with body sizes in the intermediate range can either have high or low predation rates, depending upon the community and landscape context. The importance of predatory mortality varies depending on whether populations are resident or migratory. The nonmigratory wildebeest population in the Ngorongoro experiences high predation (nearly 90% in some years; Kruuk 1972), but migratory wildebeest in the Serengeti plains experience a much lower rate of predation (Mduma, Sinclair, and Hilborn 1999; fewer than 3% of adult females die from predation). Previous studies of the major mammalian herbivores (e.g., Sinclair 1985; McNaughton and Banyikwa 1995) have some basic information on the diets of herbivore species that consists of different species or categories of vegetation.. The diet information reviewed later in this chapter furnishes a basis for the second level description of the system, in which strengths can be assigned to the feeding links.

We are interested in predicting changes in population densities that occur following perturbations, and this requires a dynamic model. If one assumes a Lotka-Volterra model of interspecific interactions, the results reviewed earlier provide enough information to construct such a model. Yodzis (1996, 1998) used this approach in his analysis of the fisheries system we have mentioned previously. However, there are a variety of reasons for believing that such a model is not an adequate description of the Serengeti system. In the first place, many of the key interactions are known to be highly nonlinear (see the following). This implies that a linearized model could at best predict the outcome of a very small perturbation. Another general difficulty in moving from simple quantification of diets and mortalities to functional relationships in dynamic models is the role of nonlethal behavioral interactions in population dynamics. Low rates of mortality need not imply small effects on fitness, as predators can significantly alter the distribution and foraging activities of their prey (Caro and Durant 1995; Bolker et al. 2003). Grouping patterns (Fryxell 1995), reproductive cycles (Mduma, Sinclair, and Hilborn 1999), and migration behavior itself (Fryxell, Greever, and Sinclair 1988) may all represent behavioral responses to predation. Because no mammalian herbivore can entirely escape predation on juveniles, the absence of predation on adults for species > 1,000 kg does not imply either a lack of direct mortality on the entire population or a lack of effect on fitness via behavioral responses in the adults. Incorporating behavior into food web models can have strong effects on system dynamics (e.g., Abrams 1999a).

FUNCTIONAL FORM OF INTERACTIONS IN THE SERENGETI FOOD WEB

Do previous studies provide a basis for transforming the web in fig. 8.1 into a dynamic model? All analyses of food web interactions are based upon an understanding of the detailed functional forms describing feeding rates as functions of species' densities, and how these rates translate into birthrates and death rates (for both consuming and consumed species). Here we adopt a simplified modeling approach, in which the populations of each species (or food web component) are implicitly assumed to be homogeneous. This allows each species to be described by a single population density. This description is consistent with some variation between individuals, provided the ecologically important properties of the population can be characterized by a mean value that is relatively constant over time. Thus we seek to describe the functional responses of the consumer species to their foods, numerical responses to different consumption rates, the nature of density dependence in the plant species, and the nature of inter- and intraspecific interference or facilitation at consumer-trophic levels. If the traits that determine interspecific interactions are adaptively variable on a short timescale, then the relationships between these traits and the densities of relevant species must also be known. Thus, an important task in determining how species are likely to respond to our standard perturbations is to assess whether previous fieldwork allows these relationships to be modeled accurately. The functional response is particularly important, and is seldom measured in ways that are appropriate for consumers in environments containing many different food types (Abrams and Ginzburg 2000). Given this general state of affairs we are led to ask—for which species in the simple Serengeti of fig 8.1 is enough knowledge available to construct a parameterized model based on realistic functional forms for feeding, density-dependence, and so on, to describe the dynamics of that species?

We begin by discussing what is known about functional and numerical responses for the best studied of the predatory species in fig. 8.1, the lion. First consider the numerical response. There is general agreement among most of the authors of the previous *Serengeti* volumes that the population growth of lions is not solely limited by the population sizes of the migratory prey species (wildebeest, zebra, Thomson's gazelle). A major increase in the wildebeest population following the eradication of rinderpest in the early 1960s was not immediately accompanied by a large increase in the number of lions on the plains (C. Packer, pers. comm.). The population of lions did increase, but after a delay, and then in jumps. Two potential explanations for this phenomenon have been proposed: (1) limitation based on dry-season food supply (Sinclair 1979); and (2) limitation based on territoriality (Packer et al. 2005). These two ideas are related, because an individual's dry-season food supply is limited to what occurs on its pride's territory. The difference is whether population limitation is caused by the food within the territory, or by the mortality associated with inability for young adults to found new prides in new territories. It is known that there are many more surviving juveniles than there are young adults that manage to establish themselves within a pride or form a new one (Hanby and Bygott 1979; Bertram 1979; Hanby, Bygott, and Packer 1995). The Ngorongoro Crater lion population has an abundant, relatively nonmigratory prey population, but young adults are still exported to other areas rather than supporting population growth in Ngorongoro (Hanby, Bygott, and Packer 1995). Whether or not mortality is caused by food supply within territories, the territorial system results in migration to "sink" habitats by many of the young adults when open territories in good habitats are not available; these individuals generally do not survive for a long period.

There are several ways to incorporate territoriality into a population dynamic model. The simplest representation of territoriality assumes that population growth depends entirely on food if territories are not filled, but is zero once territories are all occupied. Thus, if P is lion density, F denotes the food intake rate per lion (the sum of the functional responses over all prey types), B denotes the conversion efficiency of food into new surviving adults, and K_p is the lion's carrying capacity, the numerical response can be given by:

$$dP/dt = P[b(F) - d], \text{ if } P < K_p, \text{ and}$$
$$dP/dt = 0, \text{ if } P \geq K_p \tag{1}$$

(here b and d respectively denote per capita birth- and death rates). Provided that lion densities remain near K_p, only large decreases in food intake or large increases in mortality will have an impact on dynamics. This model assumes that territory size is inflexible, while we know that lion territory size in fact decreases with increasing food abundance, and that pride size within territories may also change depending on food (Packer et al. 2005). An alternative model, which allows for flexibility in the size of territory per lion, is the following expression, in which per capita birth rate is assumed to be a multiplicative function of the number of births per adult, b, (which depends on food intake) and g, the probability of survival and successful recruitment into the territory-holding adult class (which depends on current adult density):

$$dP/dt = P[b(F)g(P) - d]. \tag{2}$$

The value of g is likely to be close to 1 over a range of low lion densities, but drops rapidly as the territorial carrying capacity is approached. If K_p is the absolute maximum density of adult lions, a potential form for g is $(1 - (P/K_p)^z)$, where z is a positive exponent greater than 1. Equations (1) and (2) are still relatively gross characterizations of the actual dynamics; in particular, they ignore any effects of food supply on adult death rate, and time lags in the response of adult density to changes in food or territory availability. Hilborn and Sinclair (1979) note that adult lion survival did not seem to vary with prey density over the range that had been observed to that point, and it appears that quite low prey densities are required for reduced survival of adults. Thus, the assumption of a constant mortality rate, d, in both of these equations may be adequate. Data are currently insufficient to estimate z with any degree of accuracy, although a value much greater than 1 seems to be most consistent with the observed limited response of lion population to changing food. The value of K_p can be estimated roughly by the highest observed lion population (on the order of 3,000 individuals).

Packer et al. (2005) have analyzed long-term records of lion populations and found that population sizes since the early 1960s in both the Serengeti and Ngorongoro Crater appears to be characterized by periods of stasis in the face of gradually changing prey densities, followed by relatively abrupt shifts to significantly higher or lower levels. At least the upward shifts appear to be attributable to a combination of sufficiently high food densities, which produce large enough cohorts to allow the establishment of a new pride, and a low enough adult population that there is some possibility of squeezing in some new territories. This produces a numerical response that is intermediate between those embodied in eqs. (1) and (2). Modifying eq. (2) to reflect the saltational changes in density would require a rather complicated function with a number of fitted parameters. It is not clear whether the long-term changes are likely to differ significantly from those predicted by eq. (2), given the other uncertainties in the model.

The quantity F in eqs. (1) and (2) represents the sum of the functional responses to the different prey types, weighted by their caloric values. The function b translates this food intake into a rate of increase of the population. The simplest assumptions for these two functions are that total consumption is given by the sum of multispecies disk equations (Holling 1959; Schoener 1971), with each prey weighted by its net caloric value, and that b is a simple proportionality. These assumptions were made, for example, by Hilborn (1995). This means that the consumption rate of species i is given by:

$$
\frac{C_i N_i}{1 + \sum_{j=1}^{s} C_j h_j N_j} \tag{3}
$$

where C_i is a per capita capture rate of prey species i, s is the number of distinct prey species, N_i is the density of that prey and h_i is the handling time (including any digestive pause after consumption). It has been relatively difficult to study the form of either b or F in the field because the total food intake by lions has remained relatively similar in the face of major changes in the densities of major prey species [for example, following a die-off of wildebeest in Nairobi Park (Rudnai 1974), and during a long period of increasing wildebeest on the Serengeti Plain that began in the early 1960s (Sinclair 1979)]. However, both of these observations suggest functional responses that saturate at a low total prey density. Given the lack of information on b, and the limited variation in F, assuming proportionality to caloric intake is likely to be an adequate approximation for b. However, the form of F suffers from some additional uncertainties.

The uncertainties about F are of three types: (1) difficulty in measuring parameters in the functional responses, eq. (3); (2) uncertainty about the potential dependence of the capture rates, C_p, on population densities of prey or other predators; and (3) uncertainty about the appropriate way to incorporate seasonal variation in prey availabilities due to migration. It is useful to consider these in some detail, since these uncertainties arise in many attempts to quantify food web models.

The first set of parameters we consider are handling times, which have typically (Hilborn and Sinclair 1979; Hilborn 1995) been estimated based on the idea that the asymptote of the response is effectively equivalent to observed intake rates. The latter has most often (Hilborn and Sinclair 1979; Fryxell, Greever, and Sinclair 1988) been estimated from Schaller's (1972) observation that the biomass of prey killed by lions over the course of a year was 2,469 kg, or 6.764 kg per day. Although there is considerable uncertainty in this figure, it is not far from measurements based on other direct observations (Elliott, Cowan, and Holling 1977; Packer, Scheel, and Pusey, 1990; Hanby, Bygott, and Packer 1995), at least when these are averaged over the entire year. Elliott, and Cowan (1978) had somewhat higher estimates of the biomass of prey killed (3,376–3,927 kg) for an average female, with the range depending on whether small or large herbivores were the main component of the diet. As noted by Hilborn and Sinclair (1979), this method overestimates the handling times because there is a small but significant amount of time spent searching. Hanby, Bygott, Packer (1995) estimate this to be between 5 and 8% of total time. A somewhat more troubling problem is the possibility that the amount of a kill that is actually eaten will vary with prey density. This is predicted by models in which prey capture has costs, which is certainly the case for predators eating large prey that have the potential to injure the predator (Abrams 2000). It is also likely to be true of any system in which prey must be defended against an array of scavengers. Thus, a type 2 functional response need not even have an asymptote, and the assumption that average consumption represents the asymptotic prey mortality rate due to an average predator may not be valid.

The second type of parameter in the multispecies disk equation is the attack rate per unit prey density, C_i. This is quite problematic to measure in the field. Because prey densities vary spatially, it is difficult to determine the local density experienced by a predator. Smaller prey can be eaten quickly enough that the actual predation event is often not observed. The attack rate is likely to depend on a variety of internal variables (prey hunger and condition) and external variables (amount of cover for lions) that are themselves difficult to quantify. Scheel (1993) discusses the difficulty of determining whether a lion is hunting or not, given the opportunistic nature of prey capture. Elliott, Cowan, and Holling (1977) and Scheel and Packer (1995) used direct observations to estimate local densities and kill rates for a range of different prey species, but these estimates are highly uncertain. Because of the relatively large handling times of the major prey species (wildebeest, buffalo, and zebra, which make up > 80% of the diet in the Serengeti), the total intake is insensitive to the values of the attack rates. However, the mortality rates of particular prey are approximately proportional to the values of the C_i, so these are important determinants for estimating responses to some perturbations. Wildebeest and zebra seem to be subject to higher attack rates than most species of smaller or larger prey. Buffalo are too large to be caught effectively by small prides, while most gazelles are difficult to capture. Warthogs suffer high predation rates, since they lack the speed of gazelles and have a much lower probability of escape after attack (Elliott, Cowan, and Holling 1977).

Another uncertainty about the attack rates is to what extent they vary with relative abundances of prey. Scheel and Packer (1995) showed that the probability of a lion attacking a prey individual increased with local density of some prey species, although this relationship was weak, and even nonexistent for some prey species. Earlier observations of shifts in lion diets following a crash in the wildebeest population at Nairobi Park (Rudnai 1974), also suggested an ability of predators to switch to hunting for the currently most abundant prey species. Although Fryxell, Greever, and Sinclair (1988) used a relationship that approximated a step function to describe switching between migratory and nonmigratory prey species, Rudnai's (1974) observations are consistent with a much weaker form of switching. These data may also be consistent with a complete lack of switching, provided that lower densities of preferred prey lead to diet expansion to include a wider range of prey types. The tradeoffs that must underlie any form of multispecies choice have yet to be explored. Nevertheless, it is clear that some form of switching is likely to occur, and, as shown in the following, an accurate description of switching is important for assessing responses to our three perturbations.

Foraging theory suggests that a set of prey encountered in a single habitat should be ranked by energetic value divided by handling time, and lower-ranked prey should be dropped from the diet when highly ranked prey are common. There does not appear to be much difference between a lion's potential prey species in energy per unit weight. This, combined with the fact that handling time consists primarily of a digestive pause (Elliott et al. 1977) means that it is unlikely that prey differ enough in their ratios of energy content to handling time to suggest that any be dropped from the diet because of changes in the abundance of other prey. Scheel (1993) rather optimistically argued that his observations of lion diet were consistent with foraging theory, but the handling times he used did not include digestion. The inability to accurately measure availabilities of different prey also calls his conclusion into question.

The final problem in constructing the food intake function, F, is the problem of incorporating the seasonality in prey abundances produced by the annual migration; this results in a periodic large-magnitude reduction in the availabilities of wildebeest, zebra, and Thomson's gazelles during the dry season. It is clear that an adequate description must separate wet- and dry-season functional responses. Because the period of parental care in lions is approximately 2 years, production and survival of the young depend on both seasons, while functional responses in each season clearly depend only on the prey available at the time. The lack of seasonality in lion reproduction (Packer, pers. comm.) suggests that cubs of different ages have similar sensitivities to the lower food abundance implied by the dry season. In other words, survival may be reduced by the dry season, but not differentially reduced for specific age classes within the nonadult population. Because of this, it is possible to maintain the assumption of a homogeneous predator population. The two simplest possibilities for translating the wet- and dry-season intakes into a numerical response are: (1) to express b as a function of the sum of wet- and dry-season intake rates [e.g., $b(F_{dry} + F_{wet})$]; or (2) to decompose b into a wet-season and a dry-season contribution to ultimate cub survival, and to multiply these together to obtain a combined b [$= b_{wet}(F_{wet})b_{dry}(F_{dry})$] that can be inserted in eq. (1) or (2). The latter seems most appropriate if cub survival is often reduced by the lower food supply of the dry season, and wet-season consumption cannot compensate. This could lead to a set of equations with the following form, in the simple case of one major nonmigrant prey (buffalo, b) and a migrant prey (wildebeest,

w), where the latter is assumed to have a population effectively independent of lion population size:

$$
\frac{dP}{dt} = P
$$

$$
\left\{ b_{max}\left[1 - \left(\frac{P}{K_p}\right)^2\right]\left[\frac{\kappa_b b_b C_b N_b}{1 + (h_b\kappa_b b_b)C_b N_b}\right]\left[\frac{\kappa_w b_w C_w N_w}{1 + (h_w + \kappa_w b_w)C_w N_w}\right] - d \right\}
$$

$$
\frac{dN_b}{dt} = N_b\left[1 - \left(\frac{N_b}{K_b}\right)\right] - \left(\frac{C_b N_b P}{1 + C_b h_b N_b}\right) \tag{4}
$$

In this model, b_{max} is the maximum per capita birth rate of lions. The probability of survival of cubs within a season is described by another Holling disk equation with an asymptote of 1; survival probability is 0.5 when the food intake rate is $1/\kappa$. The combination of the framework given by eq. (4) with switching behavior can lead to quite complicated models in multiple prey systems. In the absence of wildebeest in the wet season, predators would be likely to switch to nonmigratory prey species for that part of the year; this would require making the consumption rate of buffalo increase with decreasing wildebeest density in this model.

It is clear that there are considerable uncertainties about the appropriate shapes of lion functional and numerical responses. These uncertainties are unfortunately much greater in the case of other predators. Hyenas are three times as abundant as lions, and are better able to travel substantial distances to obtain prey (Hofer and East 1995). However, they are also likely to have limited access to most of the migratory herds through much of the dry season, they also have territoriality, and their diet overlaps substantially with that of lions. Frequent observations of infanticide and siblicide also suggest a high level of self-limitation. The absence of adult buffalo and a greater dependence on wildebeest are the main differences in diet composition between hyenas and lions (Scheel and Packer 1995). As in the case of lions, annual intake seems likely to be relatively constant, unless prey density becomes very low. Thus, it seems reasonable to represent hyena dynamics using the same framework as for lions, but with the addition of direct interference from lions. Lions have been observed to kill hyenas and to take over their kills (Schaller 1972). While both events are believed to be relatively uncommon, their impact on hyena population dynamics is somewhat difficult to assess. The parameter uncertainties for all components of a dynamic model are greater for hyenas than for lions. We will not consider other predator species explicitly, since none are likely to have ecosystem effects of the same magnitude as lions and hyenas. Most of the smaller predators, particularly cheetahs (Laurenson 1995) are subject to mortality from lions, leading to problems of coexistence in the vertebrate predator guild (Durant 1998).

The functional responses of herbivores to vegetation and the appropriate classification of vegetation into functionally distinct categories are key issues for quantifying interactions between the first and second trophic levels. We know something about differences between the plant species consumed by different herbivores, and about herbivores' relative preferences for grasses of different heights (tall grasses are less nutritious than short grasses). These differences are overlain on a pattern of spatial heterogeneity in both plant height and species composition, which is heavily influenced by herbivore foraging, and by seasonal changes in rainfall. There is additional heterogeneity due to spatially varying physical factors and nutrient content of the soils. There are feedbacks of grazing pressure on vegetation dynamics via alteration of nutrient cycling and fire frequency (Dublin 1995; Holdo et al. 2007). Additional complications in describing the relationship between herbivores and vegetation were neatly summarized by Jarman and Sinclair in the first Serengeti volume (Sinclair and Norton-Griffiths 1979, p. 131): "The vegetation provides food, water, minerals, shade, and cover to hide in . . . [it] also provides cover for predators, obstruction to visual communication, and a lot of inedible material among which the animal must search for usable food." It is likely that the complexity of the relationships between vegetation and food contribute to the large number of coexisting herbivores. However, it is impossible to represent all of this complexity in a tractable model. Both competition and facilitation have been suggested to characterize interactions between different pairs of herbivores (McNaughton 1985; Sinclair 1995).

Previous models have reflected a simplified understanding of the herbivore-vegetation interaction in the Serengeti. For example, Hilborn (1995) assumes that grass is divided into two classes; green and tall, with dynamics dependent on rainfall, burning, and loss of grassland to other types of vegetation (woodland or cropland). Grass of a given category is divided among herbivores according to their biomass, and this amount determines survival according to a decelerating curve that approaches 100% survival. Coexistence in spite of the very broad overlap in diet is achieved by assigning herbivores to locations based on observed distribution over the 10 areas within the park. Competition occurs within each area independently of other areas. This eliminates any possibility of facilitation and does not allow for shifting distributions as the result of herbivory. The approach of Fryxell, Wilmshurst, and Sinclair (2004; see chapter 7 for details) is to represent vegetation as a single population, whose quality decreases as its biomass increases. Different species of herbivores have different maximum rates of harvesting vegetation and different gut capacity constraints, resulting in different optimum biomass levels. By allowing herbivores to select nearby patches based on their intake, a shifting mosaic of different biomasses is generated, which can generate coexistence of at least some species over ecologically relevant time periods (Fryxell et al. 2005). This approach ignores underlying physical heterogeneity, which is no doubt a major contributor to coexistence. It is therefore likely to overestimate the amount of competition between herbivores.

Interference within one or more herbivore species is possible, as many species have a rather complex social organization that involves territoriality, grouping, and/or hierarchical behavior (Jarman and Jarman 1979). It seems likely that trampling of riparian vegetation by wildebeest reduces food availability for buffalo (Sinclair 1995), so there is probably at least one case of interspecific interference. There is the possibility of facilitation due to large groups being less vulnerable to predators. In no cases that we know of have these positive or negative affects been quantified in a way that permits incorporation into a model of population dynamics. Moreover, scant attention has been paid to how vegetation structure can influence the interaction between predators and herbivores. Grass height, for instance, appears to influence the hunting success of lions (Packer, pers. comm.), so the influence of herbivores upon the biomass of vegetation can have reverberating effects upon interactions among higher trophic levels.

The final aspect of the ecosystem that is required for a model is a description of the dynamics of the vegetation. Although grazer exclosure experiments (McNaughton 1985) have provided estimates of maximum growth rates and the dependence of these on rainfall, there is a relatively poor understanding of the density dependence of grass growth or of competition between species. Most previous models (e.g., Fryxell, Greever, and Sinclair 1988; Fryxell, Wilmshurst, and Sinclair 2004) have terms that permit significant growth even when current aboveground biomass is very small, which is a reasonable assumption for grasses. Total growth in vegetation is a unimodal function of current aboveground biomass in most models. Here we do not require a more detailed description to identify at least some of the major uncertainties in ecosystem responses to perturbations.

We ran a number of simulations of the web in fig. 8.1, using variations on eq. (4) to represent the consumer-resource links, and simple models of grass dynamics, consisting of seasonal input into a short-grass stage followed by growth to a long-grass stage that eventually disappeared, either by consumption, fire, or decay. We utilized a variety of previous studies to estimate parameter values. In the interest of space, we will not here provide a detailed discussion of these simulation studies, because the results were most often dynamics that differed greatly from those actually observed over the past several decades. While some parameter combinations yielded dynamics that did not differ wildly from observations, these were usually not the choices of parameters that seemed to have the greatest independent empirical support. As a result, our analysis of potential responses to the three standard perturbations was based on qualitative analysis of a range of outcomes observed in these simulations, as well as general results for analyses of other multispecies, multitrophic-level systems that we feel could shed light on the dynamics of the Serengeti. We will also focus first on a single potential perturbation (increased lion mortality), and then sketch more briefly outcomes for the other two perturbations.

PREDICTED RESPONSES TO PERTURBATIONS

Increased Mortality of Lions

The first perturbation that we consider is a sustained increase in the mortality rate of lions. The impact of such an increase on the lion-population size largely determines the impacts on other components of the system. If the extreme form of territorial limitation embodied in eq. (1) is adopted, a change in lion mortality will not have any impact on the lion population, unless the change is so large that recruitment is insufficient to reach the territorially determined carrying capacity. Less extreme forms of territoriality, such as eq. (2), suggest that higher mortality will reduce the lion population, although the magnitude of the change is likely to be small unless the increase in death rate is substantial. Thus, moderate changes in lion mortality seem unlikely to have much effect on the rest of the ecosystem. However, there is at least one exception to this conclusion, which is predicted by eq. (4) and illustrated by fig. 8.2. If the territorial limit to lion-population size is high enough, or the rate of killing prey at low prey densities is high, then there may be an equilibrium at low prey-population density as well as two more at higher densities. If the only potential equilibrium is much less than the prey-carrying capacity, then there are limit cycles around that equilibrium. If the prey-carrying capacity is increased or the predator death rate is increased or its carrying capacity is decreased sufficiently, then there are three equilibria, but the highest prey-density equilibrium (which is stable) is the only equilibrium observed in the course of long-term dynamics. Under these circumstances, higher mortality of the predator could shift the system from an unstable equilibrium with a relatively low average predator density to a stable equilibrium with a higher predator density (see

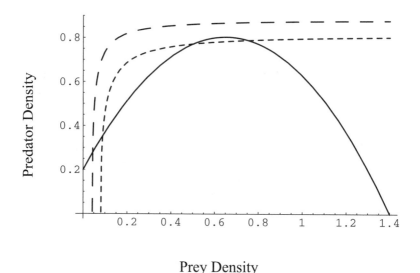

Prey Density

Fig. 8.2 Isoclines for a predator-prey system based on a territorial predator and a resident prey (see eq. 4). The solid line is the prey isocline, while the two dashed lines are alternative predator isoclines. The long-dashed line corresponds to a system with either greater intake of migrant prey or lower mortality of the predator. The only equilibrium in this case is unstable, and limit cycles are observed. The short-dashed line corresponding to a higher mortality has three equilibria, but all trajectories end up at the equilibrium with the highest prey density, which is a stable point. Sufficiently increased predator mortalities (or increased prey-carrying capacity) would only have one equilibrium on the right hand side of the diagram.

fig. 8.2). The same outcome could occur with a decreased density of the migratory herbivores, which are assumed to have a population size that is very insensitive to predator density. Because cycles in lions and their prey have not been observed, this outcome may actually be impossible in the real system; this is discussed in the following.

What happens if the increased mortality to lions is sufficient to push their population to the verge of extinction? In the case of the model illustrated in fig. 8.2, the prey only increase by a factor of 2 or more if the original equilibrium is the unstable one at low prey density. A potential empirical example of the responses of herbivore populations to the virtual elimination of lions is provided by the period of heavy poaching in the northern part of the park. Sinclair, Mduma, and Brashares (2003) document up to 10-fold increases in several antelope species during this period, and interpret them as releases from lion predation. The increase in the densities of Thomson's gazelle was close to an order of magnitude, and there were many-fold increases in warthog, oribi, and topi. This is somewhat difficult to reconcile with the ideas that: (1) lion recruitment is limited by the availability of territories and by nonmigrant prey densities, and (2) lions have strongly saturating functional responses to their major prey. Models with these properties usually have isoclines that look very similar to those in fig. 8.2, which were first illustrated by Rosenzweig and MacArthur (1963). If the low prey-density equilibrium exists, it should be cyclic.

This model has the interesting property that fertilization (increasing the prey's carrying capacity) often stabilizes a cycling system. The model has similar properties if the food intakes in the two seasons combine additively rather than multiplicatively to determine the maximum possible recruitment. The natural experiment suggests that the densities of at least some prey species are a small fraction of their carrying capacities in the presence of lions, but there does not seem to be any evidence of cycles. This paradox could be resolved in a number of ways, all of which require further research. One possibility is that there is some stabilizing process that we do not know about, which prevents cycles. In some models, behavioral or evolutionary responses of predator and/or prey have been shown to be capable of stabilizing the system (e.g., Abrams 2000). Another potentially stabilizing force is the presence of a relatively large class of senescent or diseased prey, which are caught much more easily by the predator. It is also possible that the handling times of the resident prey (other than buffalo) are small enough to eliminate the hump in the prey's isocline in fig. 8.2; this would make stable equilibria at a low prey density possible. Spatial structure in the environment also has the potential to result in at least the appearance of stability at larger spatial scales (Hoopes, Holt, and Holyoak 2005). It is currently unclear whether any or all of these phenomena account for the observed outcome of the poaching event.

There appears to be a consensus that the wildebeest population is largely unaffected by predation (Sinclair 1985, 1995; Mduma, Sinclair, and Hilborn 1999). Zebra, another migratory species, has remained surprisingly constant in population size since the 1960s, in spite of large changes in the densities of other species in the system. Models by Fryxell, Greever, and Sinclair (1988) predicted small impacts of predators on the population sizes of abundant migratory species. Migration partially decouples predator and prey dynamics. Of the species that are major components of the lion's diet, buffalo, which is a resident and constitutes a significant proportion of lion caloric intake (Scheel and Packer 1995), seems most likely to experience some change in population density as the result of a change in lion population size.

Other potential impacts of reduced lion numbers on the system include increases in competing predators and a shift in the plant community based on shifts in the numbers of different herbivores. Although competition between lions and hyenas was once thought to be virtually nonexistent

(Bertram 1979), recent work calls into question this assumption (Packer, pers. comm.), but it is difficult to put any quantitative figures on the predicted increase. Cheetahs suffer intense juvenile mortality from lions, and frequently lose kills to lion prides. Laurenson (1995) attributed much of the observed decline in cheetah numbers to lions. Thus, it is reasonable to expect an increase in cheetahs as the result of decreased lion densities. The same may be true of other small predators, which are generally less well studied. It is possible that increases in smaller predators could then cause decreases in some of the smaller prey species, such as Thomson's gazelle. There is sufficiently little known about the possible extent of increase in small predators with a decreased lion population, that models cannot contribute much to estimating these indirect effects on small herbivores. Given the likely small change in lion densities due to moderate changes in lion mortality, the likely smaller shifts in herbivore species composition, and the negligible impacts on the large migratory populations, large shifts in the plant community in response to lion mortality seem unlikely.

Increased Mortality of Wildebeest

A second potential perturbation is an increase in the mortality of wildebeest, for example, as a result of hunting or a species-specific disease. Because of the huge size of the wildebeest population, it is likely that even a several-fold decrease in the size of the population would not significantly reduce the predatory species' food intakes. Such a change would significantly reduce herbivory during the wet season, with some increase in tall grass at the expense of short grass being likely. Other herbivores that depend on short grass or on small dicots that are only exposed after grass is cropped to a short height might be harmed. Our knowledge of vegetation dynamics and herbivore responses does not seem sufficient to quantify this or to determine with much certainty whether species that consume larger amounts of tall grass (zebra and buffalo) will increase significantly. It seems likely that a sufficient reduction in wildebeest numbers would increase the amount of dry-season vegetation sufficiently to support more frequent fires, with a resulting suppression of tree regeneration (Sinclair, pers. comm.). Fire has not been included in our rough food web model, and its absence may limit our ability to predict the consequences of any perturbations with large direct or indirect effects on wildebeest populations (Holdo et al., forthcoming).

If wildebeest were reduced by an order of magnitude or more one would expect significant switching responses of predators, and this could increase predation on other prey. However, wildebeest constituted less than 30% of the caloric intake of lions over a 5-year period when wildebeest were at their peak abundance (Scheel and Packer 1995). Assuming that the functional response is described by eq. (3) and is close to its asymptote for most prey densities, there would be an approximate 40% increase in the consumption of other herbivore species if wildebeest were eliminated. It is unlikely that this increase would be spread evenly over the remaining species, but not enough is known about the nature of switching to estimate which species would be most affected. Zebras are closest to wildebeest in size and seasonal availability, and seem the most likely potential replacement for missing wildebeest in predator diets. However, there is again some inconsistency between these model predictions and recent observations, because zebra population size appears to have varied little in the face of substantial variation in many other species (Sinclair 1979, 1995; Mduma and Hopcraft, appendix, this volume).

Models of predation on two prey species that incorporate both predator self-limitation and switching suggest that positive or negative interactions between prey (via the predator) are possible, and also that alternative equilibria for a given set of parameters may occur (Holt 1977; Abrams and Matsuda 1996). Because there are potentially offsetting effects from changes in predator density and changes in predator satiation, each of these must be known reasonably accurately to predict the net effect. Accurate descriptions of switching behavior and the flexibility of territorial limitation among predators would be required to determine the sign of the change in other herbivores following a wildebeest crash. Studies leading to such descriptions are not available. If the predators do not make up for the loss of wildebeest by increased predation on other migratory species, eq. (4) suggests that a higher density of one or more nonmigratory herbivores (e.g., warthog, topi, impala, oribi) and a lower lion density might result. However, if the loss of wildebeest caused greater predation on nonmigratory species, those species are likely to decrease, in spite of any slight decrease in the lion population. Although we have limited records of other herbivore populations when wildebeest were less abundant, this was at a time when all ungulates were recovering from the rinderpest epidemic, and it cannot be used to describe the ecosystem consequences of mortality that only affected wildebeest. If fig. 8.2 is applicable, a large decline in wildebeest could result in a shift in the state of the entire system.

Thus far, we have only discussed the indirect consequences of wildebeest removal that are transmitted by predators. In fact, this is unlikely to be the only or even the most important pathway by which a change in wildebeest numbers affects other herbivores. Pathways that involve change in the plant community include the possibility of decreased exploitative competition for food, facilitation by promoting succession, and decreasing the amount of tall grass, which reduces cover from predation. As mentioned earlier, increases in fire frequency with increased fall vegetation could alter the balance between woodland and grassland. These pathways were not all modeled, because not enough is known to quantify most of them. However, the range of possible functional components discussed earlier implies something about the range of interactions.

It is possible that wildebeest removal would increase the amount of food for many other herbivore species, since wildebeest are responsible for the largest amount of removal of plant tissue. On the other hand, by keeping grass height low, wildebeest increase the quality of the remaining vegetation, and no doubt have some of the positive effects on other herbivores that were postulated by Bell (1970) in his model of grazing succession. The simplified model of grass exploitation developed in Fryxell et al. (chapter 9, this volume) predicts a net competitive effect of wildebeest on most other species. In the context of a simple spatially homogeneous system (our assumption for the model of predation discussed previously), the representation of vegetation as a single category (as in Fryxell et al., chapter 9, this volume) results in competitive exclusion of all but one herbivore species. However, coexistence of two or more species can easily be achieved if vegetation is represented by a number of size categories, with growth and herbivory moving the vegetation from one category to another (as in the model of Richards et al. 2000).

Some potential changes in vegetation could take many years, as one set of plant species slowly replaces others, so observations of past correlations between populations of different herbivores may be misleading. Natural history observations provide some guidance for resolving the question of mutualism or competition. For example, it seems likely the Thomson's gazelle population would decline in response to removal of wildebeest, since the evidence for facilitation by wildebeest grazing appears to be strongest for this species (McNaughton 1976).

Altered Productivity of the Vegetation

There are many ways that vegetative growth could be increased, and at least some of its effects on animal populations are likely to depend on whether temperature, nutrients, rain, or some other factor caused the change in productivity. Altered rainfall is responsible for most of the short-term temporal variability in productivity in the Serengeti, and has been one of the abiotic drivers incorporated into most previous models (Hilborn and Sinclair 1979; Fryxell, Greever, and Sinclair 1988, 2005; Hilborn 1995). We will follow this tradition and ask what changes might be expected in the herbivore and carnivore trophic levels in response to increased (or decreased) rainfall. The impact of increased plant growth due to greater rainfall could in theory be positive or negative. Because tall grass is poor-quality food, the total herbivore population could decrease as the result of increased productivity. Conversely, if the level of grazing is sufficient to prevent grass from escaping into the tall category, increased grass productivity should be translated into increased herbivore populations. There is good evidence that a series of years of higher dry-season rainfall increased dry-season grass production three-fold and played a role in increasing wildebeest and buffalo populations in the 1970s (Sinclair 1979). The territorial limitation of the predators means that they will not control the herbivore increase, and will exhibit a smaller proportional increase. If we regard the changes in predator populations as relatively small, we can use vegetation-herbivore models to predict approximate changes in herbivore populations in response to altered rainfall. The above observations suggest that there might be a non-monotonic response of herbivore population numbers to directional trends in production.

DISCUSSION

We have used some simple models to help illuminate the uncertainties in our predictions about consequences of potential perturbations to particular components in each level of the system. We began with a greatly simplified web that contained only a handful of the most abundant herbivore and predator species, and a grossly aggregated description of the vegetation. Developing a single model, even for this greatly oversimplified system, did not seem to be justified, because many alternative functional forms seemed equally consistent (or inconsistent) with our current knowledge. Predator-herbivore interactions in some respects appear to be better understood than are herbivore-plant interactions. Even here, potential models suggested that previous hypotheses about predator limitation may not be entirely consistent with observations.

The impacts of different herbivore species on each others' population dynamics remains largely unknown. It seems clear that spatial segregation of species is likely to play a key role in reducing competition, yet it is not clear to what extent this segregation is imposed by the physical environment or is an emergent product of the different movement rates and foraging choices of the various herbivores. The inability to carry out controlled experiments on foraging has limited our ability to develop more quantitative models of these interactions. The approach of exploring incomplete models, which we have taken here, does identify some outcomes that we can rule out, and it has served to highlight experimental studies and future modeling efforts that would contribute to an increased understanding of the web.

Are Any Outcomes Ruled Out by What We Know about the Food Web?

Even if models do not always provide a firm basis for predicting the outcomes of the perturbations we have considered, they may still set limits on the possible set of outcomes. Sometimes it is as important to know what is unlikely to occur as to know what will occur. To appreciate the extent to which our current knowledge about the fig. 8.1 web allows narrowing of the range of outcomes, we review some of the phenomena known to occur in webs similar to fig. 8.1 that are *unlikely* in that particular web. Because there is no evidence of sustained population cycles in the Serengeti system, a range of phenomena that are known from cycling systems can probably be ruled out here. Fertilization can greatly increase the bottom trophic level and decrease or cause extinction of the top level of a three-level system, due to changes in the amplitude of population cycles (Abrams and Roth 1994). Changes in cycle amplitude with changes in parameter values also greatly change the nature of the interaction between two prey that share a common predator (Abrams, Holt, and Roth 1998). Based upon the knowledge available to date, we have no reason to believe such phenomena occur to any significant extent in the Serengeti.

Surprisingly, the consequences of the types of perturbations considered here are still not completely understood in models with more than one species per trophic level, even under the simplifying assumptions that the system has a stable equilibrium, that species have fixed properties that are common to all individuals, and that there are no population cycles or environmental fluctuations. Abrams (1993) catalogued all of the potential responses to altered plant productivity in systems characterized by having linear consumer functional responses and having no direct density dependence at any of the consumer levels. Even under these somewhat restrictive conditions, if there are two species at each level, then fertilization is capable of either increasing or decreasing the total abundance of each of the levels or of any particular species, depending upon the detailed pattern and magnitudes of interactions. There has not been any similar analysis for the consequences of altering the mortality rates of one or both species on either of the consumer trophic levels. However, there have been some analyses of food webs with two species on the middle level, and one on each of the top and bottom levels (Armstrong 1979; Holt , Grover and Tilman 1994; Leibold 1996; Grover and Holt 1998; Abrams 1999b; Abrams and Vos 2003). These are likely to be reasonable approximations to six-species systems (two per level), in which the two predators are similar to each other, as are the two plants. Coexistence in this "diamond food web" requires that one of the two prey be both more vulnerable to the predator and better at exploiting the resource than is the other prey species (Armstrong 1979). Here, there is again a surprising range of responses of total trophic-level abundance to mortality applied uniformly on a particular level (Abrams and Vos 2003). One possibility is that mortality applied to predators actually increases their abundance, provided that the increase in death rate is not too large (Abrams 2002; Abrams and Vos 2003). The increase occurs because of the increase in the relative abundance of the more vulnerable prey, the resulting increase in mean growth rate of the prey population, and the saturating functional response of the predator. The saturating response moderates overexploitation of prey when the more easily captured prey species increases in abundance. A variety of other indirect effects are possible in this web. For example, increasing mortality equally on both of the herbivore species can increase the predator's population. This occurs because the mortality may favor the more vulnerable of the two prey species, resulting indirectly in an increase in food availability to the predator. These and other counterintuitive effects require a strong tradeoff among herbivores in their susceptibility to predation and their ability to subsist on low plant populations. We lack any evidence for strong tradeoffs of this sort on the herbivore level in the Serengeti. The theoretical results thus highlight the need for empirical studies focused on this question.

How Detailed a Model Do We Need?

In analyzing the uncertainties discussed earlier, we have simplified the description of the dynamics of the species represented in fig. 8.1 considerably. Thus, any question about prediction requires that we determine how the factors we have ignored might change our analysis of the system. Do we need to include age structure, spatial structure, temporal environmental variation, or more behavioral variables? The only way to answer these questions is to learn more about these additional variables and perform a theoretical analysis of their potential effects.

Chapter 9 by Fryxell et al. (this volume; see also Fryxell et al. 2005) is a first example of how an explicit representation of space can change the nature of plant-herbivore interactions. In this case, changing the temporal variation in the system can alter the outcome of competition between wildebeest and buffalo. Rules by which herbivores move toward more rewarding food patches play a major role in determining their dynamics. How these considerations alter interactions that involve predators remains to be explored.

Thus, we are unable to give a clear answer about how much detail we would need to achieve a particular level of predictive accuracy. It is clear, however, that models based on the simplest assumptions of Lotka-Volterra dynamics may not provide reasonable answers, even if we restrict ourselves to analyses based on an assumption of homogeneous populations. Many biologists leap to the opposite extreme, and assume that the only models useful in management are those crammed full of biological details. Such parameter-rich models are constrained by inadequacies in the basic data available for estimating parameters, and by difficulties in even discerning the major drivers of system behavior in complex natural systems. A major scientific and managerial challenge is working out the appropriate mapping between the resolution of desired predictions, and the level of detail embedded in theoretical models.

What about All the Missing Players?

The majority of species that occur in the Serengeti do not appear in the web in fig. 8.1. This adds another layer of uncertainty to the list that we have presented. At best, one could justify ignoring these species either because they are rare or their interactions with the rest of the web are weak. This seems a reasonable approach in at least some cases. The hope here is that we can regard these species as being driven in their dynamics by the template defined by the prime movers represented in fig. 8.1. For instance, the roan antelope is unlikely itself to drive vegetation dynamics, but its persistence will be governed by how other herbivores define available vegetation of the appropriate type. Another example is the cheetah, whose population dynamics are likely driven by interference competition with other top predators, primarily lions.

However, even if individual species are rare, their collective impact may be significant. And there is theory (McCann, Hastings, and Huxel 1998) suggesting that weak links can play a key role in stabilizing systems. Thus, it is an open question whether these simple representations of dynamics, considering just a few key players, indeed suffice to understand the entire system. One rather obvious role of rare species is that they can allow survival of their consumers through lean periods; this process is exceptionally difficult to quantify, and its role in system persistence can only be judged by long-term studies, including times of great resource scarcity. Nevertheless, our belief is that to make any headway on the issue of prediction, it is crucial to have a clear and firm understanding of the core modules (e.g., plant-herbivore interactions) onto which this proliferation of weaker interactions is placed.

There are several major categories left out of this analysis that definitely cannot be regarded as being driven by the interactors we have considered: in particular, pathogens, trees, and browsers. It is not likely to be appropriate to regard the pathogens as merely a passive source of additional mortality. Pathogen transmission depends on host densities, which are affected by food web interactions, while pathogens themselves have a large effect on those densities. (For a more detailed consideration of some possible dynamics arising at the interface of infectious disease epidemiology and community ecology see Dobson 1995; Holt and Dobson, 2006; and Cleveland et al. chapter 7, this volume). Also, the compartments shown in fig. 8.1 neglect shrubs, trees, and browsers. A full treatment of the Serengeti ecosystem is clearly important for understanding the system as a whole. Holdo et al. (forthcoming) are developing a detailed model for a web, similar to that shown in fig. 8.1, but emphasizing the interplay of fire, vegetation dynamics, and herbivory in the Serengeti.

Future Models?

Models of the Serengeti in the two previous volumes in this series have contained a number of common assumptions about the functional forms of interspecific interactions. It is not clear that these have all been appropriate, and it is time to rethink the structure of such models. This may seem like a step backward, since, unlike the simulation models in *Serengeti II* (Hilborn 1995), we are not able to produce any predictions about the future. Our current situation is not one in which we know the approximate form of the components of a predictive model, and have only to refine our estimates of parameters. We currently do not know if some important model components have been represented by the wrong functions or even left out entirely. However, our hope is that research over the next few years will allow a much more detailed specification of the structure of subsystems like the one illustrated in fig. 8.1. A series of modeling exercises building on this web should provide better insight into the array of possible futures for the Serengeti ecosystem, and help identify key linkages that should be the focus of sustained empirical study and management concern.

ACKNOWLEDGMENTS

We thank Ricardo Holdo and an anonymous reviewer for very useful comments.

REFERENCES

Abrams, P. A. 1993. Effect of increased productivity on the abundance of trophic levels. *American Naturalist* 141:351–71.

———. 1995. Implications of dynamically variable traits for identifying, classifying, and measuring direct and indirect effects in ecological communities. *American Naturalist* 146:112–34.

———. 1999a. The adaptive dynamics of consumer choice. *American Naturalist* 153:83–97.

———. 1999b. Is predator mediated coexistence possible in unstable systems? *Ecology* 80: 608–21.

———. 2000. The evolution of predator-prey interactions: Theory and evidence. *Annual Review of Ecology and Systematics* 31:79–105.

———. 2001. Describing and quantifying interspecific interactions: A commentary on recent approaches. *Oikos* 94:209–18.

———. 2002. Will declining population sizes warn us of impending extinctions? *American Naturalist* 160:293–305.

———. 2004. Trait initiated indirect effects in simple food webs: Consequences of changes in consumption-related traits. *Ecology* 85:1029–38.

Abrams, P. A., and L. R. Ginzburg. 2000. Models of predation: Prey dependent, ratio dependent or neither? *Trends in Ecology and Evolution* 15:337–41.

Abrams, P. A., R. D. Holt, and J. D. Roth. 1998. Apparent competition or apparent mutualism? Shared predation when populations cycle. *Ecology* 79:201–12.

Abrams, P. A., and H. Matsuda. 1996. Positive indirect effects between prey species that share predators. *Ecology* 77:610–16.

Abrams, P. A., B. A. Menge, G. G. Mittelbach, D. Spiller, and P. Yodzis, P. 1996. The role of indirect effects in foodwebs. In *Food Webs: Integration of Patterns and Dynamics,* ed. G. A. Polis and K. O. Winemiller, 371–95. New York: Chapman and Hall.

Abrams. P. A., and J. Roth. 1994. The responses of unstable food-chains to enrichment. *Evolutionary Ecology* 8:150–71.

Abrams, P. A., and M. Vos. 2003. Adaptation, density dependence and the responses of trophic level abundances to mortality. *Evolutionary Ecology Research* 5:1113–32.

Arcese, P., J. Hando, and K. Campbell. 1995. Historical and present day anti-poaching efforts in Serengeti. In *Serengeti II: Dynamics, management, and conservation of an ecosystem,* ed. A. R. E. Sinclair and P. Arcese, 506–33. Chicago: University of Chicago Press.

Armstrong, R. A. 1979. Prey species replacement along a gradient of nutrient enrichment: A graphical approach. *Ecology* 60:76–84.

Bell, R. H. V. 1970. The use of the herb layer by grazing ungulates in the Serengeti. In *Animal populations in relation to their food resources,* ed. A. Watson, 111–24. Oxford: Blackwell.

Bertram, B. C. 1979. Serengeti predators and their social systems. In *Serengeti: Dynamics of an ecosystem,* ed. A. R. E. Sinclair and M. Norton-Griffiths, 221–48. Chicago: University of Chicago Press.

Bolker, B., M. Holyoak, V. Krivan, L. Rowe, and O. Schmitz. 2003. Connecting theoretical and empirical studies of trait-mediated interactions. *Ecology* 84:1101–14.

Caro, T. M. 1994. *Cheetahs of the Serengeti plains: Group living in an asocial species.* Chicago: University of Chicago Press.

Caro, T. and S. M. Durant. 1995. The importance of behavioral ecology for conservation biology: Examples from Serengeti carnivores. In *Serengeti II: Dynamics, management, and conservation of an ecosystem,* ed. A. R. E. Sinclair and P. Arcese, 451–472. Chicago: University of Chicago Press.

Carpenter, S. R., and J. F. Kitchell. 1993. *The trophic cascade in lakes.* Cambridge: Cambridge University Press.

Clark J. S., R. Carpenter, M. Barber, S. Collins, A. Dobson, J. Foley, D. Lodge, M. Pascual, R. Pielke, Jr., and W. Pizer. 2001. Ecological forecasts: An emerging imperative. *Science* 293:657–60.

Dobson, A. 1995. The ecology and epidemiology of Rinderpest virus in Serengeti and Ngorongoro conservation area. In *Serengeti II: Dynamics, management, and conservation of an ecosystem,* ed. A. R. E. Sinclair and P. Arcese, 485–505. Chicago: University of Chicago Press.

Dublin, H. T. 1995. Vegetation dynamics in the Serengeti-Mara ecosystem: The role of elephants, fire, and other factors. In *Serengeti II: Dynamics, management, and conservation of an ecosystem,* ed. A. R. E. Sinclair and P. Arcese, 71–90. Chicago: University of Chicago Press.

Durant, S. M. 1998. Competition refuges and coexistence: An example from Serengeti carnivores. *Journal of Animal Ecology* 67:370–86.

du Toit, J. T., K. H. Rogers, and H. C. Biggs, eds. 2003. *The Kruger experience: Ecology and management of Savanna heterogeneity.* Washington, DC: Island Press.

Elliott, J. P., and I. M. Cowan. 1978. Territoriality, density, and prey of lion in Ngorongoro crater, Tanzania. *Canadian Journal of Zoology* 56:1726–734.

Elliott, J. P., I. M. Cowan, and C. S. Holling. 1977. Prey capture by African lion. *Canadian Journal of Zoology* 55:1811–28.

Estes, J. A., M. T. Tinker, T. M. Williams, and D. F. Doak. 1998. Killer whale predation on sea otters linking ocean and nearshore ecosystems. *Science* 282:473–76.

Fryxell, J. M. 1995. Aggregation and migration by grazing ungulates in relations to resources and predators. In *Serengeti II: Dynamics, management, and conservation of an ecosystem,* ed. A. R. E. Sinclair and P. Arcese, 257–73. Chicago: University of Chicago Press.

Fryxell, J. M., J. Greever, and A. R. E. Sinclair. 1988. Why are migratory ungulates so abundant. *American Naturalist* 131:781–98.

Fryxell, J. M., J. F. Wilmshurst, and A. R. E. Sinclair. 2004. Predictive models of movement by Serengeti grazers. *Ecology* 85:2429–35.

Fryxell, J. M., J. F. Wilmshurst, A. R. E. Sinclair, D. T. Haydon, R. D. Holt, and P. A. Abrams. 2005. Landscape scale heterogeneity and the viability of Serengeti grazers. *Ecology Letters* 8:328–35.

Gatto, M. 1991. Some remarks on models of plankton densities in lakes. *American Naturalist* 137:264–67.

Grover, J. P., and R. D. Holt. 1998. Disentangling resource and apparent competition: Realistic models for plant-herbivore communities. *Journal of Theoretical Biology* 191: 353–76.

Hanby, J. P., and J. D. Bygott. 1979. Population changes in lions and other predators. In *Serengeti: Dynamics of an ecosystem,* ed. A. R. E. Sinclair and M. Norton-Griffiths, 249–62. Chicago: University of Chicago Press.

Hanby, J. P., J. D. Bygott, and C. Packer. 1995. Ecology, demography, and behavior of lions in two contrasting habitats: Ngorongoro crater and the Serengeti plains. In *Serengetti II: Dynamics, management, and conservation of an ecosystem,* ed. A. R. E. Sinclair and P. Arcese, 315–31. Chicago: University of Chicago Press.

Hilborn, R. 1995. A model to evaluate alternative management policies for the Serengeti-Mara Ecosystem. In *Serengetti II: Dynamics, management, and conservation of an ecosystem,* ed. A. R. E. Sinclair and P. Arcese, 617–37. Chicago: University of Chicago Press.

Hilborn, R., and A. R. E. Sinclair. 1979. A simulation of the wildebeest population, other ungulates, and their predators. In *Serengeti: Dynamics of an ecosystem,* ed. A. R. E. Sinclair and M. Norton-Griffiths, 287–309. Chicago: University of Chicago Press.

Hofer, H., and M. East. 1995. Population dynamics, population size, and the commuting system of Serengeti spotted hyenas. In *Serengetti II: Dynamics, management, and conservation of an ecosystem,* ed. A. R. E. Sinclair and P. Arcese, 332–63. Chicago: University of Chicago Press.

Holdo, R. M., R.D. Holt, and J. M. Fryxell. Forthcoming. Grazers, browsers, and fire influence the extent and spatial pattern of tree cover in the Serengeti.

Holdo, R. M., R. D. Holt, M. B. Coughenour, and M. E. Ritchie. 2007. Plant productivity and soil nitrogen as a function of grazing, migration and fire in an African savanna. *Journal of Ecology* 95:115–28.

Holling, C. S. 1959. The components of predation as revealed by a study of small mammal predation of the European pine sawfly. *The Canadian Entomologist* 91:293–320.

Holt, R. D. 1977. Predation, apparent competition, and the structure of prey communities. *Theoretical Population Biology* 12:197–229.

———. 1997. Community modules. In *Multitrophic interactions in terrestrial ecosystems,* ed. A. C. Gange and V. K. Brown, 333–49. 36th Symposium of the British Ecological Society. Oxford: Blackwell Science.

———. 2002. Food webs in space: On the interplay of dynamic instability and spatial processes. *Ecological Research* 17:261–73.

Holt, R. D., and A. P. Dobson. 2006. Extending the principles of community ecology to address the epidemiology of host-pathogen systems. In *Disease ecology: Community structure and pathogen dynamics,* ed. S. K. Collinge and C. Ray, 6–27. Oxford: Oxford University Press.

Holt, R. D., J. Grover, and D. Tilman. 1994. Simple rules for interspecific dominance in systems with exploitative and apparent competition. *American Naturalist* 144: 741–77.

Holt, R. D., and B. P. Kotler. 1987. Short-term apparent competition. *American Naturalist* 130:412–30.

Hoopes, M. F., R. D. Holt, and M. Holyoak. 2005. The effect of spatial processes on two species interactions. In *Metacommunites: Spatial dynamics and ecological communities,* ed. M. Holyoak, M. A., Leibold, and R. D. Holt, 35–67. Chicago: University of Chicago Press.

Jarman, P. J., and M. V. Jarman. 1979. The dynamics of ungulate social organization. In *Serengeti: Dynamics of an ecosystem,* ed. A. R. E. Sinclair and M. Norton-Griffiths, 185–200. Chicago: University of Chicago Press.

Kruuk, H. 1972. The spotted hyaena: A study of predation and social behavior. Chicago: University of Chicago Press.

Laurenson, M. K. 1979. Implications of high offspring mortality for cheetah population dynamics. In *Serengeti II: Dynamics, management, and conservation of an ecosystem,* ed. A. R. E. Sinclair and P. Arcese, 385–99. Chicago: University of Chicago Press.

———. 1995. Cub growth and maternal care in cheetahs. *Behavioral Ecology* 6:405–409.

Leibold, M. A. 1996. A graphical model of keystone predators in food webs: Trophic regulation of abundance, incidence, and diversity patterns in communities. *American Naturalist* 147:784–812.

McCann, K., A. M. Hastings, and G. R. Huxel. 1998. Weak trophic interactions and the balance of nature. *Nature* 395:794–98.

McCarthy, M. A., L. R. Ginzburg, and H. R. Akçakaya. 1994. Predator interference across trophic levels. *Ecology* 76:1310–19.

McNaughton, S. J. 1976. Serengeti migratory wildebeest: Facilitation of energy flow by grazing. *Science* 191:92–94.

———. 1985. Ecology of a grazing ecosystem: The Serengeti. *Ecological Monographs* 55: 259–94.

———. 1992. The propagation of disturbance in savannas through food webs. *Journal of Vegetation Science* 3:301–14.

McNaughton, S. J., and F. F. Banyikwa. 1995. Plant communities and herbivory. In *Serengeti II: Dynamics, management, and conservation of an ecosystem,* ed. A. R. E. Sinclair and P. Arcese, 49–70. Chicago: University of Chicago Press.

Mduma, S. A. R., A. R. E. Sinclair, R. and Hilborn. 1999. Food regulates the Serengeti wildebeest: A 40-year record. *Journal of Animal Ecology* 68:1101–22.

Melian, C. J., and J. Bascompte. 2002. Complex networks: Two ways to be robust? *Ecology Letters* 5:705–8.

Norton-Griffiths, M. 1979. The influence of grazing, browsing and fire on the vegetation dynamics of the Serengeti. In *Serengeti: Dynamics of an ecosystem,* ed. A. R. E. Sinclair and M. Norton-Griffiths, 310–52. Chicago: University of Chicago Press.

Oksanen, L., S. D. Fretwell, J. Arruda, and P. Niemela. 1981. Exploitation ecosystems and gradients of primary production. *American Naturalist* 1108:240–61.

Packer, C., R. Hilborn, A. Mosser, B. Kissui, M. Borner, G. Hopcraft, J. Wilmshurst,

S. Mduma, and A. R. E. Sinclair. 2005. Ecological change, group territoriality and population dynamics in Serengeti lions. *Science* 307:390–93.

Packer, C., D. Scheel, and A. E. Pusey. 1990. Why lions form groups—food is not enough. *American Naturalist* 136:1–19.

Paine, R. T. 1988. Food webs: Road maps of interactions or grist for theoretical development? *Ecology* 69:1648–54.

Power, M. E., M. S. Parker, and J. T. Wootton. 1996. Disturbance and food chain length in rivers. In *Food webs: Integration of patterns and dynamics,* ed. G. A. Polis and K. O. Winemiller, 286–97. New York: Chapman and Hall.

Richards, S. A., R. M. Nisbet, W. G. Wilson, and R. G. Schmitt. 2000. Grazers and diggers: Exploitation competition and coexistence among foragers with different feeding strategies on a single resource. *American Naturalist* 155:266–79.

Rosenzweig, M. L., and R. H. MacArthur. 1963. Graphical representation and stability conditions of predator-prey interactions. *American Naturalist* 97:209–23.

Rudnai. J. 1974. The pattern of lion predation in Nairobi Park. *East African Wildlife Journal* 12:213–25.

Schaller, G. B. 1972. *The Serengeti lion.* Chicago: University of Chicago Press.

Scheel, D. 1993. Profitability, encounter rates, and prey choice of African lions. *Behavioral Ecology* 4:90–97.

Scheel, D. and C. Packer. 1995. Variation in predation by lions: Tracking a movable feast. In *Serengetti II: Dynamics, management, and conservation of an ecosystem,* ed. A. R. E. Sinclair and P. Arcese, 299–314. Chicago: University of Chicago Press.

Schoener, T. W. 1971. Theory of feeding strategies. *Annual Review of Ecology and Systematics* 2:369–404.

———. 1989. Food webs from the small to the large. *Ecology* 70:1559–89.

———. 1993. On the relative importance of direct versus indirect effects in ecological communities. In *Mutualism and Community Organization,* ed. H. Kawanabe, J. E. Cohen, and K. Iwasaki, 365–411. Oxford: Oxford University Press.

Sinclair, A. R. E. 1975. The resource limitation of trophic levels in tropical grassland ecosystems. *J. Animal Ecology* 44:497–520.

———. 1979. The eruption of the ruminants. In *Serengeti: Dynamics of an ecosystem,* ed. A. R. E. Sinclair and M. Norton-Griffiths, 82–103. Chicago: University of Chicago Press.

———. 1985. Does interspecific competition or predation shape the African ungulate community? *J. Animal Ecology* 54:899–918.

———. 1995. Population limitation of resident herbivores. In *Serengeti II: Dynamics, management, and conservation of an ecosystem,* ed. A. R. E. Sinclair and P. Arcese, 194–219. Chicago: University of Chicago Press.

Sinclair, A. R. E, and P. Arcese, eds. 1995. *Serengeti II: Dynamics, management, and conservation of an ecosystem.* Chicago: University of Chicago Press.

Sinclair, A. R. E., S. Mduma, and J. S. Brashares. 2003. Patterns of predation in a diverse predator-prey system. *Nature* 425:288–90.

Sinclair, A. R. E., and M. Norton-Griffiths, eds. 1979. *Serengeti: Dynamics of an ecosystem.* Chicago: University of Chicago Press.

Winemiller, K. O. 1990. Spatial and temporal variation in tropical fish trophic networks. *Ecological Monographs* 60:331–67.

Yodzis, P. 1988. The indeterminacy of ecological interactions as perceived through perturbation experiments. *Ecology* 69:508–15.

———. 1996. Food webs and perturbation experiments: Theory and practice. In *Food webs: Integration of patterns and dynamics,* ed. G. A. Polis and K. O. Winemiller, 192–200. New York: Chapman and Hall.

———. 1998. Local trophodynamics and the interaction of marine mammals and fisheries in the Benguela ecosystem. *Journal of Animal Ecology* 67:635–58.

———. 2000. Diffuse effects in food webs. *Ecology* 81:261–66.

NINE

Spatial Dynamics and Coexistence
of the Serengeti Grazer Community

John M. Fryxell, Peter A. Abrams, Robert D. Holt,

John F. Wilmshurst, A. R. E. Sinclair, and Ray Hilborn

Two characteristics define the Serengeti grazer community in the minds of many observers: exceptional species diversity and exceptional mobility. Twenty-eight species of large ungulates are found in the greater Serengeti ecosystem (chapter 2, appendix table A.1, this volume). Many of these species have similar resource needs and occupy similar habitats. Under these circumstances, one would normally expect to witness competitive exclusion by the dominant competitor and hence reduced diversity; yet unparalleled diversity is the name of the Serengeti game. In this chapter, we use simulation models to explore three possible explanations for this seeming paradox, all of which point to size variation among herbivores, spatial variation in resources, and behavioral responses to this variation.

One hypothesis is that no grazer is consistently at a competitive advantage. Grazers of different size tend to forage most efficiently on grass patches of different height (Wilmshurst, Fryxell, and Bergman. 2000). Species of different size may be at a competitive advantage at different times and places, but the suite of species experience similar levels of Malthusian fitness integrated over multiannual cycles (Hutchinson 1961). Such a process could only allow coexistence provided that resource availability is highly heterogeneous in both time and space, such that no species is at a perpetual advantage.

A second plausible hypothesis is that Serengeti grazers facilitate each other, rather than compete for food resources (Bell 1971; McNaughton 1976; Prins and Olff 1998). According to the grazing facilitation hypothesis, large species improve foraging opportunities for smaller species by favorably modifying plant biomass and structure. While smaller species graze in their wake, larger species move on to taller pastures. We might accordingly expect to see entrained movement patterns, with smaller species tracking the movements initiated by larger species.

A third hypothesis is that seasonal migration by the predominant species—the wildebeest—prevents competitive exclusion of other grazers. By relaxing competition during part of the year, several grazing species might be allowed to coexist on a single grass resource.

To explore the logical consistency of these alternate hypotheses, we constructed spatially explicit models of the Serengeti ecosystem. Before we can address interspecific competition, however, we need to understand the mechanistic basis for foraging by the grazing guild and how this affects rate of movement across the savannah landscape (Fryxell, Wilmshurst, and Sinclair 2004) and subsequent dynamics of the grazing system (Fryxell et al. 2005). We will start by considering the common constraints that influence rates of food intake and how these factors vary allometrically among Serengeti grazers. We then consider the spatial dynamics of one species, Thomson's gazelle, which has been studied in some detail, in order to develop a plausible behavioral model of spatial dynamics.

We then extend this model to investigate competition between Thomson's gazelles, wildebeest, and African buffalo. This group was chosen for two reasons. They all have experimentally derived grazing parameters for use in the models, somewhat reducing the uncertainties inherent in the modeling process. More importantly, these species span a wide range of body sizes, social grouping patterns, and degree of movement across savannah landscapes. In so doing, we hope to capture some of the interesting ecological contrasts that could shed light on grazer coexistence.

FORAGING CONSTRAINTS

Grazers are presented with something of a paradox. They are often surrounded by an ample quantity of food, but the quality of that food varies inversely with its abundance. For simplicity, we assume that grass abundance is primarily a function of grass height, which is reasonable for many perennial species. The reason for this maturational decline in quality is simple: as grasses grow, they must invest ever more heavily in structural compounds to support the increased weight of individual stems (Van Soest 1982). As a consequence, it doesn't necessarily pay for grazers to simply choose areas of highest resource abundance in order to feed. Nor does it make much sense

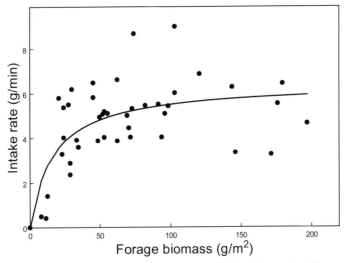

Fig. 9.1 Cropping rate (*g* dry matter per min) by Thomson's gazelles in relation to plant biomass (Wilmshurst, Fryxell, and Colucci 1999).

for grazers to simply maximize food quality, because that would imply foraging in locations with negligible food abundance. Rather, grazers should prefer to forage in locations with intermediate levels of abundance, thereby optimizing the trade-off between cropping rates and processing rates (McNaughton 1984; Hobbs and Swift 1988; Fryxell 1991).

To illustrate this fundamental trade-off, consider how cropping rates and nutritional quality vary with abundance for a typical Serengeti grazer, the Thomson's gazelle. Because the average bite size of gazelles increases with biomass, cropping rates tend to be positively associated with plant biomass (fig. 9.1). There are diminishing returns to this process, however, because gazelles have an upper limit on the size of bites they can process. Hence the cropping rate function is asymptotic, levelling off at the processing limit imposed by their mouth dimensions. This kind of instantaneous functional response is usually referred to as a type II pattern in the ecological literature (Holling 1959).

Digestibility of forage for Thomson's gazelles tends to decline with plant abundance. For example, when feeding on the grass *Cynadon dactylon,* the digestible energy content of each gram of tissue ingested declines 10% with each increase of 25 g/m² in grass abundance (fig. 9.2). This has two implications. First, gram for gram, a forager gets much lower returns on its foraging investment when grazing mature than immature grass swards. Second, the rate at which ingesta is processed through the digestive tract is inversely

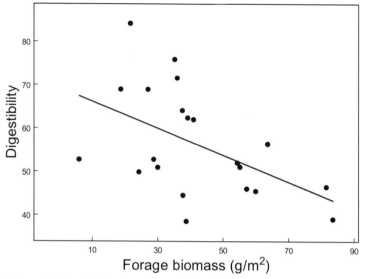

Fig. 9.2 Dry-matter digestibility by Thomson's gazelles in relation to grass biomass (Wilmshurst, Fryxell, and Colucci 1999).

related to plant abundance (Meissner and Paulsmeier 1995). If this digestive constraint falls below the cropping potential, then daily energy intake can be further compromised.

Wilmshurst, Fryxell, and Colucci 1999, measured these dual constraints for captive Thomson's gazelles fed a diet of grasses of carefully controlled quality. Their measurements suggest that gazelles realize their highest energetic returns when feeding on grass swards of 21 g/m², comparable to the abundance of grass on a freshly mown lawn (fig. 9.3). Hence, Thomson's gazelles, like many other grazers, benefit from feeding on a close-cropped lawn of regrowing grasses (McNaughton 1984; Fryxell 1991; Wilmshurst, Fryxell, and Colucci 1999).

While few studies have made the requisite measurements of both constraints, by carefully piecing together cropping rate measurements (which are relatively common in the literature) and allometric constraints on stomach capacity and digestive kinetics, one can make first order estimates of these processes in grazers of different body size (Illius and Gordon 1992; Meissner and Paulsmeier 1995; Belovsky 1997). This procedure leads to the logical conclusion that the optimal level of grass abundance increases allometrically with body mass. In other words, large species benefit best from feeding on taller swards than do smaller grazers. That result occurs because

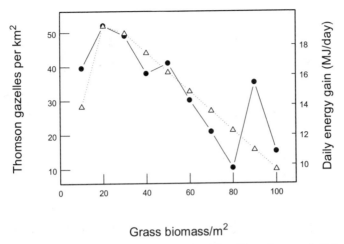

Fig. 9.3 Observed (solid line and filled symbols) versus predicted (dotted line and open symbols) densities of Thomson's gazelles, based on a daily energy-matching strategy (Fryxell, Wilmshurst, and Sinclair 2004).

larger species tend to be slightly less efficient at cropping food, evidenced by slower saturation of the cropping rate curve, as well as having a larger gut relative to their body mass. As one might predict, Serengeti herbivores of different body size tend to be found feeding on grass swards with different levels of abundance (Wilmshurst, Fryxell, and Bergman 2000).

Variation in grazer size creates a rainbow of alternate fitness optima. As a result of this allometric scaling, a heterogeneous collection of grass swards at various stages of growth might favor grazers of different size at different places at different times of year. To investigate this possibility, we need to document temporal and spatial variation in grass abundance. During the course of systematic census counts of large herbivores taken every km over a network of 220 km of roads criss-crossing the Serengeti plains, Wilmshurst (1998) monitored spatial and temporal variation in grass abundance. The routes were driven eight times per wet season between December 1994 to April 1995 and between December 1995 and June 1996, with 2 to 4 weeks between censuses. Timing of censuses was influenced to some degree by weather, because cross-country driving was sometimes impossible for a few days following rainstorms. Census routes were driven over a 3-day period.

Transects were designed to maximize coverage of a 40 × 40 km section of the Serengeti Plains, with the northwest corner at 2.535°S and 34.95°E. To ensure that routes were accurately located during each survey, established tracks were followed as much as possible, and the beginning and

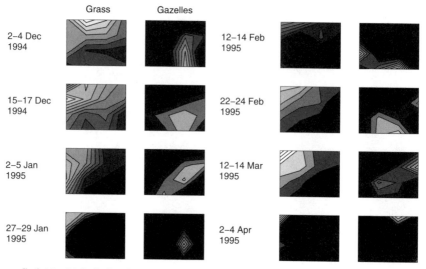

Fig. 9.4 Spatial distribution of grass biomass (left column) and Thomson's gazelles (right column) across the Serengeti Plains on eight dates during the wet season of 1994–1995, measured at two-week intervals. Relative abundance is indicated by shading, ranging from lightest (highest) to darkest (lowest). The highest grass biomass recorded was ca. 100 g/m², whereas the highest gazelle density was ca. 300 individuals/ km².

end of each transect was marked by a small pile of stones. Where tracks were absent, transects were driven cross-country, following recognizable landmarks and bearings read from a car-mounted global positioning system. Along each transect, the observer stopped at 1 km intervals to count animals and estimate grass abundance. Vehicle speed was maintained at or below 30 km/h between stops to standardize visibility and reduce disturbance of nearby animals. At each stop, all large herbivores visible within a semicircle of 200 m radius were counted, as well as noting age and sex of subsamples. After all animals had been recorded, the observer clipped all grasses and forbs in 0.25 m² randomly positioned quadrats. Dead tissue was sorted from the live component and the live component was then weighed for dry matter determination.

Wilmshurst's (1998) measurements demonstrate two important features (fig. 9.4). First, there is a rough, increasing gradient in grass abundance as one progresses from the driest (SE) corner of the system to the wettest (NW) corner of the study system. Second, after accounting for the general rainfall gradient, there is still a great deal of unexplained variation in grass abundance among locations for any given census and at any given location from one census to the next. In other words, there is enough variation in grass abundance to potentially meet the needs of several different species.

Wilmshurst's census counts suggest that Thomson's gazelles respond to this shifting mosaic of food abundance by ceaselessly redistributing themselves across the landscape (fig. 9.4). Far from being mindlessly nomadic, however, gazelle movements seem to track the shifting location of patches that are most rewarding energetically (Fig. 9.3). Further evaluation of the movement patterns suggest that gazelles tend to abandon grass patches whose energetic rewards are below the average rate of reward for the entire landscape, but tend to stay in patches that are well above-average (Fryxell, Wilmshurst, and Sinclair 2004). Once departed from a patch, gazelles tend to relocate in neighboring patches with a probability proportionate to the daily rate of energy gain (Fryxell, Wilmshurst, and Sinclair 2004). In other words, there is strong basis for modelling patch preference and probability of movement on the basis of energetic reward rates.

A SINGLE-SPECIES GRAZING MODEL

Having demonstrated that Thomson's gazelles respond to spatial and temporal variation in resource abundance by preferentially shifting among neighboring patches, we now turn to modeling the dynamics of such grazing systems. Our simulation models assume dynamic interactions among herbivores and grasses, with the local pattern of herbivore densities shaped by both demographic and behavioral responses to plant abundance (Fryxell et al. 2005). We assumed a stochastic, seasonal environment, with a single wet season alternating with a dry season of stochastically variable length. State variables in the model were updated at daily time steps. In keeping with the spatial frame of reference of the field data used to parameterize movement probabilities used in the model (Fryxell, Wilmshurst, and Sinclair 2004), the simulations were conducted on a lattice, with each grid square measuring 10 km on each side. Most of the parameters used in the model were based on empirical observations in Serengeti, the exceptions being estimates of maximum rate of herbivore population increase and mortality rates, which were estimated from allometric regressions for African artiodactyls (Western 1979).

Grass Growth Submodel

Fully parameterized models of grass growth in relation to moisture and standing biomass are scarce in the ecological literature. One of the most detailed attempts to model plant growth was Robertson's (1987) measurements on savanna grasslands in Australia, conducted as part of a decade-long collaborative study of the population dynamics of kangaroos and domesticated livestock in relation to rainfall (Caughley, Shepherd, and Short 1987). This work suggested that plant growth related curvilinearly to plant biomass, with peak growth rates at low biomass. The peak of grass growth was scaled to rainfall. The pattern of growth in the Australia study was well represented by a series of nested quadratic curves. Similar logic has been used to model grassland dynamics in the face of grazing by voles (Turchin and Batzli 2001). The logic for this relation is that a substantial amount of grass biomass occurs belowground, indeed often matching the maximum level recorded aboveground. Belowground tissue subsidizes aboveground production during periods of adequate rainfall, via carbohydrate translocation, even when there is no standing biomass aboveground. Traditional logistic formulations lack this capacity for rapid growth when there is no standing biomass and are therefore inadequate for modeling perennial grasses.

We accordingly represented the growth of Serengeti grasses by a modified logistic growth curve. Following Robertson's (1987) lead, we assumed that the daily rate of grass growth during the wet season was at a maximum at low values of vegetation biomass, declining curvilinearly with grass biomass according to the following function:

$$\frac{dV}{dt} = r_{max}[V + K(R)]\left[1 - \frac{V + K(R)}{2K(R)}\right],$$
(1)

where V = vegetation biomass and K = grass carrying capacity, both measured in dry matter g/m^2 and r_{max} is the maximum exponential rate of grass growth, expressed on an annual basis. For the growth rates observed in Serengeti, $r_{max} = 0.039$, based on McNaughton's (1985) measurements within grass exclosures. This formulation yields a series of nested growth curves whose maximum and equilibrium values are proportionate to daily rainfall R (fig. 9.5). In years with low rainfall, both the maximum rate of grass growth and the carrying capacity are markedly lower than in years of high rainfall. During the dry season, we assumed that grasses decline at a daily rate of -1.79 g/m^2, based on the intercept of the linear function $\omega(R)$.

Patterns of grass growth were based on McNaughton's (1985) measurements on the Serengeti plains. This work suggests that there is a sigmoid relationship between number of days of growth and annual rainfall. We approximated this relationship by the following function:

$$growdays(rainperyear) = 60 + \frac{300e^{0.01rainperyear}}{e^{0.01rainperyear} + e^{6.25}}, \tag{2}$$

where *rainperyear* = annual rainfall, which we assumed to average 671 mm/year in the northwestern corner and 528 mm/year in the southwestern corner of our 1,600 km² study area on the Serengeti plains, based on data from rain gauge stations 1 and 7 in table 1 of McNaughton (1985). Average rainfall levels in the other cells were interpolated from this gradient. In the stochastic simulations, variation in annual rainfall was assumed to be normally distributed, with a CV = 0.25, based on the rainfall data summarized in McNaughton (1985). During the growing season, we assumed that rain was distributed uniformly at a daily rate R = *rainperyear/growdays(rainperyear)*.

Based on data in Norton-Griffiths, Herlocker, and Pennycuick (1975), we assumed positive spatial autocorrelation in annual rainfall among adjacent grid cells of the lattice (ρ = 0.80). The annual rainfall deviate in any given cell ($z_{j,i}'$) was calculated using the random normal deviates drawn for that cell ($z_{j,i}$) and a neighboring cell ($z_{j,i-1}$), modified by the autocorrelation coefficient (ρ): $z_{j,i}' = \rho z_{j,i-1} + z_{j,i}(1-\rho^2)^{1/2}$. We started by assigning a rainfall deviate to a cell in the corner of the grid, then used the autocorrelation function and the seed deviate to calculate the deviates of the two adjoining cells in the matrix (horizontally and vertically). This process was repeated until the matrix was full.

McNaughton's (1985) measurements of biomass change within grazing exclosures indicate that maximum daily grass growth (measured in g dry matter/m² per day) is a positive function of daily rainfall R (measured in mm):

$$\omega(R) = -1.79 + 2.11R \tag{3}$$

The maximum grass biomass in McNaughton's (1985) grazing exclosures was approximately 280 g/m² in the most arid parts of the Serengeti plains during a period in which daily rainfall averaged 3.462 mm. We accordingly linked grass-carrying capacity K to rainfall through the coefficient Ψ = 280/3.462 = 80.872. In the stochastic simulations, $K(R) = \Psi R$.

Grass Consumption Submodel

Rates of consumption and energy intake by Thomson's gazelles were based on experimental data in Wilmshurst, Fryxell, and Colucci (1999). The hourly rate of energy intake was estimated by multiplying the hourly rate of consumption $[X(V) = aV/(b + V)]$ by a linear function describing the digestible energy content of forage $[Q(V) = c - dV]$, where a (the maximum hourly rate of consumption) = 380 g dry matter/h, b (vegetation biomass at which the rate of consumption is 1/2 the maximum) = 15 g dry matter/m^2, c (maximum forage quality) = 0.011201 MJ/g grass eaten, and d (the rate at which quality declines with plant biomass) = 0.05 MJ m^2 per g^2 dry matter, and V = plant biomass (g dry matter per m^2). All of these parameters were estimated from Wilmshurst, Fryxell, and Colucci's (1999) experimental trials using captive Thomson's gazelles. This yielded the following function for hourly energy intake:

$$\text{Hourly energy intake model:} \quad Y(V) = \frac{aV(c - dV)}{b + V}. \tag{4}$$

Ad libitum consumption levels may fall below the level predicted by the short-term functional response, because the consumed amount is retained longer in the digestive tract, which can become rate limiting (Fryxell 1991). Modeling this effect requires a second digestive constraint, *ad libitum* consumption as a function of plant biomass $[I(V)]$. Field measurements indicated that digestible energy content of the sward declined with sward biomass in an approximately linear fashion (Wilmshurst, Fryxell, and Colucci 1999). We accordingly modeled the digestive constraint (in MJ/day) by the function $I(V) = (e - fV)$, where e (the maximum daily intake of energy) = 22.7 MJ/day and f (the rate at which daily intake declines with sward biomass) = 0.13 MJ m^2 per g day. Daily energy gain, measured in MJ per day (fig. 9.5, panel B) was calculated by taking the lesser of the constraint functions $Y(V)t_{max}$, where t_{max} (the maximum number of hours spent foraging each day) = 9 h per day or $I(V)$:

$$\text{Daily energy intake model:} \quad Z(V) = \min\left[\frac{aVt_{max}(c - dV)}{b + V}, e - fV\right]. \tag{5}$$

The daily rate of food intake, measured in g dry matter per m^2, can be calculated similarly as the minimum of the following constraints, obtained by factoring out food quality:

$$\text{Daily food intake model:} \quad W(V) = \min\left(\frac{aVt_{max}}{b + V}, \frac{e - fV}{c - dV}\right). \tag{6}$$

Gazelles were assumed to distribute themselves among cells on a lattice with 100 km² cells, with each square cell measuring 10 km on each side. Daily dynamics in each cell ij were accordingly calculated by the following system of equations:

$$\frac{dV_{ij}}{dt} = r_{\max ij}[V_{ij} + K_{ij}(R_{ij})]\left[1 - \frac{V_{ij} + K_{ij}(R_{ij})}{2K_{ij}(R_{ij})}\right]$$
$$- N_{ij}[\Omega(V_{ij}) - \theta(V_{ij})]W(V_{ij}) \tag{7}$$

$$\frac{dN_{ij}}{dt} = N_{ij}[\Omega(V_{ij}) - \theta(V_{ij})][\Lambda(V_{ij})Z(V_{ij}) - \Gamma] \tag{8}$$

where $\theta(V_{ij})$ represents the rate of emigration from a focal cell ij and $\Omega(V_{ij})$ represents the summed rate of immigration from neighbouring cells into cell ij.

The probability of emigration $\theta(V)$ was assumed to depend on the local daily rate of energy intake $[Z(V)]$ relative to the expected rate of daily energy gain averaged over the landscape $E[Z(V)]$, following the marginal value theorem (Charnov 1976), using the following sigmoid threshold function:

$$\text{Probability of emigration:} \quad \theta(V) = \frac{E[Z(V)]^{\phi}}{Z(V)^{\phi} + E[Z(V)]^{\phi}} \tag{9}$$

with $\phi = 2.4$ (Fryxell, Wilmshurst, and Sinclair 2004). The probability of immigration to neighboring cells on the lattice was made proportionate to the rate of daily energy intake relative to that in the alternative neighbouring cells (fig. 9.5, panel D).

Demographic data on birth rates for Thomson's gazelles were not available, so we estimated these from allometric comparisons for African artiodactyls (Western 1979):

$$\Gamma(m) = (0.0000274) \times 10^{(3.18 - 0.35\log_{10}m)}, \tag{10}$$

where m = body mass in g. For Thomson's gazelles, $m = 20,000$ g, so $\Gamma = 0.0013$ per day. In the absence of data on gazelle mortality rates, we assumed that it was equal to the maximum daily reproductive rate, so $\Gamma(m)$ represents the gazelle rate of mortality, as well as the maximum daily production of young. We assumed that the maximum rate of population growth occurs when energy intake is maximized, that is, at the optimal grass biomass V_{opt}. In the case of Thomson's gazelles, $V_{opt} = 21.7$ g/m². Energy intake was accordingly translated into the per capita growth rate of the population

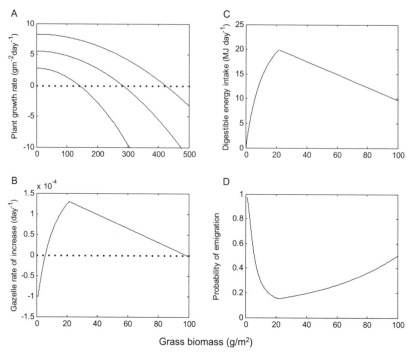

Fig. 9.5 Functional relationships used to model population dynamics of Thomson's gazelles in relation to plant biomass (Fryxell et al. 2005). (Panel A) the daily rate of grass growth for a range of daily rainfall rates (bottom curve = 1.75 mm, middle curve = 3.5 mm, and top curve = 5.25 mm rainfall per day); (panel B) the daily rate of population increase by Thomson's gazelles, (panel C) the daily rate of energy gain by Thomson's gazelles, and (panel D) the probability of emigration from a given patch. The probability of emigration was calculated on the assumption that average biomass elsewhere in the ecosystem was 100 g/m². A different curve of similar shape would apply for any different level of grass biomass averaged across the lattice.

by the function $dN/Ndt = \Lambda\, Z(V_{opt}) - \Gamma$, where N = population density of gazelles, measured in individuals per m², $Z(V)$ is the daily rate of energy intake (MJ per day), $\Lambda = 0.00013$ is the rate of conversion of daily energy intake into daily gazelle population growth, and Γ is the maximum per capita rate of gazelle increase per day, as shown in fig. 9.5, panel C.

Results of the Single-Species Model

The gazelle model generates spatial patterns in abundance very reminiscent of those observed during the biweekly censuses done in the early 1990s (fig. 9.4). Stochastic variation in rainfall at the beginning of the wet season creates a heterogeneous environmental template that the gazelle population responds to via a continuous pattern of redistribution.

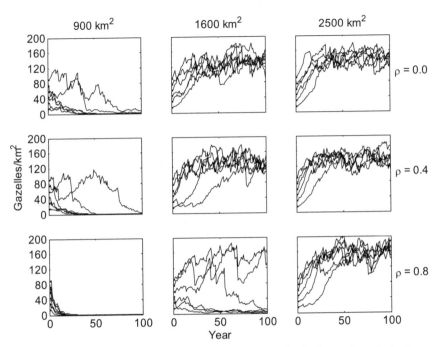

Fig. 9.6 Simulated changes in population density (individuals/km²) over time for Thomson's gazelles for 4 alternate lattice sizes: (3 × 3 cells, 4 × 4 cells, 5 × 5 cells, and 6 × 6 cells, each measuring 10 × 10 km²) and 3 alternate degrees of spatial autocorrelation in rainfall ($\rho = 0.0$, $\rho = 0.4$, $\rho = 0.8$). In each panel, 6 independent simulations are shown, each started at different initial densities of gazelles (ranging between 15 and 90 individuals/km²). Each simulation was based on a different stochastic rainfall series. Each simulation assumed that animals used an energy-matching strategy to move between cells on the lattice.

Results of the single-species model depend on the size of the savannah landscape visited by the gazelles and the degree of autocorrelation in rainfall across the system (fig. 9.6). When rainfall is highly autocorrelated among patches, then persistence of gazelles is compromised when gazelles are restricted to small landscapes. For the level of rainfall stochasticity recorded in Serengeti, Thomson's gazelles would be unlikely to persist on a savannah landscape less than 2,500 km² in size (Fryxell et al. 2005).

This conclusion has some important practical implications. It suggests that it is perhaps essential to maintain unobstructed access to large areas of grassland in order to conserve terrestrial grazers like the Thomson's gazelle. In smaller reserves, such as those in many places in East Africa, some grazers may be at appreciable risk of dying out over an extended period, at least without human intervention. It might be much easier to preserve savannah ecosystems of adequate size.

MULTISPECIES GRAZING MODEL

To extend this model to multiple species, we need to specify the functional response, movement pattern, and numerical response of each grazer to local grazing conditions. In the absence of detailed consideration of the dual constraints, we use Wilmshurst, Fryxell, and Bergman's (2000) technique of marrying direct observations of cropping rates with allometric limits on processing capacity (Meissner and Paulsmeier 1995). We focus on two species (wildebeest and African buffalo) that differ considerably in body size from Thomson's gazelles (we assumed Thomson's gazelle mass = 20 kg, wildebeest mass = 97 kg, and buffalo mass = 180 kg). In the absence of parameters for African buffalo, we used literature values recorded for yearling wood bison, which have similar body size and feeding style. Feeding rates would be much higher in adult animals, typically weighing several times more than yearlings.

We used the procedures outlined by Wilmshurst, Fryxell, and Bergman (2000) to estimate the daily rate of grass consumption by wildebeest and buffalo by the following equation:

$$W(V) = \frac{(\alpha - \beta V)V}{b + V} \tag{11}$$

where α is the maximum daily consumption rate for forage whose dry matter digestibility = 56% (the same value used for the gazelle functional response), β is the rate that maximum daily consumption declines with increasing plant biomass, and b is the vegetation biomass at which cropping rates are 1/2 the maximum level. The functional response measurements of Murray (1991) yielded the following parameter estimates for wildebeest: $\alpha_2 = 3.501$, $\beta_2 = 0.00343$, and $b_2 = 9.94$. In the absence of appropriate data for African buffalo, we used Bergman et al.'s (2001) foraging data for bison to estimate the following parameters: $\alpha_3 = 6.288$, $\beta_3 = 0.00343$, and $b_3 = 99.2$. The functional response model for each species was multiplied by digestibility to estimate digestible energy intake:

$$X(V) = \frac{(\alpha - \beta V)V(c - dV)}{b + V} \tag{12}$$

where c = maximum digestible energy content and d = rate of decline in digestible energy with increasing plant biomass. We used the same values here as for the Thomson's gazelle model. Comparison of the forage intake curves for the smaller (wildebeest) versus larger (buffalo) species show that wildebeest are more efficient at obtaining adequate energy at low vegeta-

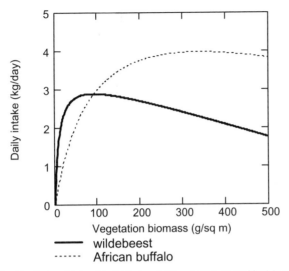

Fig. 9.7 Daily food intake by wildebeest (solid line) and African buffalo (dotted line) in relation to grass biomass.

tion abundance, probably because they have experienced strong selection for use of closely cropped swards on the Serengeti plains, whereas the larger bovid (buffalo) is far less efficient (fig. 9.7). On the other hand, buffalo should maintain positive growth at much higher levels of vegetation biomass and decline more slowly in periods of high food abundance than do smaller grazers (fig. 9.8).

We followed the same procedure as outlined earlier for Thomson's gazelle to convert the daily rate of energy intake $X(V)$ into the numerical response. This translated into the following parameter values: for wildebeest $\Gamma_2 = 7.474 \times 10^{-4}$ and $\Lambda_2 = 6.007 \times 10^{-5}$, whereas for buffalo $\Gamma_3 = 6.019 \times 10^{-4}$ and $\Lambda_3 = 6.221 \times 10^{-5}$.

Putting these pieces together, the full set of equations for the system is as follows:

$$\frac{dV_{ij}}{dt} = r_{\max ij}[V_{ij} + K_{ij}(R_{ij})]\left[1 - \frac{V_{ij} + K_{ij}(R_{ij})}{2K_{ij}(R_{ij})}\right]$$

$$- \sum_{k=1}^{3} N_{ijk}[\Omega_k(V_{ij}) - \theta_k(V_{ij})]W_k(V_{ij}) \tag{13}$$

$$\frac{dN_{ijk}}{dt} = N_{ijk}[\Omega_k(V_{ij}) - \theta_k(V_{ij})][\Lambda_k(V_{ij})Z_k(V_{ij}) - \Gamma_k] \tag{14}$$

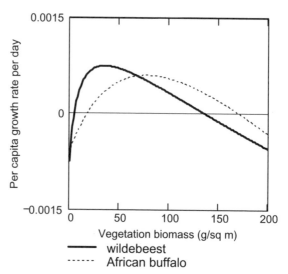

Fig. 9.8 Per capita rate of population growth (individuals per day) by wildebeest (solid line) and African buffalo (dotted line) in relation to grass biomass.

where the subscript k refers to each of the species (Thomson's gazelles = 1, wildebeest = 2, and African buffalo = 3) and i and j refer to Cartesian coordinates of the spatial lattice.

Since buffalo are resident in the tall grasslands of Serengeti, but rarely moving very far, we assumed no emigration among cells after initializing only the northern 1/2 of the grid cells. Because wildebeest and Thomson's gazelles are migratory, we shifted them to the bottom row of cells on the first day of each year and shifted them on the one hundredth day of each year to the top row of cells. Following each of these migratory movements, wildebeest shifted among cells using the same movement rules as Thomson's gazelles, as observed in Wilmshurst et al.'s (1999) analysis of radio-marked wildebeest, but with wildebeest-specific rates of energy gain used to determine the probabilities of emigration and redistribution among neighboring cells. As simulation experiments, we then turned on or off the wildebeest migration, turned on or off the nomadic interpatch movement of gazelles and wildebeest, and eliminated wildebeest altogether to observe competition between gazelles and buffalo.

Results of the Multispecies Model

Our simulations point to several interesting features. First, the complete spatially realistic model is not conducive to grazer coexistence. Wildebeest drive everything else extinct when both Thomson's gazelles and wildebeest are migratory and both wildebeest and Thomson's gazelles are capable of shifting among neighboring locations to find favorable foraging patches (fig. 9.9). By moving periodically to find suitable sites in which to forage, wildebeest rule the roost. It is clear that one advantage migratory herbivores have is access for an extended period to highly nutritious grasses growing in low rainfall conditions. This allows both wildebeest and gazelles to obtain better energetic gains and commensurately higher Malthusian fitness than buffalo. For those reasons, animals that have evolved the capacity to migrate into the short grasslands for 3 months every year ought to predominate. Wildebeest do better than gazelles because they are just as efficient at obtaining energy at low levels of vegetation abundance, but less vulnerable to death by a too-dilute diet at high levels of grass abundance.

The predominance of wildebeest extends to conditions in which

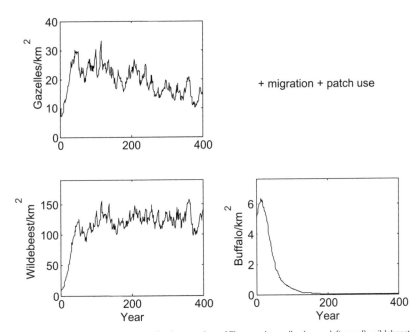

Fig. 9.9 Predicted changes in average density over time of Thomson's gazelles (upper left panel), wildebeest (lower left panel), and African buffalo (lower right panel), for a simulation model in which wildebeest are seasonally migratory, both wildebeest and gazelles shift among patches according to an energy-matching strategy, and buffalo are resident in the northern half of the ecosystem.

they and gazelles are nomadic, but not seasonally migratory, using an energy-matching strategy to distribute themselves among neighboring patches. Wildebeest, once again drive everything else extinct within 100 to 200 years (fig. 9.10). If anything, extinction is even faster than was the case for the fully migratory system. Once again, the advantage of wildebeest is that they can compete with gazelles at low plant abundance, yet also compete reasonably well with buffalo at high levels of grass abundance.

When we prevent wildebeest and gazelles from opportunistically moving to take advantage of better foraging conditions in neighboring patches, then buffalo tend to out-compete the others (fig. 9.11). Under these 'fixed residency' conditions, in other words, buffalo have a very viable combination of foraging and demographic attributes. They do not die out when the inevitable periods of poor food abundance occur, due either to droughts or periods of above average rainfall. One might think of this as a masterful integrative strategy. Note, too, that under resident conditions, perpetual coexistence is possible between a medium sized and large grazer, although gazelles still die out. Indeed, about the only way to ensure that gazelles persist is to eliminate wildebeest entirely, but let gazelles move adaptively across the landscape.

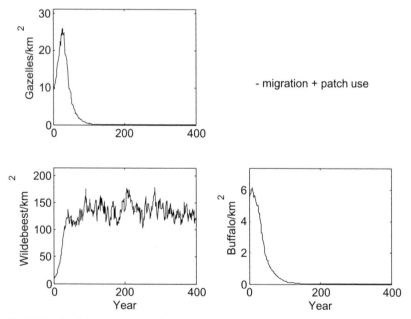

Fig. 9.10 Predicted changes in average density over time of Thomson's gazelles (upper left panel), wildebeest (lower left panel), and African buffalo (lower right panel), for a simulation model in which wildebeest are *not* seasonally migratory, both wildebeest and gazelles shift among patches according to an energy-matching strategy, and buffalo are resident in the northern half of the ecosystem.

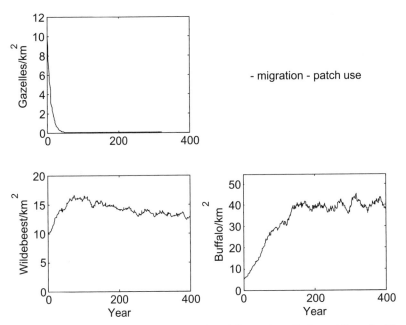

- migration - patch use

Fig. 9.11 Predicted changes in average density over time of Thomson's gazelles (upper left panel), wildebeest (lower left panel), and African buffalo (lower right panel), for a simulation model in which wildebeest, gazelles, and buffalo are resident in the northern half of the ecosystem.

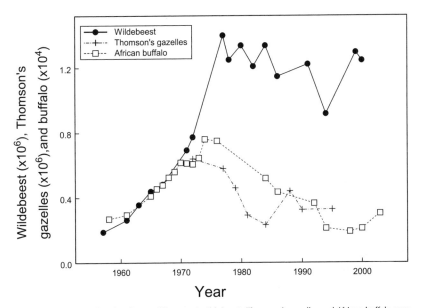

Fig. 9.12 Changes in the abundance of Serengeti wildebeest, Thomson's gazelle, and African buffalo over the past 4 decades, based on aerial census estimates.

DISCUSSION

Our models lead to three general conclusions. First, spatial processes appear to be important in allowing mobile grazer species to cope with interannual and seasonal variation in rainfall typical of savannah ecosystems. Second, neither migration, differences in body size, nor adaptive patterns of landscape use appear sufficient to allow coexistence of even three grazers, let alone 28 species. Third, temporal variation in resources, mediated via rainfall, is also inadequate to explain species coexistence.

Both African buffalo and Thomson's gazelles are ultimately losers in the competitive battle head-to-head with wildebeest. Behavioral biocomplexity is apparently not enough to ensure persistence, at least given the relatively limited behavioral repertoires we have incorporated into the model. Given the set of rules in our complete model, wildebeest have access to a strategy that is unavailable to the other herbivores, and are therefore able to play the competitive game more effectively. This suggests two things. First, we may need to look beyond mobility, at least at the crude level modeled here, to explain the exceptional diversity of the Serengeti grazer guild as a persistent phenomenon. Second, it suggests that release of wildebeest from limitation well below its resource-based carrying capacity, due to the eradication of rinderpest, may have inadvertently unleashed a slow, but nonetheless inevitable course of competitive exclusion (see fig.1.1).

Population trends for both buffalo and Thomson's gazelles show a consistent downward trend following the growth and equilibration of the wildebeest population at roughly a million animals (fig. 1.1). These patterns are consistent with the competitive exclusion hypothesis. Given the limited time frame, however, it is hard to know for sure. The spatially realistic models suggest it might take decades, if not centuries, for competitive exclusion to occur. Given the climatic, sociological, and ecological changes that may well occur over the next century, it is somewhat premature to conclude that grazer extinction is inevitable. It is nonetheless a possibility that needs serious consideration. On the other hand, if the models are correct in predicting the long-term demise of Thomson's gazelles and African buffalo relative to wildebeest, it begs a profound question: how have these species persisted for the past 2 million years?

Field studies over the past decade lend credence to the notion that Serengeti grazers make adaptive choices about where they feed. More importantly, we may be beginning to be able to predict the pattern of spatial distribution that would arise as a natural consequence of such adaptive decision making. It may not be long before one may reasonably hope to predict the shifting mosaic of both resources and a wide suite of herbivores, even in complex ecosystems like Serengeti.

Our work thus far suggests that there may be mechanistic similarities in the foraging mechanics of terrestrial foragers of different size that may lend themselves to comparative, allometric approaches. On the other hand, we know that all grazing species have some foraging preferences that are not accommodated by such simple models. For example, we have not modeled here selective grazing for particular plant parts, for which there is strong evidence of species-specific differences (Jarman and Sinclair 1979). In principle, one could include such detail into these spatially realistic models, but it would slow down the simulations considerably.

Balanced against this overall buoyant tone of predictive optimism, we still do not have a very clear idea about the biological features that permit coexistence in a speciose, yet similar, guild of terrestrial herbivores. It may be that we have failed to capture the crucial spatial scale at which these interactions take place. Any of the three hypotheses outlined at the beginning of this chapter may be important at much finer scales than we have examined here. If so, it will take much more intensive field measurements of the spatial distribution of resources and consumers and finer-scale models of individual decision-making to demonstrate the logic underlying this hypothesis. Alternatively, it may be that exclusive resource use by each species allows them to coexist. This could be achieved via tissue selection (Bell 1971; Jarman 1974) or perhaps specialization for different species or types of vegetation, such as graminoids versus dicots (Belovsky 1986a, 1986b, 1997).

Another topic that is obviously missing in our spatial analysis is the role of predators, both in shaping herbivore spatial patterns and as an agent of demographic change. Much has been learned about foraging by predominant predators of the grazer guild. For example, an exciting line of recent work has shown that there is a strong allometric pattern in the degree of predation risk for herbivores of different sizes (Sinclair, Mduma, and Brashares. 2003). Small herbivores can be attacked successfully by virtually every carnivore in the ecosystem, so they experience much higher predation rates than the largest species, available to only the largest carnivores. It is plausible to think that allometric trade-offs between predation risk and foraging efficiency might allow coexistence that would be impossible without predation as a mediating process.

We also know that lions exhibit patterns of adaptive diet selection, based on the energetic profitability of alternate prey groups (Scheel 1993). Because these lion-foraging models are based on encounters with different prey groups, and do not span a wide range of prey densities, it is tricky translating the behavioral pattern in simple trophic models, such as the cell-lattice models demonstrated here. We also know that hyenas have a spatial clan structure that responds to seasonal and multiannual fluctuations in prey abundance (Hofer and East 1993a, 1993b; Höner et al. 2005). In the absence of information on foraging rates, numerical responses by hyenas, while absolutely important, are of limited value. Hence, precious little about the magnitude of predation rates can be reliably modeled at present. We are hopeful, however, that the current generation of predator-prey studies underway in Serengeti and elsewhere in savannah ecosystems will help remedy this obvious deficiency. This is an obvious priority area for further empirical as well as theoretical work.

ACKNOWLEDGMENTS

We wish to thank the Natural Sciences and Engineering Research Council of Canada (NSERC), the Wildlife Conservation Society, the Frankfurt Zoological Society, and the National Science Foundation (NSF) for their support of our field and modeling programs in Serengeti over the years. We also warmly thank our Tanzanian colleagues for the many ways that they contribute to this long-term program. Without their unflagging support, none of this would be possible. We also thank two anonymous referees as well as Mark Ritchie for their comments on an earlier draft of the manuscript.

REFERENCES

Bell, R. H. V. 1971. A grazing ecosystem in the Serengeti. *Scientific American* 225:86–93.

Belovsky, G. E. R. 1986a. Optimal foraging and community structure: Implications for a guild of generalist grassland herbivores. *Oecologia* 70:35–52.

———. 1986b. Generalist herbivore foraging and its role in competitive interactions. *American Zoologist* 26:51–69.

———. 1997. Optimal foraging and community structure: The allometry of herbivore food selection and competition. *Evolutionary Ecology* 11:641–72.

Bergman, C. M., J. M. Fryxell, C. C. Gates, and D. Fortin. 2001. Ungulate foraging strategies: Energy maximizing or time minimizing? *Journal of Animal Ecology* 70:289–300.

Caughley, G., N. Shepherd, and J. Short, eds. 1987. *Kangaroos: Their ecology and management in the sheep rangelands of Australia*. Cambridge: Cambridge University Press.

Charnov, E. L. 1976. Optimal foraging, the marginal value theorem. *Theoretical Population Biology* 9:129–136.

Fryxell, J. M. 1991. Forage quality and aggregation by large herbivores. *American Naturalist* 138:478–98.

Fryxell, J. M., J. F. Wilmshurst, and A. R. E. Sinclair. 2004. Predictive models of movement by Serengeti grazers. *Ecology* 85:2429–35.

Fryxell, J. M., J. F. Wilmshurst, A. R. E. Sinclair, D. T. Haydon, R. D. Holt, and P. A. Abrams. 2005. Landscape scale, heterogeneity, and the viability of Serengeti grazers. *Ecology Letters* 8:328–35.

Hobbs, N. T., and D. M. Swift. 1988. Grazing in herds: When are nutritional benefits realized? *American Naturalist* 131:760–64.

Hofer, H., and M. L. East. 1993a. The commuting system of Serengeti spotted hyaenas: How a predator copes with migratory prey. I. Social organization. *Animal Behaviour* 46:547–57.

———. 1993b. The commuting system of Serengeti spotted hyaenas: How a predator copes with migratory prey. II. Intrusion pressure and commuters' space use. *Animal Behaviour* 46: 559–74.

Holling, C. S. 1959. The components of predation as revealed by a study of small-mammal predation of the European pine sawfly. *Canadian Entomologist* 91: 293–320.

Höner, O. P., B. Wachter, M. L. East, V. A. Runyoro, and H. Hofer. 2005. The effect of prey abundance and foraging tactics on the population dynamics of a social, territorial carnivore. *Oikos* 108:544–54.

Hutchinson, G. E. 1961. The paradox of the plankton. *American Naturalist* 95:137–45.

Illius, A., and I. Gordon. 1992. Modelling the nutritional ecology of ungulate herbivores: Evolution of body size and competitive interactions. *Oecologia* 89:428–34.

Jarman, P. J. 1974. The social organisation of antelope in relation to their ecology. *Behaviour* 48:215–67.

Jarman, P. J., and A. R. E. Sinclair. 1979. Feeding strategy and the pattern of resource partitioning in ungulates. In *Serengeti: Dynamics of an ecosystem,* ed. A. R. E. Sinclair and M. Norton-Griffiths, 130–63. Chicago: University of Chicago Press.

McNaughton, S. J. 1976. Serengeti migratory wildebeest: Facilitation of energy flow by grazing. *Science* 191:92–94.

———. 1984. Grazing lawns: Animals in herds, plant form, and coevolution. *American Naturalist* 124:863–86.

———. 1985. Ecology of a grazing ecosystem: The Serengeti. *Ecological Monographs* 55: 259–94.

Meissner, H. H., and D. V. Paulsmeier. 1995. Plant compositional components affecting between-plant and animal species prediction of forage intake. *Journal of Animal Science* 73:2447–57.

Murray, M. G. 1991. Maximizing energy retention in grazing ruminants. *Journal of Animal Ecology* 60:1029–45.

Norton-Griffiths, M., D. Herlocker, and L. Pennycuick. 1975. The patterns of rainfall in the Serengeti ecosystem, Tanzania. *East African Wildlife Journal* 13:347–74.

Prins, H. H. T., and H. Olff. 1998. Species-richness of African grazer assemblages: Towards a functional explanation. In *Dynamics of tropical communities,* ed. D. M. Newbery, H. H. T. Prins, and N. Brown, 448–90. Oxford: Blackwell.

Robertson, G. 1987. Plant dynamics. In *Kangaroos: Their ecology and management in the sheep rangelands of Australia,* ed. G. Caughley, N. Shepherd, and J. Short, 50–68. Cambridge: Cambridge University Press.

Scheel, D. 1993. Profitability, encounter rates, and prey choice of African lions. *Behavioral Ecology* 4:90–97.

Sinclair, A. R. E., S. Mduma, and J. Brashares. 2003. Patterns of predation in a diverse predator system. *Nature* 425:288–90.

Turchin, P., and G. Batzli. 2001. Availability of food and the population dynamics of arvicoline rodents. *Ecology* 82:1521–34.

Van Soest, P. J. 1982. *Nutritional ecology of the ruminant.* Corvallis, OR: O. and B.

Western, D. 1979. Size, life history, and ecology in mammals. *African Journal of Ecology* 17: 185–204.

Wilmshurst, J. F. 1998. Foraging behaviour and spatial dynamics of Serengeti herbivores. PhD diss., University of Guelph.

Wilmshurst, J. F., J. M. Fryxell, and C. M. Bergman. 2000. The allometry of patch selection in ruminants. *Proceedings of the Royal Society of London, Series B—267:34549.*

Wilmshurst, J. F., J. M. Fryxell, and P. E. Colucci. 1999. What constrains daily intake in Thomson's gazelles? *Ecology* 80:2338–47.

Wilmshurst, J. F., J. M. Fryxell, B. P. Farm, A. R. E. Sinclair, and C. P. Henschel. 1999. Spatial distribution of Serengeti wildebeest in relation to resources. *Canadian Journal of Zoology* 77:1223–32.

TEN

Dynamic Consequences of Human Behavior in the Serengeti Ecosystem

Christopher Costello, Nicholas Burger, Kathleen A. Galvin, Ray Hilborn, and Stephen Polasky

The Serengeti is one of the premiere natural ecosystems in the world. Expansive populations of migratory wildebeest and zebra and the lions, hyenas, and cheetahs that prey upon them, coupled with an immense diversity of other ungulates and a wide range of habitats, birds, and other flora and fauna are all responsible for its uniqueness. Yet, people have been central to the evolution of the Serengeti, and will likely play an expanding role in its future. Human populations are growing in areas surrounding Serengeti National Park. Increasing tourism to the park and travel throughout the area will increase human impacts. Because the majority of threats to the ecosystem are human caused, it is necessary to include humans in models of the long-run dynamics of the ecosystem. The model developed in this chapter integrates human behavior into a dynamic representation of the Serengeti ecosystem and builds on a growing literature of human-wildlife interactions in the Serengeti ecosystem (Barrett and Arcese 1998; Hofer et al. 2000; Loibooki et al. 2002; Damania, Milner-Gulland, and Crooks 2005).

It is now widely recognized that the future of the Serengeti depends on people and the management decisions that affect people. The Serengeti National Park (SNP) management devotes nearly its entire budget to managing people—primarily tourists and the associated roads, lodges, and campsites required to support them. Unlike many parks in southern Africa, SNP does not supply water holes, build fences, or cull populations; the only significant management activity of SNP that is not "people management" is controlled burning early in the dry season. In parts of the Serengeti ecosystem outside of SNP, people play a more important role. The Maasai are an integral part of the Ngorongoro Conservation Area and the Loliondo Game Control Area to the east of SNP. Agricultural communities are an integral part of open areas to the west of SNP. The influence of people who live around the park extends within the park itself in important ways. Managing interactions with nearby communities involves antipoaching activities and other restrictions. For example, in an effort to improve relations with nearby communities, Tanzanian National Parks (TANAPA) recently instituted a community outreach program that works with local economic development projects.

The interaction between villagers, parks, and wildlife is a recurring issue in protected-area management around the world, and has been the subject of considerable study and debate, especially in Africa (e.g., Hofer et al. 2000). Recognizing that protected areas restrict villagers' access to resources—and often provide few, if any, direct benefits—there is a growing movement to provide either cash or community services to these villages from park revenues. Protected areas are viewed as depriving local people of resources and/or land; transfer payments from the protected areas compensate them for their losses and are intended to encourage local people to protect the park and its resources. However, because human expansion often leads to human/wildlife conflict, SNP—and protected areas in general—face decisions regarding the allocation of resources between law enforcement activities to prevent poaching and encroachment and transfer payments to local communities to provide services. Only limited tools and guidelines exist to help parks managers make decisions about this resource allocation choice.

In this chapter we develop a model of the interactions between local communities, wildlife, and management options in the Serengeti. Our purpose is to provide a framework for analyzing management decision making and to evaluate different strategies for improving the lives of local people while preserving the park's wildlife. The model explicitly incorporates both human effects on the ecosystem and human responses to ecosystem changes. The model can also be used to trace the effects of changes in biophysical conditions (e.g., climate-related change, drought, and disease), changes in government policy (e.g., antipoaching monitoring and enforcement, land use regulations), or changes in economic conditions (e.g., prices of agricultural commodities, tourism demand). The model builds on similar dynamical bioeconomic models (e.g., Barrett and Arcese 1998; Damania, Milner-Gulland, and Crooks 2005), and empirical models (e.g., Hofer et al. 2000; Loibooki et al. 2002) from the recent literature. Fundamentally, this will be a time-allocation model, in which representative agents allocate their scarce labor resources to various activities to maximize an objective (DeJanvry, Fafchamps, and Sadoulet 1991).

The modeling approach is intended to be parsimonious. This model includes only a few human activities—those that have the greatest interaction with the environment (grazing, agriculture, hunting). Similarly, we model only a few ecological processes, namely those that have the greatest interaction with humans (amount of forage, size of wildlife populations). It is not intended to capture the full extent of complex biological and social interactions in the Serengeti system. Instead, our focus is largely on the human dimension of the Serengeti ecosystem, and we incorporate relatively simple ecosystem feedbacks and impacts for dynamic scenarios. The other chapters in this section focus primarily on complementary components of the ecosystem.

We begin by describing the joint human-ecosystem model, including the model of household decision making, production and consumption, and how human population, livestock numbers, and wildlife populations evolve through time. Section 3 explains the parameter values used in our model. To illustrate the model's application, we provide some results of several hypothetical scenarios in section 4, leaving more detailed exploration to future chapters. We conclude in section 5 with brief summary comments.

THE MODEL

Model Overview

We model households living near the western border of Serengeti National Park. Households within the modeled area are located in one of nine regions. The northern district, "Tarime," consists of regions 1, 2, and 3, the middle district, "Bunda/Serengeti," consists of regions 4, 5, and 6, and the southern district, "Sukuma," consists of regions 7, 8, and 9. There are many different ethnic groups in each district; however, the Kuria are dominant in the north, there are several mixed ethnic groups in the middle, and the Sukuma occupy much of the region south of the western corridor of Serengeti National Park. In each district, lower-region numbers refer to areas closer to the park. The north-south dividing lines are drawn at 5 kilometers, 15 kilometers, and 25 kilometers' distances from the park's boundary. The location of a household in a region determines the household's distance to the park, soil productivity, weather conditions, and access to wildlife populations.

Households living in particular locations, influenced by certain cultural norms, and owning certain assets (e.g., livestock), make decisions about which mix of activities to pursue for the purposes of providing food for the household and for accumulating wealth. The returns for various activities depend on the state of the ecosystem, the abilities of the household, and the assets owned by the household. For example, the expected returns from hunting depend on the size of the wildlife population (and on other variables). Human activities affect the condition of the ecosystem (e.g., hunting lowers wildlife populations), and in turn, the condition of the ecosystem in the next period will affect the next-period decisions made by households. While the model is explicitly dynamic and captures interactions between human actions and ecosystem conditions, we assume that household decisions are myopic with respect to dynamic ecosystem feedbacks.

Figure 10.1 shows a flow chart that illustrates one period of the model. Each period represents 1 year. At the start of a period there is an initial resident wildlife population in each region, an initial resident wildlife population in the park, and an initial migratory wildlife population. Each household begins the period with the amount of land, initial number of livestock, and size of the household inherited from the previous period. The size of the household determines the household labor supply. Each household chooses how to allocate its labor supply among four activities: agricultural production, livestock grazing, hunting within the local region (where the household is located), and hunting in Serengeti National Park. Production of crops, livestock, and bushmeat (harvest of wildlife) by the household is determined by the amount of household labor applied to each activity, household assets (land and livestock), wildlife populations, weather, and location-specific parameters.

We assume that households require a minimum amount of food input for consumption (crops and meat) for each period, which is based on the size of the household. If their production activities result in production of more food than is required for consumption, the household will accumulate wealth in the form of livestock and may increase in size. Should a household fail to produce enough food to meet its consumptive needs, the household must sell some livestock. If the household has sold off all of its livestock and still cannot meet its nutritional needs, then a portion of the household migrates out of the area and household size shrinks. Household size grows when the household has a successful year, resulting in an increase in its livestock (wealth).

The final step in each period involves updating variables that will set the initial conditions for the ensuing period. Both wildlife populations

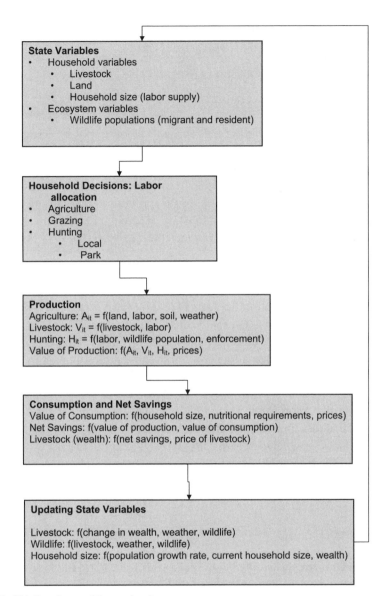

Fig. 10.1 Flow diagram of the pseudocode.

and livestock have growth functions that depend upon forage conditions, where forage depends on both weather and grazing pressure.

Household Labor Allocation Decisions

In period t, household i allocates its available labor supply, L_{it}, across four activities: agriculture, grazing, hunting locally, and hunting in Serengeti National Park. Each household is located in a region, indexed by j, $j = 1$, $2 \ldots, 9$, where $j(i)$ refers to the region in which household i is located. Denote labor allocated by household i in time period t to agriculture as L_{it}^A, grazing, L_{it}^G, local hunting, L_{it}^j, and hunting in Serengeti National Park, L_{it}^P. The household is constrained by its total labor supply so that

$$L_{it} = L_{it}^A + L_{it}^G + L_{it}^{j(i)} + L_{it}^P$$

The most efficient allocation of a household's labor in a given period depends on factors such as the spatial distribution and abundance of wildlife, number of livestock, amount of land for crops, and household size. In principle, the process of deciding how a household should allocate its labor in a period involves comparing the returns that would be generated under different allocations of that household's labor, subject to the constraint that labor allocation to each activity is non-negative and that total labor allocation across sums to household labor supply. Households choose to allocate labor to whichever of the alternatives yields the highest expected returns to the household, which occurs when marginal returns for activities are equalized.

Production

Production of crops, livestock, and harvest of wildlife by the household is determined by the amount of household labor applied to each activity, household assets (land and livestock), wildlife populations, weather, and location-specific parameters. We describe the production function for each activity in the following.

Agriculture

Production of agricultural crops depends on labor and land inputs, soil quality, weather, and crop damage from wildlife. Let the amount of land that household i can use to plant crops in period t be denoted X_{it}. We assume a Cobb-Douglas production function in labor and land, with scaling parameters reflecting weather and soil conditions. We define the following notation: λ_j is a scaling constant that reflects crop productivity in region j, θ_{jt} is a (mean 1) weather shock for region j in time period t, and β is the Cobb-Douglas production function exponent. If wildlife are present, resulting crop loss is netted out of production. In the extreme case, in which the entire crop is consumed by wildlife, production goes to zero. We define λ as the scaling coefficient of wildlife crop damage and W_{jt} as the wildlife population in region j at time t. The agricultural production function for household i in time period t is given by:

$$A_{it} = \gamma_{j(i)}\theta_{j(i)t}(L_{it}^A)^\beta(X_{it})^{1-\beta} - \lambda W_{j(i)t}$$

if this expression is positive, and $A_{it} = 0$ otherwise.

Grazing

The grazing production function assumes that a fixed amount of labor is required to maintain a given livestock herd. Let V_{it} be the number of livestock controlled by household i at the beginning of period t. We assume that a larger herd requires more labor to tend it, but at a diminishing rate. A household with V_{it} livestock units requires $(V_{it}/4)^{0.5}$ units of labor to maintain the herd. Therefore, we assume that the labor allocation toward grazing is a one-to-one function of the livestock size: $L_{it}^G = (V_{it}/4)^{0.5}$. One unit of labor, therefore, can tend to four head; two units of labor can tend to 16 head. This is the authors' empirical estimate.

Hunting

A household located in region j can hunt in region j and/or in Serengeti National Park. The tradeoff for a given household is that hunting in the park is usually more productive than hunting locally. However, hunting in the park carries a higher probability of detection, and traveling to the park is costly in terms of time. Let ϕ_j be the (instantaneous) probability of getting caught per unit of labor spent hunting in region j, and let ϕ_p be the probability of getting caught per unit labor spent hunting in the park. Given the assumption of a constant chance of getting caught each time unit, the cumulative probability of getting caught while spending L_{it}^j units of labor hunting in region j is: $\Pi_j(L_{it}^j) = 1 - \exp(-\phi_j L_{it}^j)$. The probability of getting caught while spending L_{it}^P units of labor hunting in the park is $\Pi_p(L_{it}^P) = 1 -$

$\exp(-\phi_p L^p_{it})$. The monetary fine if a person is caught hunting illegally is F. Let the catchability coefficient of wildlife, which is the proportion of the wildlife population that can be harvested per unit of labor hunting, be Q_j in region j and Q_p in the park . Then, the expected harvest of wildlife in region j is:

$$H^{j(i)}_{it} = Q_{j(i)}L^{j(i)}_{it}W_{j(i)t}[\exp(-\phi_{j(i)}L^{j(i)}_{it})]$$

Hunting in the park requires the hunting members of the household to travel to the park. Letting d_j be the distance (measured in labor units) from region j to the park, the actual time spent hunting in the park is $L^{j(i)}_{it}(1-d_{j(i)})$. The expected harvest of wildlife by household i in Serengeti National Park is:

$$H^P_{it} = Q_P L^P_{it}(1-d_{j(i)})W_{Pt}\{\exp[-\phi_P L^P_{it}(1-d_{j(i)})]\}\,.$$

The Value of Production

Each household can produce as many as three valuable products: agricultural crops, livestock, and bushmeat (wildlife harvest). Let p^A_t, p^V_t, and p^H_t be the price of agricultural crops, livestock, and bushmeat at time t. The total value of production (revenue) of household i living in region $j(i)$ at time t is given by:

$$R_{it} = p^A_t A_{it} + p^V_t V_{it} + p^H_t(H^{j(i)}_{it} + H^P_{it})$$

Consumption and Change in Wealth

All of the production activities of the household are focused toward maximizing wealth, net of nutritional requirements. In the model, we assume that each person must meet certain nutritional constraints in each period in order to subsist. These constraints are modeled as a minimum consumption of meat (m) and agricultural crops (c) per unit labor per time period. Meat consumption can derive either from harvested livestock or from bushmeat. We assume that a household will choose to reach its nutritional constraint on meat in the least expensive manner, typically consuming bushmeat rather than harvesting livestock. The total required expenditure for household i to meet nutritional needs is given by:

$$E^N_{it} = L_{it}[p^A_t c + \mathrm{Min}(p^V_t, p^H_t)m]$$

In addition to expenditures to meet nutritional needs, the household must also pay for any monetary fines from illegal hunting, should a member of the household be caught poaching. The expected value of fines is given by

$$E_{it}^F = F\{\Pi_{j(i)}(L_{it}^{j(i)}) + \Pi_p[L_{it}^P(1 - d_{j(i)})]\}$$

Finally, the household also has expenses for other purchased goods, taxes, and other fees, denoted by E_{it}^T.

The net balance for household i at the end of the period is therefore expressed as the revenue from production minus the costs of consumption, fines, and other expenses, as follows:

$$B_{it} = R_{it} - E_{it}^N - E_{it}^F - E_{it}^T$$

The household's objective is to maximize net revenue (B_{it}) in each period, subject to all relevant constraints. If B_{it} is positive in a period, t, then the value of household production in that period exceeds all expenses and the household will have an increase in wealth. If this expression is negative it means that household expenses exceed the value of household production and there will be a decrease in household wealth.

In many East African societies, wealth is held in the form of livestock, and that is what we assume in this model. Decreases in wealth result in the household's owning less livestock in the following period, whereas increases in wealth result in the household's increasing its holdings of livestock. Accordingly, the state-transition equation for livestock is expressed as follows:

$$V_{it}^{End} = Max\left(\phi, \frac{B_{it}}{p_t^V}\right)$$

where V_{it}^{End} is the size of household i's livestock herd at the end of period t. This expression accounts for the possibility that the household might end up with negative assets in a given period, and hence have to sell off all of its livestock.

Updating State Variables

There are three state variables for each household: one is fixed (land), and two are updated each period: household size and livestock. We assume that the land controlled by household i does not change over time, so that $X_{it+1} = X_{it}$. Population growth is measured by a constant, μ, but to incorporate the role of nutrition, a household only grows when it has sufficient wealth to obtain food. A household can grow, remain the same size, or shrink, depending on available resources. In good years, when there is positive wealth accumulation for the household ($V_{it}^{End} > V_{it-1}^{End}$), household size expands at a population growth rate $\mu > 1$, so the household-size transition equation is:

$$L_{it+1} = \mu L_{it}.$$

When the household has negative wealth accumulation during the period ($V_{it}^{End} < V_{it-1}^{End}$) but still has a positive amount of livestock ($V_{it}^{End} > 0$), there is no change in household size: $L_{it+1} = L_{it}$. If the household cannot meet its nutritional needs and has exhausted its wealth ($V_{it}^{End} = 0$), it will first opt to stop paying taxes or other fees ($E_{it}^T = 0$). If the household still cannot meet its nutritional needs, migration occurs, and household size shrinks. In that event, the household-size transition equation is:

$$L_{it+1} = \frac{R_{it} - E_{it}^F}{E_{it}^N} L_{it}$$

Livestock and wildlife reproduction depend on environmental conditions. Livestock growth depends on the condition of forage, which in turn depends on weather and the amount of livestock and wildlife consuming forage. Let r_v be the intrinsic growth rate of livestock, θ_{jt} the weather shock in region j during time period t, and K_j the ungulate-carrying capacity of region j for a year with average weather. Define the total number of livestock being grazed in region j during period t as $V_{jt} = \Sigma_{i \in j} V_{it}$. The livestock-transition equation (livestock growth) is a logistic of the following form:

$$V_{it+1} = r_v V_{it}^{End}\left[1 - \frac{V_{j(i)t} + W_{j(i)t}}{\theta_{j(i)t} K_{j(i)}}\right] + V_{it}^{End}$$

as long as this expression is positive, and zero otherwise.

Wildlife population growth also depends on the condition of forage, which in turn depends on weather and the amount of livestock and wildlife consuming forage. There are two types of wildlife populations—resident and migrant—with the migrant population being by far the largest. Resident wildlife populations reside in each of the nine regions; an additional population resides in the park near the western boundary. Define W_{jt}^R as the resident population of wildlife in region j at time t, and W_{Pt}^R as the resident population of wildlife in the park. The migratory herd spends part of the year (the wet season) in the interior of the park, and migrates through the area near the western boundary of the park during the dry season. (We do not explicitly model seasons here. Recognizing that many variables [e.g., farming and hunting returns] vary by season, adding seasons to the model increases complexity without providing comparable insight. However, where possible, the parameter estimates outlined in section 3 reflect averaging of seasonal variation.)

Define W_t^M as the migratory population in time t. Let a_j be the fraction of the migratory herd that spends time in region j, and let a_p be the fraction of the migratory herd that spends time in the boundary region of the park. We can define the wildlife population in region j as:

$$W_{jt} = W_{jt}^R + a_j W_{jt}^M$$

The wildlife population in the park is:

$$W_{Pt} = W_{Pt}^R + a_p W_{Pt}^M$$

Let r_w be the intrinsic growth rate of wildlife. In the absence of hunting, the wildlife-transition equation for the resident population in region j is

$$W_{jt+1}^R = r_w W_{jt}^R \left(1 - \frac{V_{jt} + W_{jt}}{\theta_{jt} K_j} \right) + W_{jt}^R$$

In the absence of hunting, the wildlife-transition equation for the resident population in the park is

$$W_{Pt+1}^R = r_w W_{Pt}^R \left(1 - \frac{W_{Pt}}{\theta_{Pt} K_P} \right) + W_{Pt}^R$$

Finally, in the absence of hunting, the wildlife-transition equation for the migratory population is

$$W_{jt+1}^M = r_w W_t^M \left(\frac{1 - W_t^M}{\theta_{mt} K_m} \right) + W_t^M$$

where θ_{mt} is the weather shock in the interior of the park where the migratory herd is present and K_m is the wildlife-carrying capacity for an average year in the interior of the park.

PARAMETERIZING THE MODEL

While the purpose of this chapter is to present and explain the model's analytics, we would like to illustrate its implementation by parameterizing it with reasonable data. Unfortunately, defensible data are scarce. In most cases, there are neither time series nor cross-sectional data available from which to statistically estimate parameters. Instead, we have relied upon case studies and expert opinion to roughly parameterize the model. Below, we discuss generally how we obtained parameters, and at the end of this section, we provide a summary of the specific parameter estimates. We stress that the data used to parameterize this model are rough; consequently, we provide only plausible approximations, not reliable estimates. Table 10.1 summarizes all of the parameter estimates we use for our base-case simulation runs.

Consumption and Other Household Data

Population Growth

We adopt a relatively conservative population growth rate of 2.9%, following Barrett and Arcese (1998), which is based on Tanzania census data. Typical household size in our simulations ranges from about 5 to 12 individuals.

Region in which Household Lives

Of the total population located within 30 km or less from the protected area boundary 33% of the population lives 10 km or less from boundary, 39% live between 10 and 20 km, and 28% live 20 to 30 km from the boundary (Emerton and Mfunda 1999).

Consumption of Crops

Assuming a 1,500k/day diet, and that 70% comes from crops, an average adult will consume about 2,000 gm of raw maize per day. A child's consumption is estimated at ¾ of an adult's. Total consumption for a typical seven-person family will be about 4,400kg/yr, which represents six labor units in our model. Thus, the per-labor unit crop requirement is about $c = 730$ kg per labor unit per year.

Consumption of Meat

We assume an average household of seven people will consume approximately 1 TLU per year. This figure is consistent both with our estimate from crop consumption and with estimates put forth in Barrett and Arcese (1998). A seven-person household represents six labor units in our model, which amounts to $m=0.17$ TLU per labor unit per year.

Livestock Production

As described in section 2.3, a household must contribute $(V_{it}/4)^{0.5}$ units of labor to maintain its livestock herd. This implies that one unit of labor is required to maintain a livestock herd of size four. It also implies that as the herd size increases, more labor is required to maintain it, but at a decreasing rate (e.g., doubling the herd to eight head would require only 1.4 units of labor).

Prices for household goods, including livestock, crops, and bushmeat, as well as other household expenditures, are from Kauzeni and Kiwasila (1994) and Loibooki et al. (2002). Table 10.1 provides the price estimates on a per-animal basis.

Agricultural Data

Each household receives an initial allocation of agricultural land based on estimates obtained from the literature (Mfunda and Emerton 1999; McLoughlin 1971). For maize production, each household produces an average of 1,200 kg/ha (Mfunda and Emerton 1999), which we adopt as a measure of productivity.

We assume a Cobb-Douglas functional form for the crop production function, where the exponent is $\beta = 0.5$. This estimate is the authors', and implies that one unit of household labor devoted to agricultural production could produce about 1,500 kg of maize in an average year. Production also depends on weather conditions in each time period, θ.

Grazing Data

Livestock populations are determined by a household's initial livestock, the growth rate of livestock, and the carrying capacity for all large animals in its own region. We assume that grazing activities do not affect crop produc-

Table 10.1 Parameter estimates

Parameter	Estimate	Units	References
Household Data			
Agricultural land owned	1.5	Hectares/ household	Mfunda and Emerton 1999; Larson 1974; Mcoughlin 1971
Initial labor per household	7	Hours/household	Von Rotenhan 1968
Initial labor allocation	25,25, 25,25	Percent to each activity	Von Rotenhan 1968; Loibooki *et al* 2002
Human population growth rate	2.9	Percent	Barrett and Arcese 1998
Scaling coefficient on crop damage	0.01	Crop units destroyed / animal	Authors' estimate
Cobb-Douglas coefficient	0.5	n/a	Authors' estimate
Consumption Data			
Required consumption of crops	730	kg/person/year	Authors' calculation
Required consumption of meat	1.875	TLU/person/year	Adams, 1975
Price of crops	300	Shillings/kg	McLoughlin 1971; Larsen 1991
Value of livestock (averaged over all livestock)	171,000	Shillings	Kauzeni and Kiwasila 1994; Loibooki *et al* 2002
Value of bushmeat	630,000	Shillings	Kauzeni and Kiwasila 1994
Grazing Data			
Initial livestock owned	5	Animals	McLoughlin 1971; Mfunda and Emerton 1999
Intrinsic growth rate of livestock	45	Percent	Tanzania Ministry of Agriculture and Food Security (2003, on-line)
Weather shock in zone j, period t[1]	1	Current rainfall / average rainfall	Mean value
Hunting Data			
Travel time to park[2]	1, 3, 9	Hours	Campbell *et al.* 2001; Cambell and Hofer 1995
Catchability coefficient in zones	.0001	% population harvested per L	Authors' estimate, based on Loibooki *et al.* (2002) and Hofer *et al.* (2000)

Table 10.1 (continued)

Parameter	Estimate	Units	References
Catchability coefficient in park	.000001	% population harvested per L	Authors' estimate, based on Loibooki et al. (2002) and Hofer et al. (2000)
Detection parameter[3] for getting caught hunting in park	1	Detection/ unit labor	Campbell et al.2001; Cambell and Hofer 1995
Detection parameter for getting caught hunting in region j	0.2	Detection/ unit labor	Campbell et al. 2001; Cambell and Hofer 1995
Fine if caught hunting illegally	100,000	Shillings	Campbell et al. 2001; Cambell and Hofer 1995
Initial migratory population	1,300,000	Animals	Campbell and Hofer 1995; Hofer 1996
Initial wildlife population in region j	2000	Animals	Campbell and Hofer 1995; Hofer 1996
Carrying capacity of region j	10,000	Animals	Authors' estimate
Initial resident wildlife population in park	15,000	Animals	Campbell and Hofer 1995; Hofer 1996
Carrying capacity in park	1,300,000	Animals	Based on current wildebeest estimate
Intrinsic growth of resident pop. of region j	50	Percent	Campbell and Hofer 1995; Hofer 1996

1. We have suppressed the randomized component of the weather shock in this chapter.
2. For zones (1, 4, 7), (2, 5, 8), and (3, 6, 9), respectively.
3. See footnote 1 for an explanation of this detection parameter.

tion. Based on estimates from the literature, each household receives an initial allocation of five livestock. Livestock grow at a logistic growth rate of 0.45/year.

Livestock are also affected by weather through foliage shocks, similar to agriculture production, discussed earlier.

Hunting Data

Households in each region can acquire meat either by hunting in region j or hunting in the park. Hunting success will depend on three important factors. Most obviously, it will depend on the amount of time spent hunting (i.e., the household's labor allocation to that activity). But it will also depend (linearly) on the wildlife population and on the catchability coefficient. The catchability coefficients in each zone and in the park are difficult to estimate. Based on Hofer et al. (2000) and Loibooki et al. (2002), as well as on the authors' calculations, we have arrived at the following estimates of catchability: $Q_j = 0.0001$ and $Q_p = .000001$. If the wildlife populations were equalized across space (including in and out of the park), this would imply that hunting outside the park is more lucrative. However, the larger animal populations in the park counteract this effect (at least in part). Furthermore, in order to hunt in the park, a household must spend time traveling to it, and faces a higher probability of being caught. Initial wildlife populations are from Campbell and Hofer (1995) and Hofer (1996) and estimated at 15,000 in the park and 2,000 in each region j. The migratory population is considerably larger, at 1,300,000, which approximates the current size of the wildebeest herd.

Travel Time to Park

At an estimated walking speed of 5 km/hr, a resident of the middle of regions 1, 4, and 7 could reach the park in 1 hour. Travel time from the middle of regions 2, 5, and 8 is 3 hours; travel from the middle of regions 3, 6, and 9 requires an estimated 5 hours of continuous walking.

Bushmeat Harvested in Park

We derive the number of animals caught from Campbell and Hofer (1995). The same authors provide an estimate for the detection parameter, which describes the likelihood of being caught while hunting in the park and in region j of 1 and 0.2 respectively. (The probability of being caught has an exponential functional form; consequently, we use a detection parameter that does not necessarily lie between 0 and 1. For example, the probability of getting caught in *zone j*—with a detection parameter of 0.2 and an initial labor allocation of 1—is 18%. A detection parameter of 1 results in a 63% chance of being caught.)

The fine for being caught hunting in the park is 100,000 shillings (US$ 94) (Campbell, Nelson, and Loibooki 2001; Cambell and Hofer 1995).

Computational Details

The model was programmed into MATLAB. For any household i, $i = 1, \ldots, I$, and any starting assets, V_{i0}, X_{i0}, and L_{i0}, the model determines the optimal allocation of time across agriculture, grazing, hunting locally, and hunting in Serengeti National Park in each period. Within 1 year, the model determines the cumulative effects of the time-allocation decisions of all households and updates the household-specific state variables. The model is run through time until a steady state is achieved (which typically takes <50 periods). The model pseudocode is as follows:

1. Input parameter values and starting values for state variables.
2. Loop over all I households.
3. Loop over all time.
4. Generate function files that determine the one-year payoffs to agriculture, livestock production, and hunting from an arbitrary allocation of L_{it} units of labor.
5. Generate function files that determine the expenditures of the household for an arbitrary allocation of time.
6. Generate function files that determines the net assets held by household i at the end of the year as a function of an arbitrary time allocation.
7. Numerically maximize net assets held by household i by choosing the optimal allocation of time.
8. Update assets held by household, and increment time by one unit.
9. End time loop.
10. End I loop.

MODEL RESULTS

Using the parameter estimates outlined in section 3, we compute initial model results. For illustrative purposes and to provide a baseline, we first run the model to steady state. We will then illustrate some nonequilibrium dynamics (i.e., comparative dynamics) and comparative statics by running several experiments and summarizing the results graphically. More detailed exploration of these (and other) scenarios of interest are left to future chapters.

Steady State

The model provides some interesting results. Human population grows rapidly from approximately 63,000 to 90,000, and reaches a steady state after about 35 years. Over this period, representative households residing in each zone are shifting their optimal labor allocations among the four activities: agriculture, livestock, hunting locally, and hunting in the park. By the time steady state is reached, the labor allocations stabilize, but because some zones are farther from the park than others, these labor allocations differ across zones. Figure 10.2 shows the steady-state labor allocation for zone 1 (near the park), zone 2 (medium distance from the park), and zone 3 (far from the park).

In steady state, households residing in zone 1 find it optimal to divide their labor across all 4 activities—where approximately half of the household labor is devoted to hunting of some sort, and half is devoted to non-hunting income-generating activity. As we move farther from the park, these allocations change, and at least three interesting results obtain. First, consistent with intuition, the percentage of labor allocated to agricultural production increases as we move farther from the park, reflecting the increase in relative returns to labor for that activity. Second, the percentage of labor spent hunting in the park is relatively insensitive to the distance from the park, at least in this simulation. The third, noteworthy result is that there are no livestock in zone 3 in steady state. At first this seems counterintuitive, because the relative returns to grazing should be higher as one moves farther from the park. However, in our simulations this effect is outweighed by the fact that wealth is stored, in our model, as cattle. Villages farther from the park (in zone 3) are at a significant disadvantage (relative to zones 1 and 2), and are therefore much less wealthy. In our model, as a household loses wealth, its cattle herd shrinks. In these simulations, by

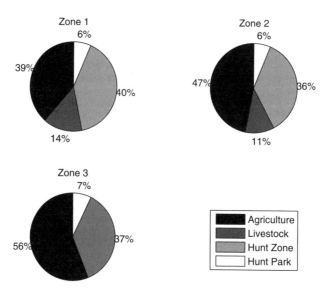

Fig. 10.2 Labor allocation in steady state.

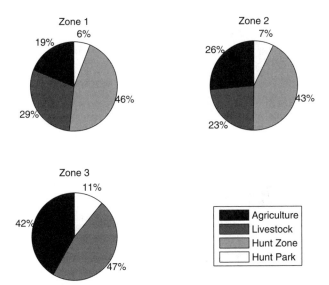

Fig. 10.3 Income-source distribution in steady state.

about year 30, households in zone 3 are just barely at the subsistence level, and therefore have no cattle (i.e., wealth) holdings.

While Figure 10.2 provides an interesting picture of the labor allocation in steady state, it does not inform us about which activities are generating the most income for the household. Figure 10.3 depicts the steady state distribution of the sources of household earnings. In all zones, around half of the household's income is generated through hunting, while the percentage of income from agriculture increases with distance from the park, from 19% near the park to 42% far from the park.

Some Comparative Dynamics

Because the human and ecological interactions adjust dynamically in our model, we are in a position to explore some comparative dynamics. While we will leave more detailed analysis to further chapters, we examine here some of the dynamics associated with a prolonged shock to productivity (e.g., a structurally shifted weather event). The procedure we adopt is to run the model, as previously described, to steady state, and then we apply a prolonged, 20-year shock to the parameter θ, which affects wildlife production, agricultural production, and cattle production. The mean value of θ from the base case simulations is θ = 1. The shock we apply reduces it by

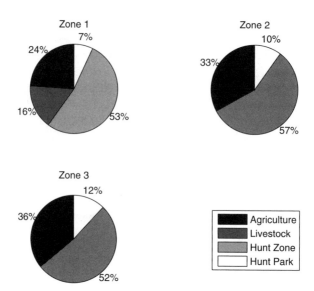

Fig. 10.4 Income-source distribution following a prolonged productivity shock.

20%, to θ = 0.8. Doing so causes all variables in the model to adjust from their steady state values, and the model reaches a new steady state (with θ = 0.8) within about 15 years.

What is the effect of this prolonged productivity shock on the optimal labor allocation? Figure 10.4 replicates figure 10.3, but at the new steady state with the productivity shock. Near the park, a household's labor allocation toward agriculture increases, but because of a reduction in wealth, its required allocation to tend livestock shrinks. Localized hunting pressure increases in all zones as a result of the weather shock.

Simply graphing the distribution of income from each source does not inform us about the ultimate impact of the productivity shock. When wildlife populations, cattle stocks, and agricultural production are all negatively impacted for a sustained period of time, human populations that depend on them will shrink. In our simulation, the human population shrinks quite rapidly, achieving a new steady state of approximately 85% of its preshock size after about 12 years (figure 10.5).

Wildlife populations are also affected, but for different reasons. First, depending on their relative densities, they may be more relied upon by humans, and so may be harvested at a faster rate. Second, forage is reduced, which effectively reduces their carrying capacity. The way we have modeled the shock, this occurs for all animals both inside and outside the park. Figure 10.6 shows the wildlife populations over time, as a percentage of

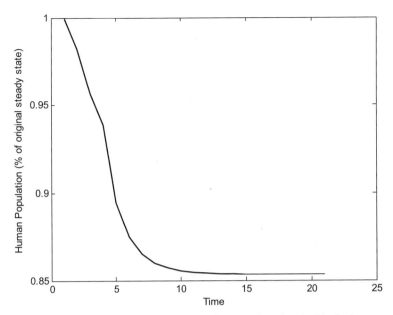

Fig. 10.5 Human population (% of original steady state) during a prolonged productivity shock.

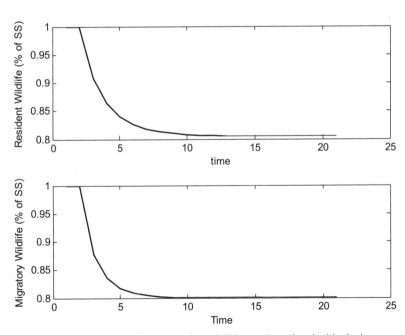

Fig. 10.6 Wildlife populations (% of original steady state) during a prolonged productivity shock.

the base-case steady-state levels, during the prolonged productivity shock. Both resident and migratory wildlife attain new steady state values that are approximately 80% of their preshock levels, about 10 years after the shock is first applied.

An Increase in the (Local) Probability of Detection

Results from the previous section illustrate how some of the variables of interest respond over time to a structural shift in a particular environmental variable. We can also use this model to explore *comparative statics*. One question that has arisen in the literature is: what are the effects of a change in antipoaching monitoring activity? While this question will be explored in more detail later, we will attempt a preliminary answer within the context of our base-case simulation model.

In particular, we would like to know how increased monitoring outside the park affects labor allocation in steady state. We model increased monitoring outside the park as an increase in the parameter ϕ_j. The base case value is $\phi_j = 0.2$, which implies that if one unit of labor specializes in hunting outside the park, the probability of detection is 18%. We increment that parameter up to a value of 0.7, which implies a probability of detection of 50%. For each value of ϕ_j, we save the steady-state labor allocation percentages for a representative household residing in zone 1. The optimized labor allocations for a zone 1 household, as a function of ϕ_j, are graphed in figure 10.7.

Several interesting results obtain. Increasing the probability of detection outside the park makes hunting outside the park relatively more expensive, in expectation, and we thus see a decline in the percentage of time allocated to hunting locally (from about 40% to about 15%, in these simulations). This raises the relative profitability of labor allocated to the main alternative—agricultural production; we see a consequent increase in labor allocated to agriculture (from about 40% to about 75%). However, because the increased monitoring effectively reduces income, the cattle herd size drops, as does the amount of time required to tend the herd. This leads to a reduction in the percentage of time allocated to livestock. In fact, for a high enough probability of detection (around 50%), the zone 1 households are so severely impoverished that their cattle herds drop to zero in this simulation, and so no time is allocated to their maintenance. Finally, and somewhat counterintuitively, the percentage allocation of hunting in the park has actually increased. This result is not so surprising upon closer inspection, because the probability of detection has not been increased there, so

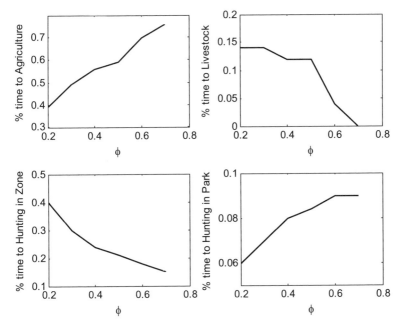

Fig. 10.7 Labor allocation in zone 1 as a function of the detection parameter.

hunting in the park has become relatively more profitable than hunting outside the park.

CONCLUSIONS, CAVEATS, AND EXTENSIONS

The model developed in this chapter synthesizes a set of rules that describe human behavior as functions of a small set of critically relevant state variables and parameters. These rules determine the spatial allocation of and level of effort dedicated toward several activities (agriculture, grazing, and illegal hunting). Embedded in this model are yield-effort relationships that determine, for example, expected illegal harvest in every location.

This example is meant to be illustrative of some of the capabilities of this type of modeling; the modeling framework is fairly flexible. For example, it is straightforward to include more household types and more geographic regions. We have thus far excluded the important impacts of congestion (what if everybody decides to hunt on one site) and endogenous prices. We have taken wildlife populations and forage quality as exogenous variables. In its current form, the model incorporates some simple ecosystem feedbacks, but the focus has been on the human components of the system. The utility of this model could be extended by including more meaningful ecosystem feedbacks. Versions of this model will be employed in subsequent chapters.

REFERENCES

Adams, C. 1975. *Nutritive value of American foods in common units.* Agriculture handbook no. 456. Washington, DC: United States Department of Agriculture.

Barrett, C. B., and P. Arcese. 1998. Wildlife harvest in integrated conservation and development projects: Linking harvest to household demand, agricultural production, and environmental shocks in the Serengeti. *Land Economics* 74(4): 449–65.

Campbell, K., V. Nelson, and M. Loibooki. 2001. Sustainable use of wildland resources, ecological, economic and social interactions: An analysis of illegal hunting of wildlife in Serengeti National Park, Tanzania. Final technical report to DFID, Animal Health and Livestock Production Programmes. London: Department for International Development.

Campbell, K., and H. Hofer. 1995. People and wildlife: Spatial dynamics and zones of interaction, In *Serengeti II: Dynamics, management, and conservation of an ecosystem,* ed. A. R. E. Sinclair and P. Arcese, 534–70. Chicago: University of Chicago Press.

Damania, R., E. J. Milner-Gulland, and D. J. Crookes. 2005. A bioeconomic analysis of bushmeat hunting. *Proceedings of the Royal Society, Series B.* 272:259–66.

DeJanvry, A., M. Fafchamps, and E. Sadoulet. 1991. Peasant household behavior with missing markets: Some paradoxes explained. *Economic Journal* 101:1400–17.

Hofer, H., K. Campbell, M. East, and S. Huish. 2000. Modeling the spatial distribution of the economic costs and benefits of illegal game meat hunting in the Serengeti. *Natural Resource Modeling* 13(1): 151–77.

Kauzeni, A., and H. Kiwasila. 1994. *Serengeti regional conservation strategy: A socio-economic study.* Dar es Salaam: Institute of Resource Assessment.

Larsen, J. S. 1991. *Agriculture in Sukumaland, Tanzania.* Copenhagen: Den.Kgl. Veterinaer-og Landbo hojskole, Okonomisk Institut.

Loibooki, M., H. Hofer, K. Campbell, and M. East. 2002. Bushmeat hunting by communities adjacent to the Serengeti National Park, Tanzania: The importance of livestock ownership and alternative sources of protein and income. *Environmental Conservation* 29(3): 391–98.

McLoughlin, P .F. 1971. *An economic history of Sukumaland, Tanzania, to 1964: Field notes and analysis.*

Mfunda, I., and L. Emerton. 1999. Making wildlife economically viable for communities living around the Western Serengeti, Tanzania. In *Evaluating Eden series.* Working Paper no.1. Dar es-Salaam: Tanzanian Bureau of Statistics:

Von Rotenhan, D. 1968. Cotton farming in Sukumaland. In *Smallholder farming and smallholder development in Tanzania,* ed. H. Ruthenberg, 51–85. Munich: Springer-Weltforum.

ELEVEN

Human Responses to Change:
Modeling Household Decision Making in Western Serengeti

Kathleen A. Galvin, Steven Polasky, Christopher Costello, and Martin Loibooki

Many people around the world are directly dependent on the land for their livelihood. Thus, the structure and condition of the environment is vitally important for peoples' land use decisions. Human and environment interactions are interactive, with people making economic decisions based on ecological and climatic variability and on economic, demographic, social, and policy conditions. Human populations around Serengeti National Park engage in land uses that include livestock production, agriculture, wildlife use, and wage labor, to various degrees. People in this region have to deal with uncertainty in rainfall and markets and in changing demographic and policy conditions. They are from several different ethnic groups, and they differ in how time and resources are allocated to various livelihood activities. Also, they live in close proximity to protected areas, which provide opportunities and constraints to livelihood strategies.

For various reasons associated with climate, policy, economy, or demographic changes, people can become vulnerable—that is, they are at a high risk of negative economic outcomes as a result of uncertain events that overwhelm the adaptations they have in place. Decision making under uncertainty is a very complex process, one where people must respond to the uncertainty within their context of economic, social, policy, and ecological constraints and opportunities. Vulnerability occurs due to their inability to cope adequately with those changes. Reducing one's vulnerability to shocks and stresses is an indication of resilience. There is a close association between people's assets status and the resources that people can draw upon in times of hardship. And further, empowering poor people to influence policy and institutions which affect one's options is another important basis for resilience. In this chapter we address the issue of vulnerability modeling of people living in the western part of the greater Serengeti ecosystem.

In this chapter we (1) place the decision making and livelihood strategies of people who live on the west side of Serengeti National Park within a conceptual framework of adaptation and vulnerability, (2) look at livelihood strategies of those people who live on the west side of Serengeti National Park, (3) use a household, individual-based model to address scenarios of change that may affect household livelihoods, and (4) address the model implications for human vulnerability and resilience. The scenarios chosen include (1) one of increasing climate variability, through changes in average annual precipitation and changes in variability of precipitation; (2) one that addresses the change in market prices for maize and for bushmeat.

CONCEPTUAL FRAMEWORK

The anthropological and sustainable development literature of decision making under uncertainty asserts that decision makers are enmeshed in sociocultural dynamics to the degree that personal decisions are often considered cultural norms rather than conscious decision processes. Low-level mechanisms become a normal way of life, often not even recognized as risk-buffering activities, and other strategies are reserved for more dire situations.

Adaptation is defined as activities or adjustments in social and/or economic systems made in response to actual or expected disturbance effects (Smit and Pilifosova 2001; Smith et al. 1996). These adjustments are intended to reduce the vulnerability of society to disturbances or changes in the system (Kates 2000). Adaptive capacity is the ability to cope with impacts of variability and change (Smit and Polifosova 2001). Capacity varies among regions and socioeconomic groups, in that those with the least capacity to adapt are generally the most vulnerable to variability and change impacts. Issues of policy, growing populations, and low agricultural and livestock production, among others, contribute to adaptive capacity and ultimately, vulnerability (Finan and Nelson 2001; Lamb 1995; Little, Hussein, and Coppock 2001). Many countries in Africa are dependent on climatic resources, and because of growing populations and lower technological capabilities, they generally have lower adaptive capacity (Downing et al. 1997; Magistro and Roncoli 2001). Most adaptations to climate and other sorts of variability are sociocultural (that is, changes in management), usually a series of reactive responses to an event such as drought (Galvin et al. 2001; Little, Hussein, and Coppock. 2001).

Vulnerability is defined as the likelihood that an individual or group will be exposed to and adversely affected by circumstances (Cutter 2001). It can also be defined as the characteristics of individuals or groups in terms of their capacity to anticipate, cope with, resist, and recover from the impacts of environmental change (Vogel 1998). Human populations can become vulnerable for various reasons, such as in changes in the environment (e.g., water available for irrigation or tree cover), changes in human and wildlife populations, climate change, diseases, conflicts, changes in environmental and social policy, and so forth. Impacts of these changes are felt unequally throughout a community or region (Galvin et al. 2001). The severity of impacts experienced will depend on which resources are available to a given group or individual within a community. Although various communities may face similar risks, they may not be equally vulnerable. Livelihood assets possessed by the rural poor can be used to assess vulnerability and external factors, such as those listed earlier, influence the levels of assets and the ways the assets can be used (Carney 1998). Due to the heavy dependency on natural resources, changes in the environment such as the timing and duration of rainfall or the distribution and numbers of wildlife can affect vulnerability.

Vulnerable people generally have a variety of alternatives to increase their adaptability and decrease their risk in times of stress and shock (Kasperson 2001). However, new and persistent environmental or political pressures, such as economic liberalization, can limit choices that have traditionally been available (Eriksen and Silva 2003; Leichenko and O'Brien 2002). Human groups may already be vulnerable, placing them at even greater risk should a natural disaster, such as a flood or long-term drought or trade liberalization occur, resulting in changed prices for local commodities.

Our central thesis is that household decisionmaking under uncertainty functions within a complex system of human-environment interactions, with many socioeconomic, policy, and institutional forces also playing an important role (fig. 11.1). At the core are households and communities who make decisions based on environmental conditions and a number of socioeconomic factors, which derive from several levels of social organization, including the local, regional, and beyond. The decision-making process is also influenced by specific land uses and household goals and is conducted under the different forms of severity, duration, and form of the perturba-

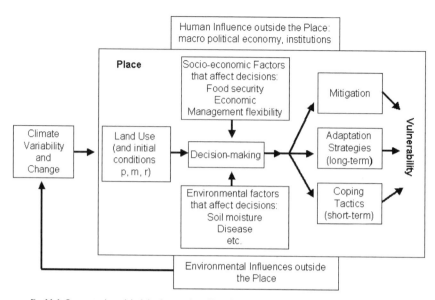

Fig. 11.1 Conceptual model of the factors that affect decision making under uncertain conditions.

tion. All these factors determine when, how, and in what form adaptation and vulnerability occurs.

WESTERN SERENGETI CULTURES AND LAND USES

The western Serengeti region of Tanzania includes many districts and numerous ethnic groups, and has a unique relationship with the Serengeti National Park (SNP). The people who inhabit these lands are generally small-scale farmers and agropastoralists who live in villages and are dependent on their natural resource base for their livelihoods. Previous studies suggest that, to varying degrees, most people rely on agriculture, pastoralism, hunting, fishing, and trading (Campbell, Nelson, and Loibooki 2001). Several of the factors of vulnerability mentioned previously apply to these populations. High in-migration, high endogenous population growth, changing market conditions, and climatic variability affect to various degrees the livelihood strategies of those who live along the western border of Serengeti National Park. Agriculture, relied upon the most, is constrained by frequent drought, lack of farm inputs that are too expensive for communities and individuals to own, lack of extension services, and destruction of crops by wild animals and fires.

Included in the factors that affect vulnerability of populations in the western Serengeti region of Tanzania is the HIV/AIDS epidemic. Although not included as one of the variables in our model, HIV/AIDS does and will continue to affect people's ability to adapt to change. The National AIDS Control Programme (NACP) estimated that in 1999 10% of the adult population in Tanzania had AIDS. The World Bank projects that AIDS will reduce GDP growth in Tanzania from 3.9% to 2.8–3.3% growth, making AIDS the "single greatest threat to Tanzania's future" (UNDP 1999, 10). Compared to the national average of 365 reported cases per 100,000 people, the Mara Region—encompassing most of the western Serengeti—had 119 reported cases per 100,000 people. This was the lowest percentage of all the regions in Tanzania. It is not certain if the low number of cases in the mostly rural Mara Region are due to a lesser amount of *actual* cases or if people simply have less access to hospitals and/or are less inclined to be tested for the disease. To the northwest of the SNP are the Tarime, Serengeti, Musoma, and Bunda districts. Directly to the west of the SNP (near the Grumeti Game Reserve) are the districts of Magu and Bariadi, while directly southwest of the SNP (at the Maswa Game Reserve) are the Maswa and Meatu districts (fig. 11.2). The Tarime, Serengeti, Musoma, and Bunda districts (north and northwest of the Grumeti Game Reserve) are composed of a mix of ethnic groups, including Ikoma, Shashi, Zangita, Wajaluo, Ikizu, Isshenyi, Zanaki, Jita, Nandi, Nata, Tatura, Sizaki, Kisii, Sinicha, Shenyi, Haematic, Luo, and the Kuria. While there are some Sukuma living in these districts, the Sukuma people tend to occupy primarily the Maswa, Magu, Bariadi, and Meatu districts to the west and southwest of the SNP, often referred to as Sukumaland. They are, or at least were, the largest single ethnic group and largest tribal group in Tanzania (Brandstrom 1979). Though the Kuria live throughout the Serengeti, Musoma, and Bunda districts, the largest number of Kuria live in the Tarime district, which borders Kenya. The Kuria (combined with the other, smaller ethnic group in the region, the Luo) make up approximately 90% of the Tarime district population (Tobisson 1986).

Many of the villages in the western Serengeti region border (or are directly on the border of) the SNP. With such a diversity of ethnic groups, and a large geographic distribution of groups to the west of SNP, there is a diversity of land uses in this region. Many villages utilize the resources inside and around the SNP's western border, such as bushmeat and firewood, for fuel as sources of income. Other allocation of labor to economic activities by households and village vary and includes subsistence farming, cash-crop cultivation, hunting, pastoralism, and other activities such as trading, fishing, and cattle raiding. The majority of people living in the western Serengeti region are agropastoralists, combining agriculture with

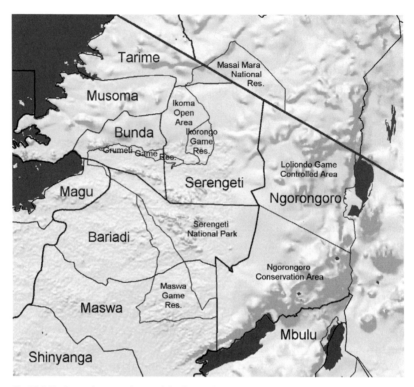

Fig. 11.2 Districts and protected areas of the Greater Serengeti Ecosystem.

some level of cattle keeping. The following sections discuss these land use activities.

Agriculture

Agriculture to the west of the Serengeti is almost always mixed with cattle keeping or other livelihood activities. A number of groups are primarily subsistence farmers, while others combine subsistence farming and cash-crop cultivation. The subsistence crops produced include maize, millet, and cassava, and cotton as a cash crop. Maize is produced as a source of income and food for the household, while cassava and millet are produced mainly for domestic consumption. Despite the economic advantage of cotton growing, cotton cultivation is limited to villages in Bunda district where there is relatively easy market access. In Serengeti District, farther north, poor infrastructure limits access to markets and therefore minimal cotton growing (Holmern et al. 2004).

Although different groups practice agriculture to varying degrees, there are conditions which all agriculturalists face, such as in-migration, population growth, and ecological factors such as unpredictable climate, drought, and soil erosion (see chapter 2, this volume). Recent estimates indicate that 59% of Tanzanian agriculturalists fall below Tanzania's poverty line of (U.S. $150) annual per capita income. Agriculturalists are also operating in an environment of uncertain production and incomplete markets. Agriculture in this semiarid region is generally unpredictable and production is subject to shock. Commonly, agriculturalists are forced to illegally poach game meat to substitute nutritional needs (Barret and Arcese 1998).

Tarime District is home to the Kuria, who are primarily agriculturalists but who combine animal husbandry with their agricultural activities (Tobisson 1980, 1986; Hendrickson, Mearns, and Armon 1996; Fleisher 2000). The Kuria were originally pastoralists. During the British colonization, the principal goal of British colonial officials in North Mara region was to transform the Kuria into peasant cultivators. The British pressed farmers to grow whatever they needed to live on while at the same time producing a surplus of cash crops (Rwezaura 1985; Fleisher 2000).

However, there has been a reluctance to embrace cash crops for several reasons: the remoteness of their land from ports, such as Musoma, has made the marketing of their crops a problematic enterprise; geographically, the Kuria's natural market is Kenya, and separate administration of the two territories has made marketing in Tanzania more difficult; another factor is the Kuria's steadfast determination not to part with cattle. Although cattle numbers are on the decline, most Kuria convert their money into cattle, which is really the only form of wealth that counts (Fleisher 2000).

Villages west and north of the Serengeti National Park corridor but south of Tarime District practice mainly subsistence agriculture, growing maize, cassava, millet, sorghum, vegetables, beans, and sometimes rice/cotton for sale (Loibooki et al. 2002; Emerton and Mfunda 1999). They are mainly occupied by dense, smallholder agriculturalists and agropastoralists. And where people are subsistence farming, they seem to also participate in cash cropping, hunting, and trading or fishing (Loibooki et al. 2002). Ninety-eight percent of 300 individual respondents were agriculturalists: 54% were subsistence agriculturalists, 44% were chiefly livestock owners, and 8% of these individuals considered bushmeat hunting their main occupation (Loibooki et al. 2002). However, commercial agriculture is becoming more common. As agricultural pressures increase, so do the pressures on the wildlife, as wild foods are substituted for the periodic decrease in agricultural production (Emerton and Mfunda 1999).

To the south of the Serengeti corridor, the Sukuma engage in crop cultivation, which is the main source of cash income. Eighty percent of Sukuma live in rural areas and are active cultivators. The Sukuma tend livestock but keep cattle as a source of animal protein and as a source of savings. Every household has a fixed farming area, which consists mainly of potentially arable land and some grazing areas (fallow land is used to graze cattle). The Sukuma, once shifting cultivators, have moved to semipermanent cultivation with the rise in cotton farming. Historically, the British encouraged cotton as a cash crop after World War I, and the German settlers established it (Birley 1982). Following the Ujamaa policy in 1967, rice replaced cotton as the principal cash crop (Meertens 1996).

With the rapid increase in population in the western Serengeti region, there has been a further expansion of cotton and rice production. With this, soil erosion is increasing rapidly (Birley 1982). Hence, increasing population densities have resulted in decreasing agricultural land per capita and in productivity. However, there is some effort to intensify production through manuring and mulching, erosion control, and pasture improvement in some areas (Meertens 1996).

Livestock

Unlike the Maasai pastoralists who live on the eastern rim of SNP, the agropastoral societies who inhabit the west side have relatively small livestock herds. The general economic strategy of the agropastoralists, such as the Sukuma, are to invest agricultural products into livestock and to reinvest livestock into agriculture (Brandstrom 1979). Livestock is the main form of property and generally the only form of storable wealth. However, cattle are perishable property and are sensitive to natural disasters, so that fortunes can be reduced overnight (Brandstrom 1979, 1990).

Livestock are the second most significant form of livelihood in most of the villages in Bunda District. The communities there consider livestock highly valuable, especially cattle, sheep, and goats. They act as sources of food security during times of famine, when they can be exchanged for grain. Cattle are also symbols of wealth/status and are also used as draft animals, especially for plowing. Chickens are used as the main source of meat/protein, since cattle, sheep, and goats are spared for traditional rituals and for harder times such as famine. Livestock are also seen as the foundation of achieving other wealth.

Of all populations who live on the west side of SNP, the Kuria keep most of the livestock. Livestock plays an essential social and economic role in Kuria society. They are both exchanged within communities and sold on the market. Kuria in the Tarime District are heavily involved in cattle raiding (Hendrickson, Mearns, and Armon 1996; Fleisher 2000). The traditional raiding was reciprocal raiding of cattle by pastoralists motivated by the desire to replenish or enlarge family herds and amass bride wealth. They sought to put livestock to economic use (Ruel 1990). Today, market-oriented cattle raiding has been tearing apart the social fabric of Kuria communities (Smith 1992). However strong these socioeconomic relationships between cattle and people, today's cattle raiding, which is driven by the market economy, has disrupted the reciprocal and distributive nature of traditional cattle raiding. Cattle for sale in local, regional, and foreign markets encourage the slaughter of cattle rather than their circulation (Fleisher 2000).

Although there are no Kuria villages within Serengeti National Park or the adjacent game reserves (Emerton and Mfunda 1999; Kauzeni and Kiwasila 1994; Leader-Williams 1996), the Kuria, as do many other groups along the western frontier, use the protected areas as a source of subsistence products, market income (wildlife), and as land insurance in times of need. Wildlife contribution to the lives of the rural people is particularly crucial during times of hardship, when crops may fail and domestic stock die (Campbell 2006).

Hunting

There has been a decrease in the density of wildlife in zones of increased human populations and agriculture (Campbell and Borner 1995; Campbell and Hofer 1995). As a general trend, as agricultural pressures increase, so do the pressures on the wildlife (Emerton and Mfunda 1999). There are several factors that drive people to hunt illegally. Aside from stochastic agricultural conditions, there are three other factors that affect hunting in the park: they include distance from the park boundary, whether the boundary is being patrolled and enforced, and the actual topographic/vegetation features (Campbell and Hofer 1995). The suitability for hunting is highest in the central parts of the Maswa and Ikorongo Game Reserves. Areas most suitable for high poaching *inside* the park are, south of the eastern end of Grumeti GR, east of the southern end of Ikorongo GR, in the NW, and in small pockets of the western corridor. Consequently, these areas of the Serengeti ecosystem that are highly exploited are 99% of the Maswa GR, the western corridor of the protected area, and the Grumeti GR. The areas that are highly exploited by hunters are a single day's walk from settlements close to the boundary (Campbell and Borner 1995).

Since many households lack access to cash and markets, sale and consumption of bushmeat has a high local economic value (Emerton and Mfunda 1999). Local communities provide both a supply of the illegal meat of hunters and the market for their products (Campbell and Hofer 1995). Hunters generally come from villages close to protected area boundaries; they decline with increasing distance of their home to the protected area. Those people that live 0 km from protected areas have one hunter per household. Almost 36,000 people originating from west of the Serengeti are engaged in hunting 75,000 resident herbivores annually and 135,000 migratory animals annually. Mduma, Hilborn, and Sinclair (1998) suggest these numbers may be too high by a factor of three. They indicate that the maximum migrant offtake is about 40,000 and is, on average, only about 20,000.

Dry periods lead to greater wildebeest mobility and wildlife move onto unprotected areas west of park boundaries, nearer villages (Barrett and Arcese 1998). This period coincides with lack of resources at the household level (chiefly agricultural), which results in a greater number of hunters becoming active (Campbell, Nelson, and Loibooki 2001). It is during these times that migratory species are targeted. Hunters also target large resident herbivores such as African buffalo, giraffe, impala, Grants gazelle, kongoni, topi, warthog, and waterbuck (Campbell and Hofer 1995). These nonmigratory species comprise approximately 33% of those killed (Campbell and Borner 1995). Snares set up in the thickets usually catch thicket-dwelling species such as buffalo, giraffe, and waterbuck. Due to its passive and unselective nature, snaring also results in killing nontargeted wildlife species, some of them in significant numbers, such as hyena (Hofer et al. 1996).

The primary reason for illegal hunting is endemic poverty (Barrett and Arcese 1998). The majority of arrested hunters were poor young adult males that owned few or no livestock (Loibooki et al. 2002). Environmental shocks such as crop failure also induce people to poach (Barrett and Arcese 1998). It is a source of protein, a source of cash income, and is traded for other local goods such as foods, fish, and so on. Many people in the western Serengeti purchase livestock meat and bushmeat; however bushmeat is cheaper than domestic meat (Loibooki et al. 2002). Bushmeat is sold in dry form to local and distant markets. There are an estimated 52,928 people (48,436 men and 4,492 women) involved in legal bushmeat hunting. Sixty percent of households in a 23-village study sell or consume bushmeat (Emerton and Mfunda 1999). Bushmeat sales equal approximately one third of average on-farm income. And one third of households lose an average of one quarter of annual harvests to wildlife.

The previous sections show that above all else, change characterizes this linked human-environment system. We look at a set of changes, through modeling, that have potential effects on people and the livestock and wildlife upon which they depend.

THE MODEL

With use of the simulation model developed in chapter 9 we look at two different scenarios of potential change and address human responses to those changes. In the first scenario, we consider changes in average annual precipitation and changes in variability of precipitation. In the second scenario, we address the effect of changes in prices for agricultural crops and bushmeat. We look at the effects of these changes on wildlife, human, and livestock numbers.

Changes in Climate

The current structure and function of the ecosystem and the human socioeconomic system in the Serengeti are determined to a large degree by rainfall patterns. The annual wildebeest migration follows the seasonal pattern of rains, moving to rich volcanic soils in Ngonrogoro when rains turn these areas green, and moving to the wetter west and north with its greater water availability during the dry season. Likewise, human societies are structured by rainfall patterns. People living in the drier areas to the east of Serengeti National Park are pastoralists, while those living in the areas that receive greater rainfall to the west of the park are predominantly agriculturalists.

Global climate change could have a large impact on the Serengeti system if it results in significant changes in precipitation patterns (see chapter 5, this volume, for a detailed discussion of potential climate change in the Serengeti system). Here, we consider changes both in average rainfall amounts and the year-to-year variability in rainfall. Because the model at this time uses annual time steps, we cannot consider changes in the seasonal patterns of rainfall. The effects of changes in average annual rainfall are shown in table 11.1. Increases in annual rainfall from the current 830 mm/year (column 1) to 1,200 mm/year (column 2) yields increases in all dimensions. Wildebeest populations increase in similar proportion to the increase in rainfall, as does poaching. Human population increases by 87%,

Table 11.1 Predicted effects of changes in mean annual rainfall. (Figures reported are steady-state values at the end of a 50-year simulation)

	Base case (column 1)	Increase in mean rainfall (column 2)	Decrease in mean rainfall (column 3)
Mean annual rainfall (mm/yr)	830	1,200	400
Wildebeest population:			
Resident population	14,890	21,450	28,330
Migrating population	1,257,000	1,809,000	613,500
Hunting offtake:			
Resident population	55	81	5,489
Migration population	20,690	30,890	9,971
Human population	135,700	253,800	68,020
Livestock population	80,050	113,600	0

due to increased productivity in agriculture (not shown), increased returns from holding livestock, as well as increased availability of bushmeat. Though conditions are more favorable with increased rainfall, human welfare is not necessarily improved. In this system, livestock represents accumulation of wealth. Livestock numbers overall increase with the increase in rainfall, from approximately 80,000 to 113,000 animals. However, the per capita livestock holdings fall. Initially, there are approximately 0.6 livestock per person. With the increase in precipitation this falls to approximately 0.45 livestock per person. Population growth absorbs all of the increase in productivity in the system, leaving the average person no better off. If, on the other hand, there were some limit to human population growth, human wellbeing would improve with the increase in precipitation. We discuss human population growth issues further in chapter 12.

Suppose, on the other hand, that there was a dramatic decline in mean annual rainfall to 400 mm/year. In this case, there are dramatic declines in the migratory wildebeest population, the human population, and livestock. The human population is highly stressed, as demonstrated by the low population, but even more dramatically by the fact that there are no remaining livestock in the system. All wealth has gone in the struggle to make ends meet. For some households there is no way to make a living when productivity in the system falls with lower precipitation, and they leave the system. Somewhat surprisingly, the resident wildebeest population increases with lower system productivity. This occurs because the resident population is released from competition with livestock. The increase in the resident herd occurs despite the decline in productivity and the increase in human predation brought on by both increased numbers of resident wildebeest and the decline in returns from other sectors.

Another possible effect of climate change is to change the variability of rainfall. For the results in table 11.1, we ran the model with the mean rainfall every year (no variance). In table 11.2, we show results with variable rainfall. We ran the model 50 times, each with a different, randomly drawn rainfall pattern through time. We report the average of the 50 runs. In column 2 of table 11.2, we show results when the standard deviation of rainfall is set at the historical average of 176 mm/year. The average results in column 2 are similar to the results reported in column 1 with a couple of exceptions. Because of the bad years, which are similar or in some cases worse than the 400 mm/year rainfall case shown in table 11.1, livestock numbers

Table 11.2 Predicted effects of changes in the variance of rainfall. (Figures reported are averages over 50 model runs for results at the end of each 50-year simulation)

	Base case: no variance (column 1)	Moderate rainfall variance (column 2)	Moderate variance with persistence (column 3)
Standard deviation of annual rainfall (mm/yr)	0	176	176
Persistence of deviations (first-order autocorrelation)	0	0	0.5
Wildebeest population:			
Resident population	14,890	32,870	21,260
Migrating population	1,257,000	1,173,300	1,196,000
Hunting offtake:			
Resident population	55	5,125	1,896
Migration population	20,690	19,890	19,950
Human population	135,700	159,150	147,830
Livestock population	80,050	7,188	32,950

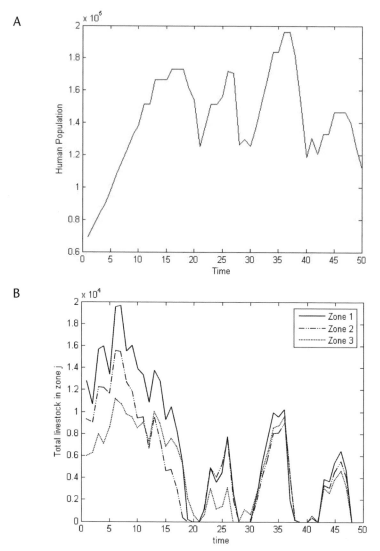

Fig. 11.3 Changes in (A) human, (B) livestock, and (C) resident and migratory wildlife populations, based on the standard deviation of rainfall of 176 mm/yr (column 2, table 11.2). Livestock numbers correspond to distance from Serengeti National Park. Zone 1 is 0–5 km, zone 2 is 5–10 km, and zone 3 is 10–25 km from the park.

tend to be quite low and are compensated by an increase in resident wildlife offtake.

We show the patterns of human, livestock, and wildebeest populations through time for a typical single 50-year simulation in fig. 11.3 (from table 11.2, column 2). Fig. 11.3, panel B shows livestock numbers relative

C

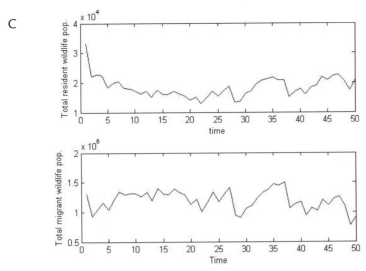

Fig. 11.3 (continued) Changes in (A) human, (B) livestock, and (C) resident and migratory wildlife populations, based on the standard deviation of rainfall of 176 mm/yr (column 2, table 11.2). Livestock numbers correspond to distance from Serengeti National Park. Zone 1 is 0–5 km, zone 2 is 5–10 km, and zone 3 is 10–25 km from the park.

to distance from the SNP. At some point in virtually all simulation runs, livestock numbers fall to zero (fig. 11.3, panel B). In bad years, livestock are sold in order to make ends meet. As discussed earlier, low livestock numbers frees the resident wildebeest population from competition for forage and the resident population increases relative to the case with no variance in rainfall, and there is a corresponding increase in hunting of resident wildebeest populations (fig. 11.3, panel C). In really bad years, or when there are multiple bad years in succession, ends cannot be met for some families and human population drops as well (fig. 11.3, panel A).

Changes in Market Conditions

In this section, we investigate changes in system dynamics if exogenously specified prices for crops or bushmeat change. The effects of changes in crop prices are given in table 11.3. Human population and livestock decrease when crop prices are doubled, while wildlife numbers change little (column 2). Increasing crop prices hurt the poor (not shown here) because they have little agricultural land and so are net purchasers of agricultural output. On the other hand, those who have agricultural land and are sellers of agricultural output are made better off by the price increase. Over-

Table 11.3 Predicted effects of changes in crop and livestock prices. (Figures reported are steady-state values at the end of a 50-year simulation)

	Base case (column 1)	Increase in crop price (column 2)	Decrease in crop price (column 3)
Crop price	300	600	150
Wildebeest population:			
Resident population	14,890	16,740	14,890
Migrating population	1,257,000	1,263,000	1,255,000
Hunting offtake:			
Resident population	55	295	56
Migration population	20,690	18,070	21,720
Human population	135,700	100,200	194,600
Livestock population	80,050	50,850	75,290

all, human population drops, because the poor tend to migrate out of the system while, at the same time, wealth (livestock per capita) increases. Because of the lower human population and greater wealth of the remaining population, there tends to be less poaching pressure on wildlife, and so wildebeest populations show a small increase. The opposite occurs with a decrease in crop prices (column 3). Human population rises, wealth (livestock per capita) falls, poaching pressure increases, and wildebeest populations show a small decline.

The effects of changes in the price of bushmeat, shown in table 11.4, have a pronounced impact on human population. When prices of bushmeat double (column 2), human population increases by 76%. This increase in population is most pronounced in the zone closest to the park (not shown). In this zone, population pressures eventually force all people in this area to sell off their livestock, which accounts for most of the decline in livestock between the cases shown in columns 1 and 2. Not surprisingly, poaching pressure is higher and wildebeest populations are lower, but not dramatically so. When bushmeat prices drop (column 3), human population remains far lower than in the other cases. Per capita wealth (livestock) is higher, as are wildebeest populations.

In reality, prices are not exogenous but are determined endogenously by supply and demand. One would expect in poor rainfall years that the

Table 11.4 Changes in bushmeat prices. (Figures reported are steady-state values at the end of a 50-year simulation)

	Base case (column 1)	Increase in mean prices (column 2)	Decrease in mean prices (column 3)
Bushmeat prices	1,000	2,000	500
Wildebeest population:			
Resident population	14,890	14,890	17,010
Migrating population	1,257,000	1,255,000	1,266,000
Hunting offtake:			
Resident population	55	57	252
Migration population	20,690	21,900	16,530
Human population	135,700	239,300	82,770
Livestock population	80,050	50,660	74,960

prices of crops, livestock, and bushmeat would all increase because of scarcity, while in years of plentiful supply prices would drop. An important improvement in the model would be to include endogenous price determination in the model. Doing so would add another layer of feedback between environmental conditions and human behavior.

DISCUSSION AND CONCLUSIONS

The vast majority of people living on the western part of the GSE are directly dependent on their environment for their livelihoods. Model results suggest they are highly vulnerable to changes in variability in climate and in prices for commodities, as shown by fluctuating human population numbers and in their wealth (livestock holdings). An increase in mean rainfall results in an increase in human population, yet the livestock-to-human ratio decreases. People are not necessarily better off—there are simply more of them. A decrease in rainfall results in a decrease in human population, and there is a concomitant reduction in wildebeest numbers. At the end of the run there are no livestock left in the system. The model runs of year-to-year variability in rainfall result in reduced livestock numbers because of bad-year reductions in their numbers. There is no agriculture in some years. Resident wildebeest increase but offtake increases, too. The result is a fluctuating human population.

The human population is also vulnerable to fluctuating prices for commodities. Increases in prices result in a decrease in human populations as the poor migrate out of the system. Livestock populations decrease as well. The consequences of a decrease in crop prices are higher human populations, increased poaching, and reduced livestock populations.

The model currently demonstrates scenarios that are all or nothing—that is, changes in climate and in prices alone push people out of the system or bring people into the system. In reality, the situation is much more complex, with people making adaptations with small economic adjustments. However, it is likely that when changes are great in magnitude or long in duration, people will have less capacity to adapt to change (Galvin et al. 2005).

People of various socioeconomic levels also have differing abilities to deal with change. The model results suggest that the rich have advantages that are lost on the poor, pushing the poor out of the system from time to time, when climate is too variable or when crop prices are too high.

In the Biocomplexity Project we will be able to refine the model, collect data in the field to parameterize it, and link it explicitly to an ecological model such as SAVANNA (cf. Ellis and Coughenour 1998). Feedbacks between the socioeconomic system and the ecological system will provide us with more realistic interactions of the two.

ACKNOWLEDGMENTS

Thanks to Craig Packer, Tony Sinclair, and the entire Biocomplexity Project team, who started out in a series of NCEAS workshops talking about the issue of human-environment interactions in the Serengeti ecosystem. Thanks also to Jennifer Sunderland and Stephanie Ewalt, whose research forms the basis of the modeling efforts. This research is supported by NSF grant DEB-0308486.

REFERENCES

Barrett, C. B., and P. Arcese. 1998. Wildlife harvest in integrated conservation and development projects: Linking harvest to household demand, agricultural production, and environmental shocks in the Serengeti. *Land Economics* 74:449–65.

Birley, M. H. 1982. Resource management in Sukumaland, Tanzania. *Africa* 52(2): 1–29.

Brandstrom, P. 1979. *Aspects of agro-Pastoralism in East Africa.* Research Report no. 51. Uppsala University, Sweden: Scandinavian Institute of African Studies.

————. 1990. Boundless universe: The culture of expansion among the Sukuma Nyamwezi of Tanzania. PhD diss., Uppsala University, Sweden.

Campbell K. L. I. 2006 Wildlife. In *Improving the husbandry of livestock kept by the poor in developing countries*, ed. A. J. Kitalyi, N. Jayasuriya, N. Owen, and T. Smith. Nottingham: Nottingham University Press.

Campbell, K. L. I., and M. Borner. 1995. Population trends and distribution of Serengeti herbivores: Implications for management. In *Serengeti II: Dynamics, management, and conservation of an ecosystem*, ed. A. R. E. Sinclair and P. Arcese, 117–45. Chicago: University of Chicago Press.

Campbell, K. L. I., and H. Hofer. 1995. People and wildlife: Spatial dynamics and zones of interaction. In *Serengeti II: Dynamics, management, and conservation of an ecosystem*, ed. A. R. E. Sinclair and P. Arcese, 534–70. Chicago: University of Chicago Press.

Campbell, K. L. I., V. Nelson, and M. Loibooki. 2001. Sustainable use of wildland resources, ecological, economic and social interactions: An analysis of illegal hunting of wildlife in Serengeti National Park, Tanzania. Final Technical Report to the Department for International Development (DFID), Animal Health and Livestock Production Programmes. London: Department for International Development.

Carney, D., ed. 1998. Sustainable rural livelihoods: What contribution can we make? Paper presented at the Department for International Development's Natural Resources Advisors Conference, July 1998. London: Department for International Development.

Cutter, S. L. 2001. A Research Agenda for Vulnerability Science and Environmental Hazards. International Human Dimensions Programme (IHDP) Update, Newsletter of the International Human Dimensions Programme on Global Environmental Change. Retrieved from http://www.ihdp.org.

Downing, T. E., L. Ringius, M. Hulme, and D. Waughray. 1997: Adapting to climate change in Africa. *Mitigation and Adaptation Strategies for Global Change* 2: 19–44.

Ellis, J. E. and M. B. Coughenour. 1998. The SAVANNA integrated modeling system: An integrated remote sensing, GIS and spatial simulation modeling approach. In *Drylands: Sustainable Use of Range Lands into the Twenty-first Century*. eds. R. Squires, and A.E. Sidahmed, 97–106. Rome, IFAD Series.

Emerton, L., and I. Mfunda. 1999. Making wildlife economically viable for communities living around the western Serengeti. Evaluating Eden series: Working paper no. 1. London: International Institute for Environment and Development.

Eriksen, S., and J. Silva. 2003. The impact of economic liberalization on climate vulnerability among farmers in Mozambique. Presentation at the Open Meeting of Human Dimensions Research Community, 16–18 October, 2003, Montreal. Retrieved from http://sedac.ciesin.columbia.edu/openmtg/docs/Eriksen.pdf.

Finan, T. J., and D. R. Nelson. 2001. Making rain, making roads, making do: Public and private adaptations to drought in Ceara, northeast Brazil. *Climate Research* 19:97–108.

Fleisher, M. L. 2000. *Kuria cattle raiders: Violence and vigilantism on the Tanzania/Kenya frontier*. Ann Arbor: University of Michigan Press

Galvin, K. A., R. B. Boone, N. M. Smith, and S. J. Lynn. 2001. Impacts of climate variability on East African pastoralists: Linking social science and remote sensing. *Climate Research* 19:161–72.

Galvin, K. A., D. S. Ojima, R. B. Boone, M. Betsill, and P. K. Thornton. 2005. NSF-funded proposal in human and social dynamics. Decision-making in rangeland systems: An integrated ecosystem-agent-based modeling approach to resilience and change (DREAMAR). Washington, DC: National Science Foundation.

Hendrickson, D., R. Mearns, and J. Armon. 1996. Livestock raiding of among the pastoral Turkana of Kenya. *IDS Bulletin* 27:17–30.

Hofer, H., K. Campbell, M. L. East, and S. A. Huish. 1996. The impact of game meat hunting on target and non-target species in the Serengeti. In *The exploitation of mammal populations,* ed. V. J. Taylor and N. Dunstone, 117–46. London: Chapman and Hall.

Holmern, T., A. B. Johannesen, J. Mbaruka, S. Mkama, J. Muya, and E. Roskaft. 2004. *Human-wildlife conflicts and hunting in the western Serengeti, Tanzania,* 1–26. Trondheim: Norwegian Institute of Natural Resources Research (NINA).

Kasperson, R. 2001. Vulnerability and global environmental change. *Newsletter of the International Human Dimensions Programme on Global Environmental Change,* Nr. 2/2001. Bonn; IHDP.

Kates, R. 2000. Cautionary tales: Adaptation and the global poor. *Climatic Change* 45:5–17.

Kauzeni, A., H. and Kiwasila. 1994. Serengeti regional conservation strategy: A socio-economic study. Dar es Salaam, Tanzania: University of Dar es Salaam, Institute of Resource Assessment.

Lamb, H. H. 1995. *Climate, history and the modern world,* 2nd ed. London: Routledge.

Leader-Williams, N. 1996. Wildlife Utilization in Tanzania. Unpublished paper presented at Workshop on the Costs and Benefits of Wildlife in Africa, Lewa Downs, Kenya.

Leichenko, R. M., and K. L. O'Brien. 2002. The dynamics of rural vulnerability to global change: The case of southern Africa. *Mitigation and Adaptation Strategies for Global Change* 7:1–18.

Little, P. D., M. Hussein, and D. L. Coppock. 2001. When deserts flood: Risk management and climatic processes among East African pastoralists. *Climate Research* 19:149–59.

Loibooki, M., H. Hofer, K. Campbell, and M. L. East. 2002. Bushmeat hunting by communities adjacent to the Serengeti National Park, Tanzania: The importance of livestock ownership and alternative sources of protein and income. *Environmental Conservation* 29:391–98.

Magistro, J., and C. Roncoli. 2001. Anthropological perspectives and policy implications of climate change research. *Climate Research* 19:91–96.

Mduma, S., R. Hilborn, and A. R. E. Sinclair. 1998. Limits to exploitation of Serengeti wildebeest and implications for its management. In *Dynamics of tropical communities,* ed. D. M. Newbery, N. Brown, and H. H. T. Prins. *British Ecological Society Symposium* 37:243–65. Oxford: Blackwell Science.

Meertens, H. C. C. 1996. Farming system dynamics: Impacts of increasing population density and the availability of land resources on changes in agricultural systems: The case of Sukumaland, Tanzania. *Agricultural Ecosystems and Environment* 56:203–15.

Ruel, M. J. 1990. Non-sacrificial ritual killing. *Man,* n.s., 25:323–35.

Rwezaura, B. A. 1985. *Traditional family law and change in Tanzania: A study of the Kuria social system.* Baden-Baden, Germany: Nomos.

Smit, B., and O. Pilifosova. 2001. Adaptation to climate change in the context of sustainable development and equity. In *Climate change 2001: Impacts, adaptation, and vulner-*

ability, ed. J. McCarthy, O. Canziani, N. Leary, D. Dokken, and K. White. Intergovernmental Panel on Climate Change. Cambridge: Cambridge University Press.

Smith, A. B. 1992. *Pastoralism in Africa: Origins and development ecology.* London: Hurst and Co.

Smith, J. B., G. V. Menzhulin, M. Campos, N. Bhatti, R. Benioff, and B. Jallow, eds. 1996. *Adapting to climate change: Assessments and issues.* New York: Springer.

Tobisson, E. 1980. Women, work, food, and nutrition in Nyamwigura village, Mara Region, Tanzania. Report no. 548. Dar es Salaam: Tanzania Food and Nutrition Centre (TFNC).

―――. 1986. *Family dynamics among the Kuria: Agro-pastoralists in northern Tanzania.* Goteborg, Sweden: Acta Universitatis Gothoburgensis.

UNDP. 1999. Tanzania Human Development Report. Washington, DC: Oxford University Press.

Vogel, C. 1998. Vulnerability and global environmental change. *Land Use Land Cover Change (LUCC) Newsletter* 3 (March): 15–19.

TWELVE

Larger-Scale Influences on the Serengeti Ecosystem: National and International Policy, Economics, and Human Demography

Stephen Polasky, Jennifer Schmitt, Christopher Costello, and Liaila Tajibaeva

Political, economic, and human demographic factors, many of which originate outside the Serengeti ecosystem, often have a profound influence on what happens inside the Serengeti ecosystem. National policies set the boundaries (literally and figuratively) on what types of human behavior are permissible in what parts of the Serengeti ecosystem. International policies also have an influence on events within the Serengeti ecosystem. Economic factors, such as the demand for products in Europe or North America, set the prices of goods and services produced in the Serengeti. In addition, the dynamics of human populations exert a powerful influence on the dynamics of the ecosystem. The effects of human demographics, economies, and policies are interconnected, making it important to consider the joint effects within a systems approach.

We begin this chapter by reviewing important national and international policies that set the institutional context for the Serengeti ecosystem. Though the greater Serengeti ecosystem contains portions of Tanzania and Kenya, we focus the discussion in this chapter on Tanzanian national policy and do not discuss national policy in Kenya. Policies determined at the national level in Tanzania define allowable human uses in various land designation categories. The Serengeti ecosystem includes many types of land use designations, including a national park, game reserves, a game-controlled area, open areas, and the Ngorongoro Conservation Area. On the international level, possibly the most important policy that affects the

Serengeti ecosystem is the Convention of International Trade in Endangered Species (CITES).

Local as well as national and international factors drive economic forces affecting the Serengeti ecosystem. Much of what people produce in the greater Serengeti ecosystem is consumed locally, but there are important international influences as well. For example, the perceived threats to the safety of international travel following the terrorist attacks of September 11, 2001, greatly reduced tourism to the Serengeti.

The final large-scale factor we examine is the influence of changes in human population on the Serengeti ecosystem. Human population growth in areas adjacent to the park drive up the demand for food, from agriculture, livestock, or hunting of wildlife, as well as the supply of labor for agriculture, grazing, and hunting. As the human population increases, more land is required for settlement, growing crops, and grazing livestock. In addition, there may be increased poaching pressures on wildlife, as well as increased likelihood of human-wildlife conflict, such as crop damage from wildlife.

In the latter part of this chapter, we use the household model developed in chapter 10 (this volume; also used in chapter 11, this volume) to simulate the impact on the Serengeti ecosystem of three potential changes in policy, economic, and demographic conditions. In particular, we explore the effects of: (1) changes in antipoaching policy, (2) the establishment of wildlife management areas, and (3) changes in human population dynamics.

POLICIES

The Serengeti ecosystem contains the largest protected grassland and savanna ecosystem in the world. National and international policies, protected-area policies in particular, have played a large role in determining the state of the ecosystem. In this section, we review important national and international policies that have had a significant influence on the Serengeti. We review these policies in the light of the ongoing debate in conservation over the relative merits of community-based conservation versus protected-area strategies.

Tanzanian National Policy

Conservation of wildlife and wild places has had a prominent place in Tanzanian national policy since independence. In 1961, President Nyerere issued the Arusha Manifesto, which stated:

The survival of wildlife is a matter of grave concern to all of us in Africa. These wild creatures amid the wild places they inhabit are not only important as a source of wonder and inspiration but are an integral part of our natural resources and of our future livelihood and well being. In accepting the trusteeship of our wildlife we solemnly declare that we will do everything in our power to make sure that our children's grandchildren will be able to enjoy this rich and precious inheritance.

The Arusha Manifesto provided the foundation for conservation policy in Tanzania. Though the manifesto predates extensive discussion of sustainability and biodiversity conservation by several decades, the meaning and importance of both is clearly expressed in this short passage.

The primary policy instrument for conservation in Tanzania has been the establishment of protected areas. Nearly 40% of the area of Tanzania has some form of protected status (International Union for the Conservation of Nature and Natural Resources [IUCN] Categories I–VI), which is one of the highest percentages of any country in the world (United Nations Envrionment Program—World Conservation Monitoring Center [UNEP-WCMC] 2004). By contrast, the comparable figure for sub-Saharan Africa, and for the world as a whole, is around 11% (UNEP-WCMC 2004).

Protected-area policies have had a profound impact on the Serengeti ecosystem. The vast majority of the ecosystem lies within some type of protected area. The most prominent protected area is Serengeti National Park, which covers a broad swath in the center of the ecosystem. In addition to Serengeti National Park, other types of protected areas within the Tanzanian portion of the Serengeti ecosystem include game reserves, a game-controlled area, and the Ngorongoro Conservation Area (NCA). Without these extensive protected areas, the Serengeti ecosystem would undoubtedly be a far different place than it is at present.

National parks have the highest degree of habitat and wildlife protection of any type of protected area. The National Park Ordinance of 1959 established regulations for governing national parks and created the Tanzania National Parks Authority (TANAPA) as the regulatory body to oversee the parks. National parks prohibit local people from residing, gathering firewood, hunting, or gathering other food within park boundaries. National parks allow some nonconsumptive human activities to occur within parks, such as photo and walking safaris, research, and educational expeditions.

Game reserves and game-controlled areas, which are under the jurisdiction of the Wildlife Division, offer considerable protection for wildlife while being somewhat less restrictive of human use than national parks. The Wildlife Conservation Act of 1974 vested authority for overseeing all wildlife within the country to the director of wildlife within the Wildlife Division. Game reserves are devoted to wildlife conservation and thus prohibit permanent human settlement, cultivation, and livestock grazing—but they allow limited hunting, with permission, as well as limited fishing and logging. There are three game reserves in the Serengeti ecosystem. The Maswa, Ikorongo, and Grumeti game reserves are located on the western border of Serengeti National Park. These game reserves provide a buffer zone between Serengeti National Park and surrounding human settlements. There is one game-controlled area within the Serengeti ecosystem, the Loliondo Game Control Area located to the east of Serengeti National Park. Unlike game reserves, game-controlled areas allow human settlement, grazing, cultivation, and licensed hunting.

The NCA was established in 1959, the same year as the National Park Ordinance. The NCA is a multiple-use area, which makes it in some ways more like a game-controlled area than a national park. The NCA was set up in an attempt to both provide for the conservation of wildlife as well as to allow Maasai to continue their pastoral livelihood. The NCA permits the Maasai to live and graze livestock within the NCA, though within the crater itself there are some restrictions on grazing and a prohibition on human settlement. Recent rules have made it illegal for non-Maasai to immigrate into the NCA. The NCA is of interest for several reasons, most notably because of its attempt to conserve wildlife as well as meet the needs of local human inhabitants. In this way, the NCA may well be one of the first examples of "community-based conservation," in which conservation activities are designed to benefit rather than exclude local human populations, though its establishment precedes the coining of this term by several decades. There remains tension, however, between the NCA Authority and the Maasai. For the Maasai, though the NCA is preferable to a National Park that would kick them out entirely, having no outside authority restricting Maasai life would be even better.

The most recent attempt to conserve wildlife and provide benefits to local inhabitants is the introduction of Wildlife Management Areas (WMAs). In 1998, after twelve years of writing, planning, and discussion, the government of Tanzania passed the Wildlife Policy of Tanzania, which authorized the formation of WMAs (Shauri and Hitchcock 1999). A WMA is an area agreed to by the minister of wildlife and local village government that is dedicated to biological natural resource conservation (Ministry of Natural Resources and Tourism 1998). Wildlife Management Areas can be set up within game-controlled areas or open areas but not in national parks, conservation areas, or game reserves. Wildlife Management Areas give villages some user rights, allowing them to participate in wildlife management and share in the benefits generated from wildlife. The Wildlife Policy of Tanzania states that the government will include relevant stakeholders in the process of deciding rules governing WMAs. However, details of how benefits are to be shared are not specified (Ministry of Natural Resources and Tourism 1998). These benefits will likely come from trophy and resident hunting, photographic safaris, beekeeping, and/or natural forest management (Mabugu and Mugoya 2001). Each WMA devises its own mechanism for generating and sharing revenue from these activities.

The Wildlife Policy of Tanzania contains provisions beyond the creation of WMAs. The goals of the wildlife policy go beyond promoting the conservation of biological diversity. The wildlife policy seeks to develop sustainable utilization of wildlife resources and raise the contribution of the wildlife sector in the country's gross domestic product (GDP) from about 2% to 5%. The wildlife policy also seeks to involve all stakeholders in wildlife conservation and sustainable utilization, and to provide fair and equitable sharing of benefits.

As part of the policy, the government outlines its role as well as the role of the private sector, nongovernmental organizations (NGOs), and local villages in conservation. Under the wildlife policy, the government distributes user rights to various stakeholders. Furthermore, the role of the government in the wildlife sector is "to provide clear policy guidelines, stimulate and promote involvement of various stakeholders, manage core wildlife protected areas, retaining ownership of wildlife resources, and see to the sector's general development" (Ministry of Natural Resources and Tourism 1998, p. 29). The wildlife policy calls on the private sector as a key player in direct investment and states that the private sector and NGOs should support the government in its endeavors under the policy (Ministry of Natural Resources and Tourism 1998). The reliance by conservationists on the private sector for investment, financing, and interest causes concern. Many caution that the local people are disadvantaged in relation to private businesses in these potential business agreements due to lack of business and legal knowledge (Mabugu and Mugoya 2001). Nongovernmental organizations or other third parties will likely be important factors in protecting local interests in WMA transactions.

A relatively small portion of the Serengeti ecosystem consists of open areas, which are areas that do not have any type of protected status. In open areas, local residents may engage in all forms of land use, including cultivation, hunting, grazing, and settlement. The regional authorities control hunting permits, granting them to resident hunters, hunting safaris, and to villages to deal with problem animals. These areas have little or no direct regulation for conservation purposes and tend to favor human uses over wildlife. The only significant open areas within the Serengeti lie in the western part of the ecosystem. Though the open areas have much higher human population densities and extensive areas of agricultural cultivation and settlement, they remain an important part of the ecosystem for wildlife. The wildebeest migration goes through these open areas almost every year and there is concern about both poaching of wildebeest in open areas (Thirgood et al. 2004) as well as damage to crops by wildlife.

Important Dimensions of Tanzanian Conservation Policy

If conserving wildlife were the primary objective facing governments, rather than one objective among many, it would be far easier to implement successful conservation policies. In developing countries such as Tanzania, education, health care, and the necessities of life are not secure for much of the population. Any conflict, real or perceived, between conservation policy and providing the necessities of life to local populations will bring heavy pressure to change conservation policy. The central dilemma of conservation policy in Tanzania, as in many other developing countries, is how to conserve wildlife within the context of a growing human population that places a high priority on the economic development necessary to reduce poverty.

There is an ongoing debate within the conservation community about this dilemma. Some conservationists have argued that conservation will only succeed in the long run if it is supported by local people. According to this line of thinking, it is necessary to have some form of community-based conservation that simultaneously addresses both conservation and the welfare of local people (e.g., Hulme and Purphree 2001; Kiss 1990; Schwartzman, Moreira, and Nepstad 2000; Wilshusen et al. 2002). When local people do not value conservation, political pressure will build to change policies that protect wildlife and habitats at the expense of the welfare of local people. Even without a formal change in policy, encroachment on protected areas and poaching of wildlife will thrive in a climate of perceived conflict between the welfare of local people and conservation objectives. In such a climate, ensuring conservation will require enforcement (e.g., guards to enforce against poaching or encroachment), which will only increase animosity of local people toward conservation. On the other hand, some conservationists have argued that protected-area strategies are the only proven strategy for conserving biodiversity and must remain the central focus of conservation strategies. These conservationists argue that community-based conservation has been largely unsuccessful to date and has diverted attention and resources toward economic development and away from conservation (e.g., Oates 1999; Terborgh 1999). Proponents of each strategy have been active in Tanzanian conservation. Conservation policy in Tanzania reflects elements of both views.

Protected-area strategies have had a longer history and more pervasive effect on the Serengeti ecosystem than has community-based conservation. Protected-area regulations set from outside the Serengeti ecosystem have dictated human actions within the ecosystem since 1921, when the British set up a big game hunting zone (McCabe, Perkins and Schofield 1992). As detailed in the previous section, protected areas, in one form or another, now cover the vast majority of the land area within the Serengeti ecosystem. Such strategies, in particular the establishment of Serengeti National Park, are a major reason for the success in maintaining the wildlife and native habitat of the ecosystem.

Protected-area policies, however, often contain the seeds of conflict with local people. Local peoples long ago lost the right to use the land and the wildlife within much of the Serengeti ecosystem. In the colonial era and the years just after independence, policies forced many people to relocate, taking away traditional sources of pasturelands, water, and food (Shauri and Hitchcock 1999). One notable exception is the NCA, which was originally part of the Serengeti National Park. The colonial powers redesignated part of the park to create a multiple-use area for the Maasai. However, as we will discuss, this attempt to combine conservation with human needs has also caused conflict. Only in open areas, which make up a small portion of the land area within the Serengeti ecosystem, are local communities in control of decisions affecting the use of land and wildlife.

The limitations on hunting are one major source of contention for local people. Hunting is banned in some areas (e.g., Serengeti National Park). Regulations allow minimal hunting with appropriate permits in game reserves and game-controlled areas, although the high permit prices and the requirement to own a firearm prohibit many local people from obtaining permission to hunt. However, illegal wildlife hunting is a major problem (Campbell, Nelson, and Loibooki 2001). Bushmeat is often the cheapest form of meat available and provides a source of income. Bushmeat is plentiful, especially during the wildebeest migration. Antipoaching enforcement is expensive and fails to provide sufficient deterrence to prevent all poaching, though it has been successful in limiting it. The presence of the Tsetse fly and of other diseases reduces the benefits of livestock ownership and has complicated a transition away from subsistence hunting in some areas (Loibooki et al. 2002).

Community-based conservation is one way to reduce conflict by aligning the interests of local communities with conservation. Community-based conservation often involves programs that aid local communities through economic development, directing revenues generated from wildlife to local communities, or giving local communities a greater say in the management of wildlife and land use. The NCA and the recently established WMAs attempt to give significant autonomy as well as the potential to keep some of the proceeds generated from wildlife-based activities to local people.

The goals of the NCA were to both conserve wildlife and allow the Maasai to continue to follow their traditional lifestyle within the NCA. The Maasai are primarily a pastoralist people, and their lifestyles tend to be consistent with conservation of wildlife. They rarely hunt for wild meat or cultivate large areas of land, thereby minimizing human-wildlife conflicts. The most recent studies of the NCA suggest that while wildlife fare well within the NCA, the Maasai have not. There are complaints that the Ngorongoro Conservation Area Authority is failing in its mission to provide adequate benefits to the Maasai (Lissu 2000). The problems that exist in the NCA may explain the indifference and sometimes hostility between the local Maasai of the NCA and government authorities.

Difficulties for the Maasai in the NCA arise because of restrictions on Maasai activities along with growth of the Maasai population. Within the NCA, there is a limit on the allowable amount of cultivation and a prohibition on wildlife hunting. The limited-cultivation allowance is constantly under threat of revocation, despite its importance as a food source for present-day Maasai. During difficult times, cultivation and wildlife hunting play a larger role in Maasai diets, and the restrictions on human activities has led to food insecurities for Maasai within the NCA (Broch-Due and Schroeder 2000). The Maasai's primary food source, livestock, also faces threats from livestock/wildebeest interactions in the NCA. The short-grass grazing areas are off limits to the Maasai during the rainy season, when wildebeest are calving, due to Malignant Catarrhal Fever, which is fatal when transmitted from wildebeest to cattle. The rainy season is the time of peak grass productivity. The Maasai within the NCA have had persistent malnutrition and food deficits. As shown in table 12.1, from 1990–1997 the Ngorongoro District (including the Maasai within the NCA) had food deficits in carbohydrates and protein every year except for carbohydrates in 1994.

Tourism, while often cited as the means to provide local benefits, creates additional problems for the Maasai. Though tourism has the potential to create jobs and economic development, in reality it has brought few jobs and little economic development to the Maasai in Ngorongoro (Broch-Due and Schroeder 2000). Tourism creates demand for land and water that may

Table 12.1 Food deficits within the Ngorongoro District 1990/1991–1996–1997

Food category	1990/91	1991/92	1992/93	1993/94	1994/95	1995/96	1996/97
1 Food production							
Maize	3,981	4,152	1,122	6,669	11,781	6,992	7,450
Millet	720	808	157	730	1,602	984	1,800
Potatoes	384	1,062	1,461	1,230	2,390	3,804	n.a.
Total	5,085	6,022	2,740	8,629	15,773	11,780	n.a.
Food beans	383	750	107	850	1,166	1,335	2,800
2 Food requirements							
Carbohydrates	13,427	13,951	14,495	15,060	15,648	16,149	16,892
Protein	2,685	2,790	2,899	3,012	3,130	3,230	3,378
3 Surplus + / deficit –							
Carbohydrates	−8,342	−7,929	−11,755	−6,431	+125	−4,369	n.a.
Protein	−2,382	−2,040	−2,792	−2,162	−1,964	−1,895	−578

Source: Table recreated from The Planning Commission (1998a).

create conflicts with the Maasai. Foreign visitors often require five-star accommodations, which includes, among other things, electricity and hot showers. The Sopa Lodge along the Ngorongoro Crater rim diverted the Oljoronyuki Stream, used by the Maasai and wildlife, for tourist needs (Lissu 2000). Another lodge along the crater rim, the Serena Lodge, has also created conflict between tourists and the Maasai. The lodge is located at Kimba on land that belongs to the Maasai village of Oloirobi. The NGO, Korongoro Integrated Peoples Orientated to Conservation, alleges that the site for Serena sits directly on a main path for livestock access into and out of the crater (Neumann 2000) as well as places unsustainable demands on the area's water supplies. Despite the money tourism brings to the government, the local people have thus far not benefited substantially from tourism in the NCA.

By the early 1990s, Tanzania began implementing other forms of community-based conservation within the Serengeti ecosystem. The governing authority for national parks, TANAPA, began setting up community policies after years of ignoring, displacing, and conflicting with local populations. Over the past 10 years, TANAPA has greatly improved relations between game wardens and the local communities. The Tanzania National Parks Authority (TANAPA) set aside funds designed to assist local communities with economic development projects, coined Support for Community Initiated Projects. The Authority uses 7.5% of its budget to fund these community projects, providing up to 75% of the total costs to new projects (Bergin 2001).

Other governmental programs have also reflected a shift toward bottom-up community-based approaches to conservation. The Wildlife Division and the Ministry of Natural Resources and Tourism set up the Serengeti Regional Conservation Strategy in 1989, now called the Serengeti Regional Conservation Project (SRCP). The SRCP based its strategy on integrated conservation and development projects. These projects strive to manage protected areas and to create buffer zones around protected areas while promoting local development (Mduma and Hando 2003). In 1993/ 1994, the SRCP set up individual projects in 10 communities, focusing on two strategies. Their main conservation strategy involved distributing game meat from wild ungulate harvests to local villages (Johannesen 2004a). By providing local villages with game meat, the authorities hoped to curb illegal hunting. Holmern et al. (2004) found that villages in the Serengeti and Bunda districts that participated in SRCP had a higher number of households participating in illegal hunting, but that their annual offtake of wildlife was lower. A second part of the strategy involved transferring tourist revenues to local authorities. In Robanda village within the

Serengeti District, the local people receive tourist revenue transfers for tourist activities created within the village through a public-private partnership with a tourist company, Sengo (Campbell, Nelson, and Loibooki 2001). Between 1996 and 2000, Robanda received a total of 21,788,244 TZ shillings (about US$31,000) from its joint partnership (Campbell, Nelson, and Loibooki 2001). However, despite the transfer of money and game meat to villages, whether SRCP has been a success is questionable. Johannesen (2004a) showed that there are no benefits to wildlife conservation when the benefits from tourism are not directly tied to reductions in illegal hunting activities. In actuality, Johannesen (2004a) found that the program may decrease conservation, because legal culling adds to predation of wildlife and increases in income from tourism revenue make people more willing to incur the financial risk associated with poaching. To benefit conservation, policies need to provide a link between reduced poaching and the benefits received; that is, cessation of game meat or tourism revenue if caught poaching (Johannesen 2004a).

A large step in the direction of community-based conservation was allowing the formation of WMAs as part of the Wildlife Policy of Tanzania, instituted in 1998. However, the policy created a basic tension in the partnership between local communities and the Tanzanian government. Although local communities receive the user rights to wildlife in WMAs, the national government retains the ownership rights to wildlife. There is some reluctance on the part of national government officials to share decision-making authority and revenue with local communities, thereby failing to provide an equal partnership for local communities in wildlife management (Shauri 1999; Shauri and Hitchcock 1999). Shauri and Hitchcock (1999, p. 8) also point out that "many of these objectives could have been implemented long ago without a new Wildlife Policy [and] one of the major shortcomings of the Tanzanian Ministries is their failure to interpret existing law to give administrative authority to take action."

Even with smooth relationships with the national government, several substantial obstacles remain before a local group can institute a WMA. There may be several local groups with different opinions on how to structure the WMA, in terms of both decision-making authority and revenue sharing. The Wildlife Policy is not available in Swahili, let alone the many native dialects of the various peoples who live within the Serengeti ecosystem, limiting knowledge of what is allowable under a WMA. Financing provides yet another hurdle for local villages. Costs of setting up a WMA can be substantial and access to financing for these costs is minimal (Mabugu and Mugoya 2001). Before a WMA can be officially recognized, local groups must complete at least eight steps (see table 12.2), culminating in a published decla-

Table 12.2 Steps for establishing a Wildlife Management Area

1.	Village assemblies meet and make a resolution to form a WMA.
2.	The resolution is sent to the District Council for ratification.
3.	Surveys of the WMA area are carried out.
4.	A village land use plan is prepared and approved by the District Council.
5.	The District Council forwards the surveys and land use plans to regional authorities.
6.	The regional authorities review the plans.
7.	The regional authorities send the plans to the minister responsible for natural resources. Upon review and full approval, the minister makes a declaration establishing the WMA.
8.	A declaration is published in the government gazette to establish the village WMA.

Source: Metcalfe et al. 1998.

ration in the government gazette (Mabugu and Mugoya 2001; Shauri and Hitchcock 1999). Communities with little existing infrastructure and local organization may have difficulty negotiating the process to form a WMA. In addition, these communities may need further substantial investment to successfully maintain a WMA once in operation (Mabugu and Mugoya 2001). Overall, the costs of setting up and maintaining a WMA may provide a prohibitive hurdle to large-scale adoption of the policy.

Currently four pilot WMA projects exist within the Serengeti Ecosystem: IKONA, ERAMATRE, Lake Natron, and Makao (see table 12.3). A fifth WMA in the Fort Ikoma Area is currently in the proposal stage. These four pilot projects are still in their infancy, but should soon provide evidence on how well the new wildlife policy is working.

A longer-running revenue sharing scheme near the Selous Game Reserve provides evidence that despite the criticisms, WMAs have the potential to alter community attitudes toward wildlife. The MBOMIPA pilot project provided 15,000,000 TZ shillings (just under US$19,000) in 1999 to the local village (Mabugu and Mugoya 2001). Evidence suggests that village attitudes toward wildlife have positively shifted because of the project (Mabugu and Mugoya 2001). Mabugu and Mugoya (2001) point out that WMAs will provide the greatest per capita benefits in sparsely populated villages and districts with abundant wildlife. Whether WMAs provide sufficient benefits to change attitudes and result in effective conservation in areas with larger populations, with different rules for allocating benefits across villages or within a village, and with different cultures and traditions, remains to be tested.

Table 12.3 WMA pilot projects in the greater Serengeti ecosystem

WMA	Location	Size (km²)	Involved communities	Natural resource uses
Ikona	Fort Ikoma Serengeti District	450	Robanda, Nyichoka Mbiso/Natta, Nyakitono/Makundusi	Wildlife, grazing, collection of indigenous tree and grass species, fish, minerals
Eramatare	Loliondo Ngorongoro District	4,500	Losoito/Malloni, Oloipiri, Olorien/ Magaiduru, Arash, Soit Sambu	Wildlife, birds, forest, stone & gravel, minerals, fish, water, bees/honey, livestock
Lake Natron	Pinyiny Ward Ngorongoro District	2,000	Collection of subvillages comprising Pinyiny Ward	Wildlife, forests, birds, caustic soda, limestone, water
Makao	Meatu and Karatu District	1,700	Makao, Sapa, Mbushi, Mang'ola, Iramba Ndogo, Mwanjoro/ Jinamo, Mwangudo, Mwabagimu	Wildlife, woodland, water, and honey

Source: International Resources Group (2000).

Tanzania is wrestling with the difficult problem of treating local communities equitably while at the same time effectively conserving wildlife. The NCA and WMAs are two attempts to address both issues. Whether these policies, or other policies that rely on community-based conservation approaches, protected-area approaches, or some combination of the two work best in the context of the Serengeti ecosystem remains an open question. Given the pressures in the system for economic development to satisfy a growing human population, there may be little time left and little margin for error to figure out the means for successful conservation.

International Treaties

Tanzania is party to numerous international agreements that affect management and conservation in the Serengeti ecosystem. These agreements emphasize the need for international cooperation in the protection of migratory, threatened, and endangered species. In its wildlife policy the government of Tanzania reaffirms its cooperation with the international

Table 12.4 International policies potentially applicable to the greater Serengeti ecosystem

---

African Convention on the Conservation of Nature and Natural Resources 1968: Signed by members of the Organization for African Unity, this convention emphasized a need for protected areas and is primarily concerned with preservation of wildlife and other natural resources.

Ramsar Convention 1971: Focuses on wetland and migratory waterfowl conservation, obligating signatories to promote wise use of wetlands. Parties to the treaty are to give wetlands special protection and create "Ramsar Sites," i.e., special wetland reserves.

Convention on International Trade in Endangered Species (CITES) 1973: Governs international trade of endangered species.

Bonn Convention 1979: Focuses on all migratory species, grouping species into conservation categories, much like the appendices of CITES.

World Heritage Listing 1979 and 1981: The United Nations Educational, Scientific, and Cultural Organization (UNESCO) focuses on preserving world/cultural heritages, imposing a legal duty on each party to do its utmost to protect listed sites. Ngorongoro and Serengeti were both added in 1979 and 1981 respectively.

Man and Biosphere Reserve 1981: Aims to perpetuate the living organisms of the earth in all their variety through collaboration between the world's governments and NGOs. Both Ngorongoro and the Serengeti were listed in 1981.

Lusaka Agreement 1994: The objective is to reduce and ultimately eliminate illegal trade in wild flora and fauna and parties are obligated to investigate and prosecute cases of illegal trade as well as provide the task force with regular information and data on illegal trade.

---

community, "committing to playing a truly international and regional role to ensure that successful conservation of wildlife is achieved both within and outside Tanzania" (Ministry of Natural Resources and Tourism 1998, p. 29). Most of these international agreements have only had marginal impact on what actually occurs in the Serengeti ecosystem, either because they are of limited applicability or because they lack adequate enforcement mechanisms. A list of international treaties that are potentially applicable is given in table 12.4. In this section we focus on perhaps the most well known and arguably most influential international policy on wildlife, the Convention on International Trade in Endangered Species of Wild Fauna and Flora (CITES).

The goal of CITES is to eliminate international trade in endangered species by removing legal demand and markets for products derived from these species. In 1973, eighty countries agreed upon CITES rules, and the treaty took effect July 1, 1975. Tanzania ratified the agreement November 29,

1979, and it took effect February 27, 1980. The agreement lists species in one of three appendices, based on their status. Appendix I incorporates species threatened with extinction and prohibits international trade in these species. Appendix II controls international trade so as to prevent nonthreatened species from becoming threatened with extinction due to trade. Appendix III provides a party to the agreement the means to enlist the support of other parties in protecting the export of species not listed in Appendix I or II.

Probably the most controversial case under CITES has been the listing of the African elephant (*Loxodonta Africana*) in Appendix I. In Tanzania, poaching for ivory steadily increased through the colonial era and after independence. At its peak, ivory poaching killed an estimated 4,000 elephants a year, largely around the Selous Game Reserve (Prins, Grootenhuis, and Dolan 2000). Within the Serengeti, poaching peaked in the late 1970s and 1980s (see fig. 2.18 in chapter 2, this volume). Between 1970 and 1986 almost 2,000 elephants either fell prey to poachers (~1,500) or migrated northward (~400–500) into the relatively more secure and protected Mara Reserve in Kenya (Dublin 1995). In 1989, CITES signatories banned all international trade in ivory. After the CITES ban, elephant numbers have continuously risen (fig. 2.18, this volume). Since the ban went into place, in 1989, four southern African countries, South Africa, Zimbabwe, Botswana, and Namibia, successfully appealed to lower the African elephant status from Appendix I to Appendix II to allow a one-time trade from their stockpile of ivory. In 1999 the four countries sold 49.5 metric tons of ivory to Japan, worth approximately US$5 million (Heltberg 2001). While Tanzania cannot sell ivory, it is allocated a quota of elephant kills under CITES. Tanzania's quota was constant from 1997 to 2002 at 50 animals, and doubled to 100 animals for 2003–2005 (CITES Secretariat 2003). Kenya has not received any quota within the last 4 years. Kenya, along with India, has been one of the main opponents of moving the elephant from Appendix I to Appendix II. In both Tanzania and Kenya the elephant remains on Appendix I.

Despite the one-time sale of stockpiled ivory, the listing of elephants under CITES remains controversial and hotly debated. Opponents of downlisting the elephant worry that allowing sales of stockpiled ivory provides loopholes for black market ivory sales. Proponents point out that the controlled sales of ivory provide badly needed conservation funds. Ivory sales in Zimbabwe have already channeled $575,000 back to local communities (Milliken 2003). Proponents also point out that in general, trade bans tend to drive up the price of the banned item and to increase the incentives for poachers (Barbier and Swanson 1990). However, the price of ivory did not in fact increase, as proponents had expected, with the ban, mostly because the ban appears to have reduced demand for ivory. Using a simulation model, Heltberg (2001) found that a trade ban will likely improve the protection of elephants from poaching if "(i) it has a large moral demand-reducing effect; (ii) it facilitates interception of smuggled goods; (iii) there is little ivory from official production piling up; (iv) it does not negatively affect law enforcement effort" (Heltberg 2001, p. 195).

Tanzania and Kenya have shown little interest in downlisting the elephant within their countries. Tanzania's new wildlife policy calls for "instituting the appropriate measure to ensure that the export of CITES species or parts derived from the forestry and fisheries sectors meet the requirements of CITES regulations" (Ministry of Natural Resources and Tourism 1998, p. 28).

ECONOMIC INFLUENCES

Tanzania is a low-income country with an economy heavily tied to agriculture and natural resources. Per capita income in Tanzania was US$290 in 2003, one of the lowest per capita income figures in the world (World Bank 2003). In 2003, Tanzania ranked 187th out of 208 countries in per capita income. Tanzania ranked 206th out of 208 countries when per capita income was measured in terms of purchasing power parity, which adjusts exchange rates so that a bundle of goods costs the same in different countries, trailed only by Malawi and Sierra Leone (World Bank 2003). In 2001, the World Bank estimated that 36% of the population of Tanzania had income below the basic needs poverty line (World Bank 2003).

Agriculture dominates the Tanzanian economy, making up 43% of GDP and employing approximately 80% of the Tanzanian workforce. By comparison, agriculture is about 25% of the economy, on average, in low-income developing countries, and around 15% in sub-Saharan African countries (World Bank 2003). Typical households spend over 60% of their income on food. At present, manufacturing (8%) and mining (2%) make up very small proportions of the economy. The other large sector of the economy is the service sector (40%), which includes tourism.

Though published economic statistics at the regional level are somewhat spotty, it is fair to say that the regions within the Serengeti ecosystem have lower per capita income and are more heavily dominated by agriculture and natural resources than is the country as a whole. The percentage of the population below the basic needs poverty line in the Mara region is 46%, 42% in Shinyanga region, and 39% in the Arusha region. These rates are among the highest in the country (fig. 12.1). In Loliondo and parts of

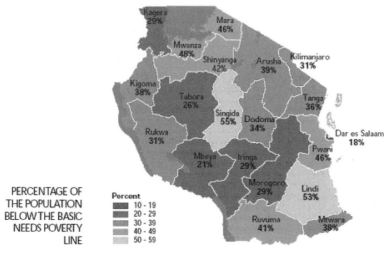

PERCENTAGE OF
THE POPULATION
BELOW THE BASIC
NEEDS POVERTY
LINE

Percent
■ 10 - 19
■ 20 - 29
■ 30 - 39
  40 - 49
  50 - 59

Fig. 12.1 Percentage of population in Tanzania below the basic needs poverty line, by region.

Ngorongoro, which tend to have low rainfall, livestock grazing dominates economic activity. Ngorongoro also receives a significant economic boost from tourism. Areas to the west of Serengeti National Park have higher rainfall totals, and small-scale agriculture dominates economic activity. The percentage of the population involved in farming, livestock, or fishing is 70% in the Mara region and 68% in the Shinyanga region, versus 62% for the country as a whole. There is virtually no industry of any significant size anywhere within the Serengeti ecosystem.

Though Tanzania is currently quite poor, there is potential for rapid economic growth and development. Unlike much of sub-Saharan Africa, Tanzania has been politically stable since independence. Political stability is a necessary ingredient for economic growth. Other ingredients for economic growth include an educated workforce, transportation and communications infrastructure, and low corruption. Over the past few years, the Tanzanian economy has in fact grown steadily and rapidly. The growth rate of the economy from 2000 through 2004 was between 5.5% and 6.3% annually (table 12.5). With increasing links to the rest of the world and improved transportation and communication, Tanzania may be on the verge of sustained rapid economic growth. Of course, much depends on continued political stability, increasing investment in the country's infrastructure, increasing investment in education, and lessening corruption.

While sustained economic growth is of great potential benefit to Tanzania's population, it could have negative environmental consequences.

Table 12.5 Per capita income and economic growth

	2000	2001	2002	2003	2004
GDP per capita, current prices (US$)	270.217	270.784	267.453	264.937	281.571
GDP, current prices (US$ billions)	9.079	9.342	9.414	9.511	10.334
GDP, annual % change, constant prices	5.6	6.1	6.3	5.5	6.3
Inflation, annual % change	6.2	5.2	4.6	5.3	5

Source: International Monetary Fund, World Economic Outlook Database (2003).

Economic growth is necessary to lift Tanzania's population out of poverty and move the country up the ranks in terms of per capita income. However, a fear among conservationists is that without proper restraint, economic growth may lead to increased pollution and environmental degradation that could threaten Tanzania's unique wildlife and ecosystems. Much depends on how the Tanzanian economy grows. If it follows a heavy industrial-manufacturing route to economic development, such as pursued by China over the past decade, environmental degradation will be a consequence. However, given its lack of an industrial base, its abundant wildlife and scenery, it is more likely in Tanzania's case that economic growth will follow a greener path. If so, economic growth could have beneficial environmental consequences. This possibility will be a reality if Tanzania becomes increasingly reliant on a "green" economy, such as ecotourism, which depends on conservation of the country's wildlife and ecosystems. Economic growth will aid rather than hinder conservation if it results in reduced pressure on the land from subsistence agriculture and shifts land use toward tourism and hunting, which depend on sustaining wildlife and ecosystems.

The fastest-growing sector of the Tanzanian economy over the past decade has been tourism. Between 1990 and 2000, revenue from tourism increased by an order of magnitude, from $65 million in 1990 to $739 million in 2000 (fig. 12.2). Tourism accounted for 16% of Tanzanian GDP in 1999. Growth in tourism stopped temporarily after the U.S. terrorist attacks in 2001, but has since resumed. There remains, however, a large, untapped potential from tourism. Tanzania has spectacular wildlife and scenery that is world renowned. The Serengeti ecosystem is the finest example of a large, relatively intact grassland ecosystem, and it contains the largest concen-

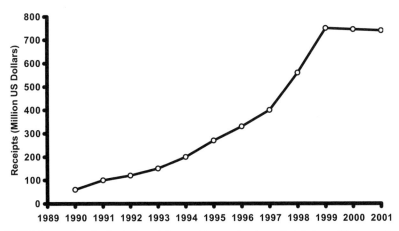

Fig. 12.2 Revenue from tourism in Tanzania increased by an order of magnitude between 1990 and 2000.

tration of large mammals remaining on earth. Yet Tanzania, despite the spectacular increase in international tourism, still earns far less from tourism than South Africa. Tanzania earned US$450 million from international tourism in 2003 while South Africa earned US$5,144 million in international tourism during the same period (World Tourism Organization 2005). Tanzania lags far behind South Africa in part because the transportation and tourism infrastructure is far better developed in South Africa than in Tanzania. For example, there are few paved roads in western Tanzania and only just recently is there a paved road linking the NCA with Kilimanjaro International Airport, near Arusha (table 12.6). For many other parks in Tanzania the roads and infrastructure are even less developed, and consequently revenues from the Serengeti and NCA effectively subsidize the budgets of the country's remaining protected areas.

Growth in tourism is not necessarily beneficial for the environment. Examples of increased tourism leading to nature being "loved to death" exist (Corfield 2000). Increases in roads and vehicle traffic can have negative impacts on wildlife and can cause erosion, air and water pollution, and mar amenities that attract tourists in the first place. Tourist lodges require water, which is already in short supply in many parts of the Serengeti ecosystem. Tourist lodges also generate garbage. Without proper disposal, baboons and other species will become problem animals.

The other controversial aspect of tourism is whether it results in benefits to local people. Receipts from Serengeti National Park go to TANAPA. Until recently there has been little link between international tourism in the park and the economy of villages outside the park. As mentioned previ-

District	Size[a] (km²)	Tarmac (km)	Gravel (km)	Earth (km)	Total (km)	Road density (km/km²)
Karatu[b]	3,300		103	278	381	0.115
Ngorongoro[b]	14,036		315	200	515	0.037
Serengeti[c]	10,942		127	326	453	0.041
Bunda[c]	2,782	39	160	406	605	0.217
Tarime[c]	3,885	81	124	595	800	0.206
Musoma Rural[c]	3,981	59	170	610	839	0.211
Magu[d]	3,070	120	125	991	1236	0.403

Source: Data compiled from The Planning Commission (1997, 1998a, 1998b).
[a]Land area for dry land only except Karatu and Ngorongoro.
[b]Arusha Profile April 1998, data = 1996.
[c]Mara Profile December 1998, data = 1996.
[d]Mwanza Profile October 1997, data = 1997.

ously, TANAPA initiated a community-based conservation project to make more of a link between villages and the park.

Hunting concessions in the game reserves and game-controlled areas within the Serengeti ecosystem have the potential to generate large amounts of revenue. In 1998, tourist hunting generated US$75,330 in Serengeti and Bunda districts, although only 9.4%, or US$7,081, of the revenues were returned to the districts (Holmern et al. 2004). As with tourism, questions and debate surround how the Tanzanian government should divide revenues generated from hunting concessions. SRCP provides money generated from wildlife cropping to communities in the western Serengeti ecosystem. In 2000, these revenues yielded each household 2,300 TZ shillings (Holmern et al. 2004), an amount less than US$3 per year. Unless communities receive more revenues, such programs are unlikely to have much influence on actions taken by local communities.

Human Demographics

The human population in the Serengeti ecosystem has increased rapidly over the past few decades, with a consequent increase in pressures on the land and wildlife. Mainland Tanzania's population has almost tripled from the first post-independence census in 1967 to the most recent census in August 2002, increasing from 11,958,654 to 33,584,607 (Tanzania National

Website 2002). Table 12.7 shows the current population numbers for the districts surrounding the Serengeti. The majority of the population lives in rural villages. Regions in the Serengeti ecosystem have experienced annual average growth rates of between 2 to 4% from 1988 to 2002. Migration patterns explain some of the differences in growth rates. In 1988, both the Arusha and Shinyanga regions had a positive net lifetime migration of 141,724 and 6,763 people, respectively. The Mara and Mwanza regions had a negative net lifetime migration of –80,431 and –33,504 people, respectively (The Planning Commission 1997, 1998a, 1998b).

Increasing population around the park is of concern for several reasons. More people mean that more food is required, which can come from increased agriculture, increased livestock, or increased hunting. Increases in any of these activities will impose some burden on wildlife, though pastoralism tends to impose fewer burdens than agriculture or hunting. Population increases coupled with increased development will likely increase demand for bushmeat (Loibooki et al. 2002). Further, population increases

Table 12.7 Human populations in the Serengeti ecosystem

Region	District	1988 Population	2002 Population	Average annual population growth for district (%)	Annual population growth for region (%)
Mara	Serengeti	113,284	176,057	3.1	
	Bunda	200,870	258,930	1.8	
	Tarime	341,146	490,731	2.6	2.50
	Musoma Urban	68,536	107,855	3.2	
	Musoma Rural	247,106	329,824	2.1	
Shinyanga	Bariadi	382,383	603,604	3.3	
	Meatu	159,439	248,214	3.2	3.30
	Maswa	221,194	304,402	2.3	
Mwanza	Magu	310,918	415,005	2.1	3.20
Arusha	Ngorongoro	68,775	129,776	4.5	4.00
	Karatu	268,129	416,316	3.1	

Sources: United Republic of Tanzania, 2002 Population Housing Census, and The Planning Commission, Mwanza (1997), Arusha (1998a), and Mara (1998b).

mean that the benefits from WMAs will be spread among more people, decreasing the amount of benefits per person. Eventually, areas around the national park will see decreasing returns to their wildlife ownership shifting the balance toward poaching and away from conservation.

Increasing human populations also threaten traditional livelihoods that require extensive land use. The per-person cattle numbers in the Ngorongoro District are decreasing, not because cattle numbers are changing, but because human populations are growing faster than cattle, effectively impoverishing the people. Decreasing cattle-per-person numbers make livestock losses to wildlife more costly and increase the necessity to find alternative forms of food (i.e., turning to illegal hunting). Furthermore, as populations throughout the country continue to increase, humans will require more land for cultivation, creating further pressure and incentives for agriculture over other forms of land use that impose fewer potential conflicts with wildlife.

APPLICATION OF THE MODEL

What does the future hold in store for the Serengeti ecosystem? Can policy influence the future trajectory of the Serengeti in a positive way? To address such questions requires going beyond describing current and past conditions. Here we use the simulation model described in chapter 10 (this volume) to consider potential effects of changes in wildlife policy and changes in assumptions about future human population growth on the dynamics of the Serengeti ecosystem.

Wildlife Policy: Antipoaching Policy and Establishment of Wildlife Management Areas

Wildlife policy has several dimensions, including whether national or local authorities have power to make decisions, who benefits from wildlife management, and how strict to enforce antipoaching policy. In chapter 10 we illustrated use of the model by analyzing a change in fines imposed on people caught poaching. Here we consider a different aspect of antipoaching policy—namely the probability of being caught poaching. In table 12.8 (column 1) we compare outcomes for the base case described in chapter 10, with a case of very lax enforcement that translates to a low probability of being caught poaching (column 2), and with a case of quite strict enforcement that translates to a high probability of being caught while poaching

Table 12.8 Predicted effects of changes in antipoaching policy
(Figures reported are "steady-state values" at the end of a 50-year simulation)

	Base case (1)	Lax antipoaching policy (2)	Strict antipoaching policy (3)
Parameter for being caught per unit of time spent poaching:			
In Serengeti Park	1.0	0.1	1.5
Out of the park	0.2	0.1	0.5
Wildebeest population:			
Resident population	14,890	13,860	20,580
Migrating population	1,257,000	1,137,000	1,284,000
Hunting off-take:			
Resident population	55	526	568
Migration population	20,690	71,440	7,982
Human population	135,700	321,100	64,650
Livestock population	74,240	50,930	42,640

(column 3). Lax enforcement of antipoaching laws increases poaching effort, hunting offtake, and results in lower wildlife populations. The hunting off-take is approximately 72,000 wildebeests in the lax enforcement case compared to just over 20,000 in the base case. Increasing enforcement of antipoaching laws increases resident and migratory populations of wildlife because it lowers hunting effort. Going from lax enforcement to strict enforcement increases the resident wildlife population by approximately 50% while the migratory population increases by about 12% (though in absolute numbers, the increase in the migratory herd is larger than in the resident herd).

What is perhaps more surprising is the effect that antipoaching policy has on the growth of the human population. In the base case, human population quickly rises to 135,000 and remains constant. At this population level, population pressures have reduced many households to the subsistence level, where they just barely get by and cannot grow further. However, with lax enforcement of antipoaching, human population rises to approximately 321,000. Reducing antipoaching effort allows more bushmeat harvest and reduces the number of people caught and fined. Both of these effects increase expected welfare of households, which in turn supports continued population growth to a much higher level. With strict enforcement of antipoaching laws, human population never rises above 80,000 and stabilizes at around 72,000. Under strict antipoaching efforts, there is less hunting and less bushmeat consumption. In addition, with stricter enforcement of antipoaching laws, there are more people caught and fined while poaching, which places additional burdens on local populations. Both of these effects push the level of human population lower than without strict antipoaching laws.

Changes in antipoaching policy also have an impact on livestock numbers. This effect arises primarily from the effect that changes in policy have on household wealth. In this model, where livestock represents the savings accounts of households, the livestock-per-person ratio is a measure of average household wealth. Total livestock numbers are highest in the base case (table 12.8, column 1) but livestock per person is highest in the strict enforcement case (column 3) and quite low in the lax enforcement case (column 2). Lax enforcement allows poorer households without many assets to make a living via poaching, resulting in increased human population, but one that has significantly lower wealth, on average. On the other hand, strict enforcement reduces the number of households that can make a living—but those that do rely more on agriculture and livestock and have somewhat higher wealth compared to the base case. An additional effect on livestock numbers comes from competition between livestock and wildlife for forage. Livestock numbers are somewhat depressed in the strict enforcement case because of greater competition, particularly with increased numbers of resident wildlife.

A different approach to wildlife policy is to give positive incentives to local communities to conserve through the establishment of WMAs rather than to prohibit hunting and punish those caught doing so. With WMAs, local communities can set hunting quotas within their local zone and hunt legally up to the quota. In addition, local communities may share in the revenue generated from trophy-hunting concessions. Table 12.9 shows the results of making these two changes in policy. Column 2 of table 12.9 shows the result when a legal quota for hunting in each zone outside the park is established. In this case, each household can legally take three wildebeests. Allowing for legal hunting up to the quota level outside the park increases hunting pressure outside the park slightly and reduces the number of people fined for poaching. This increases household food intake and wealth slightly compared to the base case. Human population stabilizes at just over 142,000, a slight increase from the base case. What we note from

	Base case (1)	Legal hunting quota (2)	Trophy hunting compensation (3)
Policy parameters:			
Legal hunting quota	None	3 wildebeest per household	3 wildebeest per household
Trophy hunting compensation (per zone)	None	None	$100,000 per zone
Wildebeest population:			
Resident population	14,890	14,890	14,900
Migrating population	1,257,000	1,254,000	1,257,000
Hunting off-take:			
Resident population	55	54	56
Migration population	20,690	22,170	20,770
Human population	135,700	142,100	138,750
Livestock population	74,240	76,950	78,380

this comparison is that allowing a modest legal harvest of wildebeest is unlikely to have much impact on the long-run dynamics of the Serengeti ecosystem. A larger effect of such a quota might occur if it were to make enforcement of illegal hunting more difficult. If enforcement of antipoaching were to decline, we might see the type of effects illustrated in table 12.8.

Similar to changing the hunting quota, distributing revenue generated by trophy hunting is unlikely to have a significant effect on long-run dynamics of the ecosystem unless compensation amounts are much larger than currently anticipated. In column 3, we show the effects of a program that distributes a portion of trophy-hunting receipts equally across all households in a zone. Suppose there are five trophy hunts in each zone each year that generate revenues of $80,000 per hunt, yielding a total of $400,000 annually. Assuming that 25% of the revenue is distributed to local communities, $100,000 per zone annually, and that there are 4,000 households in a zone, each household would receive a payment of $25 per year.

Paying compensation of this amount results in a slight increase in human population in the area. There is also a consequent slight increase in pressures on the ecosystem, as shown in column 3. If compensation were much larger, this might allow many more households to sustain themselves in the area, which would place burdens that are more serious on the Serengeti ecosystem.

Though this model does not include immigration from other areas, high levels of compensation could also be a factor increasing human population close to the park where compensation is paid. This will occur if conditions are significantly better near the park than elsewhere in the region. Drawing people toward wildlife by providing compensation based on wildlife receipts, which would tend to make conserving the wildlife more difficult, is a risk of compensation programs, as critics of community-based programs have discussed (Scholte 2003, Johannesen 2004b). Compensation programs that provide more than a token amount of compensation need to address immigration questions as well as internal population growth in order to successfully conserve wildlife and simultaneously raise the welfare of local people.

Human Population Growth

Changes in human population can have a large impact on the Serengeti ecosystem. In the simulations above (and in previous chapters), the model results show that human population is predicted to increase well above current levels, with consequent increases in the intensity of agricultural production and hunting pressures on wildlife. All of these simulations assume a type of Malthusian population dynamics in which human population grows as long as households are successful at meeting their basic needs. Eventually, however, population growth puts such pressure on what the ecosystem can provide that households can only just barely meet their basic needs. When households reach subsistence level, population growth ceases. Many parts of the world, however, have undergone a demographic transition in which human population growth has slowed, in some cases ceased altogether, as income per capita has risen. Most developed countries have very low population growth rates. Many developing countries are currently experiencing large declines in fertility rates and markedly slowing population growth rates (United Nations 2004).

The type of human population dynamics over the next few decades could potentially have a large impact on the overall dynamics of the Serengeti ecosystem. In table 12.10 we consider the effects of changes in the equation defining human population growth. We compare the base case (column 1) with cases assuming no growth in human population (column 2) and demographic transition dynamics (column 3). Comparing column 1 (with rapid population growth when conditions allow) versus zero population growth (column 2), we see that the rapid increase in human population has some influence on hunting pressure and wildebeest populations, but not as much as might at first be expected. This is due largely to the fact that antipoaching efforts prevent people from over-exploiting wildebeest populations. As discussed above, the combination of a rapid increase in population along with lax enforcement of antipoaching efforts leads to large increases in hunting pressure and a decline in wildebeest populations. Rapid population growth has a more dramatic effect on the welfare of the human population. Livestock per person, a measure of wealth, declines from 2.1 with no population growth (column 2) to 0.55 per person with rapid population growth (column 1). People are far better off in the case with low population growth. The rest of the ecosystem is also better off, though the effect, at least with effective antipoaching efforts in place, is not so dramatic.

While the goal of zero human population growth is advocated by many who care about ecosystems and wildlife, there is typically far less clarity on how this goal might be achieved. One hopeful scenario that potentially leads to a sharp decline in population growth is to assume that developing countries will undergo the same type of demographic transition that has occurred in developed countries, and that this transition occurs soon enough so that ecosystems are not overwhelmed before human population growth ceases. Column 3 of table 12.10 shows the results assuming that such a demographic transition will occur in the Serengeti region. In this scenario, we assume that population growth declines as wealth increases. With zero wealth, the population growth rate is 5%. As wealth increases, the population growth rate declines by 0.25% for each unit increase in wealth, until wealth has increased to the point where population growth falls to zero. As shown in column 3, human population grows to just above 82,000, far less than in the base case (column 1). Wealth—livestock per person—is intermediate between the case with zero population growth (column 2) and Malthusian dynamics (column 1), though closer to the zero-population-growth case. As compared to no population growth, there is more pressure on the ecosystem, because there are somewhat more people, who then hunt more and own more livestock. Except for increased livestock, human impacts on the ecosystem are less in the demographic transition case than in the base case with rapid population growth.

Table 12. 10 Predicted effects of different human population dynamics
(Figures reported are "steady-state values" at the end of a 50-year simulation)

	Base case (1)	Zero population growth (2)	Demographic transition (3)
Population dynamics	10% growth with positive wealth	0% growth	$\alpha - \beta \times$ wealth $\alpha = 0.05$ $\beta = 0.0025$
Wildebeest population:			
Resident population	14,890	14,900	14,900
Migrating population	1,257,000	1,262,000	1,259,300
Hunting off-take:			
Resident population	55	49	52
Migration population	20,690	18,600	19,760
Human population	135,700	63,000	82,230
Livestock population	74,240	132,900	125,200

CONCLUSIONS

Human demographics, policies, and economics play an important role in determining the dynamics of the Serengeti ecosystem. This chapter has examined some of the more important national and international policies affecting the Serengeti, as well as economic and demographic trends. Simple applications of the household economic model described in chapter 10 (this volume) illustrate some of the potential influences that human demographic, policy, and economic variables have on system dynamics. Of these, antipoaching policy was shown to have potentially large effects on human welfare and hunting, while the effects of WMAs, as modeled here, were quite modest. Human demography—in particular, whether there is likely to be Malthusian dynamics or a demographic transition—can also have a large impact on system dynamics.

This analysis represents only a first step in analyzing complex system dynamics that incorporate human behavior and wildlife. Fruitful areas for further research include analysis to capture other important features of the system, such as the role of expansion of agricultural lands on wildlife, and a more realistic look at immigration to, and emigration from, the region.

There are also important interaction effects, such as between climate variability, antipoaching policy, and human demographics, that may have a profound effect on long-run dynamics, for which we have only scratched the surface. Finally, there are other factors, such as cultural dynamics, that may play an important long-term role in the Serengeti. The model results point out intriguing possibilities. Further work in understanding joint human-ecosystem dynamics, along with better socioeconomic data, are necessary before we gain confidence in knowing which possibilities are in fact close to reality.

REFERENCES

Barbier, E. B., and T. Swanson. 1990. Ivory: The case against the ban. *New Scientist* (Nov.) 17:52–54.

Bergin, P. 2001. Accommodating new narratives in conservation bureaucracy: TANAPA and community conservation. In *African Wildlife and Livelihoods,* ed. D. Hulme, and M. Murphree, 88–105. Oxford: James Currey.

Broch-Due, V., and R. A. Schroeder, eds. 2000. Producing nature and poverty in Africa. Uppsala, Sweden: Nordic African Institute.

Campbell, K., V. Nelson, and M. Loibooki. 2001. Sustainable use of wildland resources, ecological, economic and social interactions: An analysis of illegal hunting of wildlife in Serengeti National Park, Tanzania. Report. Chatham, Kent: UK Natural Resources Institute (NRI).

Convention on International Trade in Endangered Species of Wild Fauna and Flora (CITES) Secretariat. Convention on International Trade in Endangered Species of Wild Fauna and Flora. Accessed November 11, 2005, from www.cites.org.

Corfield, T. 2000. The impact of tourism on the Serengeti ecosystem. Retrieved from www .serengetipark.org.

Dublin, H. T. 1995. Vegetation dynamics in the Serengeti-Mara ecosystem: The role of elephants, fire, and other factors. In *Serengeti II: Dynamics, management, and conservation of an ecosystem,* ed. A. R. E. Sinclair and P. Arcese, 71–90. Chicago: University of Chicago Press.

Heltberg, R. 2001. Impact of the ivory trade ban on poaching incentives: A numerical example. *Ecological Economics* 36:189–95.

Holmern, T., A. B. Johannesen, J. Mbaruka, S. Mkama, J. Muya, and E. Roskaft. 2004. Human-wildlife conflicts and hunting in the western Serengeti, Tanzania. Report. Trondheim: Norwegian Institute for Nature Research (NINA).

Hulme, D., and M. Murphree, eds. 2001. *African wildlife and livelihoods.* Oxford: James Currey.

International Monetary Fund. 2003. World Economic Outlook Database. Retrieved from http://www.imf.org/external/pubs/ft/weo/2003/02/data/index.htm.

International Resources Group, Ltd. 2000. The case of the Serengeti Regional Conservation Strategy Serengeti District, Arusha Region, Tanzania. Washington, DC: International Resources Group.

Johannesen, A. B. 2004a. Designing integrated conservation and development projects: Hunting incentives and human welfare with numerical illustrations from Serengeti. Working paper no 2. Trondheim: Department of Economics, Norwegian University of Science and Technology.

Johannesen, A. B. 2004b. Wildlife conservation policies and incentives to hunt: An empirical analysis of illegal hunting in western Serengeti, Tanzania. Working paper no. 3. Trondheim: Department of Economics, Norwegian University of Science and Technology.

Kiss, A., ed. 1990. *Living with wildlife resource management with local participation in Africa.* Washington, DC: The World Bank.

Lissu, T. A. 2000. Policy and legal issues on wildlife management in Tanzania's pastoral lands: The case study of the Ngorongoro Conservation Area. *Law, Social Justice & Global Development* 1 (online journal).

Loibooki, M., H. Hofer, K. L. I. Campbell, and M. L. East. 2002. Bushmeat hunting by communities adjacent to the Serengeti National Park, Tanzania: The importance of livestock ownership and alternative sources of protein and income. *Environmental Conservation* 29:391–98.

Mabugu, R., and P. Mugoya. 2001. Financing, revenue-sharing, and taxation issues in Wildlife Management Areas: Intelligent solutions for complex resource problems. Report. Dar es Salaam:Wildlife Division, Ministry of Natural Resources and Tourism and USAID/Tanzania.

McCabe, J. T., S. Perkin, and C. Schofield. 1992. Can conservation and development be coupled among pastoral people? An exaination of the Maasai of the Ngorongoro Conservation Area, Tanzania. *Human Organization* 51:353–366.

Mduma, S., and J. Hando. 2003. Alternatives to anti-poaching. Retrieved from http://www.serengeti.org/download/Antipoaching.pdf.

Metcalfe, S., B. Kaare, V. Shauri, R. A. Rugemeleza, and T. A. Lissu. 1998. Socio-legal analysis of community-based conservation in Tanzania: Policy, legal, institutional and programmatic issues, considerations and options. Report no. 1. Dar es Salaam, Tanzania: Lawyers' Environmental Action Team (LEAT).

Milliken, T. 2003. African elephants and the eleventh meeting of the conference of the parties to CITES: Trade Records of Flora and Fauna In Commerce (TRAFFIC). Retrieved from http://www.traffic.org.

Ministry of Natural Resources and Tourism. 1998. *The Wildlife Policy of Tanzania.* Dar es Salaam, Tanzania: Ministry of Natural Resources and Tourism.

National Bureau of Statistics Tanzania. 2002. Household Budget Survey 2000/01: Final report. Retrieved from http://www.nbs.go.tz/publications/index.htm.

Neumann, R. P. 2000. Local challenges to global agendas: Conservation, economic liberalization, and the pastoralists' rights movement in Tanzania. In *Preserving wildlife: An international perspective,* ed. M. A. Michael, 167–88. New York: Humanity Books.

Oates, J. F. 1999. *Myth and reality in the rain forest.* Berkeley: University of California Press,.

Planning Commission. 1997. Mwanza Region Socio-Economic Profile. Dar es Salaam, Tanzania: Dar es Salaam and Regional Commissioner's Office, Mwanza. Retrieved from http://www.tzonline.org/pdf/Mwanza.pdf.

———. 1998a. Arusha Region Socio-Economic Profile. Dar es Salaam, Tanzania: Dar es Salaam and Regional Commissioner's Office, Arusha. Retrieved from http://www.tzonline.org/pdf/Arusha.pdf.

———. 1998b. Mara Region Socio-Economic Profile. Dar es Salaam, Tanzania: Dar es Salaam and Regional Commissioner's Office, Mara. Retrieved December, 2003, from http://www.tzonline.org/pdf/Maraeng.pdf.

Prins, H. H. T., J. G. Grootenhuis, and T. T. Dolan, eds. 2000. *Wildlife conservation by sustainable use.* Boston: Kluwer Academic.

Scholte, P. 2003. Immigration: A potential time bomb under the integration of conservation and development. *Ambio* 32:58–64.

Schwartzman, S., A. Moreira, and D. Nepstad. 2000. Rethinking tropical forest conservation: Perils in parks. *Conservation Biology* 14:1351–57.

Shauri, V. 1999. The new Wildlife Policy in Tanzania: Old wine in a new bottle? Retrieved September, 2003, from http://www.leat.or.tz/publications/wildlife.policy.

Shauri, V., and L. Hitchcock. 1999. Wildlife corridors and buffer zones in Tanzania. Retrieved September, 2003, from http://www.leat.or.tz/publications/wildlife.corridors.

Tanzania National Website. 2002. Population and Housing Census. Retrieved December, 2003, from http://www.tanzania.go.tz/census.

Terborgh, J. 1999. *Requiem for nature.* Washington, DC: Island Press.

Thirgood, S., A. Mosser, S. Tham, G. Hopcraft, E. Mwangomo, M. Mlengeya, M. Kilewo, J. Fryxell, A. R. E. Sinclair, and M. Borner. 2004. Can parks protect migratory ungulates? The case of the Serengeti wildebeest. *Animal Conservation* 7:113–20.

United Nations (UN). 2004. State of World Population 2004: The Cairo Consensus at 10—Population, Reproductive Health and the Global Effort to End Poverty. New York: United Nations.

United Nations Environment Program—World Conservation Monitoring Center (UNEP-WCMC). 2004. World Database on Protected Areas. Retrieved from http://sea.unep-wcmc.org/wdpa/download/wdpa2004/index/html.

Wilshusen, P. R., S. R. Brechin, C. L. Fortwangler, and P. C. West. 2002. Reinventing a square wheel: Critique of a resurgent "protection paradigm" in international biodiversity conservation. *Society and Natural Resources* 15:17–40.

World Bank. 2003. *The world development indicators.* Washington, DC: International Bank.

World Tourism Organization (WTO). 2005. *WTO Tourism Highlights, 2005 Edition.* Retrieved from http://www.world-tourism.org/facts/menu.html.

THIRTEEN

Land Use Economics in the Mara Area
of the Serengeti Ecosystem

Mike Norton-Griffiths, Mohammed Y. Said, Suzanne Serneels,
Dixon S. Kaelo, Mike Coughenour, Richard H. Lamprey,
D. Michael Thompson, and Robin S. Reid

The northern part of the Serengeti/Mara ecosystem falls within the two Kenyan districts of Narok and Trans Mara. Within these two districts, major land use changes have been clear for quite some years. Specifically, the spread of large-scale, mechanized, and small-scale agriculture (Norton-Griffiths 1996; Homewood et al. 2001; Serneels 2001; Serneels and Lambin 2001a, 2001b), the concentration of settlements and livestock around the Maasai Mara National Reserve (Lamprey and Reid 2004), the transition of land tenure from group or communal ownership to individual ownership (Thompson 2002; Thompson and Homewood 2002) all lead to the alienation of land previously used by both resident and migratory wildlife and to the gradual loss of wildlife abundance and diversity (Broten and Said 1995; Norton-Griffiths 1995; Norton-Griffiths 1996; Otichello 2000; Said et al. 1997; Serneels and Lambin 2001a).

The importance of the Mara portion of the Serengeti ecosystem as a critical source of dry-season grazing for the Serengeti migratory wildebeest population has long been noted (Hilborn and Sinclair 1979; Mduma, Sinclair, and Hilborn 1999; Sinclair and Norton-Griffiths 1982; Sinclair 1995, Sinclair, Dublin, and Borner 1985). Indeed, with wildebeest being limited primarily by the supply of food in the dry season, any loss of dry-season habitat could have an inordinate effect on this population which is in itself the main driving force of the Serengeti ecosystem (Sinclair et al. chapter 2, this volume). Previous work (Hilborn 1995; Norton-Griffiths 1996; Sinclair 1995) indicates that losing what are called the group ranches that surround the Maasai Mara National Reserve and the Mara Triangle could cause a permanent loss of perhaps 20% of the migratory wildebeest, which in turn could trigger major changes in the Serengeti ecosystem itself.

These changes in land use are perplexing in view of the large revenues generated from wildlife tourism which should encourage investment in conservation on the part of landowners rather than land development. This chapter examines the apparent contradiction between the revenues generated by wildlife, on the one hand, and the land use changes initiated by the Maasai on the other. Our objective is to describe the economic conditions of the Mara portion of the Serengeti ecosystem within which individual economic agents—Maasai households—are making decisions about land use investment and development.

Our perspective is that outside of Protected Areas, conservation, in the sense of maintaining viable populations of large ungulates and their predators, is a matter of land use economics driven by the differential returns from the alternative land uses which are open to Maasai landowners, namely agricultural, livestock, and wildlife production, under conditions of a (mostly) free-market economy and private land tenure.

THE STUDY AREA

The circa 12,800 km^2 area selected for this study (fig. 13.1) falls within the Mara River Valley, from the highlands of the Isuria escarpment to the northwest to the Loita Hills in the southeast, and from Narok town on the Loita plains to the northeast to the Kenya/Tanzania border to the southwest. It includes two protected areas (the Maasai Mara National Reserve and the Mara Triangle, from now together referred to as the Mara Reserve) and the Group Ranches surrounding the Mara Reserve. For the purposes of this study this area is now referred to as the Mara Area.

Descriptions of the topography and vegetation of the Mara area abound (Epp and Agatsiva 1980; Dublin 1995; Serneels 2001). Of specific relevance to this study is the marked rainfall and gradient from the approximately 500 mm annual rainfall on the dry southeastern plains of Siana to the 1,200 mm annual rainfall to the northwest.

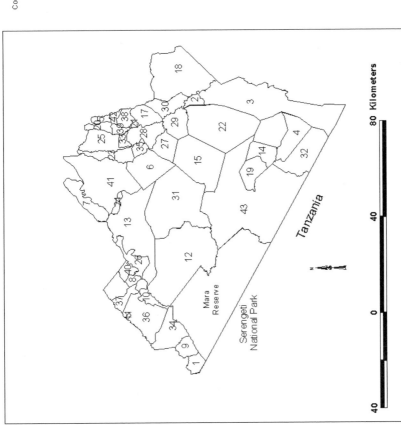

Code   Group Ranch
1 Ang'ata Barrekoi
2 Enatamatishoreki
3 Enkutoto Elang'at' enterit
4 Entakesera
5 Eorr Enkitok
6 Ewaso Ngiro
7 Ilmotiok
8 Intulele
9 Kerinkani
10 Kimintet
11 Kimintet-a
12 Koyaki
13 Lemek
14 Leshota
15 Maji Moto
16 Morijo Lota
17 Morijo Narok
18 Mosiro
19 Naikara
20 Naisoya
21 Narok Township
22 Narosura
23 Nkoben
24 Nkorrkorri
25 Nkareta
26 OlChoro-Oirowua
27 Oldonyo Orasha
28 Oleleishwa
29 Olenkuluo
30 Oletukat
31 Olkinyei
32 Olmesutye
33 Olaimutiai
34 Oloirien
35 Iloisiusiu
36 Oloonkoliin
37 Oloonkoliin
38 Olorroito
39 Oloroito
40 Olosakwana
41 Osupuko-Ololunga
42 Rotian
43 Siana

Fig. 13.1 Map of the group ranches and the Masai Mara National Reserve, Kenya.

LAND USE CHANGES IN THE MARA AREA SINCE THE 1970S

Agricultural Rents in Kenya

One of the key concepts to understanding the dynamics of agricultural land use change is that of *agricultural rents,* which are defined as the net returns to land from all agricultural activities, both large scale and small scale (Norton-Griffiths 1996; Norton-Griffiths and Butt 2006; Norton-Griffiths and Southey 1993). These net returns are strictly financial, in that they represent the *net cash returns* to landowners from their agricultural activities. They are, so to speak, cash-in-the-bank at the end of the season, or at least cash-in-the-pocket, once all the *direct expenses of production and marketing have been met.* Agricultural rents are therefore something quite different from agricultural revenues, which represent the gross returns from agricultural production before deducting the expenses of production and marketing.

Agricultural rents for Kenya have been derived from a series of aerial point sampling surveys (Norton-Griffiths 1988) dating from the 80s that covered 20 districts and some 150,000 km² of Kenya; from highland areas receiving in excess of 2,000 mm of annual rainfall to low-lying, arid areas receiving less than 200 mm (Norton-Griffiths and Southey 1993).

Some 27,500 georeferenced, 35mm large-scale (1:6,000) sample aerial color photographs were taken in the course of these surveys. Percent cover, equivalent to hectares per square kilometer, were measured from each sample photograph at an interpretation scale of 1:600 for all crops and crop combinations (Ecosystems Ltd. [ESL] 1987). Standard GIS methods were used to associate each georeferenced photograph with a wide range of physical, administrative, environmental, and other spatially organized data.

Net returns from each agricultural land use were derived from the Policy Analysis for Rural Development (PARD) crop budget methodology as applied by the Tegemeo Institute of Edgerton University, Kenya (Monke and Pearson 1989; Policy Analysis for Rural Development (PARD) 1991; Sellen 1991). For any given agricultural activity, for example, small-scale pure maize in agroecological zone UH3 (Jaetzhold and Smith 1982), all indirect and direct costs are itemized, including labor, land preparation, machinery and other inputs, and, where appropriate, costs of transport, marketing, and sales. Not included are the costs of acquiring or converting land, of land mortgages, of depreciation, or of taxes (apart from local cess where appropriate).

Any individual crop budget may be based on up to 50 farmer interviews for a given district and agroecological zone. Budgets reflect the range of variation from good, average, and bad years, from intercropping and from double cropping. In the final crop budget, gross revenues were found by multiplying yields per hectare per season by contemporary (mid 80s) farm-gate prices, and net returns were found by subtracting from them the itemized costs of production and sales. Finally, agricultural rents were derived from the individual net returns of every crop and crop combination by averaging the data from the sample photos falling within individual 5 km Universal Transverse Mercator (UTM) grid cells (6–7 samples per cell in the high-potential areas, 2–3 samples in the low-potential areas). For this study, all values were discounted to a base year of 2002 and expressed as US$ per hectare per year ($ ha$^{-1}y^{-1}$).

Agricultural rents reflect long-term net returns to land rather than the net returns in any specific year, much in the same way that climate reflects long-term averages rather than the actual weather experienced ay any specific time. Agricultural rents are closely related to mean annual rainfall, rising sharply up to 1,000 mm after which they plateau and then decline in the face of the lower temperatures found at the higher altitudes in Kenya (Norton-Griffiths and Southey 1993).

Agricultural Rents in the Mara Area

The relationship between agricultural rents and mean annual rainfalls up to 900 mm is given by Norton-Griffiths and Butt (2006) as:

$$\ln(y) = 0.836 + 0.00474X \qquad\qquad 13.1$$

where $\ln(y)$ is the natural log of the agricultural rents in $ ha$^{-1}y^{-1}$ and X is the mean annual rainfall in millimetres ($n = 958$, adjusted multiple $r^2 = 0.29$, $t = 14.98$, $p = 0.000$).

For the Mara Area (fig. 13.1), mean annual rainfalls were derived from the AWher-ACT Kenya Database (Mudsprings Inc. 1999) on a 5.5 × 5.5 km grid. A topographically dependent climate surface was fitted to the point data and interpolated at a 1 km resolution using elevation. Agricultural rents within the Mara Area were then derived by applying eq. 13.1 to each interpolated point, but were set at $240 ha$^{-1}y^{-1}$ for annual rainfalls above 900 mm, which lay outside the range of the regression line (Norton-Griffiths and Butt 2006). Finally, the surface for agricultural rents was then subdivided within a GIS environment into bands of higher and lower rents (table 13.1A).

After adjusting for the land conserved within the Mara Reserve, there is

Table 13.1 Capture of agricultural rents in the Mara Area

1A: Agricultural rents[a]

Bands ($)	< 50	50–100	100–150	150–200	> 200	Totals
Mid point ($)	25	75	125	175	249	
Hectares	27,109	336,343	191,950	202,593	530,722	1,288,717
Conserved hectares [b]	0	0	0	4,941	144,577	149,518
Available hectares [c]	27,109	336,343	191,950	197,652	386,145	1,139,199
Available rents (US$millions)	0.68	25.23	23.99	34.59	96.15	180.64

1B: Large-scale mechanized cultivation (Serneels 2002)

	< 50	50–100	100–150	150–200	> 200	Totals
1975 ha	0	0	0	0	4,002	4,002
1985 ha	0	0	837	4,451	13,867	19,155
1995 ha	0	4,808	8,998	7,459	18,849	40,114
2000 ha	0	2,540	4,691	5,459	13,382	26,072
2003 ha	0	3,526	2,907	3,030	15,432	24,895

1C: Comparison between Serneels (2000) and Africover (2000)

	< 50	50–100	100–150	150–200	> 200	Totals
Serneels 2000 ha	0	2,540	4,397	5,315	14,207	26,459
Africover large-scale area	0	9,044	20,666	12,046	31,054	72,810

1D: Total area under cultivation in the Mara Area, 2000, from Africover [d]

	< 50	50–100	100–150	150–200	> 200	Totals
Total cultivation (hectares)	0	11,634	16,019	10,832	54,011	92,496
Rents captured ($ millions)	0.00	0.87	2.00	1.90	13.45	18.22
% Rents captured	0.0	3.5	8.3	5.5	14.0	10.1

1E: Estimated total crop cover density in the Mara Area (ha km^{-2} = % cover) [e]

	< 50	50–100	100–150	150–200	> 200	Totals
1975 ha km^{-2}	0.0	0.0	0.0	0.0	3.8	1.2
1985 ha km^{-2}	0.0	0.0	2.0	8.3	12.7	6.1
1995 ha km^{-2}	0.0	5.2	16.8	13.6	17.2	12.6
2000 ha km^{-2}	0.0	2.9	9.0	10.1	12.3	8.3
2003 ha km^{-2}	0.0	3.9	5.7	5.8	14.1	7.9

Note: (a) from eq. 13.1; (b) Maasai Mara National Reserve and Mara Conservancy, (c) for use by wildlife, livestock, settlements, and agriculture; (d) corrected from eqs. 3.2 and 13.3; (e) hectares from eq. 13.4, then converted to ha km^{-2} (equivalent to % cover).

a total of 1.139 million hectares available for use and cultivation within the Mara Area. The great majority of this land remains undeveloped for agriculture. Nonetheless, the agricultural rents which are potentially available for capture (by developing the land for cultivation) are worth some $181 million. These rents are not distributed evenly but are concentrated in the areas of higher potential; for example, the highest band of >$200 ha$^{-1}y^{-1}$ covers only 34% of the land area but provides 52% of the potential rents.

Capture of Agricultural Rents in the Mara Area

The magnitude of the agricultural rents available for capture—some $460m in Narok district as a whole and $181m in the Mara Area—creates strong incentives for landowners to develop their land for cultivation. Irrespective of how good an individual might be at farming, or whether they farm the land themselves or rent it out to entrepreneurs, the incentives are there to develop the land, especially the land with high potential rents. Equally, incentives are there for outsiders to alienate the land from its traditional owners, the pastoral Maasai.

The District Agricultural Reports for Narok District suggest that this is indeed the case, for the area under cultivation has been expanding at an average rate of 8.5% pa over the last 30 years (fig. 13.2) with some 120,000 hectares under cultivation by 2001 (see also Homewood et al. 2001; Thompson 2002). Narok is no different in this respect than any other of the 19 arid and semi-arid land (ASAL) districts in Kenya, all of which report similar significant expansion of cultivation over the last 30 years (Norton-Griffiths 1998; Norton-Griffiths and Butt 2006). However, these district-level data cannot be disaggregated to the level of the Mara Area, nor do they distinguish between large-scale and small-scale cultivation.

Large-Scale, Mechanized Cultivation

Serneels (2001) mapped the spread of large-scale, mechanized cultivation (mainly wheat, farmed by contractors with short-term land leases) for the years 1975, 1985, 1995 and 2000, and again in 2003, using satellite imagery and ground control (table 13.1B). There was an initial strong growth over the 20 years from 1975 to 1995, from 4,000 ha to 40,000 ha, clearly concentrated in the areas of highest agricultural rents. Subsequently there was a marked decline to around 25,000 ha, especially in the more marginal areas of lower agricultural rents.

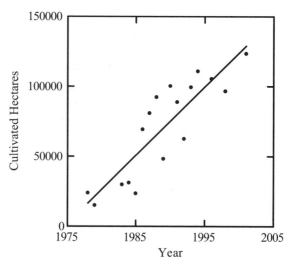

Fig. 13.2 Increase in cultivation in Narok district, 1977–2001.

This decline was due to a combination of factors, including lower rainfall, falling prices, and rising costs (as subsidies were removed), and especially the subdivision of land, so contractors were faced with negotiating leases with many individual agents (the owners of smaller land parcels) rather than with just a few agents acting on behalf of entire group ranches. The decline was accompanied by the removal of fences, to allow livestock to move back onto land that had been previously cultivated (Homewood et al. 2001; Thompson 2002; Thompson 2005).

More recently, better rainfalls and stronger prices in both 2004 and 2005 have led to an increase in the hectares under mechanized farming, but with smaller enterprises, many of which are now farmed by individual Maasai landowners themselves.

Nonetheless, despite the vagaries of climate, prices, subsidies, and tenure, it is clear that the areas of higher agricultural rents were developed sooner, more completely, and more permanently than were the areas of lower agricultural rents (Thompson, Serneels, and Lambin 2002; Thompson 2005).

Total Area under Cultivation (Africover)

There are extensive areas of small-scale cultivation within the Mara Area apart from the large-scale, mechanized cultivation mapped by Serneels—for example, along the Isuria escarpment, where some 16,000 ha were described in the mid '80s (ESL 1985). The Africover Project (Africover 2003) mapped all land uses, including both large-scale and small-scale cultivation, across the whole land area of Kenya for the base year of 2000, using a combination of satellite imagery, aerial photography, and extensive ground control. Potentially, therefore, the Africover database could be used to find the total extent of cultivation within the Mara Area. However, unlike Serneels, Said, and Lambin (2001), Africover does not map the actual hectares cultivated but rather the area within which certain types of cultivation occur. For example, a typical polygon in the Africover database might have the description *"isolated (in natural vegetation or other) rainfed herbaceous crop (field density 10–20% polygon area),"* along with an estimated area of the polygon. Clearly, the Africover areas had to be adjusted.

In the areas analyzed and mapped by Serneels (2001), the relationship with Africover (table 13.1C) was:

$$y = -221.88 + 0.373X \qquad\qquad 13.2$$

where y are the hectares of mechanised cultivation mapped by Serneels (op. cit.) and X are the hectares of large-scale cultivation as mapped by Africover ($n = 5$, adjusted multiple $r^2 = 0.95, p = 0.001$). Equation 13.2 was then used to correct the Africover estimates of large-scale cultivation in the Mara Area as a whole.

For small-scale cultivation, which Serneels did not map, the general relationship between small- and large-scale cultivation mapped by Africover in the whole Mara Area was:

$$y = 4392.78 + 0.443X \qquad\qquad 13.3$$

where y are the areas of small-scale cultivation and X the areas of large-scale cultivation mapped by Africover ($n = 5$, adjusted multiple $r^2 = 0.735, p = 0.018$). Equation 13.3 was used to estimate the hectares of small-scale cultivation from the corrected hectares of large-scale cultivation mapped by Africover.

The areas of agricultural land use for the Mara Area, including both large-scale and small-scale cultivation, were extracted from the Africover database and converted to hectares using equations 13.2 and 13.3 (table 13.1D). By the year 2000, a total of some 92,000 hectares had been cultivated within the Mara Area, capturing some $18.2 million of agricultural rents, the great majority of which are in the band of highest rents. This represents only some 10% of the total agricultural rents available for capture, so clearly the process of land conversion is still at an early stage.

A third regression (eq. 13.4) relates the total area under cultivation in the Mara Area in 2000 to the area mapped by Serneels in 2000.

$$y = 882.44 + 3.479X \qquad\qquad 13.4$$

where y are the corrected estimated areas of large and small scale cultivation mapped by Africover in 2000 and X are the areas of mechanized cultivation mapped by Serneels in 2000 ($n = 5$, adjusted multiple $r^2 = 0.98$, $p = 0.000$).

Equation 13.4 was used to estimate the total area under cultivation in the Mara Area from the areas mapped by Serneels in each of the five surveys (table 13.1B), expressed in table 13.1E as cover density (ha km^{-2}, equivalent to percent cover). All these data suggest that the areas of higher agricultural rents are being taken up sooner and more completely than are the areas of lower agricultural rents, and that the greater proportion of the land that has been converted to cultivation lies in the areas of higher agricultural rents (Serneels and Lambin 2001a, 2001b; Serneels, Said, and Lambin 2001).

Livestock Rents in the Mara Area

As with agricultural rents, livestock rents are defined here as the net returns from livestock operations expressed as \$ ha$^{-1}y^{-1}$ for the base year of 2002. Norton-Griffiths and Butt (2006) derived an empirical relationship between the net returns to livestock and mean annual rainfall (eq. 13.5) from the analysis of 14 livestock operations in Kenya, ranging from intensively managed livestock embedded within high-potential, high-rainfall agricultural areas, to extensively managed livestock in arid pastoral rangelands. The 14 studies covered annual rainfalls from 1,200 mm to 200 mm.

$$\ln(y) = -0.943 + 0.00542X \qquad\qquad 13.5$$

where $\ln(y)$ is the natural log of the livestock rents in \$ ha$^{-1}y^{-1}$ and X is the mean annual rainfall in millimeters ($n = 13$, adjusted multiple $r^2 = 0.699$, $t = 5.368$, $p = 0.000$).

Livestock rents within the Mara Area were derived by applying eq. 13.5 to the rainfall surface and then dividing into bands of increasing livestock rents. After adjusting for the areas conserved within the Mara Reserve, the livestock rents potentially available within the same 1.139 m hectares of available land amount to some \$35 million (table 13.2), 20% of the potential

Table 13.2 Livestock rents in the Mara Area

Livestock rents ($)	< 10	10–20	20–30	30–50	> 50	Totals
Mid-point ($)	9	15	25	40	70	
Hectares	179,406	281,298	288,052	287,922	252,040	1,288,718
Conserved hectares	0	0	2,651	92,662	54,206	149,519
Available hectares	179,406	281,298	285,401	195,260	197,834	1,139,199
Available rents ($millions)	1.61	4.22	7.14	7.81	13.85	34.63
Mean livestock density	23.00	24.81	31.91	31.24	40.93	30.21
% total livestock	11.99	20.28	26.47	17.73	23.53	
% land	15.75	24.69	25.05	17.14	17.37	

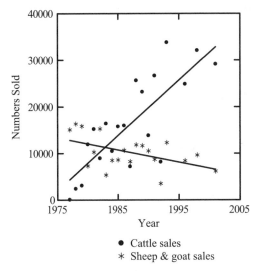

● Cattle sales
∗ Sheep & goat sales

Fig. 13.3 Livestock sales (thousands) in Narok district, 1977–2001.

agricultural rents (table 13.1A). There is a slight tendency for livestock densities to be higher in the areas of higher rents, especially in the > \$50 ha^{-1}y^{-1} band; otherwise, they are distributed evenly across the rent gradient.

Census data from the Mara Area suggest relatively stable livestock numbers over the last 30 years (Broten and Said 1995; Norton-Griffiths 1995, 1996, 1998; Lamprey and Reid 2004). However, District Livestock Reports suggest that while cattle sales are rising year after year, smallstock sales are in decline (fig. 13.3). Narok district is again similar in this regard to the 19 ASAL districts in Kenya, all of which demonstrate, over the last 30 years, a similar strong growth in cattle sales but a fall in smallstock sales, with relatively stable populations of both (Norton-Griffiths 1998; Norton-Griffiths and Butt 2006).

For smallstock, these falling sales possibly indicate an increasing use for home consumption and for local trade and barter. With cattle, however, the rising sales indicate perhaps a fundamental shift in production strategy away from extensive, subsistence production toward more intensive, market-oriented production.

Wildlife Rents in the Mara Area

In principle, wildlife rents can be defined in the same way as can agricultural and livestock rents, namely as the net returns to landowners from the provision of wildlife goods and services, expressed as $ ha$^{-1}y^{-1}$ for the base year of 2002. The situation is, however, a bit more complicated as far as wildlife rents are concerned, primarily from major distortions within the market for wildlife goods and services.

First is a policy distortion that denies to landowners highly profitable sources of wildlife rents from consumptive utilization, including sport hunting, cropping, bird shooting, and the capture and sale of wildlife. The impact of this is not just lost revenues, but lost revenues specifically in areas that cannot support wildlife viewing. Before the ban on consumptive utilization in 1977, it was possible to generate wildlife revenues throughout Narok District, for all of the land was divided into hunting blocks. In contrast, wildlife viewing today is confined to some 5% of the district around the Mara Reserve.

Further distortions arise from the actions of the tourism cartels that control the supply of clients. First, these cartels divert the greater proportion of wildlife rents away from the producers of the resource, the Maasai landowners, to tourist agents and to the providers of transport and accommodation services. As a result, landowners find it difficult to capture more than 5% to 10% of wildlife rents (Norton-Griffiths and Butt 2006). Furthermore, the market is highly inefficient, especially with respect to the flow of information between bargaining parties, resulting in marked discrepancies between the rents obtained for essentially the same services: for example, while one land parcel might receive $40 ha$^{-1}y^{-1}$ for granting access to tour operators, a second, perhaps only a few kilometers away, might receive $5 ha$^{-1}y^{-1}$. Second, by erecting barriers against local landowners becoming involved in the tourism industry, except at a trivial level (e.g., guiding), these cartels restrict the capacity of landowners to capture a greater proportion

Table 13.3 Wildlife rents in Kenya—basic statistics

	Rents	Ln(rents)	US\$ ha$^{-1}y^{-1}$
Number cases	63	63	
Median	\$5.83	1.763	5.83
Mean	\$16.96	1.365	3.92
Standard error	\$2.99	0.272	
95% CL upper		1.907	6.73
lower		0.822	2.28
Skewness	1.584	−0.414	
Kurtosis	1.640	−0.781	
SE Kurtosis	0.595	0.595	

of the available wildlife rents by, for example, becoming involved in the transport sector of the industry.

Norton-Griffiths and Butt (2006) analyzed 63 examples of wildlife rents received by landowners in Kenya (from 22 sources; Lamprey 2006; Thompson and Homewood 2002). Rents from public conservation include protected-area revenue sharing schemes with the Kenya Wildlife Service (e.g., the revenue sharing schemes around Amboseli National Park) and with county councils (e.g., the 19% of revenues disbursed to nine group ranches by the Narok County Council), and revenues from NGO and private community conservation programs. Rents from private conservation include both nonconsumptive and consumptive utilization of wildlife. Nonconsumptive utilization included concession and/or access fees with tour and lodge/camp operators (including bed night fees, local employment, etc.), and fees from simple camp sites, cultural bomas, and so on. Consumptive utilization included bird shooting and wildlife cropping, both of which are now banned by the Kenya Wildlife Service.

The statistical distribution of the wildlife rents was highly skewed, with the great majority being less than \$1 ha$^{-1}y^{-1}$ (table 13.3). Converting rents to natural logs made the distribution more normal, with median and mode acceptably close. On average, wildlife rents received by landowners in Kenya are \$5.83 ha$^{-1}y^{-1}$, with 95% confidence limits falling between \$6.73 ha$^{-1}y^{-1}$ and \$2.28 ha$^{-1}y^{-1}$.

It was clear from the data that there were marked differences between rents from different sources: for example, concession fees compared with wildlife cropping. However, after much testing, only two groups could be consistently distinguished from one another—concession and access fees on the one hand, and rents from all other sources on the other, including

Table 13.4 Wildlife rents from concession and access fees versus all others

Rents	n	Mean	SD	US$ $ha^{-1}y^{-1}$
Concession and access fees	44	2.323	1.666	10.21
All other sources	22	−0.640	1.486	0.53

Note: Pooled variance: $t = 7.053$, df $= 64$, $p = 0.000$. Rents are logged.

Table 13.5 Potential wildlife rents in the Mara Area

Mara Area	Wildlife rents	Total potential rents (US$ million)	
	$ha^{-1}y^{-1}$ ($)	Protected areas ($)	Group ranches
Highest Mara	50	7.5 m	57.0 m
Average for concession and access fees	10.21	na	11.6 m

simple campsites, cultural bomas, bird shooting, cropping, revenue sharing schemes, and so on (table 13.4). At $10.21 $ha^{-1}y^{-1}$, rents from concession and access fees are some 19 times higher than are the average rents of $0.53 $ha^{-1}y^{-1}$ from all other sources.

Although the average wildlife rents captured by landowners in the Mara are quite low, some are in fact very much higher. However, these seem to be both unreliable and highly variable, especially in response to fluctuations in tourist numbers. For example, in the good tourist years of 1999 and 2003, the wildlife rents received by the Narok and Trans Mara County Councils averaged $52 ha^{-1} (Lamprey 2006; Thompson and Homewood 2002) compared with US$11 ha^{-1} 10 years previously. Similarly, wildlife rents on some well-organized Group Ranches ranged from $35–$45 ha^{-1} in these same years, while on others it was about $2 ha^{-1}.

In general, however, the wildlife rents available for landowners, including the county councils, in the Mara area are both modest and highly variable. The total wildlife rents available on the Group Ranches could be as high as $57 million at the highest recorded rents, or as low as $11.6 million at the average rents for concession and access (table 13.5). For the protected areas, rents vary between good and bad years from $11.7m to $1.5m. This variability and uncertainty are all indicative of a poorly functioning and inefficient market. As a result, total wildlife rents on the Group Ranches compare quite unfavorably with the $35m from livestock and the $181m from agriculture (Thompson 2005).

Settlements

Lamprey and Reid (2004) mapped settlements (occupied bomas) from aerial photographs within a 1,827 km² study area to the north and east of the boundary of the Mara Reserve. Six coverages were analyzed between 1950 and 1999, with a seventh in 2003. For this study, occupied bomas were re-counted within 10 km, 20 km, 30 km, and >30 km bands from the boundary of the Mara Reserve, and converted to densities (table 13.6).

With an overall annual rate of increase of occupied bomas between 1950 and 2004 of 5.25% (from 0.02 km² to 0.26 km²), it is clear that the rates of increase are faster, and the densities higher, nearer to the Mara Reserve than further away. These same trends are also seen in the percentage of occupied bomas at different distances from the reserve (table 13.6). For example, during the 54 years between 1950 and 2004, the proportion close

Table 13.6 Density (and %) of occupied settlements around the Mara Reserve

Distance from Mara Reserve	< 10 km	10 km–20 km	20 km–30 km	> 30 km	Total
Area (km²)	413	558	450	406	1,827
1950	—	—	0.03	0.05	0.02
	—	—	(40.6%)	(59.4%)	
1961	—	0.03	0.03	0.06	0.03
		(30.5%)	(25.4%)	(44.1%)	
1967	0.01	0.03	0.06	0.10	0.05
	(6.7%)	(16.7%)	(31.1%)	(45.6%)	
1974	0.04	0.03	0.08	0.08	0.06
	(16.5%)	(16.5%)	(36.9%)	(30.1%)	
1983	0.10	0.07	0.10	0.17	0.11
	(22.2%)	(21.1%)	(22.2%)	(34.5%)	
1999	0.32	0.15	0.19	0.33	0.24
	(30.2%)	(19.7%)	(19.7%)	(30.4%)	
2003	0.34	0.23	0.30	0.20	0.26
	(29.0%)	(26.3%)	(27.5%)	(17.2%)	
Rate of increase (% pa)	9.14	4.65	4.49	2.82	5.25

to the Mara Reserve started low (< 10%) and rose rapidly, to c 30%, while far from the Mara Reserve the proportion started high (> 50%) but fell rapidly, to < 20%.

Lamprey and Reid (op. cit.) interpret the increase in occupied bomas in terms of population growth, water availability, and immigration of Maasai into the area, once certain constraints to settlement, specifically infestation by the tsetse fly, had been lifted. However, the concentration of bomas and the higher rates of increase closer to the Mara Reserve are good examples of the "Galapagos Effect," in which settlement is attracted by enhanced economic activity and opportunities—in this case, the expansion of tourism facilities, private camps, small lodges, and so on over the last 15 years around the Mara Reserve itself.

Subdivision of Land

A major and fundamental change throughout the Mara Area is the rapid transformation of land tenure from group or communal ownership to individual ownership by the subdivision of communally owned land into small parcels, each with an individual title deed (Homewood, Coast, and Thompson 2004; Thompson 2002; Thompson, Serneels, and Lambin 2002; Thompson and Homewood 2002). This process started in the Mara Area in the mid 1980s and has been riven by protracted boundary and entitlement disputes, often ending in complex court battles. Following a presidential pronouncement in the 1990s, it is now government policy that group ranches should be subdivided. The process is now some 90% complete.

Of the many factors that contributed toward the incentives to subdivide, perhaps the most pervasive was security of tenure (Homewood, Coast, and Thompson 2004): from local and central government and their "big men," from immigrant settlers, and from the possible expansion of the Mara Reserve itself. Indeed, the first group ranches to subdivide (in the northwest of the Mara Area) did so in response to perceived threats from immigrant settlers who were able to acquire land by dealing directly with group ranch committees, while nearer to the Mara Reserve there were perceived threats from the Kenya Wildlife Service (KWS), the County Council, and from the NGO conservation community to enlarge the size of the protected area.

Other incentives were more directly economic in nature, especially the evident dilution of each member's share of the communal resources in the face of population growth and the increasing number of young people at-

Table 13.7 Current status of land subdivision in a sample of 11 group ranches in the Mara Area

Number of ranches in sample	11
Total hectares	503,000
% of Mara Area	44
Decade started	
1980s	2
1990s	4
2000s	3
Not started	2
% complete (% area)	
100%	4 (22%)
> 50%	2 (17%)
< 50%	2 (41%)
Not started	3 (20%)
No. land parcels (subsample of 5 ranches)	6,521
Total Ha subdivided	222,782
Mean parcel size (hectares)	34.2
St. dev.	19.1

taining the age at which they could become registered members of a group ranch in their own right. There was also widespread dissatisfaction with the inequitable distribution of revenues from land leases for large-scale cultivation, and from tourism, with members of group ranch committees and other elites being perceived to be the main beneficiaries (Thompson 2002; Thompson and Homewood 2002).

Added to which there was a general change in lifestyles and aspirations among the communities, following better education, enhanced job opportunities, wider economic horizons, and greater exposure to the world at large, all cumulating in the wish to manage more completely their own future and economic well-being. Subdivision was seen to be the key process through which individuals could capture for themselves and their immediate family the potential land rents, be they from agriculture, livestock, or wildlife, while the clear benefits that had accrued to individual families, in terms of capturing such revenues on already-subdivided land, added further incentives.

From a sample of five Group Ranches that have completed subdivision (table 13.7), the original five land holdings under communal tenure were transformed into 6,521 individually owned land parcels of 34.2 hectares, on average (sd = 19.1ha). From this sample, once the processes of subdivision are complete, the original 38 communal landholdings within the Mara Area (fig. 13.1) of, on average, 33,000 hectares each (standard error = 5,500 ha) will be transformed into some 33,000 individually owned parcels of land. This will have, and is having, major impacts on both land use and land values.

One immediate effect of subdivision is the intensification of both agricultural and livestock production. Small parcels are not well suited to large-scale cultivation, and access to livestock resources is now becoming constrained. Mobility, especially in droughts, has become increasingly difficult, and while kinship remains critical to access scarce resources such as water, salt licks and pastures, a livestock owner now has to seek permission from many families to move to a new area.

Land values invariably rise after subdivision, both from the intensification of production and because it is now available in bite-sized chunks rather than in huge swathes. In the Mara Area, there was initially an expansion of settlement as new owners moved onto their parcels, often before they were formally adjudicated, although many new owners were slow to move, mainly for a lack of capital to develop their land and their farm. This had the side effect of forcing families that had settled but who had never been registered as members of a particular group ranch to relocate elsewhere as they were no longer welcomed or tolerated by the new landowners.

Perhaps because of this land sales became rampant, initially among the former group ranch members themselves, but later to outsiders—immigrants, tour operators, and the owners of large-scale farms. Land prices increased sharply, from $35 to $1,300 per hectare, as new investors flooded into the (previously) group ranch areas. Increased investment in land has clearly followed land subdivision, for building permanent houses and livestock pens, for clearing and preparing land for cultivation, and for water development.

One significant development has been that of neighboring landowners grouping themselves into more-or-less formal wildlife associations, especially near the Mara Reserve, to create conservation areas where, in principle, human settlements are not allowed and livestock use is regulated (Thompson 2005).

The members of these wildlife associations have three incentives to keep their land undeveloped for wildlife. First, they can capture the wildlife rents for themselves directly from tour operators rather than through group ranch committees. Second, they are creating more opportunities to establish new tourism facilities, either permanent or semipermanent. Third, the wildlife rents are providing the capital to develop other land.

IMPACTS OF DEVELOPMENT ON THE SERENGETI/MARA ECOSYSTEM

The gradual attrition of the numbers and diversity of wildlife in the Mara Area and the Mara Reserve has been noted in a number of studies (Broten and Said 1995; Norton-Griffiths 1995; Norton-Griffiths 1996; Ottichilo 2000; Said et al. 1997; Serneels and Lambin 2001a). Here we relate these losses of wildlife specifically to the economic process of capturing agricultural rents by developing land for cultivation.

Impact of Cultivation on Resident Wildlife and Livestock

Between 1977 and 2002 the Kenya Rangeland Ecological Monitoring Unit (KREMU), now the Department of Remote Sensing and Resource Surveys (DRSRS), carried out 21 aerial surveys of wildlife and livestock within the Mara Area and the Mara Reserve when the Serengeti migratory wildebeest population was clearly not present. The densities of wildebeest (the Loita population), of all wildlife (expressed in Tropical Livestock Units) and of all livestock (also expressed in TLUs) were first calculated from each of these 21 surveys within the five bands of agricultural rents used in table 13.1A before averaging within each successive decade corresponding to the Serneels' surveys. These decadal averages were then regressed against the crop cover density:

$$y = a + b^*X_1 + c^*\ln(X_2) + d^*(X_1 * \ln(X_2)) \tag{13.6}$$

where y = was the density of wildlife or livestock, X_1 was the decade (1970s, 1980s, 1990s, and 2000s), and X_2 was the natural log of the crop cover density for each decade/band (table 13.1E).

The densities of the Loita wildebeest and of all wildlife showed significant negative trends through time, and stronger negative associations with cultivation density (table 13.8). As expected, the cross-product terms (decade × ln [crop cover]) are positive, showing that both wildlife densities and crop cover are relatively higher in areas of high agricultural rents. In contrast, livestock densities show no significant trend over time, nor any significant effect from increasing cultivation.

It is clear that the Loita population of wildebeest, and all wildlife in general, are being displaced by the process of capturing agricultural rents by converting land to cultivation. Livestock show no such effects, and are being absorbed into the developing matrix of mixed agricultural and livestock land use (Homewood et al. 2001; Serneels and Lambin 2001a).

Table 13.8 Influence of crop cover on resident wildlife and livestock in the Mara Area

	Constant	Decade	ln (Crops)	Decade * ln (Crops)	r^2
Wildebeest km^{-2}	46.75***	−4.74***	−14.66***	1.62***	0.89
All wildlife km^{-2}	82.01***	−8.11***	−22.70**	2.60**	0.93
All livestock km^{-2}	81.78**	−7.06	−27.34	3.71	0.25

Note: All Wildlife and All Livestock densities as TLUs km^{-2}; decade = 1970s, 1980s, 1990s, and 2000s; Crops = crop density as ha km^{-2}; df = 14; $p < 0.01$**, $p < 0.001$***

Impact of Land Development on the Serengeti Migratory Wildebeest

The numbers of Serengeti migratory wildebeest have been monitored consistently since the mid-1960s (Mduma, Sinclair, and Hilborn 1999; Tanzania Wildlife Conservation Monitoring [TWCM] 1999). In contrast, their migratory movements have been studied in three main phases: in the early 1970s, in the course of 28 monitoring flights covering the Serengeti National Park and some of the Kenyan Maasai Mara area; since 1977, in the Mara area by the KREMU/DRSRS; and more recently through following GPS-collared individuals (Thirgood et al. 2004). For this analysis we have used the TWCM series of population censuses, the early 1970s monitoring data, and 11 of the KREMU/DRSRS censuses between 1979 and 1996, when the Mara area was flooded by migrants from the Serengeti.

The size of the migratory population grew strongly in the early 1970s, and by 1977 had established an equilibrium—the population fluctuated around a mean of some 1.3 million (fig. 13.4). In the early 1970s, when the population was low, relatively few moved into Kenya during the dry season; however, the numbers increased substantially during the early 1980s, until a precipitous drop in 1984, which never reversed. The percentage moving into Kenya dropped by 65% after 1984 (table 13.9), from an average of 0.866 million (68% of the population) to 0.307 million (27% of the population).

Of the migratory wildebeest moving to Kenya, the numbers using the Mara Reserve were high during the mid 1980s but declined sharply after 1984 (fig. 13.5), from an average of 0.612 million to 0.229 million (table 13.9). While the numbers using the Group Ranches outside the Mara Reserve have shown a more gradual decline (fig. 13.5), the numbers dropped from an average of 0.254 million before 1984 to 0.078 million after (table 13.9). Simple trend analysis confirms these impressions: from 1970 to the

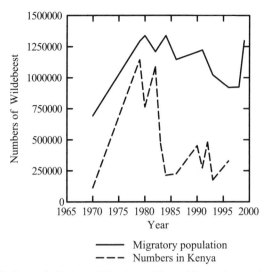

Fig. 13.4 Size of the Serengeti migratory wildebeest population and the numbers moving into Kenya each year during the dry season (1970–1999).

Table 13.9 Numbers of migratory wildebeest in Kenya 1972–1983, and rainfall 1969–2004

A. Migratory wildebeest numbers	1972–83	1984–96	df	t	p
Migratory wildebeest population	1.279m	1.139m	9	1.912	0.088
Numbers in Kenya	0.866m	0.307m	9	−4.357	0.002
Inside the Mara Reserve	0.612m	0.229m	9	−3.768	0.004
Outside the Mara Reserve	0.254m	0.078m	9	−3.759	0.004
B. Mara Reserve annual rainfall	**1969–83**	**1984–04**			
Group ranches annual rainfall (mm)	855.3	861.7	34	0.488	0.629
Mara annual rainfall (mm)	990.6	897.1	34	−1.646	0.109

Note: Pooled-variance *t* tests.

late 1990s, the densities, both inside and outside the Mara Reserve, declined significantly, as did the overall density within Kenya (table 13.10).

On the Group Ranches outside the Mara Reserve, there is absolutely no rainfall signal whatsoever that can be associated with the decline in wildebeest numbers (fig. 13.6, panels A and B; Reid et al. 2000, fig. 3; table 13.9). However, the spread of cultivation is having a clear effect, with migratory

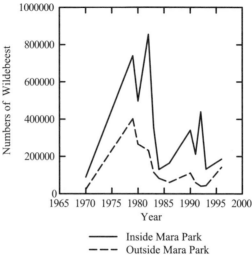

Fig. 13.5 Numbers of Serengeti migratory wildebeest using the Mara Reserve and surrounding group ranches during the dry season (1970–1996).

Table 13.10 Trends in density of migratory wildebeest in the Mara Reserve and the Mara Area 1979–1996

	Constant	Year	r^2
Densities "in Kenya"	92.3	−3.12**	0.40
Densities inside Mara Reserve	548.7	−17.82*	0.32
Densities outside Mara Reserve	33.5	−1.23***	0.45

Note: Densities in km22; df 5 10; year since 1970.

wildebeest being selectively squeezed out of the areas of high agricultural rents (> \$150 ha$^{-1}y^{-1}$) where land use development is most advanced (fig. 13.7). Multiple regression analysis confirms this impression.

$$y = 0.964 + 32.231X_1{}^{***} + 0.010X_2 - 1.129(X_1{}^*X_2)^{***} \qquad 13.7$$

where y = density of the Serengeti migratory wildebeest, X_1 is a dummy variable distinguishing between areas of high agricultural rents (> \$150 ha$^{-1}y^{-1}$ dummy variable = 1) and low agricultural rents (< \$150 ha$^{-1}y^{-1}$ dummy variable = 0) and X_1 is the year since 1970 (multiple adjusted r^2 = 0.760, n = 54).

Norton-Griffiths, Said, Serneels, Kaelo, Coughenour, Lamprey, Thompson, and Reid

A

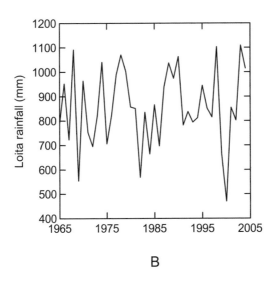

B

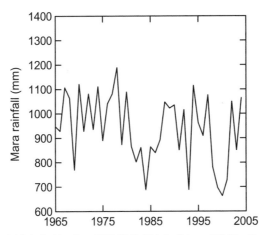

Fig. 13.6 Annual rainfalls in the Mara Area, 1965–2005: (A) Loita Plains and (B) Mara Reserve (see fig. 13.1). Constructed from PPTMAP, a computer program to create precipitation maps from point data. Input was monthly rainfalls at 20 stations in the northern Serengeti and Mara Area, 1960 to 2005. Interpolation used inverse distance weighting corrected for elevation differences between each interpolated point and the rainfall station. Numeric outputs and maps can be made for any required time period (month, years), for individual points or for masked areas. (A) was masked for the group ranches of the Mara Area east of the Isuria escarpment; (B) was masked for the Mara Reserve.

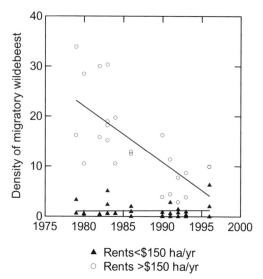

Fig. 13.7 Trends in the densities of migratory wildebeest in areas of high and low agricultural rents in the Mara Area (1977–1996).

The dummy variable X_1 is highly significant ($p = 0.000$) and expresses the higher densities of wildebeest found in the areas of high agricultural rents. The time variable X_2 is not significant on its own but shows a highly significant negative value (-1.129, $p = 0.000$) in the cross-product term. This clearly demonstrates that wildebeest densities are falling significantly in the areas of high agricultural rents.

Inside the Mara Reserve, different explanations must be sought for the drop in utilization by the migratory wildebeest. As with the group ranch areas, there is no rainfall signal to account for the change (fig. 13.6, panel B, and table 13.9), except that there are perhaps more very dry years since 1984 than in previous years. Other factors that could be important include the "ring fencing" by pastoral settlements and tourist facilities around the Mara Reserve, which, at 0.34 km^{-2}, are approaching critical values for excluding wildlife (fig. 13.8). The ever-increasing incursions of livestock, the increasing use of the Mara by tourists, poaching activities, and vegetation change (Dublin 1995; Lamprey and Reid 2004; Serneels and Lambin 2001a, 2001b) may all be playing a part as well.

In summary: use of the Mara Reserve and the Mara Area by the Serengeti migratory wildebeest population has dropped by some 65% since 1984, from 866,000 animals a year, on average, to 307,000 animals. In the absence of any evidence for a major change in rainfall, other explanations must be sought. The evidence is most clear-cut on the Group Ranches out-

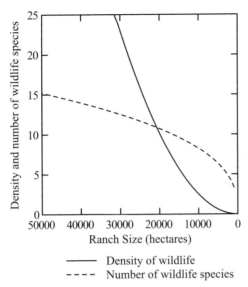

Fig. 13.8 Wildlife density (numbers km^{-2}) and diversity (number of species on each ranch) on ranches of different size, Laikipia District, Kenya, showing that both decline as ranch size declines.

side the Mara Reserve, where the drop in utilization can be associated with the explosive growth, especially in the mid 1980s, in large-scale (mechanized) and small-scale cultivation land subdivision and the concentration of both pastoral settlements and tourist facilities around the Mara Reserve. The decrease of utilization within the Mara Reserve itself requires different interpretations.

Impact of Land Subdivision

Data from ranches in Laikipia District (fig. 13.8) demonstrate clearly that both the density and diversity of large wildlife decline with ranch size. This would suggest that the subdivision of land in the Mara Area into smaller parcels is enough on its own to influence negatively the densities and diversity of wildlife. The formation of wildlife associations, where neighboring landowners cooperate by leaving their (subdivided) land undeveloped to create larger areas for capturing wildlife rents, could act against this trend.

Another potential impact of land subdivision is the proliferation of settlements, as land parcels are developed into farming homesteads. This could increase the overall density of settlements from the current 0.17 km^{-2}, reported by Reid et al. (2003), to 0.34 km^{-2}, if every land parcel were

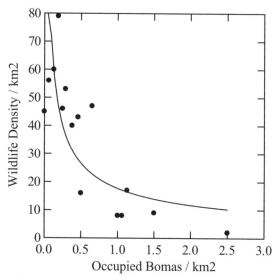

Fig. 13.9 Influence of occupied *bomas* on wildlife density (after Reid et al. 2003).

so developed (note from table 13.6, that this is the contemporary density of settlements within 10 km of the boundary of the Mara Reserve). Reid et al. (2003, section 5.1) noted that any positive relationships between pastoralists and wildlife at low densities of pastoral settlements broke down at higher densities. The relationship between the densities of wildlife and occupied bomas (fig. 13.9) reported by Reid et al. (2003, fig. 7) is:

$$\ln(y) = 4.172 - 1.438*X \qquad\qquad 13.8$$

where y is the density of wildlife in natural logs and X is the density of occupied bomas ($n = 15$, adjusted multiple $r^2 = 0.85$, $p = 0.000$). An increase in the density of occupied bomas, from 0.17 km^{-2} to 0.34 km^{-2}, would reduce wildlife densities from 50.8 km^{-2} to 39.8 km^{-2}.

Time Horizons to the Elimination of Wildlife

A simple equation relating the total hectares cultivated (as estimated in table 13.1E) to the number of years since cultivation started and the agricultural rent can be used to calculate time horizons.

$$y = 5.208 + 0.0014 * (X_1 * X_2) \qquad\qquad 13.9$$

Norton-Griffiths, Said, Serneels, Kaelo, Coughenour, Lamprey, Thompson, and Reid

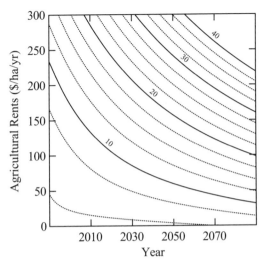

Fig. 13.10 Predicted increases in crop cover (ha km⁻²) 2010 – 2070 (from equation 13.8) as a function of agricultural rents ($U.S. ha⁻¹y⁻¹).

where y is the estimated cover density of crops (in ha km⁻²), X_1 is the number of years since crops were first grown (the first occurrence of a crop being taken as year 1) and X_2 is the agricultural rent in $ ha⁻¹y⁻¹ ($n = 21$, adjusted multiple $r^2 = 0.341$, $t = 3.367$, $p = 0.003$).

While it is never advisable to extend a regression analysis beyond the limits of the available data, the regressions of wildlife on crop cover density (table 13.8) suggest that wildebeest will be effectively eliminated at a crop density of 25% while wildlife in general will be eliminated at around 40% crop density. If the rates of land development remain the same then these critical crop cover densities will be achieved in 30 to 60 years (fig. 13.10). However, if most landowners start to develop small farms on their subdivided plots, the rate of conversion of land to cultivation and intensive livestock management, along with fencing, must inevitably become more rapid than that currently observed. The available time to the elimination of wildlife would be shorter, possibly within a decade or so. Much depends upon how quickly landowners act to optimize their net returns to land.

OPTIMAL LAND USE CHOICES FOR LANDOWNERS

The agricultural rents within the Mara Area (table 13.11) clearly outcompete those from either livestock or wildlife. The total available for capture

Table 13.11 Rents from land use in the Mara Area

	Mean \$ ha^{-1}y^{-1} (\$)	Total rents (\$m)
Agricultural rents	155.51	180.6
Livestock rents	30.40	34.6
Wildlife rents		
Mara high	50.00	57.0
Mean concession and access	10.27	11.7

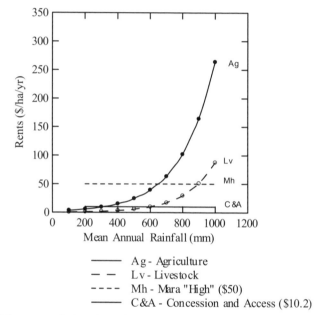

Fig. 13.11 Returns to agriculture, livestock, and wildlife in the Mara Area as a function of mean annual rainfall.

are \$181 million of agricultural rents, \$35 million of livestock rents, and anywhere between \$12 and \$57 million of wildlife rents. However, since both agricultural and livestock rents are a function of rainfall (eqs. 13.1 and 13.5), the optimal selection by a landowner will depend upon their differential returns along the rainfall gradient, compared with those from wildlife.

Comparing agricultural, livestock and wildlife rents along the rainfall gradient we see that agriculture always provides better net returns compared with livestock (fig. 13.11), while wildlife rents from concession and access fees (\$10.27 ha^{-1}y^{-1}) can offer an alternative and/or supplement to livestock rents only at lower rainfalls, below 600 mm: relatively little (14%) of the Mara Area receives less rainfall than this. In contrast, the highest wildlife rents—$50 ha^{-1}y^{-1}—seem quite competitive against both crops and livestock, and could offer landowners a genuine alternative to land subdivision and development.

These differentials, between agricultural rents on the one hand and livestock and wildlife rents on the other, go a long way to explaining the observed conversion of land to cultivation, especially in areas of higher agricultural rents. However, other factors must also be taken into account, especially rainfall, for these differential returns will vary in normal versus drought years.

Norton-Griffiths and Butt (2006) modelled the optimal choice for landowners between agricultural, livestock, and wildlife production in the pastoral areas of Kajiado District, Kenya. Parameters included agricultural and livestock rents as a function of rainfall (eqs. 13.1 and 13.4), drought years were defined as one standard deviation below average rainfall, livestock-wildlife interactions were taken into account (wildlife reduce livestock returns by 30%), and wildlife rents were set at the average for access and concession fees. When applied to the Mara Area (table 13.12) it is clear that no single production system, for example, only agriculture, does as well in terms of net returns as mixed production systems. Furthermore, defining the "optimal" selection of production systems as that which gives the highest net returns in both normal and drought years, an agricultural and livestock system where livestock returns are maximized by eliminating all wildlife are best in the wetter areas, whereas a livestock and wildlife production system is best in the drier areas. Dry areas are absent from the Mara Area, and landowners are clearly opting for the agricultural and livestock system with the elimination of wildlife.

This optimal selection depends upon the relative returns to the different production systems, and it is clear that the selection would change if the relative differentials between them were to change. In southern Africa, for example, where returns to agriculture are falling (Amalgamated Banks of South Africa [ABSA] 2004), landowners are opting instead for livestock: wildlife or even wildlife-only production systems.

In figure 13.12 we model the critical wildlife rents at which the livestock:wildlife production system becomes the optimal choice in terms of providing the highest net returns in both normal and drought years. Also shown are the contemporary average wildlife rents of $10.2 ha^{-1}y^{-1} from access and concession fees and the highest Mara Area rent of $50 ha^{-1}y^{-1}. This high wildlife rent is effective for influencing land use choices at around 600 mm—650 mm of rainfall, roughly where the wildlife associations and simi-

Table 13.12 Optimal[a] landuse strategies in the Mara Area

Land uses	Rainfall 700 mm ($)	300 mm ($)
Agriculture only, no livestock or wildlife		
Normal year	3.6	7.6
Drought year[b]	26.8	2.6
Agriculture with livestock, no wildlife		
Normal year	89.2	10.6
Drought year	34.3	3.9
Livestock only, no wildlife [c]		
Normal year	25.6	3.0
Drought year	7.5	1.3
Livestock with Wildlife		
Normal year[d]	27.5	12.2
Drought year	15.3	11.1
Wildlife only		
Average rents	10.2	10.2

Note: (a) optimal choice is that giving best net returns in both normal and dry years; (b) drought year defined as one SD below normal rainfall; (c) net returns to livestock are 48% higher when wildlife are eliminated (Norton-Griffiths and Butt 2006); (d) livestock returns reduced by 30% in the presence of wildlife.

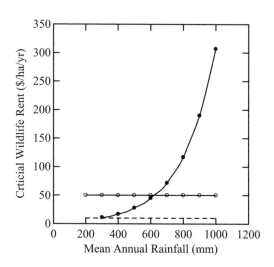

● Critical wildlife rent
○ Mara "high" ($50)
+ Concession & Access ($10.2)

Fig. 13.12 Critical wildlife rents as a function of rainfall at which the livestock:wildlife option becomes optimal.

lar initiatives are currently struggling for survival in the Mara Area. Wildlife rents of between \$100–\$150 ha^{-1}y^{-1} are required for wildlife to survive on significant areas of the Mara Area (fig. 13.12).

CONCLUSIONS: ECONOMIC DRIVING FORCES

We see four main economic driving forces that are influencing the land use decisions of the pastoral Maasai in the Mara Area. These are (1) the differential returns to land uses and production systems (2) the incentives to subdivide land, (3) the macro- and microeconomic conditions that are in turn influencing these differential returns, and (4) policy and market distortions with respect to the provision of wildlife goods and services.

Differential Returns to Land Uses

The differential returns to agricultural, livestock and wildlife production in the Mara Area (table 13.11) are so great that agricultural returns overwhelm those from either livestock or wildlife. These differentials create incentives to develop rather than conserve land, and alone are adequate to explain the contemporary patterns of land use change. In the areas of higher agricultural rents, land is being developed for agricultural and livestock production and resident and migratory wildlife are being eliminated. In areas of lower agricultural rents, livestock:wildlife production systems maintain a precarious foothold.

Incentives to Subdivide Land

The most important incentives to subdivide land are first, security of tenure—from in-migration and from land alienation by political elites, government, or even NGOs (Homewood, Coast, and Thompson 2004); second, to avoid the dilution of the value of communal resources from population growth; and finally to capture the benefits of agricultural, livestock, and wildlife production at the family level rather than through group ranch committees or other agencies (Thompson and Homewood 2002). The process of subdivision in itself sets off significant changes: land values tend to rise, and along with them the values of agricultural and livestock rents, as production strategies shift from extensive to more intensive methods.

Macro- and Microeconomic Factors

From the perspective of the individual landowner, at the macroeconomic scale, domestic and international markets are expanding, there are real gains in producer prices, ever-increasing opportunities for off-farm jobs and investment, and a wider availability and choice of goods and services. At the microeconomic level, the landowner sees improved market and transport networks, improved information networks about market conditions, and improved access to financial services. All of these create additional incentives for landowners to increase returns to land by investing in land development and production.

Policy and Market Distortions

At the policy level, the perverse policy environment (since 1977) bans sport hunting and other high-value sources of wildlife revenues and, indirectly, restricts wildlife revenue generating opportunities to the small amount (ca. 5%) of the land that can support wildlife viewing. This denies many sources of wildlife rents to landowners.

Market distortions are created through the actions of the tourism cartels who create barriers to landowners entering fully into the market for wildlife rents and then divert the greater proportion of the wildlife rents away from the wildlife producers. At best landowners can trap 10% to 15% of wildlife rents, the remainder going to the service-provider side of the industry. It is interesting here to note that in southern Africa, where the market for wildlife goods and services is essentially free of major distortions, landowners and communities typically own and manage wildlife operations themselves, rather than act as simple concessionaires—as is the case in Kenya—and accordingly capture a larger proportion of wildlife rents.

OPPORTUNITIES FOR MITIGATION AND INTERVENTION

Conservation values are clearly being lost throughout the Mara Area as landowners justifiably improve their social, political, and financial status by realizing the potential of their land. The strategic problem to be addressed is how to maintain such conservation values in the face of these major and fundamental changes in land use and land tenure. We see two major areas for positive interventions. First, one can improve wildlife rents so that they become more competitive against returns from cultivation and livestock:

second, one can support the intensification of agricultural and livestock production on land with higher agricultural potential, so as to reduce pressure on undeveloped land.

Improving Wildlife Rents

Moves to implement a policy change at government level to reintroduce high-value sources of wildlife revenues through consumptive utilization, while beneficial in the longer term, would probably have little impact in the Mara Area within the time horizons previously discussed. A policy change in Kenya of this magnitude will require a long, consultative process; it will be highly controversial and will generate heated debate. Meanwhile, the expansion of land conversion will proceed unabated. No improvement to wildlife rents in the short term will be realized from such a policy change, even if cropping and bird shooting were to be reintroduced.

The one positive feature of the subdivision process is where owners of adjacent land parcels are forming wildlife associations, in which they cooperate in keeping their land undeveloped so they can benefit from wildlife rents (Thompson 2005). These associations are now found on the (former) Koyaki, Lemek and Siana group ranches, and while they are still somewhat fluid in terms of composition and membership, they offer an excellent opportunity to strengthen the commercial relationship between the suppliers of wildlife, on the one hand, and the suppliers of clients to view wildlife on the other. Many of these members of wildlife associations have plots elsewhere on these former group ranches and use their wildlife rents as capital for farm development. This should in turn reduce, at least in the short to medium term, the pressure to develop the land currently devoted to wildlife.

This process should be encouraged and supported by all involved in the conservation and tourism industries, as it offers one of the few positive opportunities to maintain conservation values.

However, the decision to create and maintain a wildlife association is essentially an economic one, and its success will depend ultimately on the economic returns realized by the cooperating landowners. The economics of such decisions have been explored elsewhere (Norton-Griffiths 2000) and the key, as we have seen earlier, lies with the wildlife rents. The wildlife fees paid by tourism cartels to landowners must be raised significantly to make it economically worthwhile for the landowners to keep land free of development. Most of these associations are found on land with rainfalls around 700 mm, where rents of some \$100 to \$120 $ha^{-1}y^{-1}$ or more are required to keep the wildlife option sustainable in the medium to long term (fig. 13.12). To meet this target, concession areas' fees and fees for access and bed nights must all be raised, along with the wages to locally recruited staff. This will in turn require extensive negotiations with the tourism cartels, who will naturally offer strong resistance.

It is even more important for landowners to become more directly involved in the tourism industry, for only then can they capture larger proportions of the wildlife rents. An initial entry has been through the employment of local guides, trained to accepted standards—but this is trivial. Transport provides a much better opportunity, and wildlife associations should insist that only their vehicles, drivers and guides are used on their land—as already occurs on some concessions in Tanzania. This, of course, requires management skills, expertise and capital—but none of these are insurmountable. Later, the wildlife associations can enter into the ownership and provision of accommodation services, which in turn require additional skills.

Improving Agricultural and Livestock Rents

Land use in the MPSE is undergoing a rapid transformation, from pastoralism to agropastoralism, and then to settled mixed agricultural and livestock production. Such a transformation requires capital to develop farm infrastructure and new husbandry skills in intensive as opposed to extensive production, including the selection of appropriate crop mixes and livestock breeds, the adoption of new methods of crop and livestock husbandry, and the development of new marketing skills.

Today there is an almost complete lack of agricultural and livestock extension services to assist pastoralists to make this transition from extensive production on communal land to intensive production on small, individually owned parcels of land. This too often results in poor land management and poor returns, increases incentives to convert even more land, and fuels land alienation by sales to outsiders.

While in general terms the technology to effect this transition is available throughout Kenya, the transfer of this technology to these new farmers is proving to be problematic. Normally, farmers look to the government for extension services, but this need not be the case, especially today, when the government has other priorities in mind. Effective technology transfer can also be achieved by mobilizing local expertise and expertise available from the commercial suppliers of agricultural and livestock services. Whichever, here is an opportunity for a positive intervention to maintain conservation values by increasing production and profits on already-converted land, thus reducing pressures to develop unconverted land.

A second opportunity to assist in this transformation is to develop new marketing networks, marketing skills, and especially new ways of providing better market information to the producers. Remote producers achieve a ridiculously low proportion of the final end-user price for their products (Norton-Griffiths and Butt 2006) and are often at the mercy of middlemen. Breaking this stranglehold will increase returns to land.

ACKNOWLEDGMENTS

M. Norton-Griffiths is greatly indebted to his coauthors, who have generously allowed him to make use of their data and ideas, and especially to Mohammed Said, who labored so assiduously over his GIS on our behalf. We more or less bushwhacked Tony Sinclair into allowing us to put together this late entry, and his stoic patience is gratefully acknowledged.

REFERENCES

Amalgamated Banks of South Africa (ABSA). 2004. *Game ranch profitability in southern Africa (2004)*. Rivonia: The South Africa Financial Sector Forum. Accessed at www .absa.co.za.

Africover. 2003. Methodologies, applications, and downloads. Accessed at www.africover .org.

Broten, M. D., and M. Said. 1995. Population trends of ungulates in and around Kenya's Masai Mara Reserve. In *Serengeti II: Dynamics, management, and conservation of an ecosystem*, ed. A. R. E. Sinclair and P. Arcese, 169–93. Chicago: University of Chicago Press.

Dublin, H. T. 1995. Vegetation dynamics in the Serengeti-Mara ecosystem: The role of elephants, fire, and other factors. In *Serengeti II: Dynamics, management, and conservation of an ecosystem*, ed. A. R. E. Sinclair and P. Arcese, 71–90. Chicago: University of Chicago Press.

Ecosystems Ltd. (ESL). 1985. Integrated land use survey of the lake basin of Kenya. Final report to the Lake Basin Development Authority, Ministry of Energy and Regional Development. Nairobi: EcoSystems Ltd.

———. 1987. Integrated land use database for Kenya. Nairobi: Government of Kenya, Long Range Planning Unit, Ministry of Planning and National Development.

Epp, H., and J. Agatsiva. 1980. Habitat types of the Mara-Narok area, western Kenya. Technical report series no. 20. Nairobi: KREMU Department of Resource Surveys and Remote Sensing.

Hilborn, R. 1995. A model to evaluate alternative management policies for the

Serengeti-Mara ecosystem. In *Serengeti II: Dynamics, management, and conservation of an ecosystem*, ed. A. R. E. Sinclair and P. Arcese, 617–38. Chicago: University of Chicago Press.

Hilborn, R., and A. R. E. Sinclair. 1979. A simulation of wildebeest and other ungulates and their predators in the Serengeti. In *Serengeti: Dynamics of an Ecosystem*, ed. A. R. E. Sinclair and M. Norton-Griffiths, 287–309. Chicago: University of Chicago Press.

Homewood, K., E. Coast, and D. M. Thompson. 2004. In-migrants and exclusion in east African rangelands: Access, tenure and conflict. *Africa* 74 (4): 567–610.

Homewood, K., E. F. Lambin, E. Coast, A. Kariuki, I. Kikula, J. Kivelia, M. Said, S. Serneels, and D. M. Thompson. 2001. Long-term changes in Serengeti-Mara wildebeest and land cover: Pastoralism, population, or policies? *Proceedings of the National Academy of Sciences* 98:12544–49.

Jaetzold, R., and H. Schmidt. 1982. *Farm management handbook for Kenya*. Nairobi: Ministry of Agriculture.

Lamprey, R. H. 2006. Forthcoming. Profit and prestige: Game policy, tourism revenues and community conservation in Maasai Mara. *Environment, Development and Sustainability*.

Lamprey, R. H. and R. S. Reid. 2004. Expansion of human settlement in Kenya's Maasai Mara: What future for pastoralism and wildlife? *Journal of Biogeography* 31:997–1032.

Maddock, L. 1979. The "migration" and grazing succession. In *Serengeti: Dynamics of an Ecosystem*, ed. A. R. E. Sinclair and M. Norton-Griffiths, 104–29. Chicago: University of Chicago Press.

Mduma, S. A. R., A. R. E. Sinclair, and R. Hilborn. 1999. Food regulates the Serengeti wildebeest population: A 40-year record. *Journal of Animal Ecology* 68:1101–22.

Monke, E. A., and S. R. Pearson. 1989. *The policy analysis matrix for agricultural development*. Ithica and London: Cornell University Press.

Mudsprings Inc. 1999. Wher-ACT Kenya Database. Accessed at www.mudsprings.com.

Norton-Griffiths, M. 1988. Aerial point sampling for land use surveys. *Journal of Biogeography* 15:149–56.

———. 1995. Economic incentives to develop the rangelands of the Serengeti: Implications for wildlife conservation. In *Serengeti II: Dynamics, management, and conservation of an ecosystem*, ed. A. R. E. Sinclair and P. Arcese, 588–604. Chicago: University of Chicago Press.

———. 1996. Property rights and the marginal wildebeest: An economic analysis of wildlife conservation options in Kenya. *Biodiversity and Conservation* 5:1557–77.

———. 1998. The economics of wildlife conservation policy in Kenya. In *Conservation of Biological Resources*, ed. E. J. Milner-Gulland and R. Mace, 279–93. Oxford: Blackwells.

———. 2000. Wildlife losses in Kenya: An analysis of conservation policy. *Natural Resources Modeling* 13:13–34.

Norton-Griffiths, M., and B. Butt. 2006. The economics of land use change in Loitokitok Division, Kajiado District, Kenya. LUCID working paper no. 34. Nairobi: International Livestock Research Institute. Retrieved from www.lucideastafrica.org.

Norton-Griffiths, M., and C. Southey. 1993. The opportunity costs of biodiversity conservation: A case study of Kenya. CSERGE working paper GEC 93-21. Norwich, England: University of East Anglia.

Ottichilo, W. K. 2000. Wildlife dynamics: An analysis of change in the Masai Mara ecosystem of Kenya. PhD diss., ITC, The Netherlands.

Policy Analysis for Rural Development (PARD). 1991. Farm budgets in selected districts of Kenya. Policy Analysis for Rural Development working paper series 14. Nairobi: Edgerton University.

Reid, R. S., M. E. Rainy, J. Ogutu, R. L. Kruska, K. Kimani, M. Nyabenge, M. McCartney, M. Kshatriya, J. S. Worden, and L. Ng'ang'a et al. 2003. *People, wildlife and livestock in the Mara ecosystem: The Mara Count 2002.* Nairobi: International Livestock Research Institute. Accessed at www.maasaimaracount.org.

Said, M. Y., W. K. Ottichilo, R. K. Sinange, and H. M. Aligula. 1997. Population and distribution trends of wildlife and livestock in the Mara Ecosystem and surrounding areas: A study on the impacts of land-use on wildlife and environmental indicators in the East African Savannah. Nairobi: Ministry of Planning and National Development, Department of Resource Surveys and Remote Sensing.

Sellen, D. 1991. Representative farms and farm incomes for seven districts in Kenya. Research Training in Agricultural Policy Analysis Project. Nairobi: USAID.

Serneels, S. 2001. Drivers and impacts of land-use/land-cover change in the Serengeti-Mara ecosystem: A spatial modelling approach based on remote sensing data. PhD diss., University of Louvain, Belgium.

Serneels, S., and E. F. Lambin. 2001a. Impact of land-use changes on the wildebeest migration in the northern part of the Serengeti-Mara ecosystem. *Journal of Biogeography* 28:391–407.

———. 2001b. Proximate causes of land-use change in Narok District, Kenya: A spatial statistical model. *Agriculture, Ecosystems and Environment* 85:65–81.

Serneels, S., M. Y. Said, and E. F. Lambin. 2001. Land cover changes around a major east African wildlife reserve: The Mara ecosystem, Kenya. *International Journal of Remote Sensing* 22 (17): 3397–3420.

Sinclair, A. R. E. 1995. Serengeti past and present. In *Serengeti II: Dynamics, management, and conservation of an ecosystem,* ed. A. R. E. Sinclair and P. Arcese, 3–30. Chicago: University of Chicago Press.

Sinclair, A. R. E., H. Dublin, and M. Borner. 1985. Population regulation of the Serengeti wildebeest: A test of the food hypothesis. *Oecologia* 65:266–68.

Sinclair, A. R. E., and M. Norton-Griffiths. 1982. Does competition or facilitation regulate migrant ungulate populations in the Serengeti? A test of hypotheses. *Oecologia Berlin* 53:364–69.

Tanzania Wildlife Conservation Monitoring (TWCM). 1999. *Status and trends of migratory wildebeest in the Serengeti ecosystem.* Arusha: Tanzania Wildlife Research Institute.

Thirgood, S., A. Mosser, M. Borner, S. Tham, G. Hopcraft, T. Mlengeya, M. Kilewo, E. Mwangomo, J. Fryxell, and A. R. E. Sinclair. 2004. Can parks protect migratory ungulates? The case of the Serengeti wildebeest. *Animal Conservation* 7:113–20.

Thompson, D. M. 2002. Livestock, cultivation and tourism: Livelihood choices and conservation in Maasai Mara buffer zones. PhD diss., University of London.

———. 2005. Valuing land use options in the Maasai Mara. Draft report of the 2004 Socio-Economic Survey, *Better policy and management options, for pastoral lands.* Nairobi: International Livestock Research Institute.

Thompson, M., and K. Homewood. 2002. Entrepreneurs, elites and exclusion in Maasailand: Trends in wildlife conservation and pastoral development. *Human Ecology* 30: 107–38.

Thompson, M., S. Serneels, and E. F. Lambin. 2002. Land-use strategies in the Mara ecosystem Kenya: A spatial analysis linking socio-economic data with landscape variables. In *Remote sensing and GIS applications for linking people, place and policy,* ed. S. J. Walsh and K. A. Crews-Meyer. Boston: Kluwer Academic.

FOURTEEN

Propagation of Change through a Complex Ecosystem

Ray Hilborn, A. R. E. Sinclair, and John M. Fryxell

The Serengeti ecosystem is of special importance both because of the unusual diversity of wildlife, birds and other taxa, and because it is one of the few large ecosystems that has been well studied, largely unaltered by human agriculture and forestry, and has been subject to a wide range of perturbations. The perturbations have been both natural and anthropogenic, and provide us with an opportunity to examine how naturally functioning ecosystems respond to both natural and direct human-induced perturbation. This lets us explore whether current ecological theory of ecosystem behavior has any direct applicability to understanding such large systems.

The structure of this chapter is as follows. We begin with a review of the major perturbations seen in the Serengeti; much of this was discussed in chapter 2 of this volume. We then review the theories available to understand Serengeti dynamics. These theories are largely ecological, but we include some economic and social theories, as appropriate, to understand the interaction of humans and the natural components of the Serengeti ecosystem. For each theory, we look at the Serengeti to see the extent to which this theory appears to provide a useful tool for understanding how the Serengeti has behaved. Part of ecological theory, and general systems theory, is the concept of emergent properties. These are qualitative behaviors that complex systems display and are often the best descriptors of systems. We review different types of emergent properties that have arisen in theory, and look within the history of the Serengeti for evidence of these same properties.

Many of the earlier chapters in this volume have dealt with specific models of components of the system, and we compare the models we have seen, the theories and emergent properties of the theories, and the dynamics of the Serengeti. Do some of these models capture ecological theory, emergent properties, and Serengeti behavior? Then we compare how the Serengeti has responded to perturbations with other similar complex systems, looking for congruence and exceptions. Finally, we draw conclusions about how the Serengeti has responded to perturbation and the role of theory and models in understanding this response.

HISTORICAL PERTURBATIONS IN THE SERENGETI

The Serengeti-Mara ecosystem, comprising approximately 25,000 km^2, can be seen from a range of perspectives: the physical environment, the vegetation, the animals, and the humans. In this section we review the most significant or interesting perturbations or changes that have occurred in each.

The Physical Environment

Physically, the Serengeti ranges from the flat plains to the southeast, with low rainfall, to the rolling hills of the woodlands in the northwest, with higher rainfall. The system is intersected by a series of rivers running from the northeast to the southwest, draining the higher-rainfall areas in the northwest with all the rivers flowing into Lake Victoria in the west. The physical landscape of the Serengeti has remained unchanged over the last few thousand years; the only significant perturbation has been the slight cooling and increase in rainfall described by Ritchie in chapter 6 (this volume).

The dry-season drought of 1993 is a strong and well-studied perturbation. Rainfall during the dry season was 25% of the lowest previously recorded, and directly led to the death of 30% of the wildebeest population and perhaps 40 to 50% of the total park populations of large mammals. During the same period, the canine distemper virus outbreak killed 50% of the lion population, possibly because lions themselves were more vulnerable to the virus due to the low protein and fat content of their diet. The wildebeest and lions recovered rapidly from the 1993 drought and by 2000 were back at previous levels of abundance. However, buffalo have recovered only in those areas not subject to illegal hunting, and the recovery is very slow, only half-way back to their original numbers.

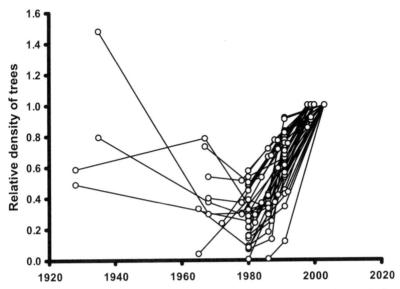

Fig. 14.1 Change in tree density (from Sinclair et al. 2007). Connected dots represent the same site from photographs.

The Vegetation

The vegetation pattern broadly follows rainfall, with the short-grass plains in the southeast giving way to taller grasses as we move northwest, and then finally savannah and woodland. The changes in vegetation in the last 100+ years have been dramatic—from the turn of the century up to the 1980s the density of trees in the woodlands steadily declined. After 1980, there has been a spectacular increase in tree density on the Tanzanian side of the border (fig. 14.1), while on the Kenyan side of the border within the Mara Reserve, the decline of the tree density has continued.

The change in tree density is ascribed to two causes. First and foremost, the increase in wildebeest in the 1970s reduced the amount of standing grass in the dry season, and this led to reduced frequency and intensity of fires and higher survival of seedlings. Second, there was a decline in elephant abundance in the 1980s due to poaching in Tanzania, whereas effective antipoaching efforts in Kenya protected elephants. Elephants eat very small trees in grassland (Dublin, Sinclair, and McGlade 1990), and the removal of elephants in Tanzania provided an opportunity for many tree seedlings to survive. The lack of elephants and reduced fires in Tanzania led to an explosion in young trees, while there has been little, if any, tree recruitment in Kenya. The landscape in Tanzania has been dramatically transformed, from what was often open savannah with a large tree every few hundred meters, to what is now in many places almost impenetrable shrub and woodland.

In the grassland there have also been significant changes, with the change in fire intensity and grazing causing the long grass to expand and the plant communities to change significantly on the plains. While much less dramatic to the human eye than the explosion of small trees in the woodlands, the grassland changes have had important impacts on grassland communities.

There has been a long, slow erosion of the riverine forest, discussed in more detail in chapter 2. The width of the Mara River has increased, washing away some longstanding forests. The peak flows in recent years have been higher and the low flows lower, likely caused by deforestation in the upper watershed in Kenya. Other causes of decline of riparian forests are thought to be changes in fire frequency, some of this caused by extensive poacher use of the riverine areas for cover and snaring.

Wildebeest and Ungulates

The decline in wildebeest numbers with the introduction of rinderpest and the recovery following its removal is perhaps the defining perturbation of the Serengeti (fig. 2.10, this volume). Few if any large ecosystems have been monitored over a period coinciding with such a large increase in the most significant species.

The change in wildebeest abundance has been felt throughout the ecosystem. As already mentioned, the higher abundance has led to more grazing pressure on grasses, subsequently less fuel for dry-season fires, and lower fire frequency and intensity, facilitating an increase in trees since 1980 in Tanzania. The resultant change in vegetation has affected both plant communities' composition and animals that depend on these communities, with such unexpected consequences as lions being able to capture prey more effectively in the woodlands due to increased cover (Hopcraft, Sinclair, and Packer 2005). The direct change in abundance of wildebeest has provided substantial opportunities for increases in lions and hyenas (Packer et al. 2005), with consequent negative impacts on their other prey and competitors, such as cheetahs (Durant 1998).

The increase in wildebeest has affected human populations as well, with legal and illegal hunting providing an estimated 30,000 wildebeest per year to communities adjacent to the north and west of the park. These areas have seen stronger population growth than communities located farther from the park boundary. A secondary impact of the profitability of illegal hunting for wildebeest has been the increased illegal catch of other resident prey.

The other ruminant to increase substantially since the 1960s is buffalo, although numbers of this species have declined considerably since the 1980s. Buffalo appear to be heavily affected by poaching in the northwest of SNP. Other ungulates have changed much less, and evidence for competitive interaction with wildebeest is weak.

Elephants

Elephant numbers have gone through two fluctuations in the past 60 years. They increased from very low abundance in 1950s to modest densities by 1970. They then declined from poaching, reaching a low point in 1988, before recovering in Serengeti (fig. 2.18). They had a major effect on tree cover in the savannah woodlands, as discussed previously.

Carnivores

Lion and hyena numbers have increased considerably as wildebeest numbers increased, while cheetahs have apparently remained constant, and wild dogs have almost totally disappeared from the ecosystem. There are no long-term records for most of the other smaller carnivores. The abundance of lions on the plains and in the adjacent woodlands has been studied intensively since the 1960s and shows increases, disease-induced declines, and recovery (Packer et al. 2005). The lions show an interesting pattern of apparently stable states, with apparent jumps between these states (fig. 1.2).

Humans

The nineteenth-century outbreak of rinderpest and the consequent death of most livestock led to a retreat of human settlement on the edges of the park, reduction in burning, and an increase in woodlands and tsetse flies (Ford 1971). These events effectively set the stage for establishment of the protected status of most of the ecosystem because there was no agriculture in the now-protected areas in the mid-twentieth century. Since the 1950s, the population in Tanzania has been growing rapidly (see fig. 1.1), currently at 2.5–3.3% per annum (chapter 12, this volume). The impact on the Serengeti ecosystem has been several-fold. Increasing population has meant increasing pressure on encroachment of agriculture into the ecosystem, and the boundaries of protected areas south of the park (Maswa game reserve) has been repeatedly adjusted as agriculture has taken over land previously in the game reserve. Increased human population size has also led to increased pressure for illegal harvesting of wildlife, and the consequent necessity of a substantial portion of the Serengeti National Park budget going to antipoaching activities.

Disease and Pathogens

Disease and pathogens have been perhaps the defining force in shaping the last 150 years of the Serengeti. We have seen how the outbreak of rinderpest and reduction in cattle and wild ruminants have shaped the vegetation, large mammal, predator, and human population dynamics. Elimination of rinderpest has done just the opposite, allowing for rebuilding ruminant, predator, and human populations. While the importance of disease in human history has been recognized for a long time (Diamond 1999; McNeill 1967), the role of rinderpest in shaping the Serengeti ecosystem is the most dramatic and best-studied impact of disease and pathogens on large ecosystems.

The role of other diseases is increasingly recognized, especially in the carnivores. The canine distemper virus (CDV) outbreak in lions in 1994 is the best known, but CDV is also thought to have been largely responsible for the ultimate disappearance of wild dogs. As in rinderpest, where livestock were the reservoir that maintained the disease in the wild animals, domestic dogs are thought to be the cause of infection of both CDV and rabies among the wild carnivores (chapter 7, this volume).

No story of African ecology would be complete without recognizing trypanosomiasis and its host, the tsetse fly, and the interaction between livestock, human occupation, burning, woodlands, and the maintenance of tsetse fly populations. The link between the rinderpest outbreak in the nineteenth century and the subsequent abandonment of agriculture near the current protected areas is a testament to the power of the tsetse fly (chapter 2, this volume).

Finally, while the role of AIDS in Africa has received considerable attention, it remains unclear what its impact on the Serengeti ecosystem will be. At present, the overall human population size adjacent to the park appears to be only slightly affected, but it is certainly possible that the next decades will see a significant reduction in human population, or the impoverishment of communities and increased pressure on wildlife.

ECOLOGICAL, SOCIAL, AND ECONOMIC THEORIES

Early ecological theory developed from the study of simple communities. The several ecological principles that developed from this theory enable us to dissect an animal community into its parts. These principles involve food chains and the food cycle, size of food niches, and the pyramid of numbers (Elton and Miller 1954, cited in Kikkawa and Anderson 1986).

Food Webs: Trophic Cascades, Bottom-up Processes, and Functional Redundancy

One of the great organizing principles of ecology, identified by C. S. Elton, is the analysis of food webs. At one level, food webs allow ecologists to describe the community based on who eats whom, which in turn provides convenient definitions for groups of species, such as herbivores, carnivores, and scavengers. Beyond simple description, food webs, combined with the laws of thermodynamics, allow prediction about the standing stock and energy flow through communities. Elton also identified this organizing principle and named it the now-familiar "pyramid of numbers." The simple theory of energy flow makes some useful predictions about community structure—namely, that no matter what perturbation might be inflicted upon a community, energy flow through primary production will always be much greater than energy flow through herbivores, which will in turn be much greater than energy flow through carnivores. Similarly, in terrestrial environments the standing stock of primary producers should be much larger than the standing stock of herbivores, which will be much larger than the standing stock of carnivores. While these relationships seem so obvious to modern observers as to be not worth mentioning, we should not forget that this theory provides some powerful predictions. Admittedly, these predictions are not going to be very exciting, nor do they provide any particularly novel insights into how complex systems respond to perturbation.

There are, however, a number of derivative theories arising from food web theory that are of much more interest in analyzing the response of the Serengeti to perturbations. The first body of theory is subsumed under the general names of "bottom-up control" or "top-down control," and concerns whether ecosystems are largely governed by limitation on the primary productivity (bottom-up control) or if the ecosystem is structured by impacts of consumers on their prey species (top-down control). Examples of top-down control that have been demonstrated most effectively include structuring of lake food webs by changes in the top predators, and this effect has been termed *trophic cascades*. (Carpenter and Kitchell 1993). In our view these are not general competing theories but theoretical frameworks, useful for understanding the behavior of different ecosystems. Some ecosystems are controlled by bottom-up forces. For instance, in marine-upwelling systems, systematic changes in ocean currents can cause many-fold changes in primary production and thus change in total abundance throughout the food web. In other ecosystems, and indeed possibly in the same ones, changes in abundance of top predators (through fishing, for instance), can dramatically change the community composition.

In the Serengeti, as in many semiarid terrestrial ecosystems, dry-season rainfall is a strong bottom-up controlling factor. The primary production of the grasses in the dry season is directly proportional to the amount of rainfall, and this is the limiting factor on wildebeest abundance (Mduma, Sinclair, and Hilborn 1999). Top-down control may be much more important for many of the other ungulates. The increase in lions and hyenas, which was caused by higher wildebeest numbers, may be causing declines in resident prey numbers (chapter 2, this volume).

The removal of elephants and the subsequent increase in tree seedling survival and tree density in the woodlands can be seen as a classic case of a trophic cascade. The tree predator (elephant) was removed by a top predator (humans) through harvesting. Elephant prey escaped from top-down control, and the nature of the ecosystem has been transformed from open savannah to thick, brushy woodland.

A second body of theory, derived from food webs, is *functional redundancy*. The basic premise is that many different species may perform similar roles in a community. Therefore, the loss of some species can be made up by substitute species, with little impact on ecosystem function (Walker 1992). It has long been recognized that communities are often organized into "guilds" of species that perform similar roles (Root 1968). Much study has been devoted to understanding how many species, with such overlapping niches, coexist. Functional redundancy turns the tables and asserts that when there are many species with similar ecological niches, several of them may be redundant to the functioning of the ecosystem, and if one species is eliminated another species will take its place.

A defining characteristic of the Serengeti is species diversity, with the number of species in different trophic groups (the ungulates, for instance) the highest seen anywhere in the world. Walker (1992) argues that this redundancy of trophic roles has important consequences for the stability of the ecosystem, and that ecosystems with high functional redundancy will be much more resistant to perturbation than ecosystems without high functional redundancy. For example, the overlap in diet between wild dogs and other larger predators (lions, hyenas, cheetahs) means that the loss of wild dogs from the system would have no consequences on other species. One of the emergent properties discussed later is *resilience*—the ability of systems to absorb and utilize change to maintain their essential structure. Functional redundancy contributes to system resilience by providing buffers to perturbation of individual species.

Advances in food web theory are highly relevant to complex systems such as the Serengeti. McCann, Hastings, and Huxel (1998) showed that food web stability can be substantially enhanced in diverse systems if there are substantial amounts of weak coupling. The high diversity of ungulates, carnivores, and birds means there are a very large number of weak trophic interactions. This stabilization by weak trophic links can come about through predator switching (Kondoh 2003) or through predator choice among different prey with space-use partitioning (Fryxell and Lundberg 1998; Lawlor and Maynard-Smith 1976; McCann, Rasmussen, and Umbanhower 2005). The addition of spatial heterogeneity to the food webs is particularly important in considering the Serengeti. The net result of the newer food web theory suggests that the Serengeti has a broad range of trophic mechanisms that buffer the system from perturbation.

Keystone Species

Not all species within a community have equal functional roles; while some species may be members of a guild with many similar species, other species may have unique and important roles. Paine (1966, 1980) introduced this concept originally, showing that intertidal communities were structured by the presence or absence of a predatory starfish. Since that time, the term *keystone species* has been defined as species with a greater role in maintaining ecosystem structure and function than one would predict based on their abundance or biomass, although a species that dominates its ecosystem in biomass can also be considered keystone.

Wildebeest are an obvious keystone species because they provide so much prey biomass to the large carnivore community, and they affect fire intensity through grazing. However, elephants can also be thought of as keystone, shaping the vegetation community. And of course, rinderpest is perhaps the prime example, shaping many aspects of the entire eco-system.

There are two major problems with the keystone concept. First, there are operational problems with identifying keystone species. We need to define which parameter that we measure—abundance, biomass, species composition, or something else. We need to specify the degree of change in the community expected from losing a keystone species. Communities are open ended, and we must state how far into the food web we should trace species' impacts. Thus, the impacts of top predators may be traced only as far as the herbivores and plants, or through other indirect links to more distant herbivores, detritivores, protozoas, even microbes (Mills et al. 1994; Power, Dietrich, and Finlay 1996, Sinclair and Byrom 2006).

Holling Resilience: Multiple States, Resilience, Slow and Fast Variables, and Panarchy

In the late 1960s and early 1970s ecologists started borrowing methods and approaches from engineering, operations research, and general systems theory to describe and evaluate ecological systems. The classic paper by C. S. Holling (1973) described possible qualitative forms of population and ecosystem behavior and popularized the concept of "multiple stable states" as one form of behavior. Up to that point, most ecological theory had focused on conditions of population stability; that is, whether a system returns to its natural equilibrium after a perturbation, or whether it is unstable, and some element of the system disappears when the system is perturbed. Holling suggested that ecosystems are rarely in any equilibrium, that change is the normal condition for ecosystems, and that normally functioning ecosystems move naturally over a range of conditions. Further, he introduced the concept of *resilience*—that movement within the range of natural conditions is in fact essential to maintaining the proper functioning of the ecosystem. Holling suggested that attempts to hold ecosystems constant—for instance, by fire suppression in western U.S. forests—cause the system to lose resilience, so that when a fire does occur, the higher fuel loads lead to intense fires and tree death that would not have occurred in a naturally functioning forest, where more frequent and less intense fires reduce fuel load and do not lead to tree death (Holling 1980).

Some aspects of the Serengeti can be viewed as a system with a stable equilibrium that is subject to perturbation. Since the elimination of rinderpest, we can argue that the natural equilibrium for wildebeest is at about 1.5 million animals, and if perturbed away from this equilibrium (as in the 1993 drought), the population will return (as it did). Managers of the Serengeti system have not attempted to maintain the Serengeti in any particular state, analogous to fire suppression in U.S. western forests. However, in other systems, like Kruger National Park in South Africa, intensive management in the form of fences, culling, and drilling of wells for water supply has led to other dramatic changes in the ecosystem that can best be thought of as collapsing bounds of stability or loss of resilience (du Toit, Rogers, and Bigg 2003).

Another important element of Holling's theories is the existence of multiple regions of stability. He suggests that a system will naturally have multiple regions that are stable. The system will remain within a region unless pushed out of it by an unusual event, whereupon it moves to a different region. Holling's classic example is forest insect pests, which were often held at low levels by avian or mammalian predation, but if for some reason the insects escaped the regulation by predators, the system moved into a new state of insect outbreak, with high insect densities. The wildebeest-rinderpest system can be thought of as having two states or regions; with rinderpest, wildebeest abundance is low, without rinderpest it is high.

A further contribution of Holling to ecosystem theory is the concept of *slow* and *fast variables.* Holling recognized that within an ecosystem some elements run on slow time scales, and rates of change are on the order of decades or even centuries, whereas other processes operate on fast time scales and may be at a pseudoequilibrium with respect to the slow variables. Holling's examples of slow variables were often forest trees, with the insect community acting as a fast variable in the context of the slowly changing forest (Ludwig, Jones, and Holling 1978). Holling's work evolved into a more general theory, called *panarchy,* that integrates the concepts of resilience and slow and fast variables (Gunderson and Holling 2002). In the Serengeti, tree and elephant numbers are clearly slow variables, and within the historical record there is no indication of stability or equilibria, but rather long periods of possibly cyclic changes in abundance. Bird species such as starlings or hornbills, whose mean abundances depend on the presence of trees, fluctuate in number according to rain. These would be fast variables.

Multiple stable states, slow and fast variables, and resilience are all closely interconnected. Thus, variables that appear to be in multiple states with respect to some slow variables may in fact be limit cycles on the longer time scale. Holling's frameworks provide powerful tools for understanding the changes in the Serengeti. First and foremost, it appears that the Serengeti is a highly resilient system, as it has absorbed a wide range of perturbations that have been inflicted on it. The concepts of slow and fast variable change provide an excellent way to understand the slow tree cycle and the more rapid response of herbivores and carnivores to these changes.

Optimal Foraging Theory and Economic Equivalent

When asked why he robbed banks, the famous American bank robber of the 1920s, Willy Sutton, is said to have replied, "Because that is where the money is." This is an individual application of the general economic theory of distribution in a market system: people and resources go where there is "profit." The underlying assumption is that individuals or firms are making rational choices in order to maximize profit, and will chose the activity or location that provides the most profit. This economic theory predicts that the profitability of different activities in space should be the same if there are no constraints on movement between locations and there is perfect information about alternative opportunities.

The same theory emerged in the study of optimal foraging in ecology (Fretwell and Lucas 1970). The assumption is that individuals distribute themselves in space in order to maximize their fitness, and so long as they can move freely, and can determine fitness in different locations, the net result should be that fitness is the same in all areas where individuals are found. Fretwell named the resulting distribution the *ideal free distribution,* recognizing that it is an ideal and that the assumptions of perfect information and free movement will rarely be completely true.

Fryxell, Wilmshurst, and Sinclair (2004; see chapter 9, this volume) apply the concept of ideal free distribution to the behavior of Thomson's gazelle, demonstrating that they distribute themselves so as to maximize net energetic intake by striking a balance between the high volume, low quality of the denser grasslands, and the high quality, low volume of recent growth. Cheetahs also distribute themselves to balance between competing pressures, in their case between competition with larger carnivores (lions and hyenas) and prey density. Thus, cheetahs are usually found in areas of generally low prey abundance, where lions and hyenas are rare, but where there are still enough prey to make a living (Durant 2000).

The structure of these theories provides a basic framework for understanding the spatial distribution of species in the Serengeti. The theory predicts that if herbivores have the ability to move, they will be found where the food is, and would seasonally follow food availability as it changes with the season. The theory would predict that carnivores are found where their prey are. The basic predictions are predicated on the freedom of movement and knowledge of alternative opportunities, and much of behavioral ecology is devoted to showing how behavior prevents such freedom of movement. The evolution of territoriality can be thought of as a way to prevent freedom of movement and the equalization of fitness in all areas.

Importance of Spatial Structure

While naturalists and ecologists have always recognized the spatial heterogeneity of natural ecosystems, a consistent theory of spatial heterogeneity was slower to emerge. The early experimental work of Gause (1936), who found that laboratory predator-prey systems were unstable unless a refuge for prey was provided, highlighted the role of spatial structure in communities. Subsequent work (Huffaker 1958) demonstrated in the laboratory that unstable predator-prey systems could be made persistent if enough spatial heterogeneity was provided. Theoretical work showed that the balance between local extinction and recolonization was the primary determinant of persistence (Hilborn 1975).

There is growing recognition in the ecological community that the persistence of many species is dependent upon such spatial heterogeneity. For instance, Holt (2002) showed that periodic immigration of keystone consumers (like wildebeest) can stabilize food web interactions (Fryxell et al., chapter 9, this volume). Chapter 9 shows that persistence of mobile grazers depends on having a large enough area to accommodate temporary periods of inadequate resources. McCann, Rasmussen, and Umbanhower (2005) show that the larger the spatial scale, the greater the stabilizing effect of consumers on food webs. All this theory supports the idea that the sheer size and spatial heterogeneity of the Serengeti contributes to its persistence and resilience.

EMERGENT PROPERTIES AND BIOCOMPLEXITY

Complex systems are made up of many interacting components, and one of the themes of general systems theory has been that there are properties of complex systems that emerge from the interaction of the many components. These *emergent properties* come in many ways from biological complexity. Emergent properties are closely related to the concept of *biocomplexity,* which has been defined as arising from "properties emerging from the interplay of behavioral, biological, chemical, physical, and social interactions that affect, sustain, or are modified by living organisms, including humans." The system may exhibit emergent or unexpected properties (i.e., behavior of the whole is often not predictable based on a study of the component parts) (Michener et al. 2001 page 1018).

Overview of Modes of System Behavior

The simplest type of emergent property of ecosystems is the overall systems behavior. Such behaviors include:

- Stable equilibria, where the system will return to the same condition when perturbed.
- Unstable equilibrium, where the system does not return to the previous equilibria and some system component will usually disappear.
- Stable limit cycles, in which there is a long-term cycle. The system will gradually return to this cycle when perturbed away from it.
- Multiple stable states, where a system moves from one state to another when perturbed far enough.
- Regions of stability (domains) where the system does not rest at an equilibrium, but moves within a range of states, and when perturbed far enough may move into another region of stability.
- Chaotic behavior, in which there is apparently random behavior of the system over time, although the apparent randomness is in fact the result of very predictable rules.
- Thresholds: specific boundaries beyond which the system behavior changes dramatically.

These modes of behavior are not incompatible; for example, thresholds are an integral part of regions of stability and multiple stable states. Similarly, within some of the regions of stability there may be stable limit cycles, or even stable equilibria. Furthermore, in a spatially large and trophically complex system, one kind of behavior may be imbedded within another. We can easily imagine that insects or bacteria, which have been relatively little studied, might be exhibiting several of these modes of behavior within a plant community that is, itself, exhibiting another behavior.

Other types of emergent properties include spatial or size patterns. For instance, fractals, the characteristic spatial structure, are considered an emergent property. Specific size structures, such as the pyramid of numbers, are another type of emergent property. In the next section we review a number of emergent properties that have been seen in other systems and explore their relevance to the Serengeti.

Stable Equilibria

A strikingly stable equilibrium is seen in the relationship between wildebeest population size and dry-season rainfall. Simple population dynamics theory suggests that the wildebeest are limited by food supply, and when perturbed, the population will return to the carrying capacity set by their food (Hilborn and Sinclair 1978). The drought of 1993 was a reasonably strong perturbation, moving the wildebeest away from the equilibrium; when dry-season rainfall returned to normal, the wildebeest recovered reasonably quickly to the equilibrium (fig. 2.13; see Mduma, Sinclair, and Hilborn 1999).

Limit Cycles

We have identified two potential examples of limit cycles in the Serengeti: the elephant-tree cycle (Dublin, Sinclair, and McGlade 1990), and the cycles of pathogen outbreaks, particularly rabies and canine distemper (chapter 7, this volume). The long cycle of tree buildup, elephant increase, tree decline, and subsequent elephant decline that has been seen in other African ecosystems, such as Tsavo (Corfield 1973), has not been documented in the Serengeti, and it can be argued that the elephant-tree cycle is really a better example of multiple stable states. However, the long time lags in both tree recruitment and growth and elephant maturity provide the basic ingredients for cyclic behavior. The elephant-tree cycle could also be viewed as a long-term, transient dynamic, subject to periodic human influences such as ivory hunting.

Cyclic outbreaks are a common pattern in many disease systems. After an outbreak much of the population is resistant, and it is not until the proportion of the population that is resistant declines enough to provide a significant new vulnerable host population that an outbreak can occur again (Anderson and May 1991; Woolhouse, Taylor, and Haydon 2001). We know there are periodic outbreaks of both rabies and CDV that come into the wild populations from domestic animals, suggesting these are cycles, as predicted from other disease studies and disease models. It is also possible that these cycles are not self-sustaining, but that the diseases would naturally die out and are only maintained by periodic introductions from outside. For example, no rabies outbreaks have been found in the park since 2000 (S. Cleaveland, pers. comm.).

Multiple Stable States And Bounds of Stability

There is a growing literature on multiple stable states (Gunderson 2000; Janssen, Anderies, and Walker 2004; Knowlton 2004; Kuijper et al. 2003;

Nystrom and Folke 2001; Scheffer et al. 2001; van Nes and Scheffer 2005), and what Holling had originally speculated is now accepted to be a common feature of ecosystems.

The Serengeti provides a number of examples of what appear to be multiple stable states. Packer et al. (2005) describe how the total lion abundance has moved rapidly between different equilibria (fig. 1.2). Sinclair (1995 and chapter 2, this volume) argues that the elephant-tree cycle is really an example of multiple stable states rather than a long-term cycle.

Holt et al. (chapter 8, this volume) showed models exhibiting multiple stable states. They state, "If the territorial limit to lion population size is high enough or the rate of killing prey at low prey densities is high, then there may be an equilibrium at low prey-population density as well as two more at higher densities. If the only potential equilibrium is much less than the prey-carrying capacity, then there are limit cycles around that equilibrium. If the prey-carrying capacity is increased or the predator death rate is increased or its carrying capacity is decreased sufficiently, then there are three equilibria, but the highest prey-density equilibrium, which is stable, is the only equilibrium observed in the course of long-term dynamics."

Thresholds

Thresholds emerge from many ecological models; they may delimit the bounds of multiple stable states or critical values below which some species will go extinct. Cleaveland et al. (chapter 7, this volume) cite several examples of thresholds. "African wild dogs require helpers to successfully rear pups, so pup survival rates drop below replacement levels at a threshold pack size (Courchamp, Clutton-Brock, and Grenfell 2000)." "In the past, these dog populations appeared unable to sustain large-scale epidemics, but the rapid growth of the human population may have permitted the disease to reach its persistence threshold" (Cleaveland et al. chapter 7, this volume).

"Human activities may also affect wildlife infection dynamics. For example, cultivation and urbanization generally result in reduced biodiversity, which may favor opportunistic species such as foxes, jackals, mongoose, and raccoons. These species may attain such high densities that they exceed the threshold for maintaining infectious diseases such as rabies, allowing persistence of the disease in areas previously free of endemic infection" (chapter 7, this volume)

As discussed earlier, elephant abundance in the Serengeti was reduced below a threshold where trees were able to escape from the combined mortality of fires and elephant predation. The multiple stable states emerging from the predator-prey model of Holt et al. (chapter 8, this volume) has numerous thresholds. The numeric value of all of these thresholds in the models depends upon a broad range of other model parameters, but it is the existence of the thresholds that emerges from the models and the nonlinearity in system response near the thresholds that is of the most interest.

Spatial Heterogeneity

Spatial heterogeneity is both a property of the physical landscape and an emergent property of the ecosystem. Anderson et al. (chapter 5, this volume) describe in detail the physical heterogeneity of rainfall, topography, and soils that directly leads to heterogeneity in the plant communities— heterogeneity driven by the physical system rather than emerging from complex ecosystem dynamics.

The physical heterogeneity has important consequences for the ecosystem. Fryxell et al. (chapter 9, this volume) showed that spatial heterogeneity either slows or prevents the competitive exclusion of other grazers by wildebeest. The spatial food web theory discussed earlier provides an explanation both for the resilience of the system and for how weak trophic interactions can lead to the emergence and maintenance of spatial heterogeneity. In contrast, trees/grass coexistence in savannas could be maintained, because woody vegetation is more susceptible to high-intensity fires compared to quickly recovering herbaceous vegetation (Higgins, Bond, and Trollope 2000).

Physical heterogeneity is augmented by heterogeneity resulting from plant-herbivore-carnivore dynamics. For instance, the lion and hyena aggregation in areas of high prey abundance produces low prey-abundance areas suitable for cheetahs, which lose most of their prey to lions and hyenas when they are abundant.

Thus far none of the analysis of disease in the Serengeti has explicitly modeled spatial heterogeneity, although it is strongly suspected that the connectivity between patches may be an important factor. For instance, domestic dog density in some villages may be high enough to maintain diseases, which then are transmitted to wild populations or smaller villages.

Wildebeest grazing generates a mosaic of short and long grass that leads to a cascade of events, including (1) patchiness of burning and burn intensity, with effects on the distribution of herbs, (2) a consequent mosaic of insect abundance and bird communities, and (3) differential utilization by topi and Thomson's gazelle, which prefer grazed patches, and buffalo and waterbuck, which prefer ungrazed patches.

The spatial and temporal pattern of resources and resource users, whether herbivores, carnivores, or humans, is an emergent property of the system that results from the resource landscape and the interaction between resources and their users. Emerging from optimal foraging theory in ecology and from economic theory is the concept of the *ideal free distribution* in ecology that says that animals should distribute themselves in space so that the fitness in all locations is the same (Fretwell and Lucas 1970). A similar theory emerges in economics, suggesting that individuals should distribute themselves in space so that the profitability is the same in all areas. If any area has a higher profitability than another then some individuals will move to the best area until the profit in that area is that of other areas (Gordon 1953).

This theory predicts the distribution that should be seen in space. Spatial patterns predicted by theory emerge from studies in the Serengeti. Studies of ungulate distribution (Wilmshurst, Fryxell, and Bergman 2000; Wilmshurst et al. 1999; Fryxell, Wilmshurst, and Sinclair 2004) show that they array themselves in space in order to maximize their fitness, generally preferring sites where the nutritional quality and digestibility are better than in locations where forage quantity is the highest. The effect of the individual behavior distorts the distribution from a true ideal free distribution. This is because of limitations in information and lack of perfectly free mobility, so that some individuals are found in less-than-ideal places. However, the theory does provide a general explanation for the spatial distribution. Cheetahs have been shown to do most of their hunting not where prey are most abundant, but where they have the highest probability of capturing prey and not having it stolen by lions or hyenas. Thus, cheetahs are found primarily on the margins of prey distribution rather than where prey are the most concentrated, because prey concentrations are where the lions and hyenas are also found (Durant 2000).

The Migration

Perhaps the most interesting feature of the Serengeti is the large-scale migration. While migration is quite common in animals, and the migration of wildebeest and zebra in the Serengeti is not large in terms of either number of animals or distance traveled compared to many bird migrations, it is the largest remaining migration of terrestrial mammals since the decimation of the North American buffalo.

We argue that the migration is an emergent property of the system that results from the spatial and temporal pattern of rainfall, nutritional content of the grasses, and the social behavior of the primary predators. The wildebeest and zebra do not need to migrate: there are resident herds of both species and Thomson's gazelle in the western corridor, and certainly wildebeest could survive in the woodlands year round. However, the migration provides access to the seasonally available high nutrition of the short-grass plains, and limits predator numbers by making their biomass seasonally unavailable to lions and hyenas. The Serengeti could not support nearly two million wildebeest and zebra if they did not migrate.

Resilience

Resilience is an emergent property of systems, and has recently been described by Folke et al. (2004) as " the capacity of a system to absorb disturbance and reorganize while undergoing change so as to retain essentially the same function, structure, identity, and feedbacks. . . . . Hence, resilience reflects the degree to which a complex adaptive system is capable of self-organization (versus lack of organization or organization forced by external factors) and the degree to which the system can build and increase the capacity for learning and adaptation."

The Serengeti ecosystem appears to be very resilient. The substantial history of perturbations, described earlier, have all been absorbed, and the Serengeti still carries on as one of the premier natural ecosystems on earth. Certainly the increase in abundance of many species since the 1960s is without precedent for large terrestrial ecosystems. Several factors contribute to the resilience of the Serengeti. The species diversity and resultant functional redundancy seem obvious. Even when the wildebeest were reduced by rinderpest to a fraction of their current abundance the essential properties of the ecosystem appeared to be intact, and Serengeti, even then (1950s), attracted worldwide attention as a natural marvel.

Spatial heterogeneity and geographic scale is another factor promoting resilience. Loss of wildlands surrounding the Mara Reserve in Kenya, and in the Maswa Reserve to the south, have been absorbed presumably because of the scale and heterogeneity of the remaining habitats. However, the lost habitat in both places lie at the edge of the system and were not intensively used by wildlife. Loss of the entire Mara or Maswa could prove a much more serious perturbation. The physical scale of the entire ecosystem must be considered. The basic theory of island biogeography and species diversity shows that size matters, and in terms of protected areas, the Serengeti ecosystem is large.

The major human perturbations that have been absorbed are the introduction of rinderpest, elephant poaching, illegal harvesting for meat of a range of species, and encroachment. Each of these factors has had significant and detectable impacts, but none has appeared to alter the functioning of the system. Perhaps we have been lucky; rinderpest did not drive the major ruminants to very low numbers, elephant poaching, both in the nineteenth and twentieth centuries, was not so severe the elephants could not recover reasonably rapidly when poaching was reduced, and the level of bushmeat harvest in the past has not been so extreme as to eliminate any species. Rhino is perhaps the only species where resilience has been exceeded and numbers may be trapped at a new and very low level.

In comparison with other natural areas the Serengeti remains relatively undisturbed. Fences have not been built, bore holes have not been drilled to provide supplemental water, major predators have not been eliminated, antipoaching activities have restricted the level of harvest, and the geographic integrity of the migration is largely intact. In comparison to Yellowstone or Kruger National parks the Serengeti ecosystem is reasonably pristine.

However, the pressures are mounting. There is growing conflict between elephants and agriculture to the west and south, human population density is increasing and adding pressure to encroachment and meat harvesting, and diseases transmitted through domestic animals appear to be growing in frequency.

LESSONS FOR OTHER COMPLEX ECOSYSTEMS SUBJECT TO CHANGE

The lessons we have learned regarding the behavior of a complex ecosystem from the Serengeti are not necessarily unique. Hughes et al. (2005) summarized current knowledge about ecosystem behavior: "Emerging theories and new multi-disciplinary approaches point to the importance of assessing and actively managing resilience; that is, the extent to which ecosystems can absorb recurrent natural and human perturbations and continue to regenerate without slowly degrading or unexpectedly flipping into alternate states. The capacity of ecosystems to regenerate following disturbance depends on sources of resilience that operate at multiple scales. The concepts of alternate states, resilience and scale are also increasingly prevalent in economics and social science, and in developing theory for linked social-ecological systems" (380).

This description is similar to what we have described for the Serengeti, yet Hughes et al. were describing large *marine* ecosystems, not a terrestrial ecosystem dominated by a migration. From this perspective the Serengeti provides no particularly new lessons. What does emerge are the following lessons:

1. *Preserve the basic physical structure of the habitat.* As the bumper stickers say, "habitat is the key to wildlife," and preservation of the basic habitat is the underlying theme of protected areas management.

2. *Do not try to hold the system in one state.* Serengeti is largely an unmanaged ecosystem—the only major "management" is protection from poaching, grazing, and encroachment. Unlike some other parks, there are no culling programs, no water or food supplementation, no fencing. The Serengeti has been fortunate that park managers have let nature take its course and have let the system move through a range of states. The evidence and theory suggest that this change promotes the resilience and basic integrity of the system.

3. *Do not extract too much.* The Serengeti has provided a sustained yield of tens of thousands of wildebeest annually through illegal harvesting (Mduma, Hilborn, and Sinclair 1998). But clearly there are limits to exploitation. Rhinos are almost completely gone, and elephants were greatly reduced during the peak of the ivory trade in the 1980s.

CONCLUSION

The Defining Characteristics of the Serengeti

The uniqueness of the Serengeti is seen in the size and scale of the migration, the diversity of species (particularly mammals and birds), and the scale and diversity of the habitat, which remains in a wild state. None of the enormous natural and anthropogenic perturbations have significantly altered any of these three characteristics. We argued earlier in this chapter that the habitat and species diversity contributes to the resilience of the Serengeti and thus provides a positive feedback to support the maintenance of the defining characteristics.

While different chapters have provided models of some of the simple components of the system, we must recognize the limited scope of these models in relation to the multiple taxonomic, trophic, and spatial diversity of the ecosystem. We have no models of the birds, reptiles, amphibians, insects, soil micro-flora and fauna—our modeling attention has been confined to large mammals, their primary food, and the human system. Who knows what biocomplexity would emerge from these other components of the ecosystem, and how these neglected elements might interact with the parts of the system where most attention is focused?

We have described a range of ecological theories, emergent behaviors, and how the data and models available relate to these theories. Did those theories prove useful, and do they provide predictions or a framework for understanding how such a complex ecosystem works?

Food web theory, and the associated topics of trophic cascades, bottom-up forcing, and functional redundancy, certainly has some power, especially in relation to the basic laws of trophic dynamics. This theory tells us that predators will be less abundant than herbivores and that the biomass of herbivores is limited by the food available and their conversion efficiency. While we are so familiar with the trophic structure of ecosystems that these predictions seem trivial, we should not ignore the basic utility of the trophic view. The associated theories of bottom-up versus top-down structure does not seem to provide any predictions, rather more a framework for understanding how the particular life histories of species might result in system behavior.

The more recent theory of food webs, suggesting weak trophic interactions and spatial heterogeneity supporting resilience, is consistent with our basic observations of the ecosystem and our intuition about the impact of the diverse food webs. While falling short of making any real quantitative predictions, we certainly feel that this work does provide useful insights to the functioning of the system.

It is not clear if the concept of functional redundancy provides much power or utility. One might have expected the reduction in wildebeest due to rinderpest to lead to the replacement of wildebeest by another ungulate with similar habitat preferences, but again perhaps the species would need to be a ruminant to take advantage of all that grass. When wildebeest numbers were lower, no species came forward to consume all that uneaten grass—it remained to burn or decompose through another trophic path. One can argue that the functional redundancy in many guilds contributes to the resilience of the system, but we have not yet seen strong evidence for this.

Of all the ecological theories examined, keystone species seems to provide the least power. From the perspective of the Serengeti, the concept of keystone species provides a convenient label for key species such as wildebeest and elephants, but little more.

Optimal foraging does provide both specific predictions in relation to spatial distribution of some species, and a powerful analytic framework for posing testable hypotheses. Within this book the theory has been used with success in the analysis of ungulate distribution (chapter 9) and as a framework for modeling human household behavior (chapters 10, 11, and 12, by Costello et al., Galvin et al., and Polasky et al., respectively).

Holling's concept of resilience and multiple stable states seems the most attractive of the theories. It provides a framework for explaining what has happened and a qualitative predictive model for nonlinear response and switching between states. Several of the modeling chapters, based upon simple dynamic models with realistic parameter estimates, provide examples of multiple stable states.

However, the power of Holling's theories lies not in specific quantitative predictions such as emerge from food web or optimal foraging theory, but in qualitative descriptions. Holling's theory does not predict to what extent the bounds of an attractor region will contract when a system is held rigidly at an equilibrium, but rather it predicts that this will happen, and one should build a specific model of the dynamics that may indeed indicate what will happen. Holling's theory predicts that change is natural and essential, that attempting to maintain a system in one state is dangerous, and that periods of growth and collapse are part of the evolution of ecosystems.

SUMMARY

The Serengeti is a large, mostly pristine ecosystem that has seen many natural and anthropogenic changes and yet still retains its essential character. As such, it is one of the most positive examples of conservation in the world, and is a treasure for the entire planet. As a testing ground for ecological theory, the Serengeti provides a number of challenges. The ability of theory to grapple with the high diversity of species and spatial complexity of the Serengeti is limited, but we do see emerging in the ecological literature concepts and approaches that seem to fit well with the Serengeti experience. Food web theory, resilience, and optimal foraging theory all provide frameworks and explanations that support observations from the Serengeti. At present none of these theories and their associated models have reached levels of precision that could provide quantitative prediction, but the actual mathematical modeling of trophically and spatially complex systems is in its infancy.

The challenge for ecologists is to develop further the mathematical models that explicitly consider the trophic and spatial complexity of the Serengeti. The modern tools of GIS systems and agent-based models may provide quantitative predictions for such complex systems.

REFERENCES

Anderson, R. M., and R. M. May. 1991. *Infectious diseases of humans: Dynamics and control.* Oxford: Oxford Science Publications.

Carpenter, S. R., and J. F. Kitchell, eds. 1993. *The trophic cascade in lakes.* Cambridge: Cambridge University Press.

Corfield, T. 1973. Elephant mortality in Tsavo National Park, Kenya. *East African Wildlife Journal* 11:339–68.

Diamond, J. 1999. *Guns, germs and steel: The fates of human societies.* New York: W.W. Norton.

Dublin, H. T., A. R. E. Sinclair, and J. McGlade. 1990. Elephants and fire as causes of multiple stable states in the Serengeti Mara woodlands. *Journal of Animal Ecology* 59: 1147–64.

Durant, S. M. 1998. Competition refuges and coexistence: An example from Serengeti carnivores. *Journal of Animal Ecology* 67:370–86.

———. 2000. Living with the enemy: Predator avoidance of hyaenas and lions by cheetahs in the Serengeti. *Behavioral Ecology* 11:624–32.

du Toit, J. T., K. H. Rogers, and H. C. Bigg, eds. 2003. *The Kruger experience.* Washington, DC: Island Press.

Elton, C. S., and R. S. Miller. 1954. The ecological survey of animal communities: With a practical system of classifying habitats by structural characters. *Journal of Ecology* 42: 460–96.

Folke, C., S. Carpenter, B. Walker, M. Scheffer, T. Elmqvist, L. Gunderson, and C. S. Holling. 2004. Regime shifts, resilience, and biodiversity in ecosystem management. *Annual Review of Ecology Evolution and Systematics* 35:557–81.

Ford, J. 1971. *The role of trypanosomiases in African ecology.* Oxford: Clarendon.

Fretwell, S. D. and H. L. Lucas. 1970. On territorial behaviour and other factors influencing habitat distribution in birds. I. Theoretical development. *Acta Biotheoretica* 19: 16–36.

Fryxell, J. M., and P. Lundberg. 1998. *Individual behavior and community dynamics.* London: Chapman and Hall.

Fryxell, J. M., J. F. Wilmshurst, and A. R. E. Sinclair. 2004. Predictive models of movement by Serengeti grazers. *Ecology* 85:2429–35.

Gause, G. F. 1936. Further studies of interaction between predators and prey. *Journal of Animal Ecology* 5:1.

Gordon, H. S. 1953. An economic approach to the optimum utilization of fishery resources. *Journal of the Fisheries Research Board of Canada* 10:442–57.

Gunderson, L. H. 2000. Ecological resilience—in theory and application. *Annual Review of Ecology and Systematics* 31:425–39.

Gunderson, L. H., and C. S. Holling. 2002. *Panarchy: Understanding transformations in human and natural systems.* Washington, DC: Island Press.

Higgins, S. I., W. J. Bond, and W. S. W. Trollope. 2000. Fire, resprouting and variability: A recipe for grass-tree coexistence in savanna. *Journal of Ecology* 13:295–99.

Hilborn, R. 1975. Effect of spatial heterogeneity on persistence of predator-prey interactions. *Theoretical Population Biology* 8:346–55.

Hilborn, R., and A. R. E. Sinclair. 1978. A simulation of wildebeest, other ungulates, and their predators in the Serengeti. In *Serengeti: Dynamics of an ecosystem,* ed. A. R. E. Sinclair and M. Norton-Griffiths, 287–309. Chicago: University of Chicago Press.

Holling, C. S. 1973. Resilience and stability of ecological systems. *Annual Review of Ecology and Systematics* 4:1–23.

———. 1980. Forest insects, forest fires, and resilience. In *Fire regimes and ecosystem properties,* ed. H. Mooney, J. M. Bonnicksen, N. L. Christensen, J. E. Lotan, and W. A. Reiners, 445–64. Washington, DC: USDA Forest Service General Technical Report.

Holt, R. D. 2002. Food webs in space: On the interplay of dynamic instability and spatial processes. *Ecological Research* 17:261–73.

Hopcraft, J. G. C., A. R. E. Sinclair, and C. Packer. 2005. Planning for success: Serengeti lions seek prey accessibility rather than abundance. *Journal of Animal Ecology* 74: 559–66.

Huffaker, C. B. 1958. Experimental studies on predation: Dispersion factors and predator-prey oscillations. *Hilgardia* 27:343–83.

Hughes, T. P., D. R. Bellwood, C. Folke, R. S. Steneck, and J. Wilson. 2005. New paradigms for supporting the resilience of marine ecosystems. *Trends in Ecology and Evolution* 20: 380–86.

Janssen, M. A., J. M. Anderies, and B. H. Walker. 2004. Robust strategies for managing rangelands with multiple stable attractors. *Journal of Environmental Economics and Management* 47:140–62.

Kikkawa, J., and D. J. Anderson, eds. 1986. *Community ecology: Pattern and process.* Melbourne: Blackwell Scientific Publications.

Knowlton, N. 2004. Multiple "stable" states and the conservation of marine ecosystems. *Progress in Oceanography* 60:387–96.

Kondoh, M. 2003. Foraging adaptation and the relationship between food-web complexity and stability. *Science* 299:1388–91.

Kuijper, L. D. J., B. W. Kooi, C. Zonneveld, and S. Kooijman. 2003. Omnivory and food web dynamics. *Ecological Modelling* 163:19–32.

Lawlor, L. R., and J. Maynard-Smith. 1976. The coevolution and stability of competing species. *American Naturalist* 110:79–99.

Ludwig, D., D. D. Jones, and C. S. Holling. 1978. Qualitative analysis of insect outbreak systems: The spruce budworm and forest. *Journal of Animal Ecology* 47:315–32.

McCann, K. S., A. Hastings, and G. R. Huxel. 1998. Weak trophic interactions and the balance of nature. *Nature* 395:794–98.

McCann, K. S., J. B. Rasmussen, and J. Umbanhower. 2005. The dynamics of spatially coupled food webs. *Ecology Letters* 8:513–23.

McNeill, W. H. 1967. *Plagues and peoples.* Garden City, NY: Anchor Press/Doubleday.

Mduma, S. A. R., R. Hilborn, and A. R. E. Sinclair. 1998. Limits to exploitation of Serengeti wildebeest and implications for its management. In *Dynamics of tropical communities,* ed. D. M. Newbery, H. T. T. Prins, and N. D. Brown, 243–65. Oxford: Blackwell Science.

Mduma, S. A. R., A. R. E. Sinclair, and R. Hilborn. 1999. Food regulates the Serengeti wildebeest: A 40-year record. *Journal of Animal Ecology* 68:1101–22.

Michener, W. K., T. J. Baerwald, P. Firth, M. A. Palmer, J. L. Rosenberger, E. A. Sandlin, and H. Zimmerman. 2001. Defining and unraveling biocomplexity. *Bioscience* 51: 1018–23.

Mills, E. L., J. H. Leach, J. T. Carlton, and C. L. Secor. 1994. Exotic species and the integrity of the Great Lakes. *BioScience* 44:666–76.

Nystrom, M., and C. Folke. 2001. Spatial resilience of coral reefs. *Ecosystems* 4:406–17.

Packer, C., R. Hilborn, A. Mosser, B. Kissui, M. Borner, G. Hopcraft, J. Wilmshurst, S. Mduma, and A. R. E. Sinclair. 2005. Ecological change, group territoriality, and population dynamics in Serengeti lions. *Science* 307:390–93.

Paine, R. T. 1966. Food web complexity and species diversity. *The American Naturalist* 100:65.

———. 1980. Food webs: Linkage, interaction strength and community infrastructure. *Journal of Animal Ecology* 49:667–85.

Power, M. E., W. E. Dietrich, and J. C. Finlay. 1996. Dams and downstream aquatic biodiversity: Potential food web consequences of hydrologic and geomorphic change. *Environmental Management* 20:887–95.

Root, R. B. 1968. The niche exploitation pattern of the blue gray gnatcatcher. *Ecological Monographs* 37:317.

Scheffer, M., S. Carpenter, J. A. Foley, C. Folke, and B. Walker. 2001. Catastrophic shifts in ecosystems. *Nature* 413:591–96.

Sinclair, A. R. E. 1995. Equilibria in plant-herbivore interactions. In *Serengeti II: Dynamics, management and conservation of an ecosystem.*, ed. A. R. E. Sinclair and P. Arcese, 91–114. Chicago: University of Chicago Press.

Sinclair, A. R. E, and A. Byrom. 2006. Understanding ecosystems for the conservation of biota. *Journal of Animal Ecology* 75:64–79.

Sinclair, A. R. E., S. A. R. Mduma, J. G. C. Hopcraft, J. M. Fryxell, R. Hilborn, and S. Thirgood. 2007. Long term ecosystem dynamics in the Serengeti: Lessons from conservation. *Conservation Biology* 21(3): 580-90.

van Nes, E. H., and M. Scheffer, M. 2005. Implications of spatial heterogeneity for catastrophic regime shifts in ecosystems. *Ecology* 86: 1797–1807.

Walker, B. H. 1992. Biodiversity and ecological redundancy. *Conservation Biology* 6:18–23.

Wilmshurst, J. F., J. M. Fryxell, and C. M. Bergman. 2000. The allometry of patch selection in ruminants. *Proceedings of the Royal Society of London, Series B–Biological Sciences* 267:345–49.

Wilmshurst, J. F., J. M. Fryxell, B. P. Farm, A. R. E. Sinclair, and C. P. Henschel. 1999. Spatial distribution of Serengeti wildebeest in relation to resources. *Canadian Journal of Zoology* 77:1223–32.

Woolhouse, M. E. J., L. H. Taylor, and D. H. Haydon. 2001. Population biology of multi-host pathogens. *Science* 292:1109–12.

FIFTEEN

Who Pays for Conservation? Current and Future Financing Scenarios for the Serengeti Ecosystem

Simon Thirgood, Charles Mlingwa, Emmanuel Gereta,
Victor Runyoro, Rob Malpas, Karen Laurenson, and Markus Borner

Tanzania has one of the world's most extensive networks of protected areas, with approximately 25% of the country's 1 million km² managed for conservation in parks and reserves. This remarkable achievement has been attained despite the fact that Tanzania is amongst the world's poorest countries, ranking seventeenth lowest in gross national income (GNI), second-lowest in purchasing power parity (PPP), and with an average per capita income of only ~$300 per annum. The conservation of Tanzania's natural resources was made a national priority in a presidential declaration by Mwalimu Julius Nyerere shortly after independence in 1961, in what is now known as the Arusha Manifesto. With this high degree of political support, the protection of Tanzania's extensive wilderness areas became government policy. Tanzania has continued to show a strong commitment to conservation and the maintenance of protected areas is increasingly seen as an important component of development because of the economic significance of wildlife tourism.

Tanzania's tourism sector has recorded strong growth since the international border with Kenya reopened in 1984. Tourist expenditures rose from $65 million in 1990 to $725 million in 2001, accounting for 12% of Tanzania's gross domestic product (GDP; World Bank 2002). The first National Tourism Policy, in 1991, provided objectives and strategies to increase the involvement of the private sector in the tourism industry (Ministry of Natural Resources and Tourism 1991). This policy was revised in 2000 to assist efforts to promote the national economy by encouraging the development of high-quality, low-impact tourism that is culturally and socially accept-able, ecologically friendly, and environmentally sustainable (Ministry of Natural Resources and Tourism 2000). The policy aims to increase tourism contribution to GDP to 25% by 2010, by which time the annual number of tourists will be approximately one million. Tanzania's tourism industry has benefited from the political stability of the country in relation to its main competitors in the African wildlife tourism market—Kenya, Zimbabwe, and South Africa. However, the volatility of the industry was demonstrated by the slump in tourism throughout East Africa following the September 11, 2001, terrorist attacks in the United States, although this effect was only temporary and tourism expenditures have continued to rise.

Despite the importance of wildlife tourism and hence natural resources for the economy, conservation is low in the priorities for the Tanzanian government's budget allocation. The Ministry of Natural Resources and Tourism (MNRT), the principal ministry for the wildlife and conservation sector, received some $46 million in 2004, only 1.4% of the government's annual budget, and was dwarfed by other ministries, such as Education and Health. Indeed, Tanzania is exceptional in a global context in that the two main government conservation agencies—Tanzania National Parks and the Wildlife Division—currently generate income for the Tanzanian Treasury rather than receive a government subsidy. It is worth stressing the point that Tanzania uses its wildlife resources as a source of income rather than a target for subsidies as this is crucial to understanding the financing options that are available to pay for conservation. Based on a protected-area network of 250,000 km^2, of which national parks comprise around 50,000 km^2 (in-cluding the 8,288 km^2 multiple-use Ngorongoro Conservation Area), game reserves comprise around 100,000 km^2, and game-controlled areas and for-est reserves make up the remaining 100,000 km^2, government expenditure on natural resource management and tourism in protected areas is ~$184 km^{-2} per annum. However, these figures exclude donor support and private investment in the wildlife sector and, as we show in the following, these additional resources can be considerable. To put these figures in context, recently published global analyses suggest that the costs of field-based ter-restrial conservation vary enormously, from < $0.1 to > $1,000,000 km^{-2} yr^{-1} (Balmford et al. 2003). The costs of effective protected areas in Africa range from $130 to > $5,000 km^{-2} yr^{-1}, with typical costs of $\cong$ $1,000 km^{-2} yr^{-1}. Such figures are, however, difficult to interpret, as they are complicated by questions of scale and efficiency in protected-area management. None the less, on this basis it is clear that overall government expenditure on pro-tected areas in Tanzania is toward the low end of the continental scale.

The Serengeti ecosystem is one of Tanzania's most important biological resources and is also the jewel in the crown of Tanzania's wildlife tourism industry. Serengeti, with the other parks of the northern circuit—Manyara, Tarangire, and Kilimanjaro—accounted for 85% of all visitor days in the national park system in 2001 (Tanzania National Parks Authority [TANAPA] 2002). The Serengeti is best known for supporting the largest herds of migratory ungulates in the world. Two million wildebeest, zebra, and gazelle migrate annually from the short-grass plains of Ngorongoro and Serengeti to the woodlands of the Mara, and these migratory herds and other resident ungulates support around 10,000 hyenas and lions, making the Serengeti one of the most important protected areas for mammals in Africa (Fjeldsa et al. 2004). The Serengeti forms the world's largest intact savannah ecosystem and is a living laboratory for the study of large-scale ecological processes—as outlined in the chapters of this volume and two previous books on the ecosystem (Sinclair and Norton-Griffiths 1979; Sinclair and Arcese 1995). Given the national and global importance of the Serengeti in terms of biodiversity conservation and economic development, it is clearly imperative to develop sustainable financing mechanisms that will guarantee the continued existence of this unique area.

This chapter focuses on the financial structure supporting conservation activities in the Tanzania sector of the Serengeti ecosystem. We start by giving an introduction to the management authorities responsible for the different components of the protected-area network that together comprises the Tanzanian Serengeti. We then give a detailed budget breakdown of revenues and expenditures for the national park and surrounding protected areas for the period 1998–2002, to assess whether financial returns from tourism and other forms of wildlife utilization are sufficient to finance conservation in the Tanzanian sector of the ecosystem or whether external financial support is required. As part of this analysis we also assess the financial returns to local communities resident in the multiple-use buffer areas adjacent to the core protected areas from both consumptive and nonconsumptive use of wildlife. This latter analysis is particularly relevant to the ongoing debate in Tanzania and elsewhere in Africa about community-based conservation and the distribution of financial benefits from wildlife and protected areas to local people (Hulme and Murphree 2001; Adams et al. 2004; Walpole 2006). In the final part of the chapter we look forward to the future and make a cautious attempt to predict how conservation activities in the different components of the Tanzanian Serengeti might be financed in the coming decade and how this may impact local communities.

The Tanzanian sector of the Serengeti ecosystem can be divided into five management categories, each of which is subject to different forms of natural resource use and different conservation finance regimes. These management categories are administered by three different government organizations, all of which are ultimately under the jurisdiction of the Ministry of Natural Resources and Tourism (MNRT). The management categories are briefly summarized here (see also fig. 2.1, this volume, for a map of the Serengeti-Mara ecosystem):

1. Serengeti National Park (SNP) is managed strictly for conservation and wildlife tourism and allows no consumptive use of wildlife or human settlement other than park staff, tourism staff, and researchers. Serengeti National Park is administered by the parastatal organization Tanzania National Parks Authority (TANAPA) which, while coming under the jurisdiction of the MNRT, has it own board of directors, and has the power to raise revenue and set its own budgets. The Tanzania National Parks Authority pays business tax to the Tanzanian Treasury.

2. Ngorongoro Conservation Area (NCA) is a multiple-use area that combines natural resource protection with development for the indigenous Maasai. Permitted land use includes wildlife tourism, pastoralism and limited cultivation, and permanent settlement. The NCA is administered by the parastatal organization Ngorongoro Conservation Area Authority (NCAA) which comes under the jurisdiction of the MNRT. As with TANAPA, the NCAA has its own board of directors, has the power to raise revenues and set budgets, and pays business tax to the Tanzanian Treasury.

3. Maswa, Grumeti, and Ikorongo game reserves (GR) are managed primarily for conservation through the consumptive use of wildlife by tourist hunting, although wildlife tourism is becoming increasingly important. There is no legal settlement, pastoralism, or cultivation within the game reserves. The three reserves fall under the jurisdiction of the Wildlife Division (WD) which is an agency of the MNRT. The WD receives a budget directly from the government, and returns all of its income to the Tanzanian Treasury, with the exception of the Selous Game Reserve in southern Tanzania, which has a local revenue retention scheme.

4. Loliondo Game Controlled Area (GCA) is a multiple-use area to the east of the SNP within which wildlife resources are under the jurisdiction of the WD. Loliondo is settled by Maasai pastoralists, who increasingly practice cultivation. There is a single tourist hunting concession and several wildlife and cultural tourism operators who work in partnership with individual Maasai villages.

5. Wildlife Management Areas (WMAs) are a new initiative in community-based natural resource management currently being established in Tanzania. They are based in part on former GCAs and are typically adjacent to large protected areas. Wildlife Management Areas are managed by community-based Authorized Associations and fall under the jurisdiction of the WD. There are plans for four WMAs in the Serengeti ecosystem: Loliondo and Natron to the east, Ikoma to the west, and Makao to the south.

With such structural complexity in the institutions responsible for conservation management in the Serengeti ecosystem, it comes as no surprise that each of these categories has a different financing system. The main source of internal revenue for the SNP and NCA is wildlife tourism. Tourist-gate fees account for ~80% of internally generated income. Other tourism-related sources, such as vehicle entry, camping, filming, aircraft landing, hotel concession fees, and other commercial enterprises account for ~15% of income (SNP 2002; NCAA 2003). In contrast, the main sources of internal income for the Maswa, Grumeti, and Ikorongo GR are hunting-block concession fees and trophy fees. Income from nonconsumptive use of wildlife in the GR is limited, although this situation may change with the development of wildlife tourism in the Grumeti and Ikorongo GR. The Loliondo GCA currently generates income through hunting-block concession fees and trophy fees and also local agreements between rural communities and tour operators providing cultural and wildlife tourism. Loliondo has been designated to become a WMA, and it is anticipated that revenues will be generated through a combination of consumptive and nonconsumptive use of wildlife.

Conservation in the Tanzanian Serengeti also receives external finance from donors. Long-term support has been provided by the Frankfurt Zoological Society (FZS), which has been active in the Serengeti since 1958. FZS currently provides ~$1 million per annum to support a variety of conservation activities throughout the Serengeti, including resource protection, ecological monitoring, veterinary intervention, tourism development, and community-based conservation. The Danish International Development Agency (DANIDA) has provided long-term support to Maasai pastoral development in the NCA. Other donors have injected large amounts of money over shorter time periods. Notable among these is the European Union (EU), which provided ~$10 million to SNP between 1998 and 2001 for infrastructure, and the Norwegian Agency for International Development (NORAD) which, in conjunction with the World Conservation Union (IUCN), provided ~$5 million between 1985 and 1995 for the development and implementation of the Serengeti Regional Conservation Strategy

(Mbano et al. 1995) and the Ngorongoro Conservation and Development Project (Thompson 1997). Less tangible and more difficult to quantify are the resources that have been provided to support research activities in the Serengeti. This research, typically externally funded and until recently conducted primarily by expatriate scientists, has contributed to conservation through improving the evidence base for management, but is difficult to quantify in terms of financial inputs.

CONSERVATION FINANCE IN THE SERENGETI ECOSYSTEM

In the following sections we provide a breakdown of income and expenditure during the period 1998–2002 for the main management components of the Tanzanian Serengeti. In the case of SNP and NCA, these data are extracted from the published annual reports of TANAPA and the NCAA, and we categorize income and expenditure using the reporting structures of these institutions. Financial information from other sectors of the ecosystem is less readily available, and we have used a variety of published and unpublished sources, including personal communications from protected-area managers and other stakeholders.

Serengeti National Park

Internally generated income in the 14,763 km^2 SNP during the 5 years 1997/1998 to 2001/2002 averaged $5,230,000 per annum and showed an increase over time (table 15.1). Income was primarily derived from tourist-gate fees, with a relatively small contribution from other sources. All income derived in the Serengeti is centrally managed by TANAPA, which as a parastatal pays tax to the Tanzanian Treasury. With the exception of Serengeti and Kilimanjaro, Tanzania's national parks do not generate sufficient income to cover their expenditures. Income from these two parks is, therefore, used to subsidize the rest of the TANAPA network of national parks.

Less than 40% of average annual income of the SNP is re-allocated as TANAPA expenditure budget to the park (table 15.1). The allocated budget to SNP during the 5 years 1997/1998 to 2001/2002 averaged $1,950,000 and, in parallel with income, showed a steadily increasing trend. These figures may not reflect expenditure, as budget allocations may, in some cases, be unrealized, although this discrepancy is difficult to quantify. Looking at SNP expenditure budget allocations by TANAPA departments, administration, which includes all staff salaries, receives the largest budget, at around

Table 15.1 Annual revenues and expenditures (USD) for Serengeti National Park for the financial years 1997/1998 to 2001/2002. Data from TANAPA annual reports.

Years	Revenue	Expenditure	Exp/Rev (%)
1997/98	4,631,247	1,314,139	28.4
1998/99	4,521,689	1,927,288	42.6
1999/00	5,119,417	1,993,281	38.9
2000/01	6,040,291	2,172,791	36.0
2001/02	5,845,775	2,352,301	40.2

Table 15.2 Annual budget allocation (USD) to Serengeti National Park Departments for the financial years 1997/1998 to 2001/2002. Data from TANAPA annual reports.

Department	1997/98	1998/99	1999/00	2000/2001	2001/02	%
Protection	55,855	57,000	95,564	114,953	134,342	5.8
Community	110,042	110,042	156,676	185,770	214,864	9.8
Ecology	77,576	79,167	132,727	159,657	186,586	15.6
Tourism	232,727	237,500	398,182	478,969	559,757	24.1
Works	99,297	101,333	169,891	204,360	238,830	10.3
Finance	794,497	1,399,246	1,135,805	1,144,035	1,152,264	34.4
Total	1,369,994	1,984,288	2,088,845	2,287,744	2,486,643	

50% of total expenditure, and resource protection, which includes anti-poaching operations, receives the least, with around 5% of total budget allocation (table 15.2).

Donor support for SNP provides a significant supplement to the TANAPA expenditure budget and must be included to calculate the real financial cost of managing the park. Donor support can be difficult to quantify, particularly where multiple agencies are involved and support may be as capital investment in infrastructure as well as annual contributions to running costs. In the case of SNP during the period of analysis, this process was simplified by the presence of two main donors—the European Union and the Frankfurt Zoological Society. European Union support to SNP was primarily for infrastructure such as roads and buildings and totaled ~$10 million over the period 1998–2001. Financial support to the SNP from FZS over the 6 years from 1998 to 2003 totaled $4,460,000, with an average contribution of $740,000 per annum (table 15.3). This support was allocated across the TANAPA departments in different proportions to the TANAPA expenditure budget allocations, with the largest proportion (40%) going

Table 15.3 Frankfurt Zoological Society financial support (USD) to the Serengeti Ecosystem 1998–2003

Department	SNP	NCA	WD	Total
2003				
Protection	357,675	143,000	118,100	618,775
Communities	35,625		106,563	142,188
Ecology	235,327	50,000		285,327
Tourism	67,000			67,000
Works/Transport	50,625	109,125	28,750	188,500
Administration	48,125			48,125
Total	**794,377**	**302,125**	**253,413**	**1,349,915**
2002				
Protection	341,125	137,500	174,000	652,625
Communities			99,900	99,900
Ecology	215,650	39,600	58,900	314,150
Tourism	77,800			77,800
Works/Transport	156,250			156,250
Administration	43,300			43,300
Total	**834,125**	**177,100**	**332,800**	**1,344,025**
2001				
Protection	337,800	186,700	181,100	705,600
Communities			107,000	107,000
Ecology	248,000	42,000	42,000	332,000
Tourism	25,600			25,600
Works/Transport	132,900			132,900
Administration				0
Total	**744,300**	**228,700**	**330,100**	**1,303,100**
2000				
Protection	372,300	185,700	143,400	701,400
Communities			95,800	95,800
Ecology	223,400			223,400
Tourism	20,600			20,600
Works/Transport	187,000			187,000
Administration	16,400			16,400
Total	**819,700**	**185,700**	**239,200**	**1,244,600**
1999				
Protection	136,000	151,300	166,700	454,000
Communities			63,300	63,300
Ecology	278,300	82,900		361,200
Tourism				0
Works/Transport	313,100	33,500		346,600
Administration				0
Total	**727,400**	**267,700**	**230,000**	**1,225,100**

Table 15.3 (continued)

Department	SNP	NCA	WD	Total
1998				
Protection	192,400	172,000	222,400	586,800
Communities			58,600	58,600
Ecology	142,200	35,500		177,700
Tourism			12,000	12,000
Works/Transport	190,700			190,700
Administration	14,000			14,000
Total	**539,300**	**207,500**	**293,000**	**1,039,800**

Note: Data from FZS audited accounts.

to resource protection. A conservative estimate indicates that donor support to SNP averaged around $3 million per annum during 1997/1998 to 2001/2002, although the EU project is now complete, thus considerably reducing donor support. The total amount invested in managing the SNP was therefore around $5 million per annum during 1997/1998 to 2001/2002, a figure notably close to the income generated by the park through tourism.

The current (2004) expenditure budget in SNP is ~$4 million, with ~$3 million coming directly from TANAPA and ~$1 million from international donors (Justin Hando, SNP Warden-in-Charge, pers. comm.). With the area of SNP totaling some 14,763 km², this represents a current investment in conservation of ~ $270 km^{-2} per annum. To put this in context, figures for protected areas of comparable size, such as Kruger National Park, are ~ $1,800 km^{-2} per annum (Gus Mills, South Africa National Parks, pers. comm.), and Yellowstone National Park are ~ $3,000 km^{-2} per annum (Glenn Plumb, United States National Park Service, pers. comm.).

Ngorongoro Conservation Area

Internally generated income in the 8,285 km² NCA during the 5 years 1998/1999 to 2002/2003 averaged $5,890,000 per annum and showed a steady increase over time, reflecting the current strength of the wildlife tourism sector in Tanzania (table 15.4). As with the SNP, approximately 80% of income was derived from tourist-gate fees and vehicle entry fees, with relatively small contributions from other sources. The Ngorongoro Conservation Area Authority (NCAA) is a parastatal organization which retains all income generated in the NCA. The NCAA pay tax to the Tanza-

Table 15.4 Annual revenues and expenditures (USD) for the Ngorongoro
Conservation Area for the financial years 1998/1999 to 2002/2003. Data from NCAA
annual reports.

Years	Revenue	Expenditure	Exp/Rev (%)
1998/99	5,156,056	5,163,414	100.1
1999/00	4,161,825	3,396,570	81.6
2000/01	6,027,707	5,397,195	89.5
2001/02	6,213,415	5,475,968	88.1
2002/03	7,892,010	4,718,934	59.8

nian Treasury and contribute funds to support community development projects in the NCA. Through mutual agreement, the NCAA also contributes funds to other partner organizations, such as the Tanzania Wildlife Research Institute, Mweka College of African Wildlife Management, and Pasiansi Wildlife Training Institute.

More than 80% of the revenues generated within the NCA were re-allocated as expenditure budget during the period 1998/1999 to 2002/2003. However, there was considerable variation between years, with the ratio of expenditure to revenue dropping sharply to 60% during 2002/2003 (table15.4). Given constant taxation levels on revenues it is not immediately clear how the remaining revenues were spent in this last financial year. As with SNP, these figures may not reflect actual expenditures, as budget allocations may in some cases be unrealized, although again this is difficult to quantify. From examination of the NCAA expenditure budget allocations by department (table 15.5), it is clear that salary costs included in the administration department receives the largest portion of the budget, averaging 59% of total expenditure over the period 1998/1999 to 2002/2003. Resource protection and ecological monitoring received a combined budget of 5% of total expenditure. One further notable trend is that the relatively high budget allocation to community development (averaging $460,000 during 1998/1999 to 2001/2002) was cut to $68,060 in 2002/2003, although this may explain the discrepancy in revenues and expenditures in 2002/2003 noted above.

The NCA also receives external support from international donors, but this is more limited in financial terms in relation to overall expenditure than is the case with SNP. The main donor during 1998 to 2003 was the Frankfurt Zoological Society, which provided a total of $1,370,000, with an average contribution of $230,000 per annum (table 15.3). Most of this support was to the Resource Protection Department (71%), with specific focus

Table 15.5 Annual budget allocation (USD) to Ngorongoro Conservation Area Authority Departments for the financial years 1998/99–2002/03. Data from NCAA annual reports.

Department	1998/99	1999/00	2000/2001	2001/2002	2002/2003	%
Protection	160,093	45,823	192,885	216,240	324,353	4.9
Community	473,999	235,671	444,917	694,290	68,060	10.0
Ecology	37,525	7,068	33,067	108,993	127,120	14.9
Tourism	143,208	147,010	388,032	393,566	354,963	7.4
Works	1,326,372	800,596	1,094,346	988,156	1,105,393	27.7
Finance	3,022,217	2,160,402	3,243,948	3,074,723	2,739,045	35.1
Total	5,163,414	3,396,570	5,397,195	5,475,968	4,718,934	

on the Rhino Protection Unit in the Ngorongoro Crater. In this restricted context the external financial support greatly exceeds the NCA budget allocation for protection of the Crater rhino population, and this donor support is believed to be essential for its survival (Mills et al. 2006). As noted above, DANIDA also has a major long-term program in the livestock and development sectors, but these are peripheral to the main subject of this chapter. Combining NCAA budget allocations with donor support, the total expenditure in the NCA was around $5 million per annum during 1998/1999 to 2002/2003. This figure is close to what is internally generated in the NCA through tourism.

Maswa, Grumeti, and Ikorongo Game Reserves

The financing situation of the game reserves (GR) is fundamentally different from the SNP and NCA. Income is derived almost entirely from hunting-block concession fees and trophy fees. The most detailed financial information is available for the Maswa GR. This reserve of 2,897 km² was established in 1962 and currently contains three hunting blocks operated on 5-year leases by different companies—Robin Hurt Safaris Tanzania, Tanzania Game Tracker Safaris, and Tanzania Big Game Safaris. Hunting revenues for Maswa GR for the period 1997 to 2001 averaged $240,000 per annum (William Ngowa, WD Warden-in-Charge, pers. comm.). Game Reserves are managed by the WD and all revenues are retained by the Tanzania Treasury. Although retention schemes whereby revenues generated within the reserves are retained locally for management purposes are in place in the Selous Game Reserve in southern Tanzania (Baldus, Kibondi, and Siege

2003), there is currently no such retention scheme in Maswa or any of the other northern game reserves. Less than 2% of annual income generated in Maswa is re-allocated from the WD as expenditure budget to the reserve, giving a total of ~ $5000 (William Ngowa, WD Warden-in-Charge, pers. comm.). This figure does not, however, include the salaries for game scouts, which are paid by central government, bringing the annual government expenditure in Maswa GR to ~$20,000, or just under 10% of income generated through hunting.

Maswa GR receives external financial support from two sources. The Frankfurt Zoological Society contributed a total of $500,000 over the period 1997 to 2001, averaging ~$100,000 per annum, although this had been reduced to ~$50,000 by 2004 (table 15.3). This money was primarily directed toward resource protection and essential infrastructure such as ranger posts. Additional external funding for Maswa comes from the private sector. The Friedkin Conservation Fund (FCF) is a locally registered NGO that is associated with the hunting company Tanzania Game Trackers Safaris. The Friedkin Conservation Fund employs and equips a private ranger force that operates in conjunction with WD rangers in resource protection duties. Similar resource protection within and adjacent to the game reserve is undertaken by rangers employed and equipped by the Cullman and Hurt Community Wildlife Project (CHCWP), a locally registered NGO that is associated with the hunting company Robin Hurt Safaris Tanzania which is the concession holders for the southern hunting block in Maswa and the adjacent Makao Open Area. During the 1990's, it was estimated that the private sector was investing $100,000 per annum in and adjacent to Maswa GR (Maswa Game Reserve 1994), and it is believed that similar sums are currently being invested, although details are not available. Expenditure in Maswa may therefore be in the region of $200,000 per annum, of which 90% is sourced from donors or the private sector. It is notable that this expenditure is similar to the income obtained by the government from hunting fees.

The 605 km^2 Ikorongo GR and 420 km^2 Grumeti GR were established in 1994 when they were upgraded from Game Controlled Areas. Each reserve contains one hunting block, and these have been leased as a single concession together with the adjacent Ikoma Open Area since they were established. Hunting revenues to the WD from the two reserves totaled ~$200,000 per annum during the period 1997 to 2001. The internal expenditure budget from the WD to the two reserves has averaged $20,000 per annum, including recurrent costs and salaries for game scouts (J. Katalihwa, WD Warden-in-Charge, pers. comm.) thus comprising about 10% of revenues generated from hunting. External financial support to the Ikorongo and Grumeti game reserves during the period 1997–2001 came from the Frankfurt Zoological Society, which contributed a total of $500,000, or $100,000 per annum, focused primarily on resource protection and essential infrastructure (table 15.3). Additional support to these two reserves has been provided by the Serengeti Regional Conservation Strategy, financed by the Norwegian Agency for International Development (NORAD). This is currently limited to a small game-cropping scheme that has been operational since 1993. This scheme has been shown to be uneconomic in terms of improving local livelihoods (Holmern et al. 2002) and at the time of this writing is effectively nonfunctional. Total expenditure in Ikorongo and Grumeti during 1997–2001 was, therefore, ~$120,000 per annum, of which 80% was donor funded. This expenditure was thus around 50% of the revenues retained by the government from hunting concession fees.

An important addendum to the financing situation in Ikorongo and Grumeti is that the hunting concession changed hands in 2002, resulting in significantly increased investment in the area. The new concession holder is Grumeti Reserves, which has employed expatriate technical advisors, trained and equipped a large ranger force, developed a community conservation program, and brought in consultants to conduct wildlife surveys and develop management plans. Grumeti Reserves plans to diversify the income generated within the reserves through the development of elite wildlife tourism and consequent reduction in tourist hunting. The development plans are ambitious and will require capital investment in the region of $15 million. The operating budget for 2003 for this relatively small corner of the Serengeti ecosystem was $3,194,000 (Rian Labascagne, Grumeti Reserves Manager, pers. comm.). These resources, comparable to the budget available to SNP to manage an area 10 times the size, were distributed across salaries for expatriate and local staff, capital expenditure on vehicles and other equipment, operating costs for helicopters, fixed-wing aircraft and other vehicles, community programs, and concession and trophy fees. It is worth noting here that the increased investment in resource protection in Ikorongo and Grumeti are believed to have had a beneficial impact throughout the Serengeti ecosystem in terms of reducing poaching (Justin Hando, SNP Warden-in-Charge, pers. comm.).

Loliondo Game Controlled Area

The Loliondo GCA is a 4,500 km² multiple-use area to the east of SNP and NCA. It is inhabited by around 40,000 Maasai, who are primarily pastoralists, although cultivation is increasing in some areas. Loliondo forms an integral part of the Serengeti ecosystem; its wildlife resources are under the jurisdiction of the WD. Loliondo contains one hunting block, which has been leased since 1992 to the Otterlo Business Corporation (OBC), a hunting company based in the United Arab Emirates. The Otterlo Business Corporation has established a hunting camp to the south of Wasso near the border of the SNP and has been active in a variety of development activities in the area, in collaboration with the district council and local residents. It is estimated that concession fees and trophy fees paid to the WD by OBC are $250,000, although details are not available. Currently, this money is distributed between the WD, the Tanzanian Treasury, and the Ngorongoro District Council, with no direct payments being made to the Maasai villages in Loliondo. However, if redirected, this income could represent a substantial economic force, capable of compensating villages for the opportunity costs of hunting and the direct costs of human-wildlife conflict. It has been suggested, however, that other funds, as well as development gifts, have been given to some of the Maasai villages by OBC, although the sums involved and the manner in which they were distributed are not known.

The other sources of revenue in the Loliondo GCA are the local agreements between safari operators and a number of Maasai villages that border the SNP. These operators take clients on walking safaris, run permanent tented camps, or both. The nature of the contracts is that a specified area of the village land, usually between 20 and 100 km², is leased to the safari operator for exclusive use in exchange for either a fixed monthly payment or a bed-night fee. Fixed payments have typically been in the range of $50–$250 annually per km² leased, and bed-night fees are usually between $5 and $15 (Gary Strand, Director, Wildlife Explorer, pers. comm.). Safari operators who currently operate in Loliondo include: Conservation Corporation Africa, Dorobo Tours, Wildlife Explorer, and Nomad. These operators are at the top end of the safari industry and actively promote ethical tourism, with benefits to local communities. The economic value of wildlife tourism in Loliondo in term of total revenues generated by the safari operators is unknown, but is almost certainly greater than the $250,000 per year generated from the OBC hunting concession. Calculating the benefits that local communities derive from wildlife tourism is complex, as cash benefits may be offset by the opportunity costs of restricted access to pastoralism and agricultural development and the direct costs of conflict between wildlife and people (Emerton 2001; Thirgood, Woodroffe and Rabinowitz 2005; Walpole and Thouless 2005). An assessment of the costs and benefits of wildlife to the Maasai inhabiting the communal group ranches to the north of the Maasai Mara National Reserve (MMNR) in Kenya suggests that the economic returns of agriculture far outweigh those from nonconsumptive use of wildlife (Norton-Griffith 1995; Norton-Griffith and Southey 1995) thus explaining the rapid agricultural intensification in the area (Serneels and Lambin 2001a, 2001b). In the case of Loliondo, it has been suggested that seven Maasai villages earn a total sum of $100,000 per annum from tourism (Nelson 2004). These economic benefits flow directly to villagers, whereas hunting revenues filter through various levels of government with corresponding depreciation in value before they reach local communities.

A detailed assessment of the economic benefits from wildlife tourism is available for Ololosokwan village in the northwest corner of Loliondo, bordering the SNP to the west and the MMNR in Kenya to the north (Nelson 2004). The total population of this village is around 4,000 and the village received land titles to around 460 km^2 in the early 1990s. Ololosokwan currently has two sources of tourism income. The first is a permanent lodge built by Conservation Corporation Africa and governed by a 15-year lease. Ololosokwan is paid $25,000 annually for use of a 100 km^2 concession around the lodge, in addition to a $3 fee per bed night. In 2002/2003, revenue to the village from this lodge was $37,640. Ololosokwan also maintains a campsite on village land adjacent to Klein's Gate into SNP that is used by various tour operators. Operators using this campsite pay $20 per bed night to the village and $10 per bed night to Ngorongoro District Council. In 2002/2003, Ololosokwan earned $18,066 from this campsite in addition to $2,723 in individual wages. Total earnings for the village from wildlife tourism in 2002/2003 were thus around $55,000 (Nelson 2004). While this village is exceptional in Loliondo in terms of revenues from tourism, it does suggest that there is great potential for community-based conservation in this area.

Government expenditure on conservation in the Loliondo GCA appears to be negligible. The WD has a wildlife officer stationed in Loliondo and has a small ranger force for regulating the activities of the hunting company. Investment by the private sector in conservation or development in Loliondo is diffuse and difficult to quantify. The OBC deploy a private ranger force within the hunting concession but the financing and activities of this force are unknown. Several of the safari operators finance community-based conservation and development activities. Notable among these in this local context is the Community Resources Trust (CRT), a local nongovernmental organization closely associated with Dorobo Tours and financed in part by surcharges to their clients, who are active in community land rights and land use planning. The financial input of this and other nongovernmental organizations in Loliondo to conservation are unknown, although undoubtedly they have played a major role in empowering local communities.

Direct revenues collected by TANAPA and NCAA from wildlife tourism in the core protected areas of SNP and NCA averaged $11 million per annum during the period 1998–2002. Revenues increased throughout this time and, even allowing for the global impact on tourism of the September 11, 2001, terrorist attacks in the United States, tourist income has continued to rise, with income in 2004 expected to be approximately $15 million. These figures represent only the revenues collected by TANAPA and NCAA and do not include the unknown but considerably larger sum that the Serengeti and Ngorongoro contribute to the wider Tanzanian economy through the tourism industry and associated businesses.

The combined conservation expenditure in SNP and NCA averaged $10 million per annum during the period 1998–2002. This period coincided with a major European Union project to develop infrastructure in the Serengeti, thus inflating expenditure. Current expenditure in the two core protected areas in 2004 is ~$8 million per annum, of which ~$1 million is provided by the Frankfurt Zoological Society. Donor funding is disproportionately important to conservation in Serengeti and Ngorongoro, as government funds are focused primarily on salaries and allow little capital expenditure on vehicles and other equipment. The discrepancy between income generated from tourism in the two core protected areas and government conservation expenditure is large; however, these funds flow to the treasury to finance the other needs of Tanzanian society, such as education and health, to NCAA to support Maasai pastoral development, and to TANAPA to support other national parks. This latter point is important, as Serengeti and Kilimanjaro are the only Tanzanian national parks with sufficient tourist income to be self-financing. The Tanzania National Parks Authority as a whole is a profit-making organization, and Tanzania is exceptional in either a developing- or developed-country context, in that the protected areas contribute income to central government rather than receive a financial subsidy.

Assessing the balance between income and expenditure in the buffer zones around the Serengeti is difficult, as numerous stakeholders are involved and financial details are often difficult to obtain. Consumptive use of wildlife in Maswa, Grumeti, Ikorongo, and Loliondo generated approximately $1 million per annum in hunting concession and quota fees payable to the WD during the period 1998–2002. This figure represents only those revenues collected by the WD and does not include the additional taxes paid to the government by the hunting companies nor the added value to the Tanzanian economy that hunting clearly provides through the provision of jobs and associated businesses (Leader-Williams, Kayera, and Overton 1995; Leader-Williams and Hutton 2005). Baldus and Cauldwell (2005) estimate that gross income from hunting in Tanzania in 2001 was $27.6 million, of which $10.5 million was accrued directly by the WD through concession and quota fees. The key point here, however, is that relatively little of the revenues generated through consumptive use of wildlife in the Serengeti buffer zones appears to trickle down to local communities.

Details on income generated through nonconsumptive use of wildlife in the Serengeti buffer zones during 1998–2002 are sketchy, and from the limited information available appear to be largely restricted to the Loliondo GCA, where wildlife tourism generated up to $100,000 per annum for local communities (Nelson 2004). This latter figure represents village income rather than profits accrued by the tourism companies and associated businesses and taxes paid to central government. It is also important to note that this income flows directly to local communities through the payment of bed-night fees and the rental of tourism concessions on village land. During the period under review (1998–2002), the western buffer zones of the Serengeti were managed primarily for consumptive use and community benefits from wildlife were limited. As noted previously, this situation is changing dramatically with the development of wildlife tourism in Ikorongo and Grumeti GR, although the mechanisms for community involvement and community benefit sharing are yet to be established.

Government expenditure on conservation in the Serengeti buffer zones during 1998–2002 was minimal. Less than 10% of the approximately $1 million per annum received by the WD from hunting concession and quota fees was reinvested back into the game reserves and game controlled areas in either resource protection or community conservation. To some extent, this shortfall was met by the donor community, with the Frankfurt Zoological Society in particular investing up to $200,000 per annum in resource protection in the game reserves. Additional resource protection has been provided by ranger forces associated with the hunting companies; however, the economic value of these contributions is difficult to assess at present. Overall, however, it is clear that the economic returns from the consumptive use of wildlife could easily finance the current levels of investment in conservation management in the Serengeti buffer zones.

The financial costs and benefits of wildlife to local communities living in or adjacent to the Serengeti buffer zones are more difficult to assess. Studies of human-wildlife conflict—in particular, crop raiding and livestock predation—have only relatively recently been initiated in the Serengeti, and at the time of this writing it is not possible to quantify the financial impact of wildlife. While there is evidence from a few villages in Loliondo that financial returns from wildlife tourism to local communities can be significant in a local context, it is not clear that these outweigh the opportunity costs of forgoing agricultural development. Elsewhere in the ecosystem, community benefits from the legal use of wildlife, as opposed to illegal hunting, appear to be minimal. The lack of local benefit sharing is one of the key drivers behind the development of community-managed WMAs which are considered in more detail in the following.

FINANCING A FUTURE SERENGETI

Predicting future economic trends is a difficult task in the best of times. This is exacerbated in the current context because financing of the Serengeti ecosystem is very dependent upon tourism, and the terrorist attacks of September 11, 2001, in the United States demonstrated how sensitive global tourism can be to external events. Global trends in wildlife tourism have, however, shown a consistent increase over the past 50 years (Walpole and Thouless 2005) and as outlined above, Tanzania has shown strong growth in the tourism sector over the last 20 years. While it is anticipated that tourism will continue to increase in Tanzania (Ministry of Natural Resources and Tourism [MNRT] 2000) it is difficult to predict to what extent this will occur and what its impact will be on the Serengeti ecosystem. Given wider financial circumstances in Tanzania and the need for wildlife tourism to contribute to the national economy, it is unlikely that tourism revenues will ever meet the full economic needs of the ecosystem. International donors and the private sector will almost certainly continue to be important sources of funding for biodiversity conservation in the Serengeti. To date, donor funding has primarily been project-driven; however, new financing mechanisms, such as Conservation Trust Funds (Bayon and Deere 1998, 1999; Lambert 2000, 2002) are currently under discussion. A further new development emerging over the past decade that will have a major impact on financing in the Serengeti is the increasing partnership between the private sector and local communities in the buffer zones (Bergen 2001; Nelson 2004). Local communities adjacent to protected areas have been authorized by the Tanzanian government to take responsibility for their wildlife resources through the formation of community-based WMAs. Communities will have the legal right to manage both the consumptive and nonconsumptive use of wildlife and will share revenues with central government. In theory, this approach will go some way toward redressing the financial balance for local communities, who currently incur most of the costs but receive few of the benefits of living adjacent to wildlife and protected areas. In the following sections we make a cautious attempt to predict how conservation activities in the different sectors of the Serengeti may be financed in the coming decade.

Serengeti National Park

Tourism will almost certainly continue to be the most important source of revenue for the SNP. Tourism revenues for SNP were ~$7 million in 2004 and it is anticipated that this will continue to increase (Justin Hando, SNP Warden-in-Charge, pers. comm.). Increased tourism does, however, bring concerns of environmental impact, and certain areas, such as the central Serengeti around Seronera, may have reached their carrying capacity, at least during the high season. Tanzania's national tourism policy is to provide high-quality, low-impact tourism with low volume but high yield in terms of revenue (Ministry of Natural Resources and Tourism [MNRT] 1998). With this policy, environmentally sensitive areas should be utilized within limits of acceptable use, incorporating restrictions on numbers of tourist vehicles and bed nights. Limits of acceptable use have been recently established for the SNP through the new General Management Plan (Tanzania National Parks Authority [TANAPA] 2005).

While tourism revenues from SNP are anticipated to continue to increase, it is clear that, as at present, a large proportion of these revenues will be retained by central government to use for social and economic development or used by TANAPA to support the rest of the national parks' estate. The Tanzania National Parks Authority currently invests less than 50% of revenues raised within the SNP back into the SNP. Tanzania National Parks Authority expenditure budgets for SNP have increased over the past decade but are still insufficient to cover essential conservation activities. This problem is exacerbated by the large proportion of the expenditure budget that is taken up with salary and other recurrent costs and a corresponding lack of funds for capital investment. A conservative estimate of the annual cost of managing the SNP in 2004 is ~$4 million, of which ~$3 million is provided by TANAPA (Justin Hando, SNP Warden-in-Charge, pers. comm.). International donors currently provide ~$1 million per annum, focused on essential activities such as resource protection, ecological monitoring, and veterinary intervention. It appears likely that this funding gap will have to be met by the international conservation community for the foreseeable future if the current high standards of professional wildlife management conducted by TANAPA are to be maintained. This raises the question of who should really finance the conservation of the Serengeti? Is this a Tanzanian responsibility, or is it a responsibility of the global community? We return to this issue at the conclusion of this chapter—but it is worth restating here that the Tanzanian government has consistently seen its wildlife resources as central to economic development.

Alternatives to direct donor funding include the development of a Serengeti Trust Fund. Trust funds have been established in many developing countries over the past decade as a way of providing long-term funding for protected areas (Bayon and Deere 1998, 1999; Lambert 2000, 2002). Trust funds are typically legally independent institutions managed by independent boards of directors. Many trust funds have a permanent endowment that has been capitalized by grants from national governments and international donors such as the Global Environment Facility (GEF). The main attraction of trust funds for international donors is their ability to reliably manage and allocate donor funds over a long period of time. Trust funds are typically formed through broad consultative processes, and are governed by a mixed public/private board of directors composed of representatives of different stakeholder groups. They are designed to have credible and transparent operational procedures, accountability, and sound financial management practices. Trust funds therefore may be able to attract new donor funding in cases where donors might otherwise be concerned about giving their money to a government agency. Furthermore, the assets of a trust fund are almost always managed and invested by outside financial institutions, either inside or outside of the country—so as to provide income for the specific duration and specific purposes of that particular Trust fund. Trust funds can also be more than just financial mechanisms—they can also serve as a valuable forum where diverse stakeholders—such as national and local government agencies, NGOs, the private sector, and international donors—come together on a regular basis to discuss and sometimes resolve important conservation issues. In the case of a Serengeti Trust Fund, initial discussions have been conducted between the main actors—the government conservation agencies, international conservation NGOs, and major international donors such as the GEF—and it is anticipated that a formal structure will be in place by the time of publication of this volume.

Ngorongoro Conservation Area

Wildlife tourism is also expected to continue to be the main economic driver in the NCA. Tourist revenues continue to show an upward trend and are expected to exceed $8 million in 2004. Tourism in the NCA is primarily focused on the Ngorongoro Crater, and there is evidence of the negative environmental impacts of tourism. At present there are no formal restrictions on the numbers of tourist vehicles visiting the crater, which regularly exceed 100 vehicles per day during the peak season. A recent workshop on the conservation and management of the Ngorongoro black rhino population recommended increasing the cost of visiting the crater, setting limits of 50 tourist vehicles per day, and encouraging the diversification of tourist activities to other attractions in the NCA, including Empakai, Olduvai, and Ndutu, and a wider range of activities, including walking and horseback safaris (Mills et al. 2006). It is anticipated that these recommendations will be formalized within the ongoing revision of the General Management Plan for the NCA (F. Manongi, Mweka Wildlife Management College, pers. comm.). While the NCAA pays corporation tax to the Tanzanian Treasury and has an obligation to provide development support to resident Maasai pastoralists, it is not burdened with the responsibility of financing other protected areas in Tanzania. Given the $8 million income from tourism, the NCAA has the potential to be self-financing in terms of conservation management. Despite this considerable income and an expenditure budget in excess of $5 million, international donor support is still necessary to provide protection to the endangered population of black rhino within the Ngorongoro Crater and to address other pressing conservation and development issues in the protected area. This anomaly is difficult to explain and presents an uncomfortable dilemma for the international conservation community.

The Game Reserves

Tanzanian Government policy is to retain a strong tourist hunting industry and as such commercial hunting will remain the primary economic activity within the great majority of Game Reserves (Leader-Williams, Kayera, and Overton 1995). Recent estimates indicate that revenues from tourist hunting to the WD exceeded $10 million in 2001 (Baldus and Cauldwell 2005). With the exception of the Selous Game Reserve, which has developed a 50% revenue retention scheme for management and community conservation (Baldus, Kibonde, and Siege 2003), Game Reserves in Tanzania receive little government investment. Resource protection activities are often undertaken by private ranger forces associated with hunting companies. Whilst this might be acceptable in principle, in practice wildlife management is in the hands of commercial operators who range from extremely professional and committed hunter-conservationists to entrepreneurs simply interested in quick profit (Baldus and Cauldwell 2005). Maswa GR and the adjacent Makao Open Area are no exception to the above rule and suffer from minimal investment from the WD despite the $250,000 per annum revenue raised from tourist hunting. The Maswa and Makao hunting concessions are currently held by reputable hunting companies with strong commitments to conservation. The current situation is no guarantee for the future however and there is a good case to be made for the development of a revenue retention scheme similar to Selous GR that would increase the local capacity of the WD to manage the Reserve.

The financing situation of the northern Game Reserves—Grumeti and Ikorongo—is in a period of considerable flux. As noted above the hunting concession changed hands in 2002 resulting in a marked increase in investment. The current concession holders have largely stopped hunting—although they continue to pay the full concession and hunting quota fees of $215,000—and are developing the area as an elite wildlife tourism destination. The reported intention is to run the concession as a commercial profit-making enterprise as opposed to an act of conservation philanthropy. This will involve the development of luxury camps and lodges, improvements to roads and other infrastructure, increased resource protection and habitat management and other interventions such as translocation to increase wildlife populations (Rian Labascagne, Grumeti Reserves Manager, pers. comm.). The initial investment is in the region of $15 million and the recurrent costs at $3 million per annum are comparable to the entire SNP. The concession holders are also investing considerable sums in the adjacent community and attempting to facilitate international donor support for development activity in the region. Whilst early signs are promising, at this stage it is too early to judge the commercial success of the wildlife tourism enterprise, its effect on conservation throughout the Serengeti Ecosystem, and its wider development impacts in the region.

Wildlife Management Areas

Perhaps the most exciting and challenging new conservation initiative in the Serengeti ecosystem is the development of community-based Wildlife Management Areas (WMAs). Tanzania's Wildlife Policy (Ministry of Natural Resources and Tourism [MNRT] 1998) advocates devolving managerial responsibility for wildlife to local communities to enable these people to economically benefit from natural resources and to create incentives for conservation. Wildlife Management Areas will empower communities to take responsibility for the management of wildlife resources on village lands through the formation of legally recognized Authorized Associations (AAs). More recently, the Ministry of Natural Resources and Tourism has released the Wildlife Management Areas Regulations, which provide WMAs with their legal framework and include detailed regulations for their management (MNRT 2002). Income generated from either consumptive or nonconsumptive use of wildlife will be retained by the AA, who will pay income tax to the Tanzanian Treasury and spend the remaining revenue on administration, village game scouts, and community development.

The Wildlife Management Area Regulations (MNRT 2002) describe 15 pilot areas for the trial establishment of WMAs in Tanzania, including two—Loliondo and Ikona—in the Serengeti ecosystem. The Frankfurt Zoological Society was formally invited by the Tanzanian government in 2002 to act as facilitator for the formation of these two WMAs—essentially to provide logistic, technical, and legal assistance in the recognition of land rights, development of village land use plans, formation of legally recognized Authorized Associations with formal constitutions, and training of community game scouts. Community leaders from two further areas— Makao Open Area to the south and Natron Open Area to the east—approached the Frankfurt Zoological Society in 2003 with requests to assist their communities in similar developments. This activity was approved by the WD, and a similar range of activities are planned for all four WMAs. Progress to date in the development of the WMAs has been mixed (Joe Ole Kuwai, Frankfurt Zoological Society, pers. comm.). WMAs are controversial with communities and other stakeholders, particularly in Loliondo, where there is already a well-established partnership between some villages (e.g., Ololosokwan) and commercial tourist operators (Nelson 2004). The situation in Loliondo is further confounded by long-standing rivalries and land disputes between the seven different Maasai villages that make up the proposed WMA, and this led to violence in 2004. This is unfortunate, as Loliondo, and to a lesser extent Natron and Makao, have great promise for successful community-based conservation programs, with the desirable combination of high-density wildlife populations and low-density human populations. It is clear that WMAs will go through many difficulties before they can contribute to their joint goals of conserving wildlife in the Serengeti buffer zones and contributing to the economic development of park-adjacent communities. It is equally clear, however, that more equitable benefit-sharing will be essential to ensure support for conservation in these communities, who bear most of the costs of conservation (Thirgood, Woodroffe, and Rabinowitz 2005).

CONCLUSIONS

The Serengeti ecosystem is without doubt a globally significant biological resource and a priority for international conservation efforts. Furthermore, within Tanzania and more widely in East Africa, it is a major economic driver that is central to the increased tourism revenues of the last decade. For both of these reasons it is clearly essential that sustainable financing mechanisms are put in place to ensure appropriate management.

Assessment of income and expenditure within the core protected areas of the SNP and the NCA during the period 1998–2002 indicates that revenues accrued by government exceed expenditure by government agencies with the remaining funds being used for conservation and development activities elsewhere in the country. Shortfalls in expenditures in both protected areas are currently covered by international donor support for infrastructure and resource protection. Conservation and management activities within the core areas are thus dependent upon continued external support from donors or alternative financing mechanisms such as conservation trust funds. Given wider economic conditions in Tanzania, and the importance that wildlife is given by government in driving development, it seems unlikely that these economic conditions will change markedly within the next decade. This raises the interesting and nontrivial question as to who bears the responsibility for financing conservation of such a globally important biological resource as the Serengeti? Clearly the Tanzanian state benefits greatly from the Serengeti in terms of driving tourism and economic development—although park-adjacent communities who bear most of the costs of conservation currently receive few of those economic benefits. However, one can argue that the global community benefits from the continued presence of the natural spectacle of the Serengeti and should therefore contribute to the cost of its conservation. These are philosophical questions that are beyond the scope of this chapter—however, the hard reality is that external financial support will continue to be necessary to maintain the core protected areas in their current favorable conservation status.

Conservation financing mechanisms in the buffer zones around the Tanzanian sector of the Serengeti ecosystem are complex and in a period of considerable flux. Assessment of income and expenditure within the Game Reserves during 1998–2002 indicates that revenues from commercial hunting far exceed expenditure by the WD for conservation and management. To some extent these shortfalls were covered by additional investment by international donors and the private sector; however, this was highly variable between game reserves. Private investment in the Ikorongo GR and Grumeti GR has recently increased and there has been a change in management emphasis from tourist hunting to high-quality wildlife tourism. At present it is too early to judge the wider effects of this investment throughout the Serengeti ecosystem, although early indications of its impact on resource protection are promising.

A notable trend over the past decade has been the development of commercial partnerships between private safari operators and local communities inhabiting the buffer zones adjacent to the Serengeti. These partnerships are best developed in the Loliondo GCA, where in some individual cases they bring significant direct economic benefits to communities. Equitable benefit sharing from the consumptive and nonconsumptive use of wildlife will be essential for the development and maintenance of support for conservation from park-adjacent communities who bear most of the costs of conservation. Community-based natural resource management initiatives are currently being formalized in Tanzania through the development of legally recognized WMAs in four areas around the Serengeti. While this process has not been without problems and controversy, it offers hope of the successful reconciliation of the two aims of community conservation—that of improving economic livelihoods in park-adjacent communities while protecting biodiversity.

ACKNOWLEDGMENTS

Simon Thirgood was based at the Africa Regional Office of the Frankfurt Zoological Society in Serengeti while writing this chapter and at the Macaulay Institute in Aberdeen while revising it. Reviews by Tony Sinclair and two anonymous referees greatly improved the chapter.

REFERENCES

Adams, W. M., R. Aveling, D. Brockington, B. Dickson, J. Elliott, J. Hutton, D. Roe, B. Vira, and W. Wolmer. (2004). Biodiversity conservation and the eradication of poverty. *Science* 306:1146–49.

Baldus, R., and A. Cauldwell. 2005. Tourist hunting and its role in development of wildlife management areas in Tanzania. *Proceedings of the 6th International Game Ranching Symposium*. Paris, July 6–9, 2004.

Baldus, R., B. Kibonde, and L. Siege. 2003. Seeking conservation partnerships in the Selous Game Reserve, Tanzania. *Parks* 13:50–61.

Balmford, A., K. Gaston, S. Blyth, A. James, and V. Kapos. 2003. Global variation in terrestrial conservation costs, conservation benefits, and unmet conservation needs. *Proceedings of the National Academy of Sciences* 100:1046–50.

Bayon, R., and C. Deere. 1998. *Financing biodiversity conservation: The potential of environmental funds.* Gland, Switzerland: IUCN.

———. 1999. *Environmental funds: Lessons learned and future propects.* Gland, Switzerland: IUCN.

Bergin, P. 2001. Accommodating new narratives in a conservation bureaucracy: TANAPA and community conservation. In *African wildlife and livelihoods: The promise and performance of community conservation,* ed. D. Hulme and M. Murphree, 88–105. Oxford: James Curray.

Emerton, L. 2001. The nature of benefits and the benefits of nature: Why wildlife conservation has not economically benefited communities in Africa. In *African wildlife and livelihoods: The promise and performance of community conservation,* ed. D. Hulme and M. Murphree, 208–26. Oxford: James Curray.

Fjeldsa, J., N. Burgess, S. Blyth, and H. de Klerk. 2004. Where are the major gaps in the reserve network for Africa's mammals? *Oryx* 38:17–25.

Holmern, T., E. Roskaft, J. Mbaruka, S. Y. Mkama, and J. Muya. 2002. Uneconomical game cropping in a community-based conservation project outside the Serengeti National Park, Tanzania. *Oryx* 36:364–72.

Hulme, D., and M. Murphree. 2001. *African wildlife and livelihoods: The promise and performance of community conservation.* Oxford: James Curray.

Lambert, A. 2000. Environmental funds: Much more than financial mechanisms. In *Proceedings of DFID Seminar on Economic Instruments for Environmental Mananagement,* 63–75. Cuiaba, Brazil, 22–23 March 2000.

———. 2002. Financing global sustainability: A proposal for multilateral environmental agreements. In *Ensuring sustainable livelihoods: Challenges for governments, corporates and civil society at Rio + 10,* 92–117. Teri Research Centre, India, Nov. 2002.

Leader-Williams, N., and J. M. Hutton. 2005. Does extractive use provide opportunities to offset conflicts between people and wildlife? In *People and wildlife: Conflict and coexistence,* ed. R. Woodroffe, S. Thirgood, and A. Rabinowitz, 140–61. Cambridge: Cambridge University Press.

Leader-Williams, N., J. Kayera, and G. Overton. 1995. *Tourist hunting in Tanzania.* Dar es Salaam, Tanzania: Department of Wildlife.

Maswa Game Reserve. 1994. *Maswa Game Reserve General Management Plan 1994–1998.* Serengeti Regional Conservation Strategy and Maswa Game Reserve.

Mbano, B., R. Malpas, M. Maige, P. Symonds, and D. Thompson. 1995. The Serengeti Regional Conservation Strategy. In *Serengeti II,* ed. A. R. E. Sinclair and P. Arcese, 605–16. Chicago: University of Chicago Press.

Mills, A., P. Morkel, P. Amiyo, V. Runyoro, M. Borner, and S. Thirgood. 2006. Managing small populations in practice: Black rhino in the Ngorongoro Crater, Tanzania. *Oryx* 40:319–23.

Ministry of Natural Resources and Tourism.1998. *The Wildlife Policy 1998.* Dar es Salaam, Tanzania: Government Printer.

———. 1991. *Tourism Policy.* Dar es Salaam, Tanzania: Government Printer.

———. 2000. *Tourism Policy.* Dar es Salaam, Tanzania: Government Printer.

———. 2002. *The Wildlife Conservation (Wildlife Management Areas) Regulations.* Dar es Salaam, Tanzania: Government Printer.

Nelson, F. 2004. *The evolution and impacts of community-based ecotourism in northern*

Tanzania. Drylands Issue Paper No. 131. International Institute for Environment and Development, London, UK.

Norton-Griffith, M. 1995. Economic incentives to develop the rangelands of the Serengeti: Implications for wildlife conservation. In *Serengeti II,* ed. A. R. E. Sinclair and P. Arcese, 588–604. Chicago: University of Chicago Press.

Norton-Griffith, M., and C. Southey. 1995. The opportunity costs of biodiversity conservation in Kenya. *Environmental Economics* 12:125–39.

Serneels, S., and E. Lambin. 2001a. Land cover changes around a major African wildlife reserve: The Mara ecosystem (Kenya). *International Journal of Remote Sensing* 22: 3397–3420.

———. 2001b. Proximate causes of land-use change in Narok District, Kenya. *Agriculture, Ecosystems and Environment* 85:65–81.

Sinclair, A. R. E., and P. Arcese, eds. 1995. *Serengeti II.* Chicago: University of Chicago Press.

Sinclair, A. R. E., and M. Norton-Griffith, eds. 1979. *Serengeti.* Chicago: University of Chicago Press.

Tanzania National Parks Authority (TANAPA). 2002. *Maximizing revenues in Tanzanian National Parks: Towards a better understanding of park choice and nature tourism in Tanzania.* Unpublished report.

———. 2005. *Serengeti National Park General Management Plan 2005–2015.* Dar es Salaam: Tanzanian National Parks.

Thirgood, S., R. Woodroffe, and A. Rabinowitz. 2005. The impact of human-wildlife conflict on human lives and livelihoods. In *People and wildlife: Conflict and coexistence,* ed. R. Woodroffe, S. Thirgood, and A. Rabinowitz, 13–26. Cambridge: Cambridge University Press.

Thompson, D. M., ed. 1997. *Multiple land-use: The experience of the Ngorongoro Conservation Area, Tanzania.* Gland, Switzerland: IUCN.

Walpole, M. J. 2006. Partnerships for conservation and poverty reduction. *Oryx* 40: 245–46.

Walpole, M. J., and C. R. Thouless. 2005. Increasing the value of wildlife through nonconsumptive use? In *People and wildlife: Conflict and coexistence,* ed. R. Woodroffe, S. Thirgood, and A. Rabinowitz, 122–39. Cambridge: Cambridge University Press.

World Bank. 2002. *Tourism in Tanzania: Investment for growth and diversification.* Washington, DC: World Bank.

SIXTEEN

Integrating Conservation in Human and Natural Ecosystems

A. R. E. Sinclair

Conservation of biota and their resources is a human enterprise, a human construct embedded in a human context. It has developed because humans have taken over some 90% of the world and are sequestering most of its resources, so that explicit efforts are now required to provide resources for the remaining wild species and prevent their extinction (Pimm et al. 1995). Conservation in the greater Serengeti ecosystem is an example of this enterprise, as emphasized by the previous chapters. However, maintaining conservation in the long term requires integration of the needs of humans and the rest of the biota. Such merging lies at the heart of sustainable development.

There are two paradigms for conservation; namely, community-based conservation for sustainable use, and protected-area conservation (PAC) (Brandon, Redford, and Sanderson 1998; Terborgh 1999; Bruner et al. 2001; Terborgh *et al.* 2002; Child 2004). Community-based conservation (CBC) is usually linked to improving human welfare and poverty reduction, often through integrated conservation and development projects (ICDPs). Adams et al. (2004) summarize the various approaches to CBC and PAC, and emphasize that there is confusion in achieving the two goals of conservation of biodiversity and reduction of poverty. As a consequence, many ICDPs have had minimal success for conservation (Barrett and Arcese 1995; Hackel 1999; Songorwa, Buhrs, and Hughey 2000; Newmark and Hough 2000; Adams and Hulme 2001, Ferraro and Kiss 2002). Adams et al. (2004) disentangle four positions in the continuum, from strict protection to pure development:

1. *Poverty and conservation are separate realms.* Trying to combine them is inefficient and compromises conservation. Success is measured in terms of biodiversity, not human welfare.
2. *Poverty is a constraint on conservation.* Poverty necessarily results in exploitation and often elimination of resources, and hence biodiversity. Pragmatically, conservation contributes to poverty reduction through outreach programs that provide net benefits to local peoples so as to counteract exploitation.
3. *Conservation should not compromise poverty reduction.* Conservation should at least not undermine the livelihoods of the poor. There is a moral obligation to improve human welfare through economic benefits.
4. *Poverty reduction depends on maintaining functioning ecosystems.* Conservation of fully functioning ecosystems will provide the poor with improved livelihoods. Here CBC focuses on sustainable consumptive exploitation of wildlife species to provide economic benefits.

These different positions are not mutually exclusive. Because different species require different methods for conservation, any particular policy and approach will not be able to address all conservation needs. Thus, these several policies address different needs, and none are sufficient alone to represent all fully functioning ecosystems. Nevertheless, these four positions emphasize the link between community-based conservation and human welfare through poverty reduction and sustainable development.

SUSTAINABLE DEVELOPMENT

Historically, it was the concerns of developing countries attending the United Nations Conference on the Human Environment in Stockholm, in 1972, that led to the process of integrating conservation and development. The first milestone was the publication of the *World Conservation Strategy* by IUCN (1980). The strategy had three objectives:

1. The global maintenance of life support systems for humans and their environment.
2. The preservation of genetic diversity as an insurance against loss in both domestic and wild species, and as an investment for future use.
3. The sustainable development of species and ecosystems for human use (Adams 1998).

Over the next two decades, sustainable development was pursued in developing countries through Integrated Conservation and Development Projects (ICDPs), culminating in the United Nations Conference on Environment and Development at Rio de Janeiro, in 1992, termed the "Earth Summit" (Holmberg, Thomson, and Timberlake 1993). This Earth Charter set out a list of principles for international development and environmental cooperation, and in particular the Biodiversity Convention, signed by 152 countries. Commitment to conservation and protection of biodiversity were central themes.

However, a new direction for sustainable development emerged at the World Summit on Sustainable Development in Johannesburg, in 2002, a direction that ignored the role of protected areas and emphasized economic rationalism, both inside and outside protected areas (Sanderson 2004). The summit "conceded that protected areas had to be justified by economic and social criteria, not conservation or ecological integrity", contrary to the objectives of the World Conservation Strategy (Sanderson 2004). Given these conflicting objectives, it is important to understand the roles of protected-area conservation and community-based conservation.

THE ROLE OF PROTECTED-AREA CONSERVATION

The IUCN has set out nine different roles for protected areas (table 16.1), ranging from complete protection of biodiversity to supporting local peoples. Protected areas that exclude or minimize the effects of modern human activity, as set out by the IUCN (McNeely and Miller 1984; Wright and Mattson 1996; Burgess et al. 2004), are essential, for several reasons.

1. *Protection of endangered species, especially those threatened by humans.* Numerous examples exist worldwide. For example, the Wood Buffalo National Park in northern Canada is the only location in the world that can provide the endangered whooping crane (*Grus americana*) with an area in which the species can breed naturally.
2. *Protection of species that cannot coexist with humans.* Obvious examples of such species are large carnivores, large ungulates such as elephants, those that require intact ecosystems, such as interior forest species (many birds, large primates) or species that require fragile ecosystems, such as wetlands (many plants, insects, amphibians), or undisturbed breeding sites.
3. *Minimizing the loss of biodiversity.* Exploitation, particularly of tropical forests, is the main cause of extinction (Pimm et al. 1995; Pimm 1998). Until

we understand the roles of the full complement of species (biodiversity) it is wise to preserve intact representative portions of all ecosystems. At least these areas provide the baseline (see the following) with which to compare disturbed systems concerning the function of biodiversity.

4. *Maintaining natural processes.* Jope and Dunstan (1996) cite, as an example, undisturbed catchment areas that are required to maintain water supply for humans. Woodlands are required to maintain watertables and waterflow. The classic example is the present salinization of Australian agricultural areas following the clearing of eucalypt woodlands in the early twentieth century (Hobbs 1993; McFarlane, George, and Farrington 1993). Nutrient cycling and pollution buffering are other essential services provided by protected areas.

5. *Preserving a gene bank.* Although most species have no current economic value, we realize that they could act as an important genetic reservoir for future use. The classic example is that of the aquatic microorganism *Thermus aquaticus,* capable of living in hot pools at Yellowstone National Park at temperatures extreme enough to kill most other organisms. Thomas Brock discovered this species in 1966, but its usefulness was not realized until 1985, when Kary Mullis developed the Polymerase Chain Reaction using the *Taq* polymerase, named after the microorganism from which it came (Biel 2004).

6. *Maintaining ecological baselines.* There is the need to provide ecosystem baselines undisturbed by humans in order to understand the impacts humans are having on ecosystems. We need control areas where human interference is kept to a minimum (Sinclair 1983, 1998; Peterson 1996; Arcese and Sinclair 1997; Sinclair and Byrom 2006).

Baselines allow us to distinguish between natural or long-term disturbance and local human impacts, in order to predict the potential outcomes of current and future disturbance events (Willis et al. 2005). Without such baselines we can misinterpret causes of change. The great famines of northern Africa in the past 30 years have been ascribed to unfortunate and unavoidable droughts. Undoubtedly, there is a long-term climate change leading to greater climatic variability and lower productivity (Hulme 2001; Thuiller et al. 2006). However, this is only part of the problem. Other evidence on long-term overgrazing (e.g., Wu, Thurow, and Whisenant 2000), derived from experimental baselines where grazing was kept under control (Sinclair and Fryxell 1985) indicated that the prevailing rainfall was clearly sufficient for livestock needs and pointed to overgrazing as the likely cause of the famine. (Overgrazing occurs when the plant community composition changes due to the impacts of top-down herbivory, usually to species

Table 16.1 Categories of protected areas as outlined by IUCN (Wright and Mattson 1996)

1. *Scientific reserve/ strict nature reserve:* These areas possess some outstanding ecosystems, features, and/or species of flora and fauna of national scientific importance; they often contain fragile ecosystems or areas of particular importance to the conservation of genetic resources.
2. *National Park:* Relatively large areas that contain representative samples of major natural regions, features, or scenery, where plant and animal species, geomorphological sites, and habitats are of special scientific, educational, and/or recreational interest.
3. *Natural monument/natural landmark:* One or more specific natural features of outstanding national significance, which, because of their uniqueness or rarity, should be protected; these features are neither the size nor the diversity of features that would justify their inclusion as a national park.
4. *Nature conservation reserve/ managed nature reserve/wildlife sanctuary:* These are specific sites or habitats whose protection is essential to the continued well-being of resident fauna or migratory fauna of national or global significance.
5. *Protected landscape or seascape:* Natural or scenic areas found along coastlines and lake and river shores, sometimes adjacent to visitor-use areas or population centers, with the potential to be developed for a variety of recreational uses.
6. *Resource reserve/ Interim conservation unit:* These areas normally comprise an extensive, relatively isolated and uninhabited area having difficult access, or include regions that are lightly populated yet may be under considerable pressure for colonization or resource development.
7. *Natural biotic area/ Anthropological reserve:* Natural areas where the influence or technology of modern humans has not significantly interfered with or been absorbed by the traditional ways of life of the inhabitants.
8. *Multiple-use management area/ Managed resource area:* Areas designed to provide sustained production of or access to water, timber, fauna, pasture, marine products, and outdoor recreation.
9. *Biosphere reserve:* Areas containing unique communities with unusual natural features, examples of harmonious landscapes resulting from traditional patterns of land use, and examples of modified or degraded ecosystems that are capable of being restored.
10. *World heritage site:* Areas that are of true international significance.

of inedible herbs or even bare ground, and subsequently that community does not return to the original one when the herbivory is removed; i.e., there is a change in state.)

In essence, the baseline highlighted a change in ecosystem function, namely a drop in the resilience of the system to fluctuations in climate. Thus, baselines have allowed us to understand that both long-term trends in climate and top-down grazing effects of herbivores were necessary to create these events, while neither cause was sufficient on its own.

COMMUNITY-BASED CONSERVATION

Protected-area conservation has been challenged on philosophical grounds (Birch 1990; Cronon 1995; Adams 2001; Adams and Mulligan 2003) or from historical viewpoints. If the premise is that protected areas are samples of pristine nature then there is plenty of evidence to show that most if not all areas of the world have changed over the past millennia (Adams 2003). Indeed, the densest currently uninhabited Congo forest is now known to have been inhabited by humans some 2 to 3 millenia ago (Willis, Gillson, and Brncic 2004). "Pristineness" is not, however, the premise for protected areas.

Nevertheless, protected areas cover less than 10% of the global terrestrial surface, and it is unlikely this value will increase much more. Thus, from standard species-area relationships, something less than 50% of the terrestrial biota will ever be found in protected areas. Furthermore, many protected areas are not large enough to maintain viable populations in isolation from the surrounding matrix (see also Rodrigues et al. 2004). Therefore, if we are to ensure the persistence of the other half of the biota we must look to human-dominated ecosystems; that is, we employ community-based conservation.

Community-based conservation through consumptive exploitation of natural resources has been forcefully promoted in southern Africa, sometimes to the exclusion of other ideas (Child 1996, 2004; Jones and Murphree 2004). The rationale is that conservation is only justified if local peoples benefit by improved wealth (Hulme and Murphree 2001; Murphree 2004). This is, of course, only one of several equally important justifications outlined in table 16.1 for the greater good of mankind. At the same time, local peoples do not have to employ consumptive exploitation to obtain benefits from conservation (Walpole and Thouless 2005; Leader-Williams and Hutton 2005).

In many cases around the world, implementation of community-based conservation has lead to problems and the policy has not always been successful (Agrawal and Gibson 1999; Adams 2001; Murray Li 2002; Conley and Moote 2003; Lybbert and Barrett 2004; McShane and Wells 2004; Homewood, Coast, and Thompson 2004). Murombedzi (2003) concludes that such programs in Africa benefit local authorities but not individuals in the community.

Thus, as currently practiced, community-based conservation has some fundamental problems that limit its effectiveness. First, CBC favors only those species that are either useful to or tolerant of humans, and the latter are often those least in need of protection. Other species that are detrimental to humans, such as large carnivores, are often excluded.

Second, CBC relies on species providing some value, usually economic value and often through commercial exploitation, so as to advance human welfare. Yet even with "useful" species where economic value has been assessed, it rarely exceeds 10% of the potential agricultural value. Thus, there is a strong incentive to replace the few useful wild species with domestic species (Norton-Griffiths et al., chapter 13, this volume; Emerton 2001; Bond 2001), a trend contrary to the need for maintaining a full complement of species.

Third, the CBC justification for consumptive exploitation of wildlife, especially in Africa, relies on short-term, static measures of per capita profit. There has been little, if any, consideration of five trends: (1) population increase will necessarily mean per capita wealth will decline unless consumption of the resource increases (Polasky et al. chapter 12, this volume); (2) there is an inherent psychological expectation of increasing wealth—people are not content to receive the same income for their whole life—so that even with a constant population, consumption must increase to meet this expectation. These two trends are multiplicative; (3) for several reasons, the productivity of the resource is likely to decline, such as shrinking total area and overexploitation of the remaining area. Thus, the system is unlikely to be sustainable in the long term (Barrett and Arcese 1995); (4), human-dominated systems, where CBC must be practiced, result in a distorted biota that may not be sustainable in the long term. The population declines of common bird species in Britain, attributed both to a decline of insect food due to changes in agriculture and to an increase in predation, are a reflection of this distortion of the biota. Their society has been forced to subsidize the ecosystem (Tapper, Potts, and Brockless 1996; Chamberlain et al. 2000; Robinson et al. 2004); (5) the vast majority of the biota has no assessed economic value *at the margin,* and hence little conservation effort is spent on them. The value of individual species is unknown (Sedjo 2000; Simpson 2000), although there are global estimates for all species (Costanza et al. 1997; Moran, Pearce, and Wendelaar 1997; James, Gaston, and Balmford 2001).

In summary, community-based conservation is essential for the long-term sustainability of the biota. There is an urgent need to find ways of counteracting the destabilizing trends discussed above while maintaining the full complement of species. Innovative ways of providing benefits to peoples without distorting the biota or causing extinctions are now being explored (Borgerhoff Mulder and Coppolillo 2005; Breitenmoser et al. 2005). Impacts on human lives have to be assessed (Thirgood, Woodroffe, and Rabinowitz 2005) and reduced (Osborn and Hill 2005). In the end, the costs of conservation must be met by those who benefit the most, mainly those in the developed nations. These nations must devise adequate benefit programs for local communities (Getz et al. 1999; Sinclair, Ludwig, and Clark 2000; Balmford and Whitten 2003; Nyhus et al. 2005; Woodroffe, Thirgood, and Rabinowitz 2005)—and by "adequate" I mean that people should be better off from the CBC compensation than if they had adopted other forms of resource use with no conservation. Standardized methods across ecosystems of monitoring progress toward conservation and sustainability are needed (Balmford et al. 2005; Green et al. 2005; Stem et al. 2005).

INTERACTIONS OF PROTECTED AREAS AND HUMAN-DOMINATED ECOSYSTEMS AROUND SERENGETI

The approach in this volume has used a variety of models to predict outcomes from the dynamics of plant and animal populations, natural communities, ecosystem processes, human ecosystems, and human welfare. Different types of models have been used to meet different ends, and Hilborn has compared these in chapter 14. We have used likely changes in the system as the starting point to understand the dynamics within the natural system (e.g., the food web dynamics of Holt in chapter 8, and the behavioral dynamics of grazing systems of Fryxell in chapter 9). Similarly, perturbations are used to simulate the responses of the human ecosystems, the rebounding effects on natural system, and the consequences for human welfare (e.g., chapters 10–13 of Costello, Galvin, Polasky, and Norton-Griffiths).

Climate Change

An important conclusion that arises from the analysis of climate and habitats in the palaeontological record (Peters et al. chapter 3, this volume) is that the Serengeti region has changed radically from tropical forest to desert many times. At least within the past 10,000 years such changes occur every few thousand years or even more rapidly—perhaps every few hundred years. Thus, the Serengeti that we see today will most likely be very different in 200 years. Such conclusions are more than likely to apply to other conservation areas around the world. Therefore, the long-term strategies of conservation agencies should be planning for radical changes in their ecosystems.

For the greater Serengeti region we have used a variety of likely trends and disturbances to the system, but it is convenient to start with the results from the large-scale analyses of Olff (chapter 4) and Ritchie (chapter 6). Global climate models predict that CO_2 levels, rainfall, temperature, and nitrogen deposition will all go up. However, these predictions are not necessarily matched by observed long-term trends. For example, Ritchie found that although Lake Victoria rainfall has increased over the past 100 years (see fig. 2.8), possibly due to temperature increases over the lake surface, within the Serengeti the increase over the first 50 years has changed to a decreasing trend since 1960. At the same time, overall variability in rainfall has decreased.

Rainfall is the overriding abiotic driver in the system, so changes in precipitation could have a dominating effect. Ritchie predicts that decreasing average rainfall will reduce overall plant production. This in turn will benefit smaller herbivore species, particularly browsers. Herbivore diversity could decrease, and if so, carnivore (top-down) regulation will predominate, due to the processes observed in Sinclair, Mduma and Brashares (2003).

The decrease in rainfall variability, especially the longer period between extreme events, such as floods, has important consequences for plant regeneration. Ritchie predicts fewer pulsed regeneration events of the sort demonstrated for the main Serengeti tree species described in chapter 2 (figs. 2.15–2.16). Sharam (2005) has shown that riverine forest trees germinate only following flood seasons. The western lowland forest on the Grumeti also requires nurse trees (*A. polyacantha* stands) for successful sapling survival, and such stands appear as pulses at intervals of about 50 years. Thus, fewer flood events will reduce the rate of regeneration of forests.

Reduced rainfall and productivity will alter migration routes, taking animals outside the protected area and exposing them to human hunting. The consequences for the human system are described in the following, but for the protection of the migration, on which the whole ecosystem depends, this prognosis provides a reason for buffer areas.

Disease Transmission

One of the most noticeable disturbances in the past 30 years is episodic disease, discussed in chapter 7 by Cleaveland et al. There are essentially two aspects to disease transmission.

First, there are background enzootic parasites that reside within the wildlife—for example, malignant catarrhal fever in wildebeest, *Theileria* and *Brucella* in buffalo, *Trypanosoma* in carnivores and ungulates, and rabies in small carnivores (and dozens of other parasites). They have always been in the system; however, increasing human populations and encroachment on the native fauna have lead to increased contact between wildlife and both domestic stock and humans, with consequent effects on human livelihood strategies and welfare.

The second aspect concerns epizootics. We have observed periodic outbreaks of distemper, in particular, with consequent crashes in carnivore numbers. Rinderpest is another example, although its consequences in recent decades have been much reduced. These perturbations arise in the human ecosystem, with little affect there but with major consequences on the native biota. Thus, disease causes disturbances by being transmitted both ways between the human and the natural ecosystems.

Modeling Perturbations within the Natural System

Different modeling scenarios applied to the natural system have made predictions on trends in the major mammal populations (Hilborn and Sinclair 1979; Hilborn 1995). In chapter 8, Holt et al. focus on a generalized food web to predict the consequences of high carnivore mortality (such as might occur through disease, as above). In general, this mortality has little effect on the rest of the biota because of buffering from territorial behavior of carnivores. Indeed, territoriality creates stasis until the system has been pushed far enough to produce a jump in lion numbers, as observed by Packer et al. (2005). Thus, jumps from one state to another might be predicted. The main effect was the prediction of decreasing resident herbivores, and this has now been observed (Sinclair, Mduma, and Brashares 2003).

The second scenario involved the high mortality of wildebeest. There is little predicted effect on carnivores, but some effect on grass biomass through increasing tall grass. Indirectly, the prediction is a decline of species such as Thomson's gazelle through reduced facilitation. These predictions are consistent with historical reports of long grass in the 1930s on what are now short-grass plains, when wildebeest numbers were probably one tenth the present numbers (see chapter 2, fig. 2.13, this volume). However, as wildebeest numbers have increased, gazelle numbers have declined, suggesting that competition is the dominant process, as Holt et al. note (chapter 8, this volume).

Fryxell et al. (chapter 9) model behavioral grazing strategies as a mechanism to account for the coexistence of the highly diverse ungulate community. They conclude that the behavioral complexity is not sufficient to provide coexistence. In particular, they predict that buffalo and gazelle should decline in number. Such declines are seen in both species (fig. 2.24 for buffalo). Buffalo populations that collapsed in the 1993 drought, throughout the system, have been slowly increasing in areas where they do not overlap with the migratory wildebeest in the dry season, hence avoiding competition. However, in the Mara Reserve and northern Serengeti, where the two species overlap, buffalo have shown no increase subsequent to 1993 (fig. 2.14, this volume). One of the important issues that emerges from this chapter is that of scale—systems have to be large enough to allow the unrestricted movements of populations as they search for their optimum resource level. Restriction of area could lead to collapse of the population as it is forced to make do with unsuitable resources.

Modeling the Human-Wildlife Interaction

The model of Costello et al. (chapter 10) is used at two scales, that of the individual household by Galvin et al. (chapter 11) and that of the macrolevel of national and international agencies by Polasky et al. (chapter 12). Within Tanzania, indigenous peoples living around the park are either small-holding agriculturalists with livestock in the west or are pastoralists in the east. The most important point that emerges from the empirical research and the modeling is that *wealth is measured in terms of livestock*. These communities translate surplus agricultural production and cash income into cattle, this being the only long-term banking system for wealth that can grow. Thus, the statistic of livestock per capita is the critical measure of wealth.

We see, however, a different dynamic emerging in Kenya. In that country it is the cash value of the income derived from the land that is the driving variable (Norton-Griffiths et al., chapter 13). Incomes from agricultural and pastoral practices are outstripping those from traditional conservation practices, so that community-based conservation is losing ground.

Local peoples in the greater Serengeti ecosystem are vulnerable to many different perturbations outside their control—changes in climate, ecology, government policies, economics, and demography. In addition, seasonal variability of resources, such as rainfall and the appearance of migrating wildlife, imposes seasonality on this vulnerability. Individual households make livelihood decisions to reduce their vulnerability to unpredictable changes. Galvin et al. (chapter 11) model two such changes.

The first scenario postulates changes in average rainfall. With increased rainfall everything goes up—agricultural production, livestock, and more than anything else, human numbers. Consequently, livestock per capita—and hence wealth—goes down, producing a counterintuitive and perverse result. They conclude that the only way to increase wealth is to slow the human population growth.

The converse perturbation of a reduction in rainfall, the one actually predicted by Ritchie to occur (chapter 6), does not provide the opposite outcome. Declining production results in either the death of livestock or the sale of the remainder. People's wealth is used up, so they leave. Resident wildlife populations increase because they are better adapted to environmental conditions and they are now also released from hunting. Something analogous to these outcomes may have occurred following the Great Rinderpest of 1890. Although the perturbation was not due to lower rainfall, the near complete loss of cattle removed all resilience to seasonal or annual fluctuations in food, and people abandoned the western Serengeti. In Ethiopia, where events were better documented, such a severe famine resulted in the death of one third of the human population (Pankhurst 1966).

The second scenario modeled the doubling of food crop prices. This resulted in people having to sell off livestock and some having to leave. Consequently, livestock per capita of those remaining increased, so there was less hunting of wildlife. In contrast, if the price of bushmeat increased, human numbers doubled, livestock were sold off, and wealth declined. Thus, neither result from changes in microeconomics would have been expected, at the outset.

At the macro level, scenarios that were modeled included (1) changes in antipoaching policy, (2) the establishment of Wildlife Management Areas (WMAs), and (3) increasing human populations. In the first scenario, decreasing the level of enforcement encouraged human immigration, an increase in poaching, and small declines of wildlife, and the opposite result prevailed with increasing enforcement.

The second scenario allowed legal harvesting of wildlife in WMAs. At most levels of harvesting there was little change in either the natural ecosystem or in human populations and wealth. However, if revenues from harvesting were high (perhaps from very high harvest levels or the value of the commodity was high), then human immigration occurs, wealth declines, and poaching increases. Empirical evidence for this immigration is given in Campbell and Hofer (1995).

Rapidly increasing human populations (the third scenario) had little effect on the natural ecosystem, provided that enforcement was effective, but caused major declines in wildlife if enforcement was lax. Perversely, wealth declined, despite increased poaching, because livestock did not increase as fast as the human population. This decline in wealth is seen in other areas of Africa with rapidly increasing human numbers but stationary livestock numbers (Wu, Thurow, and Whisenant 2000), and this has resulted in an increasing consumption of bushmeat (Brashares et al. 2004).

Demographic transition, involving declining birthrates and human population growth, produced moderate impacts on the natural ecosystem in terms of poaching, and a moderate increase in the wealth of people (see table 12.9). However, it incorporated the major advantage of stabilizing the human population, and therefore, the whole system.

In general, three important conclusions emerge from these scenarios of change in the human system (chapters 10–13). First, in Tanzania and parts of Kenya, livestock are the standard measure of wealth, because this is the only asset that grows over time, survives the normal dry season when food crops dwindle, and can be used for cash to obtain food during droughts. Indeed, livestock are the means of dealing with fluctuating environmental conditions, as seen in the Turkana of northern Kenya (McCabe 2004). However, perturbations—both environmental and socioeconomic—can have negative effects on wealth, sometimes in counterintuitive and perverse ways. In Kenya, cash income is the measure of wealth. Livelihoods are assessed on the basis of competing land uses, and livelihoods may be compromised if conservation does not compensate for lost opportunities.

Second, if human populations surrounding protected areas increase too fast, from immigrants attracted, for example, by high resources in ICDPs (see Homewood, Coast, and Thompson 2004 for the Ngorongoro Conservation Area), or high bushmeat prices, or even high rainfall, then livestock per capita and hence wealth declines. These outcomes provide reasons why many ICDPs, including the Serengeti Regional Conservation Project (chapter 10 and Mbano et al. 1995), have not met their objectives (Holmern et al. 2002, Ferraro and Kiss 2002). In particular, the consumptive exploitation of wildlife as a means of increasing per capita wealth may not achieve that end but rather could produce the perverse result of a decline in wealth. Increasing human populations also increase poaching. Thus, both PAC and community-based conservation are negatively affected under these conditions.

Third, at moderate levels of harvesting and hence revenue generation, WMAs have little negative impact on the natural ecosystem while at the same time they have a positive indirect functional role. They act as buffer zones to the protected area, so reducing both the impact of humans on the natural system and wildlife nuisance effects on human systems

(Thirgood et al. chapter 15). If new settlement in the WMA can be regulated a net benefit is obtained for both Protected Area Conservation and Community-based Conservation. As Polasky et al. (chapter 12) point out, the fundamental dilemma for conservation in developing countries is how to conserve biota within the context of a growing human population that places a higher priority on economic development, necessary to reduce poverty.

The main conclusion is that unless human population increase in areas surrounding protected areas is stopped, or even reversed, the future of conservation in both the community areas and the protected areas will be seriously compromised. However, it is far from clear how zero human population growth can be achieved. Polasky et al. (chapter 12) show that population growth can decline if the same type of demographic transition that has occurred in developed countries takes place before the natural ecosystems are overwhelmed.

BOUNDARIES BETWEEN HUMAN AND NATURAL AREAS IN THE GREATER SERENGETI REGION

We have described above, and in most of the previous chapters, the way the protected area system and the human-dominated system influence each other. Inevitably these interactions also lead to negative consequences and conflicts. Human populations living on the edges of the natural ecosystem have several impacts (chapter 2). First, there are direct impacts through exploitation of resources—largely by hunting of ungulates and removal of predators by snaring, and to a lesser extent cutting of trees for building poles. Other inadvertent impacts arise from poachers setting fires next to riverine forests in order to attract wildlife into snares or within range of arrows. Such fires are the main cause of the decline of the rare and threatened forests of northern Serengeti (Sharam 2005).

Indirect effects of humans can be seen in the consequences of political events, such as the closing of the Tanzania-Kenya border during 1977–1986. This conflict led to a collapse of revenue, a decline of antipoaching activities, and burgeoning poaching (chapter 2). Second, the availability of bushmeat within the natural area encourages people to settle as close to the park as possible, creating a hard boundary and exacerbating the human-wildlife conflict. This settlement process is seen in the high rates of increase (15% per annum) along the northwest boundary of the park during the 1980s and 1990s, far above the population increase due to reproduction (3% per annum; Campbell and Hofer 1995). Third, influences of humans through deforestation and overgrazing far upstream on the Mara River in Kenya are likely having an effect on the flood patterns of the river. Consequently the width of the river is now twice as wide as it was in the 1960s (chapter 2).

Impacts of the natural area on humans are largely felt through animal populations spilling out of the protected area into agricultural areas, causing damage to crops. Wildebeest have been doing this each dry season into the wasukuma, waikoma, and wakuria areas as far back as records go, and most of the peoples living close to the protected areas subsist on this meat when it is available (C. Packer personal observation). To some extent, bushmeat obtained from killing the encroaching animals mitigates the damaging effects of encroachment. However, such benefits do not accrue when elephants or baboons pillage crops from the safety of the protected area. In the case of elephants these problems have increased, as the population has rebounded after the ivory poaching of the 1980s (chapter 2).

Hard Boundaries

Solutions to human-natural area conflicts fall into two categories: those that artificially separate the two systems through barriers and create a hard boundary, and those that form a zone of intermediate use, consistent with both areas, and so create a soft edge.

Superficially, hard-boundary solutions are tempting because they prevent incursions of wildlife into agriculture, and clearly demarcate where humans cannot go. In some cases, usually where high human density confronts high wildlife density, as along forest-agriculture boundaries, this approach may be the only solution.

However, even where they must be employed, barriers and fences create other problems, and they work intermittently at best. Fences, in particular, must be maintained. Unless they are checked, effectively daily, breaks allow large mammals to escape into agriculture, and they may not find their way back through the break. Such was the case with a fence that was to exclude moose (*Alces alces*) in the Yukon, Canada (Krebs, Boutin, and Boonstra 2001). Animals trapped on the wrong side of the fence then have to be removed. In the case of tree-climbing animals, such as baboons or large cats, fences do not work.

Indirect problems of fences occur when animals congregate at the barrier. Sometimes, they press against the fence and are killed (wildebeest died along fencelines that protected wheat farms north of the Mara Reserve in the 1980s). Lions also learned to trap wildebeest and other species against fences at Keekerok in the Mara Reserve (pers. obs.), and Canadian lynx

(*Lynx canadensis*) trapped snowshoe hares (*Lepus americanus*) against fences in the Yukon (Krebs, Boutin, and Boonstra 2001).

Fences that prevent the normal migration routes of animals can cause severe mortality and population declines, as occurred with the Kruger (Whyte and Joubert 1988; Ogutu and Owen-Smith 2003) and the Kalahari wildebeest in Botswana, when fences prevented their movement. Habitat destruction by animals patrolling the fence also accompanies the prevention of movements, a feature originally described as the "fence effect" by Krebs, Keller, and Tamarin (1969).

Soft Boundaries

The concept of a soft edge between human and wildlife activity necessarily requires a compromise on both sides. Human activities must be consistent with use of the same area by wildlife. Thus, buildings, agriculture, and fences are inconsistent with this objective. In contrast, pastoralism, tourism, and wildlife viewing, and even controlled hunting, whether for sport or commercial off-take are uses that are consistent with the idea of soft boundaries.

The main problem with this approach lies in its application. Without due care it simply results in another hard edge where it confronts agriculture. In the past these approaches, under the heading of *buffer zones,* have come to be viewed as simply a mechanism by which conservation can expand protected areas at the expense of local peoples. Indeed, the term "buffer zones" now has negative connotations among social scientists (Neumann 2004).

Soft edges, however, are the only solution along extensive boundaries. It is simply not feasible, for example, to contain the Serengeti wildebeest migration within a fence—especially because the range and hence the boundary shifts with prevailing weather and long-term climate. Soft edges, therefore, must contain the following conditions if they are to work:

1. At the interface with agriculture (or other purely human dominated systems) the impact of wildlife should be at a minimum.
2. At the other edge of the boundary zone, the impact of humans must be at a minimum.
3. To achieve the above two conditions, there has to be two gradients, one of decreasing wildlife presence from protected area to agriculture, and the other of decreasing human impact in the reverse direction.

How these gradients are imposed requires careful planning and management. Soft boundaries also have to be sufficiently wide that a workable gradient can be produced. Sport and commercial hunting on the agricultural end of the zone both reduces the densities of wildlife and provides a deterrent, due to the disturbance from hunting—many species will prefer to reside closer to the protected area, where, by design, there is less hunting. Conversely, tourism viewing and pastoralism can be promoted at the edge of the protected area. In summary, within a sufficiently wide soft boundary, there must be zones of different management strategies conscientiously applied and rigorously monitored by enforcement authorities.

COMMUNITY-BASED CONSERVATION IN THE SERENGETI ECOSYSTEM

As we describe previously, protected areas and community-based conservation areas are different mechanisms to meet conservation needs, and both are required for a global conservation strategy. Both approaches generate revenue. The degree to which this revenue is ploughed back into conservation, on the one hand, and community benefits, on the other, differs with the type of administration, but does not clearly differ with the PAC-CBC continuum (Ferraro and Kiss 2002). What does appear essential is that (1) there is community involvement, especially in planning and control, and (2) such communities have access to both natural resources and markets for CBC to be successful (Brooks et al. 2006).

Thirgood et al. (chapter 15) provide a summary of the different revenues and expenditures from the various administrations in the Serengeti ecosystem. These show a gradation of investments in community-based conservation. The Serengeti National Park earns income directly from tourism and about half of this is reinvested in the protected area. A proportion is also used for conservation projects in surrounding villages, in particular for construction of infrastructure such as roads, schools, and clinics (J. Hando, pers. comm.). In return, the villages control incursions to the park. Thus, the protected area is reaching out to the community, and social contracts are generated.

The Ngorongoro Conservation Area also expends a significant proportion of its income on infrastructure as well as development for local peoples who inhabit the area. Both the SNP and NCA administrations also pay a substantial amount of revenue to the central government, and this is not reinvested in either conservation or development. Game reserves, such as Maswa, earn income from hunting licenses. All of this revenue goes to the central government, with almost nothing of this income reinvested in the reserve or in surrounding villages. Thus, internal community-based conservation is minimal. In contrast, the Wildlife Management Areas earn income from wildlife, and nearly all of it is reinvested in the form of development for local people.

In summary, there is a range of investments in community-based conservation. The strictly protected areas, such as Serengeti National Park, play a role in community-based conservation through its outreach programs, so that the division between pure protection and community protection is not black and white. In addition, because park revenue derives from non-consumptive use, the dangers of overuse of the wildlife resource that are inherent in consumptive use are minimal though not absent—too many tourists can have negative impacts on both rare species such as cheetah and on the functioning of the natural ecosystem.

The analyses by Thirgood et al. (chapter 15) clearly show that reinvestment by the central government for development of local communities is a very small proportion of the income produced from conservation. The short-fall is made up largely by donations from philanthropic donors or international agencies. There is a concern that the presence of external funding inhibits the investment from internal agencies for the sake of community conservation. At this point the cause-and-effect relationships of these two funding processes are not clear.

CONCLUSIONS

Various types of modeling approaches have simulated perturbations to the combined natural and human systems included in the greater Serengeti ecosystem. These perturbations comprised changes in rainfall, wildebeest numbers, commodity prices, and human populations. The overall conclusion was that changes in the natural system affected human welfare, and equally, changes in the human system had impacts on the protected area.

The five conclusions that emerge from change in the human system are: (1) in Tanzania, livestock are the standard measure of wealth, because this is the only asset that grows over time. Perturbations, both environmental and socioeconomic, can have negative effects on wealth, sometimes in counterintuitive and perverse ways. (2) In Kenya, wealth is assessed through the cash income from land. In both countries, however, the loss of wealth from negative effects due to conserving wildlife (e.g., opportunity costs due to loss of other activities, damage from wildlife) need to be compensated from national and international sources. (3) If human populations surrounding protected areas increase too fast, livestock per capita and hence wealth declines. These outcomes provide reasons why many ICDPs, including the Serengeti Regional Conservation Project, have not met their objectives. (4) The consumptive exploitation of wildlife as a means of increasing per capita wealth may not achieve that end but rather could produce the perverse result of a decline in wealth. Increasing human populations also increase poaching. Thus, both the protected area and community conservation areas are negatively affected under these conditions. (5) However, at moderate levels of harvesting and hence revenue generation, community conservation areas, such as the WMAs, have little negative impact on the natural ecosystem while at the same time they act as buffer zones to the protected area, so reducing both the impact of humans on the natural system and wildlife nuisance effects on human systems.

To reduce the negative impacts of one system upon the other, soft boundaries are required. These boundary zones must be wide enough to include two gradients: one a gradient of human disturbance from high on the outside to low at the park boundary, the other a gradient of animal impacts with the opposite trend. Land uses that are inconsistent with these gradients, such as agriculture on the park boundary (thus creating a hard boundary) should be phased out.

Human benefits from the conservation of biota may include consumptive exploitation within soft boundaries close to the outside of the zone. However, consumptive exploitation is not a necessary requirement in order to provide benefits to local peoples. Benefits can accrue from nonconsumptive uses such as tourism and from payments in return for sustainable stewardship by national governments and international organizations. However, I emphasize that the costs of conservation must be met by those who benefit the most, which is mainly those in the developed nations, and these nations must devise adequate benefit programs for local communities that leave people better off from the CBC compensation than if they had adopted other forms of resource use with no conservation.

These results highlight an essential point; namely, that both the conservation of biota and the sustainability of human welfare require the combination of community-based conservation and protected-area conservation. Each is essential, but neither is sufficient alone for the conservation and sustainability of resources.

The main conclusion from these models is that unless human population increase in areas surrounding protected areas is stopped—even reversed—the future of conservation in both the community areas and the protected areas will be seriously compromised. At the same time, the most effective way to increase human wealth is to slow population growth.

REFERENCES

Adams, W. M. 1998. Conservation and development. In *Conservation science and action,* ed. W. J. Sutherland, 286–315. Oxford: Blackwell Science.

———. 2001. *Green development: Environment and sustainability in the third world,* 2nd ed. London: Routledge.

———. 2003. When nature won't stay still: Conservation, equilibrium and control. In *Decolonizing nature,* ed. W. M. Adams and M. Mulligan, 220–46. London: Earthscan.

Adams, W. M., R. Aveling, D. Brockington, B. Dickson, J. Elliott, J. Hutton, D. Roe, B. Vira, and W. Wolmer. 2004. Biodiversity conservation and the eradication of poverty. *Science* 306:1146–49.

Adams, W. M., and D. Hulme. 2001. If community conservation is the answer, what is the question? *Oryx* 35:193–201.

Adams, W. M., and M. Mulligan. 2003. Introduction. In *Decolonizing nature,* ed. W. M. Adams and M. Mulligan, 1–15. London: Earthscan.

Agrawal, A., and C. C. Gibson. 1999. Enchantment and disenchantment: The role of community in natural resource conservation. *World Development* 27:629–49.

Arcese, P., and A. R. E. Sinclair. 1997. The role of protected areas as ecological baselines. *Journal of Wildlife Management* 61:587–602.

Balmford, A., and T. Whitten. 2003. Who should pay for tropical conservation, and how should the costs be met? *Oryx* 37:238–50.

Balmford, A., P. Crane, A. Dobson, R. E. Green, and G. M. Mace. 2005. The 2010 challenge: Data availability, information needs, and extraterrestrial insights. *Philosophical Transactions of the Royal Society, London—Series B* 360:221–28.

Barrett, C. B., and P. Arcese. 1995. Are integrated conservation-development projects (ICDPs) sustainable? On the conservation of large mammals in sub-Saharan Africa. *World Development* 25:1073–84.

Biel, A. W. 2004. The bearer has permission. *Yellowstone Science* 12:5–20.

Birch, T. H. 1990. The incarceration of wilderness: Wilderness areas as prisons. *Environmental Ethics* 12:3–26.

Bond, I. 2001. CAMPFIRE and the incentives for institutional change. In *African wildlife and livelihoods,* ed. D. Hulme and M. Murphree, 227–43. Oxford: James Currey.

Borgerhoff Mulder, M., and P. Coppolillo. 2005. *Conservation: Linking ecology, economics and culture.* Princeton, NJ: Princeton University Press.

Brandon, K., K. H. Redford, and S. E. Sanderson. 1998. *Parks in peril: People, politics and protected areas.* Washington, DC: Island Press.

Brashares, J. S., P. Arcese, M. K. Sam, P. B. Coppolillo, A. R. E. Sinclair, and A. Balmford. 2004. Bushmeat hunting,wildlife declines and fish supply in West Africa. *Science* 306: 1180–83.

Breitenmoser, U., C. Angst, J-M. Landry, C. Breitenmoser-Wursten, J. D. C. Linnell, and J-M. Weber, 2005. Non-lethal techniques for reducing depredation. In *People and wildlife: Conflict or coexistence?* ed. R. Woodroffe, S. Thirgood, and A. Rabinowitz, 49–71. Cambridge: Cambridge University Press.

Brooks, J. S., M. A. Franzen, C. M. Holmes, M. N. Grote, and M. Borgerhoff Mulder. 2006. Testing hypotheses for the success of different conservation strategies. *Conservation Biology* 20:1528–38.

Bruner, A. G., R. E. Gullison, R. E. Rice, and G. A. B. de Fonseca. 2001. Effectiveness of parks in protecting tropical biodiversity. *Science* 291:125–28.

Burgess, N., J. D. Hales, E. Underwood, E. Dinerstein, D. Olson, I. Itoua, J. Schipper, T. Ricketts, and K. Newman. 2004. *Terrestrial ecoregions of Africa and Madagascar.* Washington, DC: Island Press.

Campbell, K., and H. Hofer. 1995. People and wildlife: Spatial dynamics and zones of interaction. In *Serengeti II: Dynamics, management and conservation of an ecosystem,* ed. A. R. E. Sinclair and P. Arcese, 534–70. Chicago: University of Chicago Press.

Chamberlain, D. E., R. J. Fuller, R. G. H. Bunce, J. C. Duckworth, and M. Shrubb. 2000. Changes in the abundance of farmland birds in relation to the timing of agricultural intensification in England and Wales. *Journal of Applied Ecology* 37:771–88.

Child, B. 1996. The practice and principles of community-based wildlife management in Zimbabwe: The CAMPFIRE programme. *Biodiversity and Conservation* 5 (3): 369–98.

———, ed. 2004. *Parks in transition.* London: Earthscan.

Conley, A., and M. A. Moote. 2003. Evaluating collaborative natural resource management. *Society and Natural Resources* 16:371–86.

Costanza, R., R. d'Arge, R. de Groot, S. Farber, M. Grasso, B. Hannon, K. Limburg, S. Naeem, R. V. O'Neill, J. Paruelo, R. G. Raskin, P. Sutton, and M. van den Belt. 1997. The value of the world's ecosystem services and natural capital. *Nature* 387:253–61.

Cronon, W. 1995. The trouble with wilderness: Or, getting back to the wrong nature. In *Uncommon ground: Toward reinventing nature,* ed. W. Cronon, 69–90. New York: W.W. Norton.

Emerton, L. 2001. The nature of benefits and the benefits of nature. In *African wildlife and livelihoods,* ed. D. Hulme and M. Murphree, 208–26. Oxford: James Currey.

Ferraro, P. J., and A. Kiss. 2002. Direct payments to conserve biodiversity. *Science* 298: 1718–19.

Getz, W. M., L. Fortmann, D. Cumming, J. du Toit, J. Hilty, R. Martin, M. Murphree, N. Owen-Smith, A. M. Starfield, and A. I. Westphal. 1999. Conservation—sustaining natural capital: Villagers and scientists. *Science* 283:1855–56.

Green, R. E., A. Balmford, P. R. Crane, G. M. Mace, J. D. Reynolds, and R. K. Turner. 2005. A framework for improved monitoring of biodiversity: Responses to the World Summit on Sustainable Development. *Conservation Biology* 19:56–65.

Hackel, J. D. 1999. Community conservation and the future of Africa's wildlife. *Conservation Biology* 13:726–34.

Hilborn, R. 1995. A model to evaluate alternative management policies for the Serengeti-Mara ecosystem. In *Serengeti II: Dynamics, management and conservation of an ecosystem,* eds. A. R. E. Sinclair and P. Arcese, 617–38. Chicago: University of Chicago Press.

Hilborn, R., and A. R. E. Sinclair. 1979. A simulation of wildebeest and other ungulates and their predators in the Serengeti. In *Serengeti: Dynamics of an ecosystem,* ed. A. R. E. Sinclair and M. Norton-Griffiths, 287–309. Chicago: University of Chicago Press.

Hobbs, R. J. 1993. Effects of landscape fragmentation on ecosystem processes in the Western Australia wheatbelt. *Biological Conservation* 64:193–201.

Holmberg, J., K. Thomson, and L. Timberlake. 1993. *Facing the future: Beyond the earth summit.* London: Earthscan and International Institute for Environment and Development.

Holmern, T., E. Roskaft, J. Mbaruka, S. Y. Mkama, and J. Muya. 2002. Uneconomical game cropping in a community-based conservation project outside the Serengeti National Park, Tanzania. *Oryx* 36:364–72.

Homewood, K., E. Coast, and M. Thompson. 2004. In-migrants and exclusion in East African rangelands: Access, tenure and conflict. *Africa* 74:567–610.

Hulme, D., and M. Murphree, eds. 2001. *African wildlife and livelihoods.* Oxford: James Currey.

Hulme, M. 2001. Climatic perspectives on Sahelian desiccation: 1973–1998. *Global Environmental Change* 11:19–29.

International Union for Conservation of Nature and Natural Resources (IUCN). 1980. *The world conservation strategy.* Geneva: International Union for Conservation of Nature and Natural Resources, United Nations.

James, A., K. J. Gaston, and A. Balmford. 2001. Can we afford to conserve biodiversity? *Bioscience* 51:43–52.

Jones, B. T. B., and M. W. Murphree. 2004. Community-based natural resource management as a conservation mechanism: Lessons and directions. In *Parks in transition,* ed. B. Child, 63–104. London: Earthscan.

Jope, K.L., and J.C. Dunstan. 1996. Ecosystem-based management: Natural processes and systems theory. In *National parks and protected areas: Their role in environmental protection,* ed. G. R. W. Wright, 45–62. Oxford: Blackwell Science.

Krebs, C. J., S. Boutin, and R. Boonstra, eds. 2001. *Ecosystem dynamics of the boreal forest: The Kluane project.* Oxford: Oxford University Press.

Krebs, C. J., B. L. Keller, and R. H. Tamarin. 1969. *Microtus* population biology. *Ecology* 50: 587–607.

Leader-Williams, N., and J. M. Hutton. 2005. Does extractive use provide opportunities to offset conflicts between people and wildlife? In *People and wildlife: Conflict or coexistence?* ed. R. Woodroffe, S. Thirgood, and A. Rabinowitz, 140–61. Cambridge: Cambridge University Press.

Lybbert, T., and C. B. Barrett. 2004. Does resource commercialization induce local conservation? A cautionary tale from southwestern Morocco. *Society and Natural Resources* 17:413–30.

Mbano, B. N. N., R. C. Malpas, M. K. S. Maige, P. A. K. Symonds, and D. M. Thompson. 1995. The Serengeti regional conservation strategy. In *Serengeti II: Dynamics, management and conservation of an ecosystem,* ed. A. R. E. Sinclair and P. Arcese, 605–16. Chicago: University of Chicago Press.

McCabe, J. T. 2004. *Cattle bring us to our enemies. Turkana ecology, politics, and raiding in a disequilibrial system.* Ann Arbor: University of Michigan Press.

McNeely, J.A., and K.R. Miller, eds. 1984. *National parks, conservation and development: The role of protected areas in sustaining society.* Washington, DC: Smithsonian Institute Press.

McFarlane, D. J., R. J. George, and P. Farrington. 1993. Changes in the hydrologic cycle. In *Reintegrating fragmented landscapes,* ed. R. J. Hobbs and D. A. Saunders, 147–86. New York: Springer-Verlag.

McShane, T. O., and M. Wells, eds. 2004. *Getting biodiversity projects to work: Towards more effective conservation and development.* New York: Columbia University Press.

Moran, D., D. Pearce, and A. Wendelaar. 1997. Investing in biodiversity: An economic perspective on global priority setting. *Biodiversity and Conservation* 6:1219–43.

Murombedzi, J. 2003. Devolving the expropriation of nature: The 'devolution' of wildlife management in southern Africa. In *Decolonizing nature,* ed. W. M. Adams and M. Mulligan, 135–51. London: Earthscan.

Murphree, M. W. 2004. Who and what are parks for in transitional societies? In *Parks in transition.* ed. B. Child, 217–32. London: Earthscan.

Murray Li, T. 2002 Engaging simplifications: Community-based resource management, market processes and state agendas in upland Southeast Asia. *World Development* 30: 265–83.

Neumann, R.P. 2004. A century of changing land uses and property rights in Tanzania's Selous Game Reserve. In *Beyond the arch: Community and Conservation in Greater Yellowstone and East Africa,* ed. A.W. Biel. 228–43. Yellowstone National Park, WY: Yellowstone Centre for Resources.

Newmark, W.D., and J.L. Hough. 2000. Conserving wildlife in Africa: Integrated development and conservation projects and beyond. *Bioscience* 50: 585–92.

Nyhus, P. J., S. A. Osofsky, P. Ferraro, F. Madden, and H. Fischer. 2005. Bearing the costs of human-wildlife conflict: The challenges of compensation schemes. In *People and wildlife: Conflict or coexistence?* ed. R. Woodroffe, S. Thirgood, and A. Rabinowitz, 107–21. Cambridge: Cambridge University Press.

Ogutu, J.O., and N. Owen-Smith. 2003. ENSO, rainfall and temperature influences on extreme poplation declines among African savanna ungulates. *Ecology Letters* 6:412–19.

Osborn, F. V., and C. M. Hill. 2005. Techniques to reduce crop loss: Human and technical dimensions in Africa. In *People and wildlife: Conflict or coexistence?* ed. R. Woodroffe, S. Thirgood, and A. Rabinowitz, 72–85. Cambridge: Cambridge University Press.

Packer, C., R. Hilborn, A. Mosser, B. Kissui, M. Borner, G. Hopcraft, J. Wilmshurst, S. Mduma, and A. R. E. Sinclair. 2005. Ecological change, group territoriality and population dynamics in Serengeti lions. *Science* 307:390–93.

Pankhurst, R. 1966. The Great Ethiopian famine of 1888–1892: A new assessment. Part 2. *Journal of the History of Medicine* (July): 271–94.

Peterson, D.L. 1996. Research in parks and protected areas: Forging the link between science and management. In *National parks and protected areas: Their role in environmental protection.* ed. R.G. Wright, 417–34. Oxford: Blackwell Science.Pimm, S. L. 1998. Extinction. In *Conservation science and action,* ed. W. J. Sutherland, 20–38. Oxford: Blackwell Science.

Pimm, S. L., G. J. Russell, J. L. Gittleman, and T. M. Brooks. 1995. The future of biodiversity. *Science* 269:347–50.

Robinson, R. A., R. E. Green, S. R. Baillie, W. J. Peach, and D. L. Thomson. 2004. Demographic mechanisms of the population decline of the song thrush *Turdus philomelos* in Britain. *Journal of Animal Ecology* 73:670–82.

Rodrigues, A. S., S. J. Andelman, M. I. Bakarr, L. Boitani, T. M. Brooks, R. M. Cowling, L. D. C. Fishpool, G. A. B. da Fonsecca, K. J. Gaston, and M. Hoffmann et al. 2004. Effectiveness of the global protected area network in representing species diversity. *Nature* 428:640–43.

Sanderson, S.E. 2004. A conservation agenda in an era of poverty. In *Beyond the arch: Com-*

munity and Conservation in Greater Yellowstone and East Africa, ed. A.W. Biel. 278–86. Yellowstone National Park, WY: Yellowstone Centre for Resources.

Sedjo, R. A. 2000. Biodiversity: Forests, property rights and economic value. *Conserving nature's diversity,* ed. G. C. Van Kooten, E. R. Bulte, and A. R. E. Sinclair, 106–22. Burlington, VT: Ashgate.

Sharam, G. J. 2005. The decline and restoration of riparian and hilltop forests in the Serengeti National Park, Tanzania. PhD diss., University of British Columbia, Vancouver.

Simpson, D. 2000. Economic perspectives on preservation of biodiversity. In *Conserving nature's diversity,* ed. G. C. Van Kooten, E. R. Bulte, and A. R. E. Sinclair, 88–105. Burlington, VT: Ashgate.

Sinclair, A. R. E. 1983. Management of conservation areas as ecological baseline controls. In *Management of large mammals in African conservation areas,* ed. N. Owen-Smith, 13–22. Pretoria: Haum.

———. 1998. Natural regulation of ecosystems in Protected Areas as ecological baselines. *Wildlife Society Bulletin* 26:399–409.

Sinclair, A. R. E., and A. Byrom. 2006. Understanding ecosystems for the conservation of biota. *Journal of Animal Ecology* 75:64–79.

Sinclair, A. R. E., and J. M. Fryxell. 1985. The Sahel of Africa: Ecology of a disaster. *Canadian Journal of Zoology* 63:987–94.

Sinclair, A. R. E, S. A. R. Mduma, and J. S. Brashares. 2003. Patterns of predation in a diverse predator-prey system. *Nature* 425:288–90.

Sinclair, A. R. E, D. Ludwig, and C. Clark. 2000. Conservation in the real world. *Science* 289:1875.

Songorwa, A. N., T. Buhrs, and K. F. D. Hughey. 2000. Community-based wildlife management in Africa: A critical assessment of the literature. *Natural Resources Journal* 40: 603–43.

Stem, C., R. Margoluis, N. Salafsky, and M. Brown. 2005. Monitoring and evaluation in conservation: A review of trends and approaches. *Conservation Biology* 19:295–309.

Tapper, S. C., G. R. Potts, and M. H. Brockless. 1996. The effect of an experimental reduction in predation pressure on the breeding success and population density of grey partridges *Perdix perdix. Journal of Applied Ecology* 33:965–78.

Terborgh, J. 1999. *Requiem for nature.* Washington, DC: Island Press.

Terborgh, J., C. van Schaik, L. Davenport, and M. Rao, eds. 2002. *Making parks work.* Washington, DC: Island Press.

Thirgood, S., R. Woodroffe, and A. Rabinowitz. 2005. The impact of human-wildlife conflict on human lives and livelihoods. In *People and wildlife: Conflict or coexistence?* ed. R. Woodroffe, S. Thirgood, and A. Rabinowitz, 49–71. Cambridge: Cambridge University Press.

Thuiller, W., O. Broennimann, G. Hughes, J. R. M. Alkemade, G. F. Midgeley, and F. Corsi. 2006. Vulnerability of African mammals to anthropogenic climate change under conservative land transformation assumptions. *Global Change Biology* 12:424–40.

Walpole, M. J., and C. R. Thouless. 2005. Increasing the value of wildlife through non-consumptive use? Deconstructing the myths of ecotourism and community-based tourism in the tropics. In *People and wildlife: Conflict or coexistence?* ed. R. Woodroffe, S. Thirgood, and A. Rabinowitz, 122–39. Cambridge: Cambridge University Press.

Whyte, I. J., and C. J. Joubert. 1988. Blue wildebeest population trends in the Kruger National Park and the effects of fencing. *South African Journal of Wildlife Research* 18: 78–87.

Willis, K. J., L. Gillson, and T. Brncic. 2004. How virgin is virgin forest? *Science* 305: 943–44.

Willis, K. J., L. Gillson, T. Brncic, and B. Figueroa-Rangel. 2005. Providing baselines for biodiversity measurement. *Trends in Ecology and Evolution* 20:107–108.

Woodroffe, R., S. Thirgood, and A. Rabinowitz. 2005. The future of co-existence: Resolving human-wildlife conflicts in a changing world. In *People and wildlife: Conflict or coexistence?* ed. R. Woodroffe, S. Thirgood, and A. Rabinowitz, 388–405. Cambridge: Cambridge University Press.

Wright, R. G., and D. J. Matson. 1996. The origin and purpose of National Parks and Protected Areas. In *National parks and protected areas: Their role in environmental protection,* ed. G. R. W. Wright, 3–14. Oxford: Blackwell Science.

Wu, X. B., T. L. Thurow, and S. G.Whisenant. 2000. Fragmentation and changes in hydrologic function of tiger bush landscapes, south-west Niger. *Journal of Animal Ecology* 88:790–800.

The Main Herbivorous Mammals and Crocodiles in the Greater Serengeti Ecosystem

Simon A. R. Mduma and J. Grant C. Hopcraft

Estimates of population numbers or of density are available for many of the Serengeti's larger herbivorous mammals. These were obtained by several methods. Larger ungulates have been counted repeatedly through aerial surveys, and trends in numbers have been reported in chapter 2 (this volume) and earlier papers (Grimsdell 1979; Sinclair 1979; Campbell and Borner, 1995). For other species we have estimates of density from ground transects or total counts. For many of the smaller species we have only rough estimates of density or simple descriptions of distribution.

In table A.1 we present the available information, with notes, where appropriate; scientific names are given in table A.2. The orders Proboscidea, Hyracoidea, and Tubulidentata have recently been placed in the Superorder Afrotheria.

Wildebeest
Trends in this species are given in chapter 2, fig. 2.13. The most recent aerial census of 2003 was obtained digitally. Numbers appear to have stabilized, having recovered from the drought of 1993.

Zebra
This species has remained at approximately the same level since counts began, in 1966.

Table A.1 The main herbivorous mammals in the Serengeti ecosystem

Common name	Latin name	Number	S.E.	Density (N/km²)	Notes
Wildebeest	Connochaetes taurinus	1,086,754	166,570		2006
Zebra	Equus burchelli	183,815	30,360		2003
Thomson's gazelle	Gazella thomsoni	328,620	25,053		1996
Grant's gazelle	Gazella granti	54,628	10,345		2003
Topi	Damaliscus korrigum	38,990	10,833		2003
Kongoni (Coke's hartebeest)	Alcelaphus buselaphus	16,043	3,173		2003
Impala	Aepyceros melampus	90,692	11,089		2003
Defassa waterbuck	Kobus defassa	1,186	487		(at least) 2003
Warthog	Phacochoerus aethiopicus	3,737	671		(at least) 2003
Buffalo, African	Syncerus caffer	30,276			2003
Giraffe	Giraffa camelopardalis	10,460	1,742		2003
Elephant	Loxodonta africana	2,360			2003
Eland	Taurotragus oryx	15,773	3,943		2003
Roan antelope	Hippotragus equinus	180	108		Maswa G.R. Oct. 2005
Oribi	Ourebia ourebi	7,000			1992
Bohor reedbuck	Redunca redunca			2	
Klipspringer	Oreotragus oreotragus	< 200			on kopjes
Dikdik (Kirk's)	Madoqua kirki			1	
Steinbuck	Raphicerus campestris			0.25	
Grey duiker	Sylvicapra grimmia			0.4	

				Mara & Grumeti Rivers, 2001
Hippopotamus	*Hippopotamus amphibius*	<808		
Rhinoceros	*Diceros bicornis*	13		
Bushbuck	*Tragelaphus scriptus*		0.2	Salai plains
Oryx (Fringe eared)	*Oryx beisa*	<100		on hill tops
Mountain reedbuck	*Redunca fulvorufula*	<200		Maswa G.R. Oct. 2005
Greater kudu	*Tragelaphus strepsiceros*	53	45	present in western rivers
Bushpig	*Potamochoerus porcus*			1 obs. Mara forest, 1972
Giant forest hog	*Hylochoerus meinertzhageni*			widespread, scarce
Aardvark	*Orycteropus afer*			widespread, scarce
Pangolin, Ground	*Manis temmincki*			present in Loita hills
Lesser kudu	*Tragelaphus imberbis*			on kopjes
Bush hyrax	*Heterohyrax brucei*			on kopjes
Rock hyrax	*Procavia capensis*			Mara River forest only
Tree hyrax	*Dendrohyrax arboreus*			on plains
Cape hare	*Lepus capensis*			in savanna
Crawshay's hare	*Lepus crawshayi*			Kuka Hill
Red rock hare	*Pronolagus rupestris*			on eastern plains
Spring hare	*Pedetes capensis*			widespread, scarce
Cape crested porcupine	*Hystrix africae-australis*			widespread, scarce
North African crested porcupine	*Hystrix cristata*			
Black and white colobus	*Colobus guereza*			western rivers only
Olive baboon	*Papio anubis*	7369	1456	widespread, abundant (1991)
Patas monkey, Red	*Erythrocebus patas baumstarki*			very rare, western rivers
Vervet monkey	*Cercopithecus aethiops*			widespread, abundant

Table A.2 Scientific names of larger mammals in the Serengeti ecosystem

Common Name	Latin Name
Order Insectivora	
Four-toed hedgehog	*Erinaceus albiventris*
Order Primates	
Black and white colobus	*Colobus guereza*
Olive baboon	*Papio anubis*
Patas monkey, Ikoma	*Erythrocebus patas baumstarki*
Vervet monkey	*Cercopithecus aethiops*
Greater galago	*Galago crassicaudatus*
Bushbaby, lesser	*Galago senegalensis*
Order Carnivora	
Lion	*Panthera leo*
Spotted hyena	*Crocuta crocuta*
Cheetah	*Acinonyx jubatus*
Leopard	*Panthera pardus*
African wild dog	*Lycaon pictus*
Black-backed jackal	*Canis mesomelas*
Golden jackal	*Canis aureus*
Side-striped jackal	*Canis adustus*
Bat eared fox	*Otocyon megalotis*
Striped hyena	*Hyaena hyaena*
Aardwolf	*Proteles cristatus*
Wildcat	*Felis silvestris*
Serval	*Leptailurus serval*
Caracal	*Caracal caracal*
Egyptian (Great grey) mongoose	*Herpestes ichneumon*
Banded mongoose	*Mungos mungo*
Dwarf mongoose	*Helogale undulata*
Black-tipped mongoose	*Herpestes sanguineus*
Genet, Common	*Genetta genetta*
White-tailed mongoose	*Ichneumia albicaudata*
Marsh mongoose	*Herpestes paludinosus*
Civet, African	*Viverra civetta*
Ratel, Honey badger	*Mellivora capensis*
Zorilla	*Ictonyx striata*
Palm civet	*Nandinia binotata*
Striped weasel, African	*Poecilogale albinucha*
Spotted-necked otter	*Lutra masculicollis*
Cape clawless otter	*Aonyx capensis*

Common Name	Latin Name
Order Proboscidea	
Elephant	*Loxodonta africana*
Order Hyracoidea	
Bush hyrax	*Heterohyrax brucei*
Rock hyrax	*Procavia capensis*
Tree hyrax	*Dendrohyrax arboreus*
Order Tubulidentata	
Aardvark	*Orycteropus afer*
Order Perissodactyla	
Zebra, Burchell's	*Equus burchelli*
Rhinoceros, Black	*Diceros bicornis*
Order Artiodactyla	
Giraffe	*Giraffa camelopardalis*
Buffalo, African	*Syncerus caffer*
Eland	*Taurotragus oryx*
Topi	*Damaliscus korrigum*
Kongoni (Coke's hartebeest)	*Alcelaphus buselaphus*
Wildebeest	*Connochaetes taurinus*
Impala	*Aepyceros melampus*
Defassa waterbuck	*Kobus defassa*
Bohor reedbuck	*Redunca redunca*
Mountain reedbuck	*Redunca fulvorufula*
Thomson's gazelle	*Gazella thomsoni*
Grant's gazelle	*Gazella granti*
Oryx (Fringe eared)	*Oryx beisa*
Roan antelope	*Hippotragus equinus*
Oribi	*Ourebia ourebi*
Klipspringer	*Oreotragus oreotragus*
Dikdik (Kirk's)	*Madoqua kirkii*
Steinbuck	*Raphicerus campestris*
Grey duiker	*Sylvicapra grimmia*
Hippopotamus	*Hippopotamus amphibius*
Greater kudu	*Tragelaphus strepsiceros*
Lesser kudu	*Tragelaphus imberbis*
Bushbuck	*Tragelaphus scriptus*
Warthog	*Phacochoerus aethiopicus*

Common Name	Latin Name
Bushpig	*Potamochoerus porcus*
Giant forest hog	*Hylochoerus meinertzhageni*
Order Pholidota	
Pangolin, Ground	*Manis temmicki*
Order Rodentia	
Spring hare	*Pedetes capensis*
Cape crested porcupine	*Hystrix africae-australis*
North African crested porcupine	*Hystrix cristata*
Order Lagomorpha	
Cape hare	*Lepus capensis*
Crawshay's hare	*Lepus crawshayi*
Red rock hare	*Pronolagus rupestris*

Thomson's gazelle

Trends suggest that numbers have declined once wildebeest reached their asymptote (chapter 1, fig. 1.2).

Grant's gazelle

There is little detailed information on trends. An earlier rough estimate of 50,000 in 1982 (Sinclair and Norton-Griffiths 1982) compared with the count of 54,628, (SE 10,345) in 2003 suggests no major change in that interval.

Topi

Numbers have not changed much, from 55,000 in 1971 (Sinclair 1972) compared to a recent count of 38,990 (SE 10,833) in 2003.

Kongoni

Similarly, this species has not changed in number since 1971. They may be spreading into the western corridor, where they once did not occur. However, numbers there are still very small.

Impala

This species has declined, at least since 1976 (Grimsdell 1979). However, in local areas where thickets have increased through woodland generation, densities appear to have increased.

Waterbuck

This species is confined to major rivers with permanent water. Numbers are very low and difficult to census with precision. Trends are not possible to obtain with any certainty.

African Buffalo

This species, like wildebeest, recovered after the rinderpest disappeared, but then collapsed through both poaching and drought. They are recovering slowly in the southern half of Serengeti, but not in northern Serengeti or Mara Reserve (chapter 2, fig. 2.24).

Giraffe

Giraffe appear to have declined a small amount since 1971.

Elephant

Trends for elephant are described in chapter 2, fig. 2.18. Since 1990 they appear to be quickly increasing.

Eland

This is one of the migrant species that depends on the northwestern Serengeti in the dry season. Numbers were estimated by using the Systematic Aerial Reconnaissance technique.

Roan Antelope

A small number of about 50 used to occur in northwestern Serengeti and Mara, but numbers have declined, probably to less than 10. Small herds have been seen around the Bangwesi Hill. Roan is relatively more abundant in Maswa.

Oribi

Oribi occur only in the *Terminalia* woodland with 2 m-high *Hyparrhenia* grass in northwest Serengeti. A decade ago they numbered 7,000. They are the most abundant of the small antelopes in the appropriate habitat.

Klipspringer

This species occurs on the kopjes of the woodlands but not on the plains.

Steinbuck

These occur in singles or pairs at very low density in the dry, eastern woodlands and around the edge of the plains.

Grey Duiker

Duiker occurs in broadleaved woodland, both in *Combretum* on rocky hills and in the *Terminalia* woodland. They are always in singles and at very low density.

Black Rhinoceros

Trends in this species are reported in chapter 2, fig. 2.23. Currently they number a mere 13, closely guarded by rangers.

Oryx

These occur only on the semiarid Salai plains of the NCA and Loliondo. Occasionally they wander into eastern SNP. Their numbers are very low.

Reedbuck

Two species occur. Bohor commonly occur throughout the long-grass plains, as recorded in night road transects, but they are rarely seen by day. They are also seen along rivers in daylight hours throughout the woodlands. Mountain (Chandler's) reedbuck occurs in small groups or singly on top of the highest hills—Kuka, Lobo, Nyaruswiga, Nyaroboro, and Itonjo.

Kudu

Both species occur in very small numbers at the extremes of the ecosystem. Greater kudu occur in the kopjes at the southern end of Maswa. Lesser kudu occur in the mountain thickets and forests of the Loita hills, east of the Mara Reserve in Kenya.

Pigs

Warthog are ubiquitous, though in low numbers, on the eastern plains. They appear to have dropped in number in recent years. Bushpig occur in the riverine forests but are rarely seen. M. Turner observed one Giant Forest Hog in 1972 in the riverine forest of the Mbalipali, 7 km west of Kogatende. He assumed it had wandered downstream along the Mara from the Loita forests.

Hyrax

Three species occur. Bush hyrax (*Heterohyrax*) feeds on dicots and climbs trees, whereas rock hyrax (*Procavia*) feeds on monocots and does not climb trees. Both are widespread on most but not all woodland kopjes and those at the edge of the plains. Kopjes further into the plains are without hyrax. Tree hyrax (*Dendrohyrax*) lives in the montane forests of the Mara watershed but not in the lowland forests of the Grumeti and Mbalageti.

Hares

Three species occur. Cape hares occur on the Serengeti plains, whereas Crawshay's hare lives in the woodlands (J. Flux, pers. comm.). They are indistinguishable in the field. The red rock hare is very localized, recorded so far only on top of Kuka Hill, but common where found.

Primates

Olive baboon and vervet monkeys are ubiquitous in the woodlands. They are most abundant associated with the Grumeti and Mbalageti rivers. Black-and-white colobus is also restricted to the forests of those rivers. Greater galago is also restricted to riverine forest of both north and west. Bushbabies occur throughout the woodlands.

Nile Crocodile (Reptile)

The density of Nile crocodile in the two main river systems is 0.33 and 1.24 per square kilometer in the Mara and Grumeti Rivers respectively (Games and Severre 1999).

REFERENCES

Campbell K., and M. Borner. 1995. Population trends and distribution of Serengeti herbivores: Implications for management. In *Serengeti II: Dynamics, management and conservation of an ecosystem,* ed. A. R. E. Sinclair and P. Arcese, 117–46. Chicago: University of Chicago Press.

Games, I., and E. Severre. 1999. Aerial survey for crocodiles in Tanzania, October 1999. A Report to the Director of Wildlife, Tanzania. Typescript.

Grimsdell, J. J. R. 1979. Changes in populations of resident ungulates. In *Serengeti: Dynamics of an ecosystem.* ed. A. R. E. Sinclair and M. Norton-Griffiths, 31–45. Chicago: University of Chicago Press.

Sinclair, A. R. E. 1972. Long term monitoring of mammals in the Serengeti: Census of non-migratory ungulates, 1971. *East African Wildlife Journal* 10: 287–97.

———. 1979. The eruption of the ruminants. In *Serengeti: Dynamics of an ecosystem.* ed. A. R. E. Sinclair and M. Norton-Griffiths, 82–103. Chicago: University of Chicago Press.

Tanzania Wildlife Research Institute. 2001. Aerial census of Hippopotamus in Tanzania mainland, dry seasons, 2001. TAWIRI Aerial Survey Report.

———. 2004. Aerial census in the Serengeti ecosystem, wet season 2003. TAWIRI Aerial Survey Report.

Peter A. Abrams
Department of Ecology and Evolutionary
    Biology
Zoology Building, University of Toronto
25 Harbord Street, Toronto, Ontario
M5S 3G5, Canada
abrams@zoo.utoronto.ca

T. Michael Anderson
Community and Conservation Ecology
    Group
Centre for Ecological and Evolutionary
    Studies
University of Groningen
P.O. Box 14, 9750 AA Haren
The Netherlands
t.m.anderson@rug.nl

Peter Andrews
Natural History Museum
London SW7 5BD, UK
pjandrews@btinternet.com
Fax: 44 (0) 207 942 5546

Miranda Armour-Chelu
College of Medicine, Department of
    Anatomy
Laboratory of Evolutionary Biology
Howard University, 520 W St. N.W.
Washington, DC 20059, USA
machelu@earthlink.net
Fax: 202-265-7055

Raymond L. Bernor
College of Medicine, Department of
    Anatomy
Laboratory of Evolutionary Biology
Howard University, 520 W St. N.W.
Washington, DC 20059, USA
rbernor@howard.edu
Fax: 202-265-7055

Robert J. Blumenschine
Center for Human Evolutionary Studies
Department of Anthropology, Rutgers
    University
131 George St., New Brunswick, NJ
    08901-1414, USA
rjb@rci.rutgers.edu
Fax: 732-932-1564

Raymonde Bonnefille
CNRS, Cerege, BP 80
13545 Aix-en-Provence Cedex 04
France
bonnefille@cerege.fr
Fax: +33 (0) 4 42971540 or 1595

Markus Borner
Frankfurt Zoological Society
PO Box 14935
Arusha, Tanzania
markusborner@fzs.org

Nicholas Burger
Donald Bren School of Environmental
    Science & Management
UC Santa Barbara
Santa Barbara, CA 93106, USA

Sarah Cleaveland
Centre for Tropical Veterinary Medicine
Royal (Dick) School of Veterinary Studies
University of Edinburgh, Easter Bush,
    Roslin, Midlothian
UK EH25 9RG
sarah.cleaveland@ed.ac.uk
Fax: +44 (0) 131 651 3903

Christopher Costello
Bren School of Environmental Science and
    Management
UC Santa Barbara
Santa Barbara, CA 93106-5131, USA

Michael Coughenour
Natural Resource Ecology Laboratory
Colorado State University
Fort Collins, CO 80523, USA
mikec@nrel.colostate.edu

Meggan Craft
Department of Ecology, Evolution and
    Behavior
University of Minnesota, 1987 Upper
    Buford Circle
St. Paul, MN 55108, USA
craft004@umn.edu
Fax: +1 (0) 612 624 6777

Jan Dempewolf
Department of Geography
University of Maryland
1104 LeFrak Hall, College Park, MD 20742,
    USA
dempewol@umd.edu
Fax: (301) 314-9299

Andy Dobson
Ecology and Evolutionary Biology
Eno Hall
Princeton University
Princeton, NJ 08544-1003, USA
andy@eno.princeton.edu
Fax: +1 (0) 609 258 1334

Kathleen A. Galvin
Department of Anthropology and Natural
    Resource Ecology Laboratory
Colorado State University
Fort Collins, CO 80523, USA
Kathleen.Galvin@colostate.edu
Fax: 970-491-7597

John M. Fryxell
Department of Integrative Biology
University of Guelph
50 Stone Road E, Guelph
Ontario, Canada, N1G 2W1
jfryxell@uoguelph.ca

Emmanuel Gereta
Tanzania National Parks
PO Box 3134
Arusha, Tanzania
emmanuel_gereta@hotmail.com

Katie Hampson
Ecology and Evolutionary Biology
Guyot Hall, Princeton University
Princeton, NJ 08544-1003, USA
khampson@princeton.edu
Fax: +1 (0) 609 258 1334

Terry Harrison
Center for the Study of Human Origins
Department of Anthropology
New York University
New York, NY 10003, USA
terry.harrison@nyu.edu
Fax: 914-666-3668

Richard L. Hay (1926–2006)
Department of Geosciences
University of Arizona
Tucson, AZ 85621, USA

Ray Hilborn
School of Aquatic and Fisheries Sciences
University of Washington
Box 355020
Seattle, WA 98195-5020, USA
rayh@u.washington.edu

Robert D. Holt
Department of Zoology
University of Florida
223 Bartram Hall, P.O. Box 118525
Gainesville, FL 32611-8525, USA
rdholt@zoo.ufl.edu

J. Grant C. Hopcraft
Frankfurt Zoological Society
P.O. Box 14935
Arusha
Tanzania
granthopcraft@fzs.org

Magai Kaare
Ministry of Water and Livestock
    Development
P.O. Box 322
Musoma, Tanzania
m.t.kaare@sms.ed.ac.uk
Fax: +44 (0) 131 651 3903

Dixon S. Kaelo
International Livestock Research Institute
P.O. Box 30709, Nairobi, Kenya
golekaelo@yahoo.com

T. Kimbrell
Department of Zoology
University of Florida
223 Bartram Hall
P.O. Box 118525
Gainesville, FL 32611-8525, USA
tkimbrell@zoo.ufl.edu

Richard Kock
Conservation Programs
Zoological Society of London
Regent's Park, London, NW14RY, UK
richard.kock@zsl.org

Richard H. Lamprey
P.O. Box 21472
Nairobi 00505, Kenya
lamprey@infocom.co.ug

Karen Laurenson
Frankfurt Zoological Society
P.O. Box 14935
Arusha, Tanzania
karenlaurenson@fzs.org

Tiziana Lembo
Centre for Tropical Veterinary Medicine
Royal (Dick) School of Veterinary Studies
University of Edinburgh, Easter Bush,
    Roslin, Midlothian
UK EH25 9RG
s0128864@sms.ed.ac.uk

Daniel A. Livingstone
Department of Biology
Duke University
Durham, NC 27708-0338, USA
livingst@duke.edu
Fax: 919-660-7372

Martin Loibooki
Tanzania National Parks
P.O. Box 3134
Arusha, Tanzania

Rob Malpas
Conservation Development Centre
P.O. Box 24010
Nairobi, Kenya
robmalpas@cdc.info

Curtis W. Marean
Institute of Human Origins
School of Human Evolution and Social
    Change
P.O. Box 872402
Arizona State University
Tempe, AZ 85287-2402, USA
curtis.marean@asu.edu
Fax: 480-965-1386

Simon A. R. Mduma
Serengeti Biodiversity Program
Tanzania Wildlife Research Institute
P.O. Box 661
Arusha, Tanzania
mduma@habari.co.tz

Kristine L. Metzger
Serengeti Biodiversity Program and Centre
    for Biodiversity Research
6270 University Blvd.
University of British Columbia
Vancouver, BC, Canada V6T 1Z4
metzger@zoology.ubc.ca

Titus Mlengeya
Tanzania National Parks
P.O. Box 3134
Arusha, Tanzania
tanapavet@yahoo.com
Fax: 255 28 262 1537

Charles Mlingwa
Ministry of Livestock Development
P.O. Box 9152
Dar es Salaam, Tanzania
cmlingwa@hotmail.com

Mike Norton-Griffiths
P.O. Box 15227
Langata 00509, Kenya
mng5@compuserve.com

Han Olff
Community and Conservation Ecology
    Group
Centre for Ecological and Evolutionary
    Studies
University of Groningen
P.O. Box 14
9750 AA Haren
The Netherlands
h.olff@rug.nl

Craig Packer
Department of Ecology, Evolution and
    Behavior
1987 Upper Buford Circle
University of Minnesota
St. Paul, MN 55108, USA
packer@biosci.cbs.umn.edu
Fax: +1 (0) 612 624 6777

Charles R. Peters
Department of Anthropology and
    Institute of Ecology
Baldwin Hall
University of Georgia
Athens, GA 30602-1619, USA
chaspete@uga.edu
Fax: 706-542-3998

Stephen Polasky
Department of Applied Economics
University of Minnesota
1994 Buford Ave
St. Paul, MN 55108, USA
polasky@umn.edu

Denné N. Reed
Dept. of Anthropology
University of Texas at Austin
1 University Station C3200
Austin, TX 78712, USA
Phone: (512) 471-7529
Fax: (512) 471-6535
reedd@mail.utexas.edu

Robin S. Reid
International Livestock Research Institute
P.O. Box 30709
Nairobi, Kenya
r.reid@cgiar.org

Mark E. Ritchie
Department of Biology
Syracuse University
Syracuse, NY 13244, USA
meritchi@syr.edu

Victor Runyoro
Ngorongoro Conservation Area Authority
P.O. Box 1
Ngorongoro Crater, Tanzania
ncaa_faru@cybernet.co.tz

Mohammed Y. Said
International Livestock Research Institute
P.O. Box 30709, Nairobi, Kenya
m.said@cgiar.or

Jennifer Schmitt
Department of Ecology, Evolution and
    Behavior
University of Minnesota
1987 Upper Buford Circle
St. Paul, MN 55108, USA

Suzanne Serneels
Department of Geography
University of Louvain
3 Place Louis Pasteur, 1348 Louvain-la-
    Neuve, Belgium
suzy.serneels@broederlijkdelen.be

Anthony R. E. Sinclair
Serengeti Biodiversity Program and Centre
    for Biodiversity Research
6270 University Blvd.
University of British Columbia
Vancouver, BC, Canada V6T 1Z4
sinclair@zoology.ubc.ca
Fax: 604 822 0653

Liaila Tajibaeva
Department of Economics
Ryerson University
350 Victoria Street
Toronto, Ontario
M5B 2K3, Canada

Simon Thirgood
Macaulay Institute
Aberdeen, AB15 8QH, UK
s.thirgood@macaulay.ac.uk

D. Michael Thompson
Environment Agency UK
29 Southwood Drive, Bristol, BS9 2QX
mthomp@zoom.co.uk

Lars Werdelin
Department of Palaeozoology
Swedish Museum of Natural History, Box
    50007
S-104 05 Stockholm, Sweden
werdelin@nrm.se
Fax: +46-8-5195 4184

John F. Wilmshurst
Parks Canada, 145 McDermot Ave.
Winnipeg, Manitoba
Canada, R3B 0R9
john_wilmshurst@pc.gc.ca

Page numbers in italics refer to figures.

Abiotic factors: climate, 151–54; constraint, 4; fire, 157–59; topography and soil, 154–57
Adaptation, 325, 326, 327, 328, 342; definition, 326
Adaptive capacity: definition, 326
African buffalo, 334; *ad libitum* intake, 291; competition with wildebeest, 26, 27; cropping rates, 291; demography, 291; human-wildlife conflict, 104, 127; increase, 421; maximum population growth rate, 291; nonmigrant, 255; poaching, 39; population, 481; population time series, 296; response to rainfall, 122, 123; response to soil fertility, 122, 123; rinderpest, 38–39; spatial distribution, 115, 121 trends, 503
African wild dog: rabies, 214–16, 222; vaccination, 222. *See also* Wild dog
Africover Project, 387
Agricultural fields, spatial distribution, 124
Agriculture, 204, 328, 300–32, 341, 362; change over time, 388; commercial, 331; historical development, 108–44; loss of wildlife, 379; mechanized cultivation, 385, 386; rents Kenya, 382, 383, 405–7; rents Mara, 383–88, 400; total cultivation, 386

Agropastoralist, 329, 331, 332; invasion into East Africa, 96–98
AIDS, 329; in Africa 422. *See also* HIV/AIDS
Alternate states, concept of, 436
Amboseli, 66, 192
Anthrax, 126, 210, 218
Antipoaching, 353, 368–72; policy 369, 370
Arusha, 103, 108, 109
Arusha Manifesto, 348, 349
Attack rate, of lions, 254
Authorised Associations, 465
Axum, Ethiophia, 108

Banagi, 8, 23
Bantu, 98
Baobab, 11
Bariardi district, 329
Baselines, ecological, 474, 475; theory of, 423–29, 438, 439
Bat-eared foxes, CDV, 211–13
Beans, 331
Benguela ecosystem, 241
Beta-diversity, 138, 141, 142
Biocomplexity, 429–36; of Serengeti, 3, 4
Biodiversity: ecosystem functioning, 128; importance of parks, 129; loss of, 102, 104, 107, 108, 128

Biodiversity Convention, 473
Bioeconomic models, 302
Biomass, 199–201
Biosphere reserve, 9, 475
Biotic constraint, 4
Biotic factors: browsing, 160, 161; grazing,
159, 160; herbaceous vegetation, 163,
164; human impacts, 164–67; termites,
161, 162; trees, 162, 163
Black rhinoceros, trends, 504. *See also*
Rhinoceros
Body size of herbivores, 115, 118, 119, 120,
122
Border closure, 36, 37
Bottom-up processes, 423–25, 438
Boundary: hard, 484–86; soft, 486, 487
British colonization, 331
Browsing, 160, 161; 200
Buffalo, African. *See* African Buffalo
Buffer zone, 486. 489.
Bunda, 103, 108
Bunda district, 329, 332
Bushmeat, 3, 304, 316, 326, 329, 331, 334–36,
339–41, 353, 484; harvest of, 316

Caliche, 13
Canine distemper, 3, 32, 210–14, 227–28,
230–31, 418, 422, 431, 480
Carbon dioxide, 183, 185, 195, 196, 198
Carnivores, 421
Carrying capacity, of prey, 261
Cash crops, 109; cultivation of, 329–31
Cassava, 330, 331
Catastrophic events, 98, 106
Catena, 14, 155
Cattle, 330–33; introduction into East Africa,
105; raiding, 329, 333; response to
rainfall, 120, 123, 125; response to soil
fertility, 125; spatial distribution, 124. *See
also* Livestock
Chaotic behavior, 430
Cheetah, 202, 203; ideal free distribution,
428, 434; interference competition, 268;
mange, 218; mortality from lions, 256,
262
Chickens, 332
CITES, 348, 360–62
Climate, 15, 16, 151–54, 170; and diseases,
204; dry season, 191; effect of soil fertility, 117; global models, 479; regional,
186, 199; sensitivity to, 198; variability,
189, 199, 203, 204, 325, 326, 341; wet
season, 191; woody vegetation, 194, 195

Climate change, 3, 183, 478, 479; global, 335;
since last ice age, 96
Commercialization of agriculture and
livestock husbandry, 108–9
Commodity prices, 341, 342
Community Resources Trust, 457
Community-based conservation, 352–54,
357, 471, 476–78, 481, 487, 489
Competition, 277, 480; exclusion, 293; migra-
tion, 293, 294; resident species, 294;
spatial scale, 297; effects of predators, 297
Compositional variance, 137, 138, 141–44
Congo forest, 476
Conservation Corporation Africa, 457
Conservation finance, 447, 448
Conservation policy, 352–59
Conservation Trust Funds, 460
Conservation values: Mara area, 410–13
Consumption and wealth, 308, 309
Consumption: crops, 312; meat, 313
Consumptive use of wildlife, 458
Contagious Bovine Pleuropneumonia, 220
Controllers, 137
Cotton, 330, 331, 332
Crater highlands, 9, 13, 49, 53, 188
Crocodile, trends, 505
Crop production, 313
Cropping rates: Thomson's gazelle, 279;
wildebeest, 291; African buffalo, 291
Crops, 205
Culling, 229
Cullman and Hurt Community Wildlife
Project, 454
Cultivation: cash-crop, 329–31; land use
change, 115; origin, 96
Cultural heritage, 128
Cultures, Western Serengeti, 328–35
Cushites, 98, 100

Danish International Development Agency,
donor support, 447
Data: agricultural, 313; grazing, 313; hunting,
315
Decisions, land use, 325, 326, 327
Decomposition, 202
Deforestation, 485; in Kenya, 420
Detritivores, 202
Digestibility, 279
Disease, 244, 422, 423, 531; biodiversity and,
229; control of, 223–28; cycles, 431;
effect on livestock, 106, 126; eradication,
226–27; of humans, 205; of livestock,
204; reservoirs, 222–23, 227, 230–31;

role in human historical population, 102, 104, 106, 107, 108; transmission of, 479, 480

Disturbance, propagation of, 4. *See also* Perturbation

Diversity, herbivores, 200

Domains, 430

Domestic dog, 3; CDV, 211–14, 223–26

Domestication of livestock, 95. 101, 105, 332. *See also* Livestock

Draft animals, 332

Drought, 25, 327, 328, 331; 418, 481; causing human wildlife conflict, 127; effect on herbivores, 124, 127; effect on livestock, 106; refugia, 66–68; regulation of historical human populations, 104

Dry-season refugia, 66–68, 127

Dual constraints grazing model: Thomson's gazelle, 278–280; wildebeest, 291–92; African buffalo, 291–92

Dung, 203

Early humans, 101–4

Earth charter, 473

East Coast Fever, 220

Ecological baselines. *See* Baselines: ecological

Economic: influences, 362–68; theory, 428, 429

Economic driving forces: Mara area, 409, 410

Economic factors; macro- and micro-, 410

Economics, 204

Economy: agriculture, 362; green, 364; hunting concessions, 366; the model, 368–74; tourism, 364

Ecosystem: health, 209; stability, 426, 427

Ecosystems: lessons, 436, 437

Ecotourism, 96, 100, 112, 204, 364

El Nino (ENSO), 152, 188–190, 199, 200

Eland: migration, 126; spatial distribution, 124; trends, 503

Elephant, 201; and agriculture, 436; and trees, 431; CITES listing, 361; evolution, 101; heterogeneity, 167, 168; human-wildlife conflict, 104, 126, 127; keystone species, 425, 438; numbers, 421, 431; pillaging, 485; poaching, 2, 38, 419, 421, 424; population, 32, *31,* 38; response to rainfall, 122, 123; response to soil fertility, 122; role of, 30, 31; spatial distribution, 115, 119, 121; trends, 503

Embagai, 53, 60

Emergent properties: and biocomplexity, 429–436: in ecosystems, 417, 418

Emerging diseases, 221, 226, 231; properties of, 4

Environment: drought, 418; perturbation, 418

Environmental change, 183, 198. *See also* Climate; General circulation model (GCM)

Equilibria, 430, 439

Equilibrium: stable, 259; unstable, 259, 430

Erosion, soil, 331, 332

European Union, donor support, 447

Evolution: late quaternary, 74–77; of large herbivores, 70–74; of carnivores, 70–74; Pliocene and Pleistocene, 71–74

Extinction, cascades of, 243

Facilitation (grazing), 277–78

Famine, 332, 474

Farming, subsistence, 100, 101, 103, 107, 108, 113, 128, 329, 330, 331

Fence effect, 486

Fences, 485, 486

Financing a future Serengeti, 460

Fire, 22, 23, 25, 28, 157–59, 167, 168, 170, 200, 484; and wildebeest, 23, 25; effect on green flush, 106; effects on woody species, 106; on plains, 420; prehistoric use, 79; use by early humans, 101–4, 105; use by pastoralists, 106

Fishing, 328, 329, 331

Floodplain(s), 122

Food quality for herbivores, 109, 113, 119

Food security, 205; sources of, 332

Food web, 200, 241, 242, 243, 246, *247,* 423–25, 438, 480; diamond, 267; functional forms, 249–59; pathogens, 269; Serengeti, 245–49; stability, 425, 429; time scales, 242

Foot-and-mouth disease, 220

Footprint tuff, 51, 52, 54

Forage, 310

Foraging constraints, 278

Foraging theory, optimal. *See* Optimal foraging theory

Forest, riverine, 420

Fort Ikoma, 358

Fossil remains of early humans, 100

Fractal dimension, 138

Frankfurt Zoological Society, donor support, 447

Friedkin Conservation Fund, 454

Functional: redundancy, 423–25, 438; response, of lions, 252, 253; seasonality, 255

Game Controlled Areas, conservation management, 446
Game Reserves: concession fees, 453; conservation management, 446; donor support, 454
Garusi valley, 50
Gazelles, 201, 202. *See also* Grant's gazelle; Thomson's gazelle
General circulation model (GCM), 183, 187, 192–94
Genetic diversity, 472
Geological evolution, 49–66
Getis-Ord hotspot statistic, 118–21, 124
Giraffe, 334; response to rainfall, 122, 123; response to soil fertility, 122, 123; spatial distribution, *121;* trends, 503
Glaciation period, 96
Global change, factors, 184
Global climate change, 335. *See also* Climate; General circulation model (GCM)
Globalization, 128
Goats, 332; domestication of, 105; introduction to East Africa, 105; response to rainfall, 120; spatial distribution, 122, 124
Gol Kopjes, 77
Gol Mountains, 10, 12, 13
Grant's gazelle, 8, 334; migration, 19, 110, 126; response to soil fertility, 123; response to rainfall, 123; trends, 502. *See also* Gazelles; Thomson's gazelle
Grass abundance on Serengeti plains, 282
Grass consumption, 279
Grass growth, 285; maximum growth, 285; carrying capacity, 285
Grasses, 195–200; C_3, 195
Grasshopper species, 34
Grazer population dynamics model: single species, 283–89; multi-species, 290–98
Grazing, 159, 160, 167, 168, 170, 200, 277–98, 307; lawns, 159; livestock, 313
Green flush after fire, 112
Grey Duiker, trends, 504
Group ranches, 379, 392, 395, 398–400, *401, 402,* 411; in Narok, 101, 102, 109, 113, 115
Grumechen, 13
Grumeti Game Reserve, 9, 11, 104, 112, 119, 350, 329, 333, 334, 455
Grumeti river, 8, 23, 24, 29, 40, 67, 144
Grzimek, 23

Haematic ethnic group, 329
Handling time, of lions, 253
Hard boundary, 484–86

Hares, trends, 505
Harvest, wildlife, 306. *See also* Hunting
Harvey, Gordon, 8
Herbivores, 196, 200, 203; migratory, 203; resident, 115–24; spatial aggregation of, 118–120; spatial segregation, 265; species richness, 201
Heterogeneity: agents, 151–67; definition of, 136–39; effect on herbivores, 104; elephants, 167, 168; grazing, 159; hotspots, 159, 168, 169; in time and space, 167–70; patterns, 140–51; scales, 139, 140; soils, 154; spatial, 429, 433, 438; spatial and temporal, 4; vegetation cover, *150*
Hippo, human-wildlife conflict, 104
Historical perturbations: carnivores, 421; disease and pathogens, 422, 423, 436; elephant, 421, 436; humans, 421, 422, 436; physical environment, 418; Serengeti, 418–23; vegetation, 419, 420; wildebeest and ungulates, 420, 421. *See also* Perturbation; Disturbance, propagation of
HIV/AIDS, 210, 221, 230–31, 329
Holocene pastoralists, 80, 81
Homo, scavenging and hunting, 78
Hotspots, 159, 168, 169; analysis of, 115–25
Households: agriculture, 313; antipoaching, 322, 323; assets, 304; change in wealth, 308, 309; comparative dynamics, 319–22; consumption, 304, 308, 309, 312, 313; crop consumption, 312; hunting, 315, 316; labor allocation, 306; livelihoods, 326; livestock grazing, 313, 315; livestock sale, 204; meat consumption, 313; model, 303–6; model caveats, 323–24; model computation, 316, 317; model parameters, 312–17; model results, 317–23; model variables, 309–11; population growth, 312; region, 312; steady state model, 317–19; welfare, 370
Human: activity consequences, 243; capital, 5; consumption, 308; decisions, 3; demographics, 366–68, 373–75; ecosystem model, 303; evolution, 96; impacts, 40; livelihoods, 100, 108, 128
Human population, 1, *2, 3, 36,* 325, 327, 333, 335–37, 339–42, 348, 482, 484, 489; demographic transition, 373, 374; density, 108, 114; fertility, 1; growth, 372–74; Tanzania, *363,* 366, 367, 370, 371; welfare, 204, 336; well-being, 336
Human-dominated ecosystems, 478–84
Human-environment interactions, 327

Human-natural areas, boundaries, 484–87
Humans: early, 95, 101–4, and migration, 78; and rinderpest, 421, 422, 436; food deficits, 355; population increase, 422; prehistoric, 77–81; settlement, 421; and tsetse fly, 421. *See also* Human population
Human-wildlife conflict(s), 126–27
Human-wildlife interaction, 481–84
Hunter-gatherer(s), 95–108, 126
Hunting, 125, 307, 328, 329, 331, 333–37, 339–41; and boundary zones, 487, 489; bushmeat harvest, 316; concessions, 366; early humans, 101, 102, 104; illegal, 125, 205, 125; modern humans, 101; quotas, 370; stone age, 79, 80; travel time, 316; trophy, 371
Hyena, 334; lion interference, 256; population, 19; self-limitation, 256; CDV, 211–14; numbers, 421, 424
Hyrax, species; 504; trends, 504

Ideal free distribution, 428, 434, 435
Ikizu ethnic group, 329
Ikoma ethnic group, 329
Ikorongo Game Reserve, 9, 112, 333, 350
Impala, 334; anthrax, 218; effect of agriculture, 127; response to rainfall, 115, 119, 123; response to soil fertility, 115, 119, 123; spatial distribution, 115, 121; trends, 502
Indian Ocean, 16, 186, 192; dipole, 189, 200
In-migration, 331
Integrated Conservation & Development Projects, 356
Interactions, human-environment, 327
International treaties, 359–62
Interspecific interference, 258
Intertropical Convergence Zone, 22, 104, 151, 186, 190. *See also* Rainfall
Isshenyi ethnic group, 329
Isuria escarpment, 11. *See also* Siria escarpment
IUCN, 349
Ivory poaching, 30, 37, 38, 485
Ivory trade, 30, 31, 361, 362

Jackals, CDV, 213–14
Jita ethnic group, 329

Kagera River, 69
Kajiado district, 407
Kalahari, 486

Karatu, 109
Keekerok, 485
Kenya, 197; Wildlife Service 394; human fertility rate in, 1
Keystone species, 425, 426, 438; elephant, 425, 438; predator, 243; rinderpest, 425, 526; wildebeest, 33, 34, 41, 425, 438
Khosian people, 98
Kisii ethnic group, 329
Kiteto district, 105
Klipspringer, trends, 503
Kongoni, 334; response to rainfall, 119, 122, 123; response to soil fertility, 122, 123; spatial distribution, 121; trends, 502
Kopjes, 156, 157
Kruger National Park, 242, 427, 436, 486; bovine tuberculosis in, 216
Kudu, trends, 504
Kuka, 13, 18, 40
Kuria ethnic group, 329, 331, 333

Labor allocation, 306, 318
Laetoli, 50, 51, 53, 54, 55, 56, 57; area, 58, 59; beds, 50, 51, 54, 77, 84; fauna, 59, 60; pollen studies, 60, 61; sector, 50–53, 84; vegetation, 58
Lake Ndutu, 50
Lake Eyasi, 12, 50, 66, 75
Lake Manyara, 69; National Park, 155
Lake Naivasha, 76
Lake Natron, 68, 69, 358
Lake Olduvai, 53
Lake Victoria, 9, 12, 16, 20, 24, 36, 47, *48, 49*, 67–70, 82, 98, 99, 152, 186, 188, 203, 479; as drought refuge, 66–70
Lamai, 8
Landowners, land use choice, 405–12
Landscape, 203; size effects of, 289
Land subdivision, incentives, 409; Mara area, 394–96, 403, 404, 409, 411
Land use, 219–21, 223; change, 114, 125; decisions, 325, 326, 327; method for assessment, 118; planning, 128, 129; regional trends, 109; returns, 409; strategies, 405–13; Western Serengeti, 328–35
Large mammals, evolution of, 70–74
Lattice, 283
Lemagrut, 53
Lemuta member, 57, 66
Lengai, 12, 13, 49, 51, 52, 54; eruptions of, 52
Leopard: human-wildlife conflict, 126; livestock predation, 126
Levels of social organization, 327

Limit cycles, 430, 431
Lion: and cheetah, 420; bovine tuberculosis, 216–17; attack rate, 254; canine distemper, 422; capture rate, 420, 432; CDV, 211–15; functional response, 252, 253; handling time, 253; increased mortality, 259–62; livestock predation, 126; Ngorongoro, 251; numbers, 421, 422, 424, 432; population, 19; population growth, 250, 251;
Lithology, 13, 14,110
Livelihood strategies, 34–36, 325, 326, 328, 330
Livelihoods of people, 128
Livestock, 204, 332–33, 481–82, 488, 489; diseases of, 106, 126, 210, 219–20; domestication, 95, 101, 105; impact of cultivation on, 397; introduction to East Africa, 105; numbers, 370, 371, 374, 389; pastoralism, 100, 102, 103, 105, 106, 107; population, 336, 337, 340–42; rents, 388–390, 405–7, 412, 413; role of droughts, 106; sales, *389;* wealth, 304
Locational variance, 137, 138
Loita hills, 11, 13, 380, *401*
Loita wildebeest, 397
Loliondo Game Control area, 10, 35, 101, 102, 113, 302, 455
Loliondo region, 362
Lotka-Volterra models, 245, 246, 249
Luo ethnic group, 329
Lyme disease, 229

Magu district, 329
Maize, 326, 330, 331
Malaria, 205
Malignant catarrhal fever, 126, 210, 219–21, 354
Mammals: large, evolution of, 70–74; scientific names, 500–502
Mange, 217–18
Manonga valley, 51
Mara area: conservation values, 410–13; land subdivision, 394–96, 403, 404, 409; land use change, 382–96; settlements, 393, 394, 403, 404
Mara Region, 329, 331
Mara Reserve, 435, 481, 485
Mara River, 8, 11, 23, 24, 29, 40, 47, 49, 67–69, 82, 144, 165; as drought refuge, 66–70; bank erosion, 29; width of, 420
Market access, 330
Market economy, 333

Masai (*also* Maasai), 12, 332; Mara Reserve, 8, 11, 26, 31, 35, 36, 38, 39, *48,* 67, 68, 80, 81, 99, 100, 110; 136, 165, 167, 168, 203, 379–81, 383, 385, 388, 393, 394, 396–98, 402–4; pastoralists, 99, 105, 106, 108, 111, 128
Masek beds, 54, 58, 83
Maswa district, 329
Maswa Game Reserve, 10, 24, 49, 104, 115, 119, 125, 329, 333, 350, 435, 487
Mau highlands, 68 69, 76, 82
Mbalageti river, 23, 144
Meatu district, 329
Microbes, soil, 156
Migration, 203, 278, 434, 435; and cultivation, 397; and land development, 398–403; emergent property, 434, 435; fences, 486; Grant's gazelle, 110; humans, 78; large herbivores, 110; Thomson's gazelle, 110; wildebeest, 2, 110, 111, 112; zebra, 110, 112, 124, 126; species evolution, 70, 71. *See also* Wildebeest migration
Millet, 330, 331
Missionaries, 8
Model computations, 316
Model households, overview, 303–6
Model: human ecosystem, 303; individual-based, 326; simulation, 335–41; climatic changes, 335–39; market condition changes, 339–41
Modeling, individual behavior, 4
Models: details of, 267, 268; future, 269, 270; missing players in, 268, 269
Models of predation, two species, 263
Mongooses, CDV, 213
Monsoon, Indian Ocean, 186, 188, 190
Moore, Monte, 8
Moose, 485
Mortality: and food supply, 25; of wildebeest, 25
Moru kopjes, 143
Mosonik, 68
Mount Kilimanjaro, 99
Mugumu, 12, 13, 108
Multiple states, 426–28, 430–33, 439
Musoma district, 329
Musoma, 103, 108, 331
Mwanza, 109

Naabi hill, 152, 153
Naibor soit, 66
Nairobi, 103, 108, 109

Naisuri gorge, 58
Nandi ethnic group, 329
Narok district, 12, 103, 113, 385, 389, 390
Nasera rock, 75, 77
Nata ethnic group, 329
National Centre for Ecological Analysis and
    Synthesis, 3, 4
Ndolanya beds, 50, 57, 84
Ndutu beds, 54, 58, 83
Newcastle Disease, 223
Ngaloba beds, 57
Ngorongoro Conservation Area, 8, 9, 19, 35,
    47, 53, 54, 60, 101, 102, 111, 113, 302, 335,
    350, 352, 354, 359, 363, 488, 489; con-
    servation management, 446; income and
    expenditure, 451; malignant catarrhal
    fever, 219
Ngorongoro Crater, 201
Ngorongoro Highlands, 99, 110, 152, 155, 165
Niche, of herbivores, 127
Nile crocodile, trends, 505
Nile river, 98
Nilotic people, 98, 100
Nitrogen: mineralization, 203; plant, 198,
    203; soil, 198
Nonconsumptive use of wildlife, 459
Nonmetric multidimensional scaling, *147*
Nsongezi beds, 69
Nutrients: litter, 203; plants, 196, 200; soil,
    196, 200
Nyanza, 103

Olbalbal, 52
Oldeani, 53
Olduvai Gorge, 10, 16, 50, 52, 54, 55, *56,* 58,
    78, 80, 79; fossil pollen, 65, 66; large
    mammals, 62–65; small mammals, 61, 62
Olduvai sector, 53, 54
Olduwan industrial complex, 78
Olmoti, 53, 54
Ololosokwan Village, economic benefits of
    wildlife, 457
Optimal foraging theory, 428, 429, 439
Orangi river, 144
Oribi, trends, 503
Oryx, 19; trends, 504
Otterlo Business Corporation, 456
Overgrazing, 474, 485

Panarchy, 426–28
Pastoralism, 328, 329
Pastoralists, 95–115, 331, 333, 335; in Holo-
    cene, 80, 81

Pathogens, 32, 33. 422, 423; and domestic
    dogs, 32; infectious disease of humans,
    32, 33; weather and outbreaks, 33
Patterned variance. *See* Variance: patterned
Perturbation: altered vegetation productiv-
    ity, 264, 265; increased lion mortality,
    259–62, 267; increased wildebeest
    mortality, 262–64; modeling of, 480,
    481; predicted responses of, 259–265; to
    ecosystem, 244, 245. *See also* Distur-
    bance, propagation of
Pigs, trends, 504. *See also* Warthog
Plains, volcanic ash, 49–54
Plant physiology, 195
Poaching, 38, 39, 244, 333, 335, 340, 342,
    369, 370, 484, 489; of elephant, 2, 361,
    419; of rhino, 2; of wildebeest, 2
Policy: antipoaching, 369, 370, 375; conserva-
    tion, 352–59; hunting quota, 371; market
    distortions, 410; national, 348–52
Population cycles, 266
Population estimates, large animals, 498–99
Population growth, human, 312, 328, 331,
    336
Poverty, 102, 108, 109, 128, 219–20, 223, 331,
    334, 471, 472
Precipitation patterns: changes in, 335;
    herbivore susceptibility to, 267; patterns
    of, 248, 249. *See also* Rainfall
Predator-prey system, isoclines, *260*
Predators, 201, 203
Prices, 327, 339, 340, 341; bushmeat, 340,
    341; commodities, 341, 342; crop, 339,
    340, 342; market, 326
Primary production, 109, 116, 170
Primates, trends, 505
Production, 306–8; agriculture, 306, 307;
    crop, 306; grazing, 307; hunting, 307,
    308; livestock, 306, 313; primary, 186,
    196; value of, 308
Protected area conservation, 352, 471,
    473–76, 489
Protected Areas: costs of effective protection,
    444; financing structures, 445; networks,
    443, 444

Quaternary, 74–77; extinctions, 75, 76

Rabies, 3, 32, 210–11, 214–16, 222–26, 229–31,
    284, 431, 432, 480
Rainfall, 153, 186–88, 198, 199, 418, 424, 431,
    479, 481, 482; and agriculture, 406, 407;
    and climate, 20–22; decadal oscillation

Rainfall (*continued*)
in, 21; drought, 21, 25; dry season, 21, *21;* and grass production, 424; Intertropical Convergence Zone, 22; isohyets, 16; Mara area, 399, *401,* 406, 407, *408;* models, 265; seasonality, 113; spatial autocorrelation on Serengeti plains, 285; variability in, 341; and wildlife, 407, 408;
Rainfall gradient: response of herbivores to, 110–22; response of human population density, 110–22
Reedbuck, trends, 504
Rents: agriculture, 382–88, 405–7, 412, 413; livestock, 388–90, 405–7, 412, 413; wildlife, 390–93, 405, 412
Resident herbivores. *See* Herbivores: resident
Resilience, 325, 326, 426–28, 435, 436, 439; as emergent property, 435
Resource transition, 95
Revenue retention schemes, 463
Rhinoceros, 101; lack of resilience, 436, 437; poaching of, 2, 38. *See also* Black rhinoceros
Rice, 331, 332
Rift valley, 49, 98, 111
Rinderpest, 24, 25, 28, 32, 38,47, 80, 106, 126, 164, 168, 210, 217–18, 226–28, 230–31, 421, 422, 427, 436, 482; and wildebeest, 250; keystone species, 426; shaping the ecosystem, 422
Riverine forest, 479, 484; change in, 420; decline of, 29, 30; regeneration of, 29
Roan antelope, 268; trends, 503
Robanda village, 356, 357
Ruwana river, 67

Sadiman volcano, 50, 53, 60, 68
Salai plains, 10, 15, 19 23, 49, 54
Sambu, 68
Sand dunes, 49
SAVANNA model, 342
Savanna trees, change in, 27–29
Scavenging, by early humans, 95, 101
Scenarios, household decision making, 326, 335, 342
Selous Game Reserve, 358, 361
Semivariance, 138, 148
Semivariograms, *147*
Serengeti: border closure 36, 37; abiotic change, 20–23; biotic change, 23–34; changes in savanna trees, 27–29; climate, 15–16; competition in grazers, 26; current and future change, 20–34;

decline of riverine forests, 29, 30; district, 329; environmental changes, 185, 195; eruption and regulation of ungulates, 25, 26; fauna, 18–20; fire, 22–23; Great Rinderpest, 24, 25; human influences, 34–40; landform and evolutionary change, 68–70; livelihood strategies, 34–36; pathogens, 32, 33; plains, 8, 23; poaching, 36, 38–40; rainfall and climate, 20–22; role of elephants, 30, 31; topography and geography, 12–14; wildebeest as keystone, 33, 34; vegetation, 16–18
Serengeti ecosystem, 3, 347, 350, 358, 359; conservation management, 446; development, 397–405; perturbations, 417–23, 435, 436; stability, 429, 436, 437
Serengeti-Mara ecosystem, 9, 15, 16, 47, *48,* 49, 82, 136, 379
Serengeti National Park, 9, 35–38; conservation management, 446; donor support, 449; income and expenditure, 448; western corridor, 110, 111, 113
Serengeti plains: geological evolution, 49–66; vegetation evolution, 54–66
Serengeti Protected Area, establishment, 8–12
Serengeti Regional Conservation Project, 356
Serengeti Trust Fund, 462
Seronera, 8, 13, 15, 17, 53
Settlements, 484; Mara area, 393, 394, 403, 404
Shashi ethnic group, 329
Sheep, 332; domestication, 105; introduction into East Africa, 105; response to rainfall, 120, 125; response to soil fertility, 125; spatial distribution, 120, 124
Shenyi ethnic group, 329
Shifting mosaic of grass abundance, 283
Short grass plains, 112
Simanjiro district, 105
Simulation model, 335
Sinicha ethnic group, 329
Siria escarpment, 67, 68. *See also* Isuria escarpment
Sizaki ethnic group, 329
Social organization, levels of, 327
Socioeconomic research, 3
Soft boundary, 486, 487
Soil: 154–57, 196, 200; catena, 155; erosion, 331, 332; microbes, 156; nitrogen mineralization, 156; respiration, 156
Soil fertility: determination, 116–18; response

of human population density, 110–22; response of resident herbivores, 110–22
Sorghum, 331
Southern Maasailand, 99, 113
Southern Oscillation Index, 152
Southern Sudanic people, 98
Spatial aggregation, herbivores, 118–20
Spatial segregation, herbivores, 265
Spatial variance, 136–38, 144–48
Special heterogeneity, 429, 433, 438
Species richness, herbivores, 201
Speke Gulf, 9, 12, 40, 69, 70
Speke, John, 8
Stability: equilibria, 430, 431; regions, 430; thresholds, 430
Stable equilibrium, 259, 430
Stable limit cycles, 430, 431
Stable states, 430
Steinbuck, trends, 503
Stone age hunters, 79, 80
Subsistence farming. See Farming, subsistence
Sukuma ethnic group, 329, 332
Sukumaland, 329
Sustainable development, 471, 472, 473
Systematic Reconnaissance Flight, 118, 121, 124

Tanzania Government: budget allocations, 444; Ministry of Natural Resources, 446; National Tourism Policy, 443
Tanzania National Parks, 349, 356
Tanzania National Policy, 348–52
Tanzania Wildlife Conservation Monitoring, 398
Tanzania, human fertility rate, 1
Tarime district, 329, 331, 333
Tatura ethnic group, 329
Taxonomic habitat index, 61, 62
Temperature: air, 192; sea surface (SST), 188, 189, 192, 199
Termites, 161, 162, 202
Thomson's gazelle: *ad libitum* intake, 286; body mass, 287; cropping rates, 279; decline, 26; demography, 287; digestibility, 279; energy intake, 286; ideal free distribution, 428; mange, 218; maximum population growth rate, 288; migration, 110, 126; movement rates, 287; population, 19, 480; population time series, 296; trends, 502. *See also* Gazelles; Grant's gazelle
Threats, human-caused, 301
Thresholds, 430, 432, 433

Top-down control, 423, 424, 438
Topi, 334; response to rainfall, 123; response to soil fertility, 123; spatial distribution, 121, 126; trends, 502
Topography, 154–57
Total exchangeable bases, 111, 113
Tourism, 354, 364; numbers, 37; revenue, *365;* Tanzania, 443
Trading, 328, 329, 331
Travel time, 316
Treaties, international, Serengeti ecosystem, 359–62
Trends: animals, 201; extrapolating of, 184; rainfall, 187; temperature, 194
Trona, 54
Trophic cascades, 423–25, 438
Trophic position, 104
Trypanosomiasis, 210, 222, 223, 229, 422
Tsavo, rinderpest, 226
Tsetse fly, 28; trypanosomiasis, 422
Tuberculosis, 210, 216–17
Turkana, 483
Turner, Myles, 8, 27

Ujamaa policy, 332
Uncertainty, 325; decision making under, 325, 326, 327; in rainfall, 325
Ungulates: competition, 26; eruption of, 25, 26; regulation of, 25
United Nations Environment Program, 349
Unstable equilibrium, 259, 430
Urbanization, 108–9
Utimbara, 68

Vaccination: rabies, 222–26, 228; rinderpest, 226–28
Variability: climatic, 325, 326, 341; in rainfall, 341
Variance: compositional, 137, 138; locational, 137, 138; patterned, 137, 138, 148–50; spatial, 136–38, 144–48
Vegetation: altered productivity of, 264, 265; growth of, 258; interaction with herbivores, 257; perturbation of, 419, 420; woody, 197, 199
Vegetation change, plains, 55–58
Villagers, interaction with wildlife, 302
Volcanic influence, 98, 111, 113, 117
Vulnerability, 325, 326, 328, 329; definition of, 327

Wajaluo ethnic group, 329
Wakuria people, 107

Warthog, 334; spatial distribution of, 115, 119, 121, 126. *See also* Pigs

Wasukuma, 107

Water: availability, 197, 201, 203; surface, 203

Waterbuck, 334; response to rainfall, 122, 123; response to soil fertility, 122, 123; spatial distribution, 121; trends, 503

Water-use efficiency (WUE), 196, 198

Wavelet analysis, 190–93

Wealth, 308

Weather shock, 319

Weather cycles, 190–93. *See also* Climate

Weeds, 107

Western corridor, Serengeti National Park, 110, 111, 113

White Nile, 70

Wild dog, 3, 202, 203; canine distemper, 422, 432; extinction, 20, 32; livestock predation, 126; numbers, 421, 425, 432

Wildebeest, 342; *ad libitum* intake, 291; and dry season rain, 431; and fire, 420; and grass production, 424; and lions, 420, 432; as prey for early humans, 104; competition, 26; competition with livestock, 126; cropping rates, 291; demography, 291; facilitation, 264; grazing, 34, 433; harvest, 369; increase, 419–20; independent of lions, 256, 261; in Kalahari, 486; keystone species, 33, 34, 41, 425, 438; Loita population, 397; maximum population growth rate, 291; MCF, 219; mortality, 262–64, 480; poaching, 2, 420, 421; population, 7, 19, 25, 26, 497; population time series, 296; rinderpest,

218, 226; trends, 497. *See also* Wildebeest migration; Rinderpest.

Wildebeest migration, 2, 8, 23, 24, 110, 111, 112, 124, 126, 127, 335, 398; and cultivation, 397; and land development, 398–403; and fences, 486; in Mara area, 398–403. *See also* Migration; Wildebeest

Wilderness, 108

Wildlife Conservation Act, 349

Wildlife Management Area, 350, 351, 354, 357–59, 368–72, 374, 447, 464, 488, 489; regulations, 465

Wildlife policy: antipoaching, 368–72, 375; incentives, 370; of Tanzania, 350, 351, 357

Wildlife rents, 390–393, 405–12

Wildlife, 325, 327, 332–35, 338–40; density of, 333; elimination, 404, 405; fencing, 405; impact of cultivation, 397, 404, 405; migrant, 310; pressures on, 333; population, 338, 339; resident, 310

Winds, 186, 189, 199

Woody vegetation. *See under* Vegetation

World Conservation Monitoring Centre, 349

World Heritage Site, 475

Yukon, 485, 486

Zanaki ethnic group, 329

Zangita ethnic group, 329

Zebra: migration, 110, 112, 124, 126; population, 19, 26, 261, 263; predator regulation, 26; trends, 497

Zebu, 105

Zoonoses, 106, 221